计算机科学与技术丛书

深入浅出
ASP.NET Core

周家安◎编著

清華大學出版社
北京

内 容 简 介

ASP.NET Core是微软推出的跨平台、开放源代码的Web开发框架。本书秉持“现学现用”的原则，知识讲解通俗易懂，并配有示例代码。每个示例都是针对知识点而设计的，代码量适中，功能和结构简单，便于读者参考学习和扩展改造。

全书共18章。第1章和第2章讲述ASP.NET Core应用程序的初始化过程以及运行环境的设定；第3章单独介绍ASP.NET Core应用程序的设计模式——组件化，即依赖注入技术；第4章和第5章讲述配置应用程序的方法，包括使用配置文件和选项模式；第6章和第7章讲述HTTP管道和HTTP状态存储；第8～11章涉及Web开发的重点知识：Razor页面、MVC框架、模型绑定及Web API；第12章和第13章讲述MVC框架的应用扩展，包括过滤器和标记帮助器的使用，以适应实际开发需求；第14章介绍静态文件服务，服务器可向客户端提供目录/文件的访问入口；第15章讲述路由约束；第16章和第17章讲述ASP.NET Core独有的客户端技术——SignalR和Blazor；第18章主要涉及基本的安全功能，即验证与授权。

本书适合作为高等学校、培训机构.NET课程相关的教材或.NET爱好者的参考书。对于想通过自学步入编程大门的读者，也推荐阅读本书。

图书在版编目（CIP）数据

深入浅出：ASP.NET Core / 周家安编著. —北京：清华大学出版社，2024.2
（计算机科学与技术丛书）
ISBN 978-7-302-65668-5

Ⅰ. ①深…　Ⅱ. ①周…　Ⅲ. ①网页制作工具—程序设计　Ⅳ. ①TP393.092.2

中国国家版本馆CIP数据核字（2024）第048976号

责任编辑：盛东亮　吴彤云
封面设计：李召霞
责任校对：时翠兰
责任印制：杨　艳

出版发行：清华大学出版社
网　　址：https://www.tup.com.cn，https://www.wqxuetang.com
地　　址：北京清华大学学研大厦A座　　**邮　　编：**100084
社 总 机：010-83470000　　**邮　　购：**010-62786544
投稿与读者服务：010-62776969，c-service@tup.tsinghua.edu.cn
质 量 反 馈：010-62772015，zhiliang@tup.tsinghua.edu.cn
课 件 下 载：https://www.tup.com.cn，010-83470236
印 装 者：小森印刷霸州有限公司
经　　销：全国新华书店
开　　本：186mm×240mm　　**印　张：**34.5　　**字　数：**798千字
版　　次：2024年4月第1版　　**印　次：**2024年4月第1次印刷
印　　数：1～2000
定　　价：128.00元

产品编号：098506-01

前言

PREFACE

ASP.NET Core 由微软官方推出，开放源代码并以社区为主，可以生成运行于 Windows、macOS、Linux 等操作系统的新型 Web 应用程序。ASP.NET Core 并不是 ASP.NET 的延续版本，而是经过重新设计和优化的框架。由于它是编译运行的，因此在性能上的优势明显。而且，其内部对异步任务和安全性做了大量集成工作，有些安全功能是默认启用的，如对跨站漏洞攻击的防范。经过六七个版本的迭代，ASP.NET Core 已趋向完善，对 Web 前端与其他数据访问技术的兼容性也得到极大提升。对于常规 Web 功能、微服务、移动后端以及物联网后端等应用场景，ASP.NET Core 都是不错的选择。

本书所涉及的内容针对性强，只要读者具备.NET 或 C#编程相关基础，即可通过本书快速掌握 ASP.NET Core 的关键技术。本书在每个知识点的讲解后都会附上专门的示例，方便读者将学到的知识马上付诸实践，加深印象。

ASP.NET Core 配有官方开发工具，并共享.NET SDK 工具。无论读者使用的是 Windows 还是 Linux 操作系统，都可以执行 dotnet new 命令创建 ASP.NET Core 应用项目，或执行 dotnet run 命令运行应用程序。

本书推荐使用官方提供的工具编写代码。

- Visual Studio：Windows、macOS 用户均可以使用，简称 VS。VS 是著名的集成开发环境，提供从编码、校验、生成到调试和运行的完整支持。除了代码提示功能，还有联想功能，可以根据代码上下文以及大数据汇总推断各种代码片段，极大地提高编码效率。
- Visual Studio Code：简称 VS Code 或 VSC。VS Code 是 VS 的一个分支版本，着重代码编辑功能。VS Code 通过安装扩展支持各种编程语言。理论上，只要拥有足够的扩展，VS Code 就能编写任何程序语言的代码（如 C、C++、Python、C#、Java 等）。VS Code 能运行在 Windows、Linux 及 macOS 等操作系统上，同时也支持 ARM 架构，如 Raspberry Pi OS。使用时先执行 dotnet new 命令创建 ASP.NET Core 项目，然后在 VS Code 中打开项目所在目录即可。

虽然 ASP.NET Core 可以细分出 Razor Pages、MVC、Web API、Blazor 等项目，但实际上这些功能是可以在同一个项目中实现的。ASP.NET Core 以服务容器为核心，可以组件化扩展。只要向容器注册服务类型，就能开启相关的功能，如 MVC 与 Blazor 功能可以同时启用（URL 路由不能有冲突）。

本书适合有一定.NET 或 C#基础的读者阅读，也可以作为高等学校或培训机构的辅助教材。也欢迎想了解 ASP.NET Core 的开发人员阅读本书。

由于编者水平有限，书中难免出现不妥之处，望广大读者不吝批评指正。

编 者

2024 年 3 月

目录

CONTENTS

第 1 章　初始化 ASP.NET Core 应用程序

本章要点

- ❑ WebApplicationBuilder 类与 WebApplication 类的关系
- ❑ 运行 ASP.NET Core 应用程序
- ❑ 为 Web 服务器设置 URL
- ❑ 处理应用生命周期事件

1.1　应用程序的启动过程

ASP.NET Core 应用程序在启动之前，需要经过以下步骤。

（1）创建 WebApplicationBuilder 对象实例。

（2）（可选）向 Services 集合添加所需要的服务类型。

（3）（可选）配置相关的选项类型。

（4）调用 Build 方法创建 WebApplication 实例。

（5）（可选）组建 HTTP 处理管线。

（6）启动应用程序，同时启动内置的 Kestrel 服务器。

要创建基本的 ASP.NET Core 应用程序项目，可以执行 dotnet new web 命令。例如，执行以下命令将创建一个名为 Demo 的 ASP.NET Core 项目。

```
dotnet new web -n Demo -o .
```

其中，new 命令指明即将完成的操作为创建新的应用程序项目；web 参数表示项目类型，此处指 ASP.NET Core 项目。若希望创建控制台应用程序，可以执行以下命令。

```
dotnet new console …
```

-n 参数表示待创建项目的名称，示例中为 Demo；-o 参数指定被创建项目的存放路径，上述示例中，“.”表示存放到当前工作目录下。也可以使用绝对路径，如

```
dotnet new web C:\test\mydir          // Windows
dotnet new web /home/user/apps        // Linux
```

Web 项目模板将生成以下 C#代码。

```
// 创建 WebApplicationBuilder 对象
var builder = WebApplication.CreateBuilder(args);

/* 此处可以向 Services 集合添加服务类，或配置应用程序 */

// 创建 WebApplication 对象
var app = builder.Build();

// 构建 HTTP 管道
app.MapGet("/", () =>"Hello World!");

// 启动应用程序
app.Run();
```

1.2 WebApplicationBuilder 类

WebApplicationBuilder 类没有公共的构造函数，因此不能直接实例化。只能通过 WebApplication 类的 CreateBuilder()方法获得 WebApplicationBuilder 实例。此方法是静态成员，可在代码中直接调用，如

```
WebApplicationBuilder builder = WebApplication.CreateBuilder();
```

如果 ASP.NET Core 应用程序需要通过命令行参数获取配置参数，则应该将传递给程序入口点（Main 方法）的参数继续传递给 CreateBuilder()方法，即

```
WebApplicationBuilder builder = WebApplication.CreateBuilder(args);
```

获得 WebApplicationBuilder 实例后，就可以使用它配置应用程序了，如向依赖注入容器（ServiceCollection）添加服务类、设置 Kestrel 服务器、添加配置参数的数据源等。CreateBuilder()方法在创建 WebApplicationBuilder 实例的过程中也会添加一些默认的配置，以保证 ASP.NET Core 应用程序的基本代码能够正常运行。这些默认配置包括：

（1）为内置 Web 服务器 Kestrel 设定默认值；

（2）配置应用程序的运行环境；

（3）默认添加的配置，来源有命令行参数、环境变量、appsettings.json 文件；

（4）配置默认的日志输出方案；

（5）将当前目录设置为应用的内容根目录；

（6）配置 IIS（Internet Information Services）集成，以便能在 Windows 操作系统以 IIS 服务器托管 ASP.NET Core 应用程序。

应用程序配置完成后，调用 Build()方法，创建并返回 WebApplication 实例。调用 Build()方法后，依赖注入容器将变为只读状态，无法修改。运行以下代码后会发生错误。

```
WebApplicationBuilder builder = WebApplication.CreateBuilder(args);
/* 配置应用程序 */
```

```
var app = builder.Build();

// 错误：此时 Services 无法修改
builder.Services.AddRazorPages();
```

WebApplicationBuilder 对象只是为应用程序的运行准备好相关的配置参数，而不会真正启动 ASP.NET Core 应用程序。要启动应用程序，需要调用 WebApplication 实例的相关方法。

1.3　启动应用程序

WebApplicationBuilder.Build()方法返回 WebApplication 实例后，可以适当地安排 HTTP 管线，以便能处理并响应来自客户端的 HTTP 请求，如

```
var app = builder.Build();
// 运行一个简单的中间件
app.Use(async (httpcontext, next) =>
{
    // ...
    await next();
});
// 构造一个终结点，HTTP 管线在此处结束
// 并沿着调用堆栈将响应消息回传给客户端
app.MapGet("/", () =>"Hello World!");
```

最后，调用 WebApplication 实例的 StartAsync()、Run()、RunAsync()等方法正式启动应用程序。

StartAsync()方法完成应用程序启动后会马上返回。此时，开发人员需要想办法保证应用程序能持续运行并循环处理客户端请求，否则应用程序启动之后会立即退出。当需要关闭应用程序时，应调用 StopAsync()方法。

下面的示例演示了如何按 Esc 键退出应用程序。

```
var builder = WebApplication.CreateBuilder(args);
var app = builder.Build();                          // 构建应用程序

app.MapGet("/", () =>"Hello World!");               // HTTP 管线

// 启动应用程序
await app.StartAsync();
// 循环检测用户是否按下了 Esc 键
while(true)
{
    ConsoleKeyInfo key = Console.ReadKey(intercept: true);
    if(key.Key == ConsoleKey.Escape)
    {
        // 跳出循环
        Console.WriteLine("*** 应用程序即将退出 ***");
        break;
```

```
    }
}
await app.StopAsync();                                    // 停止应用程序
```

调用 StartAsync()方法后，只要应用程序顺利启动，该方法立即返回；接着程序进入一个死循环（while 循环的条件永远都是 true，此循环会无限次地执行），在循环代码内部，等待用户的键盘输入。如果用户按下的是 Esc 键，就会跳出 while 循环。

跳出循环后，就会调用 StopAsync()方法，停止应用程序，如图 1-1 所示。

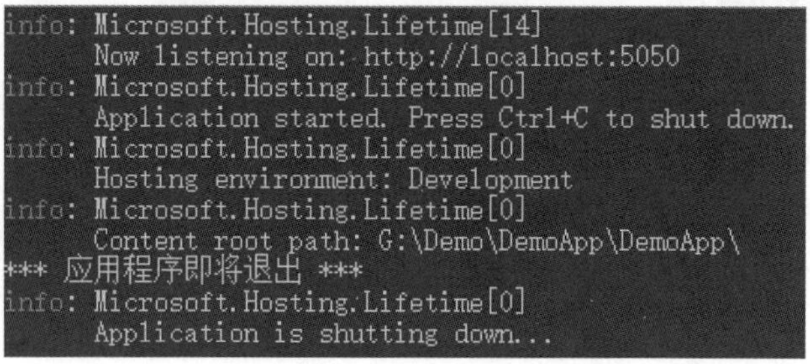

info: Microsoft.Hosting.Lifetime[14]
Now listening on: http://localhost:5050
info: Microsoft.Hosting.Lifetime[0]
Application started. Press Ctrl+C to shut down.
info: Microsoft.Hosting.Lifetime[0]
Hosting environment: Development
info: Microsoft.Hosting.Lifetime[0]
Content root path: G:\Demo\DemoApp\DemoApp\
*** 应用程序即将退出 ***
info: Microsoft.Hosting.Lifetime[0]
Application is shutting down...

图 1-1　按下 Esc 键后应用程序停止

如果调用的是 Run()或 RunAsync()方法，开发人员不需要编写额外的代码等待应用程序退出，因为 Run()或 RunAsync()方法被调用后会处于等待状态，直到应用程序退出才会返回。下面的示例演示 Run()方法的使用。

```
var builder = WebApplication.CreateBuilder(args);

/* 配置应用程序 */

// 构建应用程序实例
var app = builder.Build();

// …

// 启动应用程序
app.Run();
```

如果要调用 RunAsync()方法，就要使用异步等待。

```
await app.RunAsync();
```

应用程序运行后，可以按 Ctrl+C 快捷键退出。

1.4　使用 Host 初始化应用程序

Host（翻译为“主机”）是一种具有特定功能的服务类型，它封装并管理与当前应用程序有关的各种资源，如应用程序的配置数据、依赖注入容器，以及服务容器中注册的所有实现了 IHostedService 接口的对象。当 Host 启动时，它会在服务容器中检索所有实现 IHostedService

接口的类型实例，然后逐个调用它们的 StartAsync()方法。

1.4.1　通用主机

Host 可以用于许多典型的.NET 应用程序中（如控制台应用程序），称为通用主机（Generic Host）。

使用通用主机构建应用程序的大致过程如下。

（1）创建 HostBuilder 实例。可以直接使用该类的构造函数进行实例化，也可以调用 Host 类的 CreateDefaultBuilder()静态方法获取 HostBuilder 实例。

（2）通过调用 HostBuilder 的成员方法或扩展方法完成配置，如调用 ConfigureServices()方法向支持依赖注入的服务容器添加服务类型。

（3）调用 Build()方法创建 Host 实例。

（4）调用 Host 实例的 StartAsync()、Run()、RunAsync()等方法启动主机。

（5）（可选）调用 StopAsync()方法停止通用主机。

1.4.2　示例：简单的通用主机

下面的示例将演示一个简单的使用通用主机的控制台应用程序，并在其中启动两个服务项目。步骤如下。

（1）创建控制台应用程序项目。

（2）执行以下命令，添加 Nuget 包引用。

```
dotnet add package Microsoft.Extensions.Hosting
dotnet add package Microsoft.Extensions.DependencyInjection
```

（3）在代码文件中引入以下命名空间。

```
using Microsoft.Extensions.Hosting;
using Microsoft.Extensions.DependencyInjection;
```

（4）定义两个服务类（DemoService1 和 DemoService2），它们都实现 IHostedService 接口。这两个类将作为服务运行在通用主机上。

```
// 第 1 个服务类
public sealed class DemoService1 : IHostedService
{
    public Task StartAsync(CancellationToken cancellationToken)
    {
        Console.WriteLine("服务 1：开始运行");
        return Task.CompletedTask;
    }

    public Task StopAsync(CancellationToken cancellationToken)
    {
        Console.WriteLine("服务 1：停止运行");
        return Task.CompletedTask;
```

```
        }
    }

    // 第 2 个服务类
    public sealed class DemoService2 : IHostedService
    {
        public Task StartAsync(CancellationToken cancellationToken)
        {
            Console.WriteLine("服务 2：开始运行");
            return Task.CompletedTask;
        }

        public Task StopAsync(CancellationToken cancellationToken)
        {
            Console.WriteLine("服务 2：停止运行");
            return Task.CompletedTask;
        }
    }
```

主要是实现两个方法：StartAsync()方法在服务被启动时调用；同理，当服务被停止时则调用 StopAsync()方法。本示例中仅调用 Console.WriteLine()方法输出简单的文本，以便从控制台输出得知服务的运行状态。

（5）创建 HostBuilder 实例。

```
var hostBuilder = Host.CreateDefaultBuilder(args);
```

（6）调用 ConfigureServices()方法，将上述步骤中实现的两个服务类注册到服务容器中。

```
hostBuilder.ConfigureServices(services =>
{
    services.AddSingleton<IHostedService, DemoService1>();
    services.AddSingleton<IHostedService, DemoService2>();
});
```

AddSingleton()方法表示这两个服务类在整个应用程序生命周期内是单个实例的——它们的构造函数只调用一次。

（7）创建通用主机实例。

```
var host = hostBuilder.Build();
```

（8）启动通用主机。

```
await host.StartAsync();
Console.WriteLine("通用主机已启动，按 Enter 键退出");
Console.ReadLine();
await host.StopAsync();
```

调用 StartAsync()方法启动通用主机，随后 Console.ReadLine()方法会等待用户的键盘输入。只要按 Enter 键，ReadLine()方法会立即返回，然后调用 StopAsync()方法停止通用主机。

运行示例程序后，若看到屏幕上输出“服务 1：开始运行”和“服务 2：开始运行”，则表明托管在通用主机上的两个服务已顺利启动；此时按 Enter 键，通用主机停止，托管在其中的两个服务也会停止，并在屏幕上输出相关信息，如图 1-2 所示。

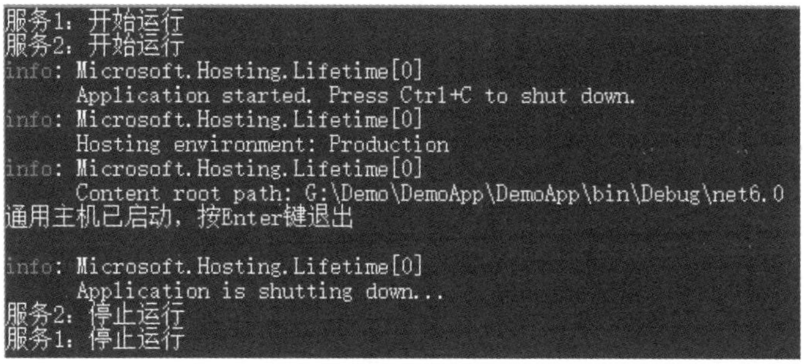

服务1：开始运行
服务2：开始运行
info: Microsoft.Hosting.Lifetime[0]
Application started. Press Ctrl+C to shut down.
info: Microsoft.Hosting.Lifetime[0]
Hosting environment: Production
info: Microsoft.Hosting.Lifetime[0]
Content root path: G:\Demo\DemoApp\DemoApp\bin\Debug\net6.0
通用主机已启动，按Enter键退出
info: Microsoft.Hosting.Lifetime[0]
Application is shutting down...
服务2：停止运行
服务1：停止运行

图 1-2　在通用主机上运行两个服务示例

1.4.3　Web 主机

对于 ASP.NET Core 项目，新版本推荐使用前文所介绍的初始化方法，即使用 WebApplicationBuilder 类型实例构建 WebApplication 对象。而使用 Web 主机仅用于对旧版本的兼容，或者同时使用通用主机和 Web 主机的项目。

Web 主机有两种使用方法。

第 1 种方法是先创建用于通用主机的 HostBuilder 实例，然后再配置 Web 主机，最后创建通用主机实例并启动主机。请思考下面的示例代码。

```
// 创建通用主机的 Builder
var hostBuilder = Host.CreateDefaultBuilder(args);
// 添加 Web 主机配置
hostBuilder.ConfigureWebHostDefaults(webhostBuilder =>
{
    // 添加服务
    webhostBuilder.ConfigureServices(services =>
    {
        services.AddControllers();
    })
    // 构建 HTTP 管线
    .Configure(appBuilder =>
    {
        appBuilder.Use(async (ctx, next) =>
        {
            ctx.Response.Headers.Add("Access-Mode", "Read");
            await next();
        })
        .UseRouting()
        .UseEndpoints(endPoints =>
        {
            endPoints.Map("/", () =>"示例应用程序");
        });
```

```
    });
});

// 构建主机实例
var host = hostBuilder.Build();
// 启动主机
await host.RunAsync();
```

要在 HostBuilder 上添加 Web 主机的配置，可以调用 ConfigureWebHost()方法，也可以调用 ConfigureWebHostDefaults()方法。上述示例调用了 ConfigureWebHostDefaults()方法，它会为 Web 主机设置常用的默认参数，如使用内置的 Kestrel 服务器、IIS 集成等。

ConfigureServices()方法用于向依赖注入容器添加服务类型，定义应用程序需要用到的功能；Configure()方法用于构建 HTTP 管线，处理客户端的 HTTP 请求。

最后，调用 Build()方法创建通用主机实例，并调用 RunAsync()方法启动主机，ASP.NET Core 应用程序开始运行。

使用 Web 主机的第 2 种方法是调用 WebHost 类的 CreateDefaultBuilder()静态方法创建 WebHostBuilder 实例，接着配置服务容器和 HTTP 管线，最后调用 Build()方法创建 Web 主机实例并运行它。示例代码如下。

```
// 创建 WebHostBuilder 实例
var webhostBuilder = WebHost.CreateDefaultBuilder(args);
// 配置依赖注入容器
webhostBuilder.ConfigureServices(services =>
{
    services.AddControllers();
});
// 构建 HTTP 管线
webhostBuilder.Configure(app =>
{
    app.UseRouting();
    app.UseEndpoints(endPoints =>
    {
        endPoints.MapControllers();
        endPoints.Map("/", () =>"示例应用程序");
    });
});

// 创建 Web 主机实例
var host = webhostBuilder.Build();
// 启动 Web 主机
host.Run();
```

除了使用 WebHost.CreateDefaultBuilder()方法，还可以直接实例化 WebHostBuilder 对象，并手动配置 Web 主机，如

```
var webhostBuilder = new WebHostBuilder();
// 使用内置的 Kestrel 服务器和 IIS 集成功能
```

```
webhostBuilder.UseKestrel().UseIISIntegration();
// 添加应用程序的配置源
webhostBuilder.ConfigureAppConfiguration(cfg =>
{
    // 命令行参数
    cfg.AddCommandLine(args);
    // 环境变量
    cfg.AddEnvironmentVariables();
    // JSON 文件
    cfg.AddJsonFile("appsettings.json");
});
```

1.5　设置应用程序的 URL

Web 服务器需要绑定一个或多个有效的统一资源定位符（Uniform Resource Locator，URL），并在这些 URL 上侦听连接。客户端（通常是 Web 浏览器）通过 URL 找到要访问的 Web 应用程序。因此，URL 可以标识应用程序的唯一性。

为 ASP.NET Core 应用程序设置 URL 有多种方法，大体上可分为两类。

（1）硬编码。即直接把要使用的 URL 写到程序代码中，其缺点是一旦修改了 URL，就需要重新编译应用程序。

（2）应用程序配置。此方法比较灵活，修改 URL 后不需要重新编译应用程序。参数的配置方案比较多，如命令行参数、环境变量、配置文件等。

1.5.1　调用 UseUrls()方法

调用 WebApplication.CreateBuilder()方法创建 WebApplicationBuilder 实例后，就可以通过它的 WebHost 属性调用 UseUrls()方法为 ASP.NET Core 应用程序指定 URL。该方法是 IWebHostBuilder 接口类型的扩展方法，由于 WebApplicationBuilder 类实现了 IWebHostBuilder 接口，所以能够调用 UseUrls()方法。

UseUrls()方法接受一个或多个字符串类型的参数，每个参数值表示一个有效的 URL。示例代码如下。

```
var builder = WebApplication.CreateBuilder(args);

// 为应用程序设置 3 个 URL
builder.WebHost.UseUrls(
"http://localhost:4252",
"http://localhost:6703",
"http://localhost:5361");

var app = builder.Build();
…
```

运行应用程序后，在浏览器地址栏中输入上述 3 个 URL 中的任意一个，均可以访问，如图 1-3 所示。

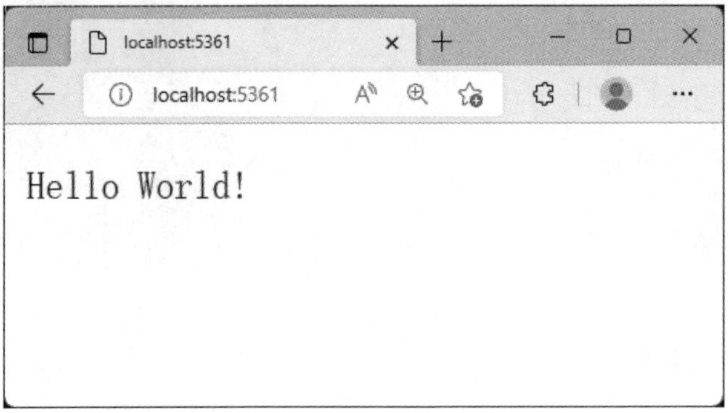

图 1-3　使用指定的 URL 访问 Web 应用

也可以使用 UseSetting()方法配置 URL，如

```
builder.WebHost.UseSetting(WebHostDefaults.ServerUrlsKey,
"http://localhost:12345;http://localhost:54218;http://*:8080");
```

WebHostDefaults 类定义了一组默认的应用配置字段，其中 ServerUrlsKey 表示 URL 的配置键，即 urls。其实，UseUrls()方法内部也是调用了 UseSetting(WebHostDefaults.ServerUrlsKey, …)方法，封装为 UseUrls()方法调用起来更简单。

1.5.2　使用 WebApplication 类的 Urls 属性

Urls 属性是一个字符串类型的集合对象，可以调用 Add()方法添加一个 URL，或者调用 Remove()方法删除一个 URL。

下面的代码通过访问 Urls 属性为应用程序添加两个 URL。

```
var builder = WebApplication.CreateBuilder(args);
var app = builder.Build();

app.Urls.Add("http://localhost:5050");
app.Urls.Add("http://localhost:6060");
…
```

Urls 属性是 WebApplication 类的实例成员，允许在调用 WebApplicationBuilder.Build()方法之后进行修改。

1.5.3　调用 Run()方法时传递 URL

在调用 WebApplication 对象的 Run()方法（或 RunAsync()方法）时，可以向 url 参数传递一个有效的 URL，该 URL 将被应用到应用程序上。此处只能设置一个 URL，而且会将应用程序之前配置的所有 URL 全部删除。

请思考以下示例。

```
var builder = WebApplication.CreateBuilder(args);
var app = builder.Build();

...

// 配置 3 个 URL
app.Urls.Add("http://localhost:8472");
app.Urls.Add("http://192.168.0.107:13055");
app.Urls.Add("http://127.0.0.1:9288");

app.Run("http://127.0.0.1:16249");
```

上述代码中，先是通过 Urls 属性添加 3 个 URL。接着，在调用 app.Run()方法时设置了第 4 个 URL。

运行示例程序后，只有 http://127.0.0.1:16249 这个地址才能访问应用程序，如图 1-4 所示，Urls 属性中添加的 3 个 URL 被删除。不管前面的代码通过何种方式设置了 URL，最终只有调用 Run()或 RunAsync()方法所指定的 URL 有效。

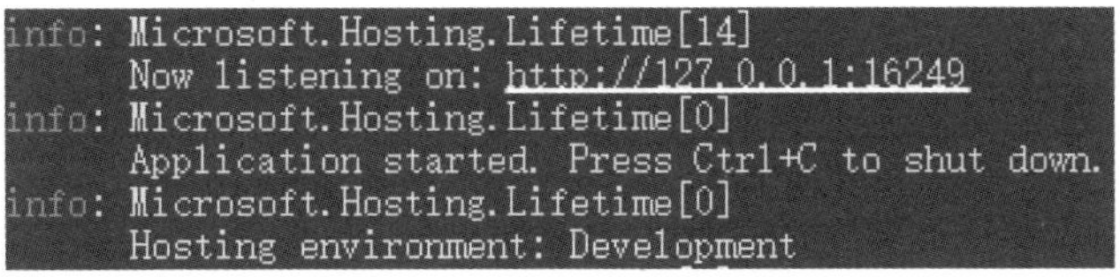
info: Microsoft.Hosting.Lifetime[14]
Now listening on: http://127.0.0.1:16249
info: Microsoft.Hosting.Lifetime[0]
Application started. Press Ctrl+C to shut down.
info: Microsoft.Hosting.Lifetime[0]
Hosting environment: Development

图 1-4　只有一个 URL 可用

1.5.4　通过 ServerAddressesFeature 对象设置 URL

ServerAddressesFeature 类（位于 Microsoft.AspNetCore.Hosting.Server.Features 命名空间）是 IServerAddressesFeature 接口的默认实现类。在应用程序代码中一般通过 IServerAddresses-Feature 接口来引用。其实，WebApplication 类的 Urls 属性、Run()方法等成员内部也是使用 Server-AddressesFeature 类设置 URL 的，只是做了一层封装而已。

ServerAddressesFeature 类公开一个集合类型的属性——Addresses，在代码中可以调用 Add()方法添加 URL。示例代码如下。

```
var builder = WebApplication.CreateBuilder(args);
var app = builder.Build();

// 获取服务实例
IServer server = app.Services.GetRequiredService<IServer>();
// 获取 Feature
IServerAddressesFeature? feat = server.Features
.Get<IServerAddressesFeature>();
if(feat != null)
```

```
{
    // 设置 URL
    feat.Addresses.Add("http://127.0.0.1:8210");
    feat.Addresses.Add("http://localhost:7466");
    feat.Addresses.Add("http://localhost:23051");
    feat.Addresses.Add("http://localhost:16620");
}

app.MapGet("/", () =>"Hello World!");

app.Run();
```

上述代码先从服务容器中取出实现了 IServer 接口的默认服务器对象(通常是 KestrelServer 类)，再从服务器对象的 Features 集合中查找出实现 IServerAddressesFeature 接口的对象。最后通过 IServerAddressesFeature. Addresses 属性添加所需 URL。

本示例演示的方法比较复杂，建议直接使用封装好的 WebApplication.Urls 属性。

1.5.5 使用命令行参数

前面几个示例使用的 URL 配置方法均属于硬编码方式，其缺点是一旦 URL 需要变动，就要修改代码并重新编译应用程序，在实际开发中不太方便。而通过传递命令行参数配置 URL 的方式则相对灵活，只要在运行应用程序时提供名为 urls 的参数即可。

在调用 WebApplication 类的 CreateBuilder()方法创建 WebApplicationBuilder 实例时，应用程序的配置源中已默认添加对命令参数的支持，因此，开发人员不需要编写额外的处理代码。

下面给出一个简单的示例。

```
// 在屏幕上打印命令行参数
Console.WriteLine("命令行参数: {0}", string.Join(", ", args));

var builder = WebApplication.CreateBuilder(args);
// 创建应用程序
var app = builder.Build();
app.MapGet("/", () =>"你好, 世界");
// 启动 Web 应用程序
await app.RunAsync();
```

其中，args 是程序入口点——Main 方法的参数，传递给应用程序的命令行参数会保存在 args 参数中。请注意，CreateBuilder()方法有多个重载，若希望接收传递给应用程序的命令行参数，就必须调用带 args 参数的重载，即

```
public static WebApplicationBuilder CreateBuilder(string[] args);
```

如果调用的是无参数的 CreateBuilder 方法，那么应用程序是不会读取命令行参数的。还可以通过 WebApplicationOptions 类引用命令行参数，它公开了一个名为 Args 的属性，可用于设置命令行参数，如

```
…
var builder = WebApplication.CreateBuilder(new WebApplicationOptions
{
    Args = args
});
// 创建应用程序
var app = builder.Build();
…
```

假设该示例生成的可执行文件为 DemoApp，那么，在执行应用程序时可以通过以下方法设置 urls 参数。

```
DemoApp --urls http://localhost:32085
DemoApp --urls=http://127.0.0.1:14520
DemoApp /urls http://*:15992
DemoApp /urls=http://127.0.0.1:8088
```

其中，http://*:15992 表示侦听本地所有 IP 地址的 15992 端口。如果要指定多个 URL，需要用分号（必须是英文符号）分隔，如

```
DemoApp --urls http://localhost:1788;http://127.0.0.1:48160
```

如果应用程序是以程序集的方式生成（即生成.dll 文件），也可以通过 dotnet 命令运行，如

```
dotnet DemoApp.dll --urls=http://localhost:44225
```

1.5.6　使用配置文件

默认情况下，ASP.NET Core 应用程序会将名为 appsettings.json（包括 appsettings.Development.json 等文件）的配置文件添加到配置源中。因此，通过该文件配置 URL 也非常方便，而且当 URL 需要变动时直接修改配置文件即可。

打开 appsettings.json 文件，加入字段名为 urls 的配置。

```
{
    …
    "urls": "http://*:59148;http://localhost:23804;http://localhost:6731"
}
```

多个 URL 使用分号（英文）分隔。每个 URL 的开头不能有空格，但结尾可以有空格。

例如，下面的配置是无效的，应用程序运行后会发生错误，因为第 2 个 URL（http://localhost:23804）的开始处存在空格。

```
"urls": "http://*:59148; http://localhost:23804;http://localhost:6731"
```

但下面的配置是有效的，因为空格出现在 URL 的末尾，在应用程序中是允许的。

```
"urls": "http://localhost:23804 ;http://localhost:6731"
```

1.5.7　使用环境变量

ASP.NET Core 应用程序项目默认会读取环境变量所提供的配置，而用于设置 URL 的环境

变量名为“<前缀>_URLS”，预设的前缀为 ASPNETCORE，即在运行应用程序前可以通过设置 ASPNETCORE_URLS 环境变量指定 URL。

下面的示例将为应用程序指定两个 URL 地址，然后运行应用程序（假设应用程序名为 DemoApp）。

```
set ASPNETCORE_URLS=http://localhost:7554;http://localhost:19265
dotnet DemoApp.dll
```

1.5.8 使用 launchSettings.json 文件

在应用程序的开发阶段，launchSettings.json 文件仅用于本地测试，通常位于 Properties 目录下。ASP.NET Core 应用项目模板默认会创建此文件，在 Visual Studio 开发环境中，可以在项目属性窗口中以图形化方式编辑 launchSettings.json 文件。

launchSettings.json 文件为 ASP.NET Core 项目配置多个运行环境，在运行应用程序时可以通过配置名称选择需要的环境。

其中，位于配置节点下的 applicationUrl 字段，可以用来配置应用程序的 URL，如

```
"<环境名称>": {
…
"applicationUrl": "http://localhost:6322",
…
      }
```

如果有多个 URL，请使用分号（英文）分隔，如

```
"<环境名称>": {
…
"applicationUrl": "http://localhost:6322;http://localhost:16378",
…
      }
```

1.5.9 Kestrel 服务器的侦听地址

ASP.NET Core 内置的 Kestrel 服务器也支持 URL 配置，不过该配置是基于 Socket 层面的，因此只能使用 IP 地址和端口号确定服务器要侦听的 URL。

KestrelServerOptions 类定义了 3 个公共方法，用来配置 Web 服务器的侦听地址。

（1）Listen()方法：指定要侦听的 IP 地址和端口号。

（2）ListenAnyIP()方法：仅指定服务器端口号，应用程序会侦听绑定到网络接口上的所有 IP 地址，包括 IPv4 地址和 IPv6 地址。该方法优先侦听 IPv6 地址，如果 IPv6 地址不可用，就侦听 IPv4 地址。

（3）ListenLocalhost()方法：指定服务器端口号，仅侦听本机地址（如 127.0.0.1）。

请看下面的示例。

```
var builder = WebApplication.CreateBuilder(args);

// 配置 Kestrel 服务器
builder.WebHost.ConfigureKestrel(options =>
{
    // 设置侦听地址
    options.ListenAnyIP(10072);
});

var app = builder.Build();
…
```

上述代码指定 Kestrel 服务器绑定到任意 IP 地址，侦听端口为 10072。

1.5.10　通过 HTTP.sys 配置 URL

与 Kestrel 一样，HTTP.sys 也是 ASP.NET Core 内部已实现的服务器类型。HTTP.sys 只能运行在 Windows 平台，不支持跨平台。

要在 ASP.NET Core 项目中使用 HTTP.sys 服务器，需要调用 UseHttpSys()方法。也可以结合 HttpSysOptions 类进行相关配置。其中，UrlPrefixes 属性用于设置服务器的 URL 地址。

通过 HTTP.sys 配置 URL 的示例代码如下。

```
var builder = WebApplication.CreateBuilder(args);
// 使用 HTTP.sys
builder.WebHost.UseHttpSys(option =>
{
    // 设置 URL
    option.UrlPrefixes.Add("http://*:12005");
});
…
app.Run();
```

其中，http://*:12005 表示侦听本机上所有地址上传入的请求，并绑定到 12005 端口。

1.5.11　PreferHostingUrls()方法的作用

当 ASP.NET Core 应用程序同时使用 IWebHostBuilder 和 IServer 接口配置 URL 时，通过 IServer 所设置的 URL 会覆盖由 IWebHostBuilder 配置的 URL。

请思考下面的代码。

```
var builder = WebApplication.CreateBuilder(args);

// 配置第 1 个 URL
builder.WebHost.UseUrls("http://localhost:6500");

// 配置第 2 个 URL
builder.WebHost.ConfigureKestrel(option =>
```

```
    {
        option.ListenLocalhost(6502);
    });

    var app = builder.Build();
    ...
```

第 1 个 URL 是通过 UseUrls()方法配置的，第 2 个 URL 则是通过 Kestrel 选项配置的。由于 Kestrel 选项所配置的 URL 具有更高的优先级，使得应用程序选择了第 2 个 URL。即应用程序会侦听 http://localhost:6502。

调用 PreferHostingUrls()方法会改变此优先级。若将 preferHostingUrls 参数设置为 true，那么通过 IWebHostBuilder 配置的 URL 的优先级会更高，并被应用程序使用；若 preferHostingUrls 参数设置为 false，则 IWebHostBuilder 所配置的 URL 会被 Kestrel 选项所配置的 URL 替代。

在上面的示例代码中加入 PreferHostingUrls()方法的调用。

```
builder.WebHost.PreferHostingUrls(true);

// 配置第 1 个 URL
builder.WebHost.UseUrls("http://localhost:6500");

// 配置第 2 个 URL
builder.WebHost.ConfigureKestrel(option =>
{
    option.ListenLocalhost(6502);
});

var app = builder.Build();
```

此时应用程序会优先使用 IWebHostBuilder 接口配置的 URL，即 http://localhost:6500。

1.6 应用程序生命周期事件

ASP.NET Core 应用程序在初始化过程中会向服务容器注册 IHostApplicationLifetime 接口类型的服务。该接口描述了几个与应用程序生命周期相关的事件，具体如下。

（1）ApplicationStarted：只有在应用程序完全启动之后才会触发。

（2）ApplicationStopping：应用程序正在停止运行，但尚未完成操作。

（3）ApplicationStopped：当应用程序正常停止（此过程未发生错误）后触发。

IHostApplicationLifetime 接口的默认实现类是 ApplicationLifetime（位于 Microsoft.Extensions.Hosting.Internal 命名空间）。调用 app.Services.GetService<IHostApplicationLifetime>方法后返回的对象正是 ApplicationLifetime 实例。不过，此处通过 IHostApplicationLifetime 接口访问 ApplicationLifetime 对象的成员即可，并不需要将类型强制转换为 ApplicationLifetime。目前，.NET 平台并不支持自定义实现 IHostApplicationLifetime 接口，因此开发人员无法将自己编写的类型添加到服务容器中（应用程序运行后会发生错误）。

下面的示例将演示应用程序生命周期事件的简单处理过程。

```
// 从服务容器中获取 IHostApplicationLifetime 服务
IHostApplicationLifetime? lifetime = app.Services.GetService
<IHostApplicationLifetime>();
if(lifetime != null)
{
    // 在应用程序启动后触发
    lifetime.ApplicationStarted.Register(() =>
    {
        app.Logger.LogInformation("应用程序已启动");
    });
    // 当应用程序正在停止时触发
    lifetime.ApplicationStopping.Register(() =>
    {
        app.Logger.LogInformation("应用程序正在停止……");
    });
    // 在应用程序完全停止后触发
    lifetime.ApplicationStopped.Register(() =>
    {
        app.Logger.LogInformation("应用程序已停止");
    });
}
```

ApplicationStarted、ApplicationStopping 等属性类型都是 CancellationToken 结构体（位于 System.Threading 命名空间），需要通过 Register()方法注册一个委托实例，才能在与委托绑定的方法中响应事件。本示例只是将文本信息写入日志记录中。在实际开发中，可根据具体情况作出响应。例如，在应用程序启动后创建一些数据文件，当应用程序停止后将这些文件删除。

运行示例程序后，会在控制台界面看到输出的日志信息，如图 1-5 所示。

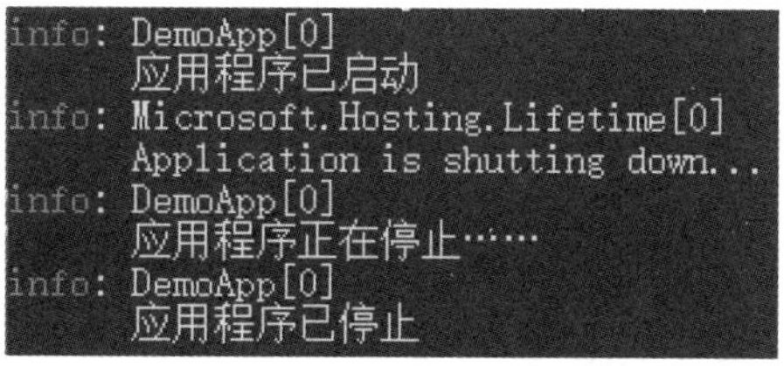
info: DemoApp[0]
应用程序已启动
info: Microsoft.Hosting.Lifetime[0]
Application is shutting down...
info: DemoApp[0]
应用程序正在停止……
info: DemoApp[0]
应用程序已停止

图 1-5　输出应用程序生命周期信息

第 2 章　运 行 环 境

本章要点

- 定义运行环境
- 适配多个运行环境

2.1　定义运行环境

ASP.NET Core 应用程序允许配置多个运行环境，并可以针对某个环境执行特定的代码。这使应用程序更易于被管理。例如，在项目开发过程中，为了能直观地发现、排查错误，可以选择启用带有详细异常信息的错误页面；一旦项目交付给客户并正式上线运行，出于安全考虑，应用程序在运行过程中将使用自定义错误页，并隐藏详细的异常信息（可以选择将详细信息写入服务器的日志中，仅把错误的摘要信息返回给客户端）。又如，在开发测试阶段，应用程序应当使用测试数据库；当项目正式投入使用后，才使用真实的数据库。这样做可有效避免因测试过程中出现的错误使数据库遭到破坏。

为了让应用程序能识别出不同的运行环境，需要为每个运行环境命名。原则上，环境名称只是一个字符串对象，开发人员可以定义任意有效的命名。当然，ASP.NET Core 应用程序预定义了 3 个运行环境，方便开发人员直接使用。这些预定义名称通过 Environments 类（位于 Microsoft.Extensions.Hosting 命名空间）的静态字段公开。

（1）Development：表示当前应用程序处于开发阶段。

（2）Staging：表示当前应用程序的开发基本完成，处于预发布阶段。

（3）Production：表示应用程序已完成开发和测试，即将投入生产环境中使用。

可以通过命令行参数和环境变量为应用程序指定运行环境。配置项的字段名为 environment，命令行参数可以使用--environment 或/environment，环境变量可以使用 ASPNETCORE_ENVIRONMENT。

下面的示例通过命令行参数将应用程序的运行环境配置为 Development。

```
dotnet demo.dll --environment=Development
```

也可以使用环境变量配置。

```
set ASPNETCORE_ENVIRONMENT=Production          // Windows
export ASPNETCORE_ENVIRONMENT=Production       // Linux

dotnet demo.dll
```

当然，也可以通过直接编写代码设置运行环境。

```
WebApplicationOptions appopt = new()
{
    // 命令行参数
    Args = args,
    // 指定运行环境
    EnvironmentName = "Shared"
};
var builder = WebApplication.CreateBuilder(appopt);
// 创建应用程序
var app = builder.Build();

...

// 启动应用程序
app.Run();
```

首先实例化 WebApplicationOptions 对象，然后通过 EnvironmentName 属性设置运行环境，最后把此 WebApplicationOptions 对象传递给 CreateBuilder()方法。该示例使用了一个自定义的环境名称——Shared。应用程序运行后，可以在输出的日志记录中查看正在使用的运行环境，如图 2-1 所示。

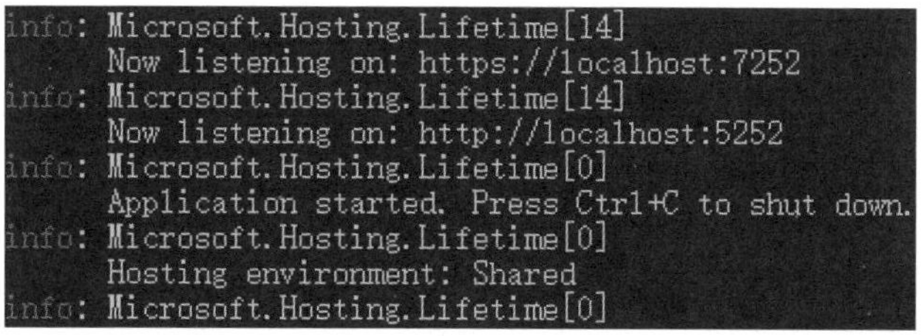
info: Microsoft.Hosting.Lifetime[14]
Now listening on: https://localhost:7252
info: Microsoft.Hosting.Lifetime[14]
Now listening on: http://localhost:5252
info: Microsoft.Hosting.Lifetime[0]
Application started. Press Ctrl+C to shut down.
info: Microsoft.Hosting.Lifetime[0]
Hosting environment: Shared
info: Microsoft.Hosting.Lifetime[0]

图 2-1　自定义的运行环境

本书推荐使用命令行参数或环境变量设置应用程序的运行环境，这样做比较灵活，切换环境不需要重新编译程序代码。

2.2　Is{EnvironmentName}扩展方法

ASP.NET Core 应用程序支持多个环境配置，在运行阶段需要根据不同的环境执行相应的代码。

WebApplication 类有一个名为 Environment 的公共属性，声明类型为 IWebHostEnvironment

接口，此接口派生自 IHostEnvironment 接口，因此它继承了 IHostEnvironment 接口的 EnvironmentName 属性。也就是说，应用程序代码通过 IWebHostEnvironment. EnvironmentName 属性可以获知当前正在使用的运行环境名称。虽然该属性定义了 set 访问器，但是不要在代码中修改它的值，那样做会导致应用程序对运行环境作出错误的判断。

程序代码可以根据 EnvironmentName 属性的值判断应用程序正在使用的运行环境。在实际使用中，调用扩展方法会更方便。这些扩展方法以 Is{EnvironmentName}的格式命名，由 HostEnvironmentEnvExtensions 类公开。具体说明如下。

（1）IsDevelopment()方法：如果当前正在使用的环境名称为 Development，该方法就返回 true，否则返回 false。

（2）IsProduction()方法：如果正在使用的环境名称为 Production 就返回 true，否则返回 false。

（3）IsStaging()方法：同上，判断当前的运行环境是否为 Staging。

（4）IsEnvironment()方法：如果所使用的运行环境名称在 Development、Production 和 Staging 之外，就需要调用该方法进行判断。

下面的示例将根据多个运行环境构建 HTTP 管线。

```
var builder = WebApplication.CreateBuilder(args);

var app = builder.Build();

if (app.Environment.IsDevelopment())
{
    app.MapGet("/", () =>"当前项目正处在开发阶段");
}
else if (app.Environment.IsProduction())
{
    app.MapGet("/", () =>"当前项目已投放使用");
}
else if(app.Environment.IsEnvironment("NoAPI"))
{
    app.MapGet("/", () =>"此版本不公开 Web API");
}
…

app.Run();
```

上述代码中，针对不同的环境调用 MapGet()方法。由于方法调用被写在 if...else 语句中，这使得 MapGet()方法在整个 HTTP 管线的构建中仅调用一次。传递给 MapGet()方法的第 1 个参数值“/”表示匹配 URL 的根地址（如 https://testhost/）。在此 if...else 语句之后如果还需要调用 MapGet()方法，那就不能再匹配根地址“/”，否则会导致该地址存在多个匹配项，应用程序无法选择执行哪个地址，进而发生错误。

在最后一个 else 分支中，由于 NoAPI 环境是自定义的名称，需要调用 IsEnvironment()方法验证当前环境。

假设通过命令行参数设置运行环境为 NoAPI，那么示例程序运行后将执行以下代码，结果如图 2-2 所示。

```
app.MapGet("/", () =>"此版本不公开 Web API");
```

图 2-2　应用程序运行在 NoAPI 环境中

2.3　多运行环境下的配置文件

ASP.NET Core 项目模板会生成一个名为 appsettings.json 的配置文件，默认的 Web 应用程序配置除了加载 appsettings.json 文件外，还会查找以 appsettings.{EnvironmentName}.json 格式命名的配置文件。这些配置文件是专用于特定运行环境的。例如，若当前运行环境为 Production，应用程序在启动时会加载 appsettings.json 和 appsettings.Production.json 两个配置文件。

下面给出一个示例。应用程序项目中包含以下 3 个配置文件。

（1）appsettings.json：适用于所有运行环境的配置文件。

（2）appsettings.Demo.json：适用于名为 Demo 的运行环境的配置文件。

（3）appsettings.Development.json：适用于名为 Development 的运行环境的配置文件。

3 个配置文件都包含一个相同的字段——my_key。

```
// appsettings.json
{
    "my_key": "通用于各环境的配置"
}

// appsettings.Development.json
{
    "my_key": "适用于开发阶段的配置"
}

// appsettings.Demo.json
{
```

```
    "my_key": "适用于 Demo 运行环境的配置"
}
```

在应用程序中读入 my_key 的值并返回给客户端。

```
var builder = WebApplication.CreateBuilder(args);
var app = builder.Build();

app.MapGet("/", () =>
{
    // 读取 my_key 配置项的值
    string cfgValue = app.Configuration["my_key"];
    return $"【my_key】的值：{cfgValue}";
});

app.Run();
```

app.Configuration 属性中存储了应用程序从各个配置源加载的配置信息（包括 appsettings.json、appsettings.Development.json 等配置文件），其数据类型实现了 IConfiguration 接口。程序代码可以通过以下索引器访问特定配置项的值。

```
[DefaultMember("Item")]
public interface IConfiguration
{
    // 支持读写某个配置项的值
    string this[string key] { get; set; }

    …
}
```

其中，参数 key 用于指定配置项的字段名，在本示例中是 my_key。

假设以 Demo 环境运行示例程序，并从 Web 浏览器访问其根 URL，将得到如图 2-3 所示的响应信息。

如果给应用程序指定的运行环境没有匹配的配置文件，默认会读取 appsettings.json 文件中的配置信息。例如，以 Test 环境运行应用程序，会得到如图 2-4 所示的响应信息。

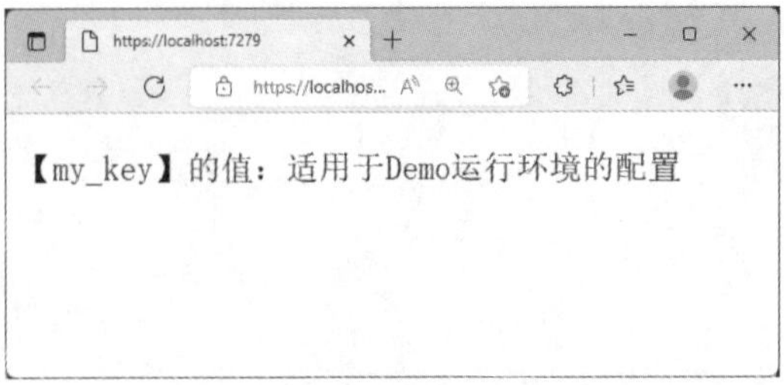

图 2-3　以 Demo 环境运行示例程序

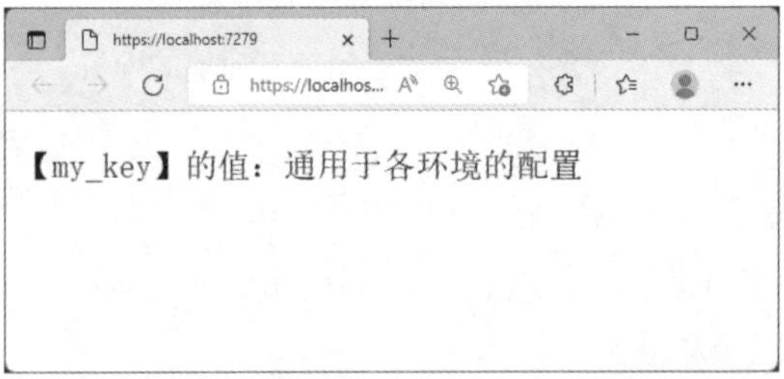

图 2-4　默认读取 appsettings.json 文件中的配置

2.4　用于环境筛选的 Razor 标记

在 ASP.NET Core 项目中，运行于服务器的 HTML 标记都会使用到 Razor 标记语法。它可以在 HTML 标记中插入 C#或 VB（Visual Basic）代码。当客户端（通常是 Web 浏览器）发出请求后，服务器会执行 Razor 标记并生成最终呈现给客户端的 HTML 代码。

HTML 标记本身无法检查 ASP.NET Core 应用程序正在使用的运行环境，因此需要借助 Razor 标记对运行环境进行分析，不同运行环境下显示相应的内容。例如，如果应用程序运行在 Development 环境下，为了便于测试，开发人员会考虑在 HTML 页面中显示一些调试信息；而当应用程序运行在 Production 环境下时，由于应用程序已投入生产环境中使用，就不再需要在 HTML 页面上显示详细的调试信息了。又如，某个应用程序即将发布到两台服务器上，运行于 A 服务器的应用程序是面向内网用户的，设置运行环境为 ENV1；而运行在 B 服务器上的应用程序是面向外网用户的，设置运行环境为 ENV2。在组织 HTML 标记时，如果运行环境为 ENV1，就显示登录用户的完整信息，并且显示公司内部通知；如果运行环境为 ENV2，就显示登录用户的简要信息，并且不显示公司内部通知。

Razor 语法使用 EnvironmentTagHelper 类（位于 Microsoft.AspNetCore.Mvc.TagHelpers 命名空间）扩展 HTML 功能，对运行环境进行筛选。该类定义了 3 个公共属性。

（1）Include：指定一个环境列表，只要当前应用程序所使用的环境在此列表中，就会呈现 HTML 内容。

（2）Exclude：其含义为“排除”，即指定一个环境列表，如果应用程序正在使用的环境在此列表中，就不会呈现 HTML 内容。其逻辑与 Include 属性相反。

（3）Names：指定一个环境列表，如果应用程序正在使用的环境在此列表中，就会呈现 HTML 内容。其处理方式与 Include 属性相同。

这 3 个属性的类型都是字符串，指定运行环境列表时，环境名称之间使用逗号（英文）分隔。例如：

```
<environment include="Development,Test">
<div>你好，世界</div>
</environment>
```

上述示例中 Include 属性指定了两个环境名称——Development 和 Test，只要当前应用程序的运行环境为 Development 或 Test，那么服务器就会将“你好，世界”输出到响应流中（呈现在 Web 浏览器上）；如果程序正在使用的环境不是 Development，也不是 Test，那么“你好，世界”将不会输出到响应流中。

下面的示例将使用 Razor Pages 呈现 HTML，并在 HTML 页面上通过 environment 标记筛选运行环境。大致步骤如下。

（1）在构建 app 前，需要向服务容器注册 Razor Pages 功能。

```
var builder = WebApplication.CreateBuilder(args);
// 添加 Razor Pages 功能
```

```
builder.Services.AddRazorPages();

var app = builder.Build();
```

（2）当 app 构建后，向 HTTP 管线中添加 Razor Pages 终结点。

```
app.MapRazorPages();
```

（3）调用 MapFallback()方法，向 HTTP 管线中添加一个“回退”终结点，当客户端所请求的 URL 无效时执行。

```
app.MapFallback(() =>"找不到指定的资源");
```

（4）开始执行 app。

```
app.Run();
```

（5）在项目目录下新建子目录，命名为 Pages。

（6）在 Pages 目录下新建文件，命名为 index.cshtml。

（7）在 index.cshtml 中输入以下内容。

```
@page
@addTagHelper Microsoft.AspNetCore.Mvc.TagHelpers.EnvironmentTagHelper,
Microsoft.AspNetCore.Mvc.TagHelpers

<html lang="zh-cn">
<head>
<meta charset="utf-8"/>
<title>Demo</title>
</head>
<body>
<environment include="Development">
<span>项目开发中……</span>
</environment>
<environment include="Production">
<span>项目已上线</span>
</environment>
</body>
</html>
```

在 HTML 文档的第 1 行必须写上@page 指令，表示此页面用于 Razor Pages。第 2 行使用@addTagHelper 指令将 EnvironmentTagHelper 类导入，其格式为“<完整类名>, <程序集名>”。EnvironmentTagHelper 类所在的程序集名称与命名空间名称相同，都是 Microsoft.AspNetCore.Mvc.TagHelpers。

当运行环境为 Development 时，将呈现一个<span>元素，包含文本“项目开发中......”；当运行环境为 Production 时，<span>元素中的文本呈现为“项目已上线”。

假设应用名称为 DemoApp，在启动应用程序时通过命令行参数指定运行环境为 Development。

```
dotnet DemoApp.dll --environment=Development
```

Web 浏览器中呈现的内容如图 2-5 所示。

按 Ctrl+C 快捷键退出应用程序，接着以 Production 环境运行。

```
dotnet DemoApp.dll --environment=Production
```

在 Web 浏览器中导航到应用程序 URL，页面呈现效果如图 2-6 所示。

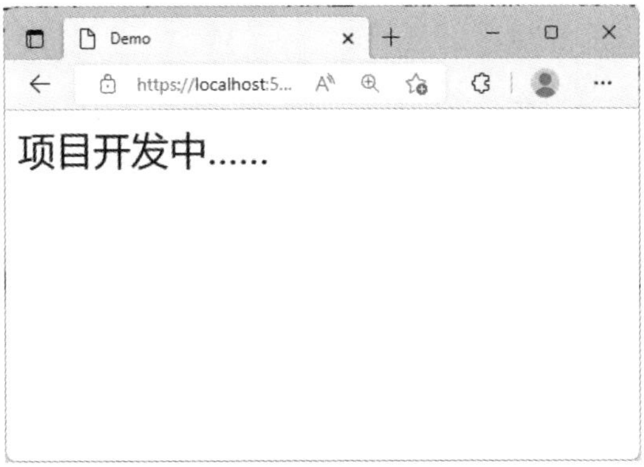

图 2-5　以 Development 环境运行

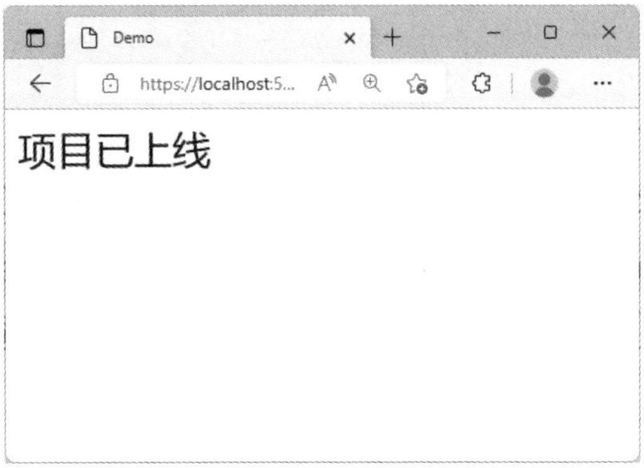

图 2-6　以 Production 环境运行

2.5　运行环境与依赖注入

被添加到服务容器中的对象都可以使用依赖注入。由.NET 运行时自动创建类型实例，并传递该实例所需要访问的对象引用。最常见的做法是通过构造函数进行注入。假设 A 类的实例中需要访问 B 类的实例，那么在 A 类的构造函数中声明一个 B 类的参数，并在构造函数内部接收 B 类实例的引用，存放到 A 类的私有字段中。最后将 A、B 两个类都注册到服务容器中。当调用者获取 A 类的实例时，服务容器会自动调用 A 类的构造函数，且自动将 B 类的实例传递给 A 类的构造函数。依赖注入是 ASP.NET Core 的一个核心功能，在后续章节中会详细阐述。

另外，像 MVC 中的控制器类，或者 Razor Pages 中的页面模型类，尽管开发人员不需要将它们放进服务容器中，但它们的实例是由.NET 运行时自动创建的，因此也支持依赖注入。

而包含应用程序运行环境信息的相关类型在程序初始化过程也会被注册到服务容器中。对于使用通用主机启动的.NET 应用程序，可以通过 IHostEnvironment 接口得知当前运行环境；而 ASP.NET Core 应用程序是在通用主机的基础上启用 Web 主机的，因此除了 IHostEnvironment 接口，还可以通过 IWebHostEnvironment 接口获取运行环境信息。这是因为 IWebHost-Environment 接口派生自 IHostEnvironment 接口，所以继承了 EnvironmentName 属性。

在下面的示例中，将定义一个名为 TestService 的类。该类公开一个 GetHubName()方法，此方法会根据应用程序当前所处的运行环境返回不同的字符串。

```
public class TestService
{
```

```
    readonly IHostEnvironment _hostEnv;

    public TestService(IHostEnvironment hostenv)
    {
        _hostEnv = hostenv;
    }

    public string GetHubName()
    {
        return _hostEnv.EnvironmentName switch
        {
            "DATA_CNT"        =>"数据中心",
            "REPORT_SVR"      =>"报表中心",
            "NEWS_GRP"        =>"新闻组",
            _                 =>"未知环境"
        };
    }
}
```

在TestService类中要先声明一个IHostEnvironment接口类型的私有字段，命名为_hostEnv，在构造函数中进行赋值。对象的引用来自构造函数参数，此参数会通过依赖注入自动获得所需对象的引用。

在 GetHubName()方法中，使用 switch 子句对 EnvironmentName 属性的值进行分析，如果当前运行环境的名称为 DATA_CNT，则返回“数据中心”；如果当前的运行环境是 REPORT_SVR，则返回“报表中心”；如果当前的运行环境是 NEWS_GRP，则返回“新闻组”；如果运行环境的名称与筛选的表达式均不匹配，则返回“未知环境”。

该示例也可以使用 IWebHostEnvironment 接口。

```
readonly IWebHostEnvironment _hostEnv;

public TestService(IWebHostEnvironment hostenv)
{
    _hostEnv = hostenv;
}
```

在项目的 Program.cs 文件（即 ASP.NET Core 应用程序的初始化代码）中，创建 WebApplicationBuilder 实例后，需要将上文定义的 TestService 类注册到服务容器中。

```
var builder = WebApplication.CreateBuilder(args);
// 向服务容器注册自定义类型
builder.Services.AddSingleton<TestService>();
…
```

AddSingleton()方法表示 TestService 类型将使用单实例模式，即在整个应用程序生命周期内，它只创建一个对象实例（仅调用一次构造函数）。

在调用 Build()方法后，需要为 app 对象添加一个绑定到根 URL 的终结点，并在处理代码中调用 GetRequiredService()方法主动获取 TestService 实例，然后调用 GetHubName()方法，将结果返回给客户端。代码如下。

```
app.MapGet("/", () =>
{
    // 获取自定义服务类的实例
    var svr = app.Services.GetRequiredService<TestService>();
    // 调用服务的实例方法
    return svr.GetHubName();
});
```

以 NEWS_GRP 环境启动应用程序，然后使用 Web 浏览器访问应用程序的根地址，服务器返回“新闻组”，如图 2-7 所示。

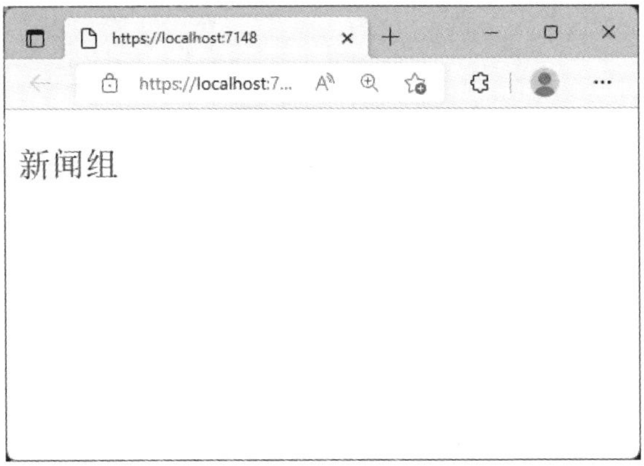

图 2-7　返回特定于运行环境的字符串

第 3 章 依赖注入

本章要点

- 服务容器与 ServiceProvider
- 服务的生存期
- 注入多个服务实例

3.1 依赖注入与服务容器

假设有 3 个类——A、B、C，A 类中需要引用 B 类的实例，B 类中需要引用 C 类的实例，比较常见的方案是通过类的构造函数传递被引用的对象。例如：

```
public class A
{
    private B _b;

    public A(B b)
    {
        _b = b;
    }
}

public class B
{
    C _c;

    public B(C c)
    {
        _c = c;
    }
}

public class C { }
```

如此一来，A、B、C 类之间就存在了依赖关系，即创建 B 类实例前要先创建 C 类实例，创建 A 类实例前要先创建 B 类实例。

```
C xc = new C();
B xb = new B(xc);
A xa = new A(xb);
```

随着需求的变化，A 类所依赖的类型可能不仅仅是 B 类，而是结构与 B 类相近的其他类型。于是，可以将 A 类对 B 类的依赖关系修改为 A 类对 ITest 接口类型的依赖，然后 B 类实现 ITest 接口。

```
public class A
{
    private ITest _t;

    public A(ITest ts)
    {
        _t = ts;
    }
}

// 公共接口
public interface ITest { }

public class B : ITest
{
    …
}
```

将来因功能需要，可能会扩展出 D 类。D 类同样实现 ITest 接口，因此在创建 A 类实例时，既可以向构造函数传递 B 类实例，也可以传递 D 类实例。

```
ITest _e1 = new B();
A var1 = new A(_e1);

ITest _e2 = new D();
A var2 = new A(_e2);
```

采用抽象类/实现类的设计模式也是可行的，如

```
public abstract class LoaderBase
{
    public abstract string GetData();
}

public class XMLLoader : LoaderBase
{
    public override string GetData()
    {
        return "XML 数据";
    }
```

```
    }

    public class JSONLoader : LoaderBase
    {
        public override string GetData()
        {
            return "JSON 数据";
        }
    }

    public class CSVLoader : LoaderBase
    {
        public override string GetData()
        {
            return "CSV 数据";
        }
    }
```

LoaderBase 是一个抽象类，包含 GetData()抽象方法。假设该类的作用是通过 GetData()方法从数据文件中加载数据。常见的数据文件可以是 TXT、JSON、XML、INI、CSV 等格式。于是，从 LoaderBase 类可以派生出具有相应功能的类。上述代码中，XMLLoader 类用于加载 XML 格式的数据，JSONLoader 类用于加载 JSON 格式的数据，CSVLoader 类则用于加载 CSV 格式的数据。

再假设有一个 ConfigReader 类，专用于读取文件数据。

```
public sealed class ConfigReader
{
    private readonly LoaderBase _loader;

    public ConfigReader(LoaderBase loader)
    {
        _loader = loader;
    }
}
...
```

ConfigReader 类依赖的是 LoaderBase 类，所以它不需要考虑被读取的数据来自什么格式的文件，具体的文件加载工作由 LoaderBase 的实现类去完成。在创建 ConfigReader 类的实例时，可以向构造函数传递 JSONLoader 实例或 CSVLoader 实例。

```
ConfigReader reader = new(new JSONLoader());
ConfigReader reader = new(new CSVLoader());
```

通过构造函数参数或属性向调用者传递它所依赖的对象，称为依赖注入（Dependency Injection，DI）。给调用者赋值其所依赖对象的过程即是“注入”。此过程是通过一种特定的管理机制主动实现的，而不是调用者直接去获得。上述示例代码编写方式只适合少量对象间的依赖注入，如果一个应用程序中存在很多类型，并且类型之间的依赖关系非常复杂，手动编写代码的方式不仅工作量大，而且混乱，也不利于维护和扩展。于是，就需要一种机制，可以自动管理类型之间的依赖关系，当调用方需要某个类型的实例时，该管理机制能够自动创建该类型的实例，自动创建该类型所依赖的类型实例并完成注入。

这种机制包括以下两部分。

（1）一个容器，即服务容器，可以将存在依赖关系的类型添加到该容器中，称为服务。

（2）自动激活实例。当调用者需要容器中某个类型实例时，服务容器会自动调用类型构造函数创建实例，并根据构造函数的参数列表自动创建和引用所依赖的类型实例。

3.1.1　ServiceCollection 类

ServiceCollection 类是.NET 内置的服务容器，它实现了 IServiceCollection 接口。依据接口的继承层次，可知 ServiceCollection 类也实现了 IList<T>、ICollection<T>、IEnumerable<T>等接口。也就是说，ServiceCollection 类的使用方法与其他的泛型集合相同。其中，类型参数 T 为 ServiceDescriptor 类。因此，被添加到服务容器中的服务类型将通过 ServiceDescriptor 对象来描述。

ServiceDescriptor 类公开了以下属性。

（1）ServiceType：表示服务类型的 Type 对象。

（2）ImplementationType：如果服务的实现方式为“接口+实现类”或“抽象类+派生类”，那么 ServiceType 属性表示服务的接口或抽象类，ImplementationType 属性表示服务的实现类。ImplementationType 属性是可选的，可以不指定。

（3）ImplementationInstance：表示服务类型的实例对象。

（4）Lifetime：表示服务对象的生存期类型。

为了使服务容器更容易访问，ServiceCollection 类提供了以下扩展方法（在 ServiceCollection-DescriptorExtensions 和 ServiceCollectionServiceExtensions 类中定义）。

（1）Add()、TryAdd()：将单个或多个服务添加到服务容器中。

（2）Replace()：替换容器中的 ServiceDescriptor 实例。

（3）RemoveAll()：删除容器所有与指定类型相同的服务。

（4）AddSingleton()、TryAddSingleton()：将服务添加到容器中，并且服务实例的生存期为单个实例。

（5）AddScoped()、TryAddScoped()：将服务添加到容器中，服务实例的生存期被限定为“作用域”（Scoped）范围内。

（6）AddTransient()、TryAddTransient()：将服务添加到容器中，服务实例的生存期为瞬时性。

上述扩展方法中，以 TryAdd 开头命名的方法表示只有当目标服务类型未在容器中注册时才会添加服务，否则不进行任何处理。

3.1.2 ServiceProvider 类

当服务类型注册到容器后，可使用 ServiceProvider 类获取服务类型的实例。服务类型的实例化过程由 ServiceProvider 类负责完成，调用代码只须告诉 ServiceProvider 类想要获取哪个类型的服务即可。

ServiceProvider 类实现了 IServiceProvider 接口。在编写代码过程中，应通过 IServiceProvider 接口类型的变量引用 ServiceProvider 实例，因为 ServiceProviderServiceExtensions 类中定义的扩展方法均面向 IServiceProvider 接口（GetService、CreateScope 等）。

ServiceProvider 类没有公共构造函数，不能直接使用 new 运算符实例化，而是调用 IServiceCollection 类型的扩展方法 BuildServiceProvider()获得 ServiceProvider 实例。

3.2 .NET 项目中的依赖注入

非 ASP.NET Core 项目在默认情况下不包含与依赖注入相关的引用，需要手动添加 Nuget 包引用。

在应用程序项目所在的目录下执行以下命令，可添加对 Microsoft.Extensions.DependencyInjection 包的引用。

```
dotnet add package Microsoft.Extensions.DependencyInjection
```

或者直接编辑项目文件，在<Project>下添加以下内容。

```
<ItemGroup>
<PackageReference Include="Microsoft.Extensions.DependencyInjection"
Version="x.y.z" />
</ItemGroup>
```

其中，Version="x.y.z"表示要引用的 Nuget 包的版本号。

在代码文件中引入相关的命名空间。

```
using Microsoft.Extensions.DependencyInjection;
```

然后创建容器实例，并调用 BuildServiceProvider()扩展方法产生 ServiceProvider 实例。

```
IServiceCollection services = new ServiceCollection();
// 添加服务
services.AddTransient<ITestService, MyService>();
IServiceProvider serviceProvider = services.BuildServiceProvider();
```

当需要访问服务实例时，可通过 serviceProvider 对象获取。

```
ITestService? theservice = serviceProvider.GetService<ITestService>();
```

GetService()方法将返回指定服务类型的实例，泛型参数用于指定要获取的服务类型，上述代码中，服务类型为 ITestService。如果服务容器中并不存在指定的服务类型，该方法将返回 null。因为在使用服务实例时需要进行 null 检查。

```
theservice?.DoSomething();
```

也可以调用非泛型的 GetService()方法，通过传递方法参数指定期望的服务类型。

```
ITestService? theservice = serviceProvider.GetService(typeof(ITestService))
as ITestService;
```

3.3 ASP.NET Core 项目中的依赖注入

在 ASP.NET Core 项目中，默认已包含对 Microsoft.Extensions.DependencyInjection 包的引用，开发人员不需要手动添加相关的 Nuget 包引用。

调用 WebApplication.CreateBuilder()方法会自动创建服务容器（ServiceCollection）实例，并向容器添加一些必要的基础服务。创建 WebApplicationBuilder 实例后，通过 Services 属性可以访问服务容器实例。

WebApplicationBuilder 类的内部封装了对 HostApplicationBuilder（位于 Microsoft.Extensions.Hosting 命名空间）实例的引用。ServiceCollection 容器的实例也是在 HostApplicationBuilder 类中初始化。以下是 HostApplicationBuilder 类的部分源代码。

```
public sealed class HostApplicationBuilder
{
    …
    private readonly ServiceCollection _serviceCollection = new();

    private Func<IServiceProvider> _createServiceProvider;
    private Action<object> _configureContainer = _ => { };
    private HostBuilderAdapter? _hostBuilderAdapter;

    private IServiceProvider? _appServices;
    …
}
```

通过 Services 属性将服务容器实例对外公开。

```
public IServiceCollection Services => _serviceCollection;
```

_createServiceProvider 字段是委托类型，用于创建 ServiceProvider 实例。当调用 Build()方法时，该委托会被执行。创建的 ServiceProvider 实例将赋值给_appServices 字段。

```
_appServices = _createServiceProvider();
```

同时，调用 MakeReadOnly()方法使服务容器（ServiceCollection）变为只读集合。之后就不能再向容器添加服务了，当然也不能删除或替换服务。

```
_serviceCollection.MakeReadOnly();
```

最后，把 HostApplicationBuilder 实例封装到 WebApplicationBuilder 类中。

```
public sealed class WebApplicationBuilder
{
    …
```

```
    private readonly HostApplicationBuilder _hostApplicationBuilder;
    …
}
```

通过 Services 属性继续对外公开服务容器。

```
public IServiceCollection Services => _hostApplicationBuilder.Services;
```

在 Build()方法中也调用了 HostApplicationBuilder 类的 Build()方法。

```
public WebApplication Build()
{
    …
    _builtApplication = new WebApplication(_hostApplicationBuilder.Build());
    return _builtApplication;
}
```

ServiceCollection 对象的传递过程如图 3-1 所示。

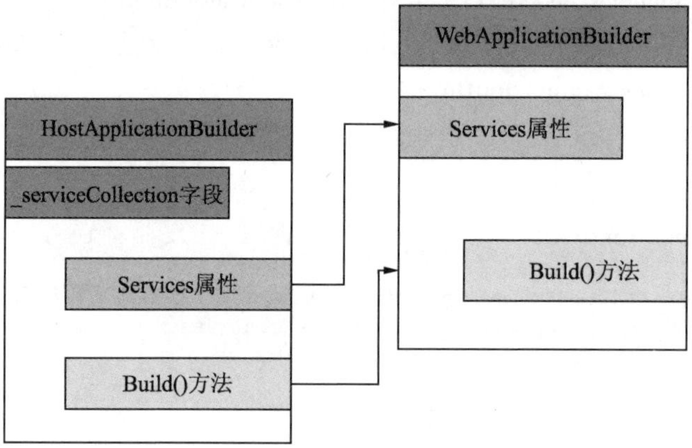

图 3-1　ServiceCollection 对象的传递过程

待 WebApplication 对象构建完成后，服务容器处于只读状态，不能被修改。此时可以访问 app.Services 属性获得对 ServiceProvider 实例的引用，再通过调用 GetService()方法获取要使用的服务实例。

```
// 获取服务实例
ILoggerFactory? logfac = app.Services.GetService<ILoggerFactory>();
// 使用服务实例
if(logfac != null)
{
    // 创建日志记录器
    ILogger logger = logfac.CreateLogger("示例日志");
    // 写入一条日志记录
    logger.LogInformation("此条记录仅用于测试");
}
```

上述代码从服务容器中获取 ILoggerFactory 类型的服务实例，再调用它的 CreateLogger()

方法创建一个用于写入应用程序日志的对象，最后调用 LogInformation()方法写入一条测试日志。

3.4　构建存在依赖关系的服务

本节将通过一个示例演示如何在应用程序中定义存在依赖关系的服务类型。

定义一个名为 DataSender 的服务类，调用它的 Sendout()方法后会将字符串数据发回给客户端。DataSender 类依赖名为 IDataWriter 的服务。IDataWriter 定义为接口，包含一个 Output()方法，可将字符串数据以特定的格式输出。

```
public interface IDataWriter
{
    ValueTask Output(string data);
}
```

本示例中还定义了两个实现了 IDataWriter 接口的类，它们的功能都是将字符串数据写入 HTTP 响应流中。其中，JsonDataWriter 类将数据以 JSON 格式写入响应流，XmlDataWriter 类则将数据以 XML 格式写入响应流。

```
public class JsonDataWriter : IDataWriter
{
    readonly HttpContext _httpContext;
    public JsonDataWriter(IHttpContextAccessor accessor)
    {
        _httpContext = accessor.HttpContext!;
        // 设置响应头
        _httpContext.Response.ContentType = "application/json;charset=
        UTF-8";
    }

    public async ValueTask Output(string data)
    {
        string res = $"{{ \"data\": \"{data}\" }}";
        // 写入响应内容
        await _httpContext.Response.WriteAsync(res);
    }
}

public class XmlDataWriter : IDataWriter
{
    readonly HttpContext _httpContext;
    public XmlDataWriter(IHttpContextAccessor accessor)
    {
        _httpContext = accessor.HttpContext!;
        // 设置响应头
        _httpContext.Response.ContentType = "application/xml;charset=
        UTF-8";
```

```
    }

    public async ValueTask Output(string data)
    {
        string odata = $"<data>{data}</data>";
        // 写入响应内容
        await _httpContext.Response.WriteAsync(odata);
    }
}
```

这两个类在结构上和实现逻辑上都很相似，区别在于 JsonDataWriter 类生成的数据为 JSON 格式，而 XmlDataWriter 类生成的数据为 XML 格式。

这两个类都依赖 IHttpContextAccessor 接口类型的服务，作用是通过该服务可以引用与当前 HTTP 通信关联的 HttpContext 对象，然后再通过 HttpContext 对象将数据写入 HTTP 的响应流中。最后，DataSender 类依赖一个 IDataWriter 类型的服务对象，其具体的实例类型可能是 XmlDataWriter 或 JsonDataWriter。

将相关的服务添加到服务容器中。

```
var builder = WebApplication.CreateBuilder(args);

// 注册服务
builder.Services.AddHttpContextAccessor();
builder.Services.AddSingleton<IDataWriter, JsonDataWriter>();
//builder.Services.AddSingleton<IDataWriter, XmlDataWriter>();
builder.Services.AddTransient<DataSender>();
…
```

因为本示例需要从服务容器中获取 IHttpContextAccessor 服务，所以要调用 AddHttpContext-Accessor()扩展方法注册默认的 HttpContextAccessor 服务。JsonDataWriter 和 XmlDataWriter 类型的服务可以同时加入容器中，在获取服务实例时，优先获取最后添加到容器中的类型。假设先添加 JsonDataWriter 服务，再添加 XmlDataWriter 服务，那么最终被注入 DataSender 类构造函数的是 XmlDataWriter 服务。

```
builder.Services.AddSingleton<IDataWriter, JsonDataWriter>();
builder.Services.AddSingleton<IDataWriter, XmlDataWriter>();
```

以下代码尝试从服务容器中获得 DataSender 实例，并调用它的 Sendout()方法。

```
app.MapGet("/", async () =>
{
    // 获取 DataSender 实例
    DataSender ds = app.Services.GetRequiredService<DataSender>();
    // 使用服务实例
    await ds.Sendout("示例数据");
});
```

得到的结果如下。

```
<data>示例数据</data>
```

可以对示例做一个扩展，定义一个选项类，用于配置数据输出的格式。

```
// 枚举
public enum Format { Json, Xml}

// 选项类
public class DataOutputOptions
{
    public Format OutputFormat { get; set; }
}
```

将选项类对象也添加到服务容器中。

```
builder.Services.Configure<DataOutputOptions>(options =>
{
    options.OutputFormat = Format.Json;
});
```

在注册 IDataWriter 服务时，可以根据 DataOutputOptions 选项类的配置决定产生 JsonDataWriter 实例还是 XmlDataWriter 实例。

```
builder.Services.AddSingleton<IDataWriter>(provider =>
{
    // 获得选项类的实例引用
    IOptions<DataOutputOptions> opt = provider.GetRequiredService
    <IOptions<DataOutputOptions>>();
    // 获取 IHttpContextAccessor
    IHttpContextAccessor accessor = provider.GetRequiredService
    <IHttpContextAccessor>();
    // 根据 OutputFormat 属性的值创建不同的类型实例
    IDataWriter datawriter = opt.Value.OutputFormat switch
    {
        Format.Json => new JsonDataWriter(accessor),
        Format.Xml => new XmlDataWriter(accessor),
    };
    return datawriter;
});
```

调用 JsonDataWriter 类或 XmlDataWriter 类的构造函数时需要注入 IHttpContextAccessor 类型的服务实例，由于此处是手动调用构造函数，所以要先从服务容器中获取 IHttpContext-Accessor 服务实例，再手动传递给构造函数。

此时运行示例程序，将得到 JSON 格式的数据。

```
{ "data": "示例数据" }
```

若需要 XML 格式的数据，可以将选项类配置为 XML 格式。

```
builder.Services.Configure<DataOutputOptions>(options =>
{
    options.OutputFormat = Format.Xml;
});
```

3.5 服务的生存期

服务注册到容器时，有 3 种生存期可供选择。

（1）单实例服务：生存期最长，实例生存期与服务容器相同。在整个容器的生存期内，服务仅创建一个实例。

（2）瞬时服务（暂时性服务）：生存期最短，每次从服务容器中获取时都会创建新的服务实例。

（3）作用域服务：从容器根部的 ServiceProvider 上创建子作用域。在该作用域的生存期内，服务只创建一个实例。当该作用域的生命周期结束时会释放服务实例。

下面将通过一个示例验证服务在不同生存期的实例化行为。代码定义一个服务类，在其构造函数中生成一个新的 GUID 值。该 GUID 可以通过 MyID 属性获得。

```
public class TestService : IDisposable
{
    readonly Guid _guid;
    public TestService()
    {
        _guid = Guid.NewGuid();
        Console.WriteLine("{0}实例化", GetType().Name);
    }

    // 获取与实例相关的 GUID
    public Guid MyID => _guid;

    public void Dispose()
    {
        Console.WriteLine("{0}实例即将释放", GetType().Name);
    }
}
```

此服务类实现了 IDisposable 接口，当实例被释放时会向屏幕输出提示信息。

首先，测试一下单实例服务的实例化行为。

```
// 创建服务容器
var services = new ServiceCollection();
// 注册服务
services.AddSingleton<TestService>();
// 构建 ServiceProvider
using(var provider = services.BuildServiceProvider())
{
    // 获取 3 次服务实例
    var x1 = provider.GetRequiredService<TestService>();
    var x2 = provider.GetRequiredService<TestService>();
    var x3 = provider.GetRequiredService<TestService>();
    // 输出 GUID
```

```
    Console.WriteLine("变量 1: {0}", x1.MyID);
    Console.WriteLine("变量 2: {0}", x2.MyID);
    Console.WriteLine("变量 3: {0}", x3.MyID);
}
```

上述代码将从服务容器中获取 3 次 TestService 实例，接着依次输出它们的 GUID 值。运行代码后发现 3 次输出的 GUID 相同，说明从容器中获取的引用为同一个服务实例，如图 3-2 所示。

接下来测试一下瞬时性服务的实例化行为。

```
// 创建服务容器
var services = new ServiceCollection();
// 添加服务
services.AddTransient<TestService>();
// 构建 ServiceProvider
using var svcPrvd = services.BuildServiceProvider();
// 获取 4 次服务实例
var x1 = svcPrvd.GetRequiredService<TestService>();
var x2 = svcPrvd.GetRequiredService<TestService>();
var x3 = svcPrvd.GetRequiredService<TestService>();
var x4 = svcPrvd.GetRequiredService<TestService>();
// 依次输出它们的 GUID
Console.WriteLine($"变量 1: {x1.MyID}");
Console.WriteLine($"变量 2: {x2.MyID}");
Console.WriteLine($"变量 3: {x3.MyID}");
Console.WriteLine($"变量 4: {x4.MyID}");
```

代码运行后，可以看到 4 个变量的 GUID 均不相同，这表明容器为服务创建了 4 个实例，如图 3-3 所示。

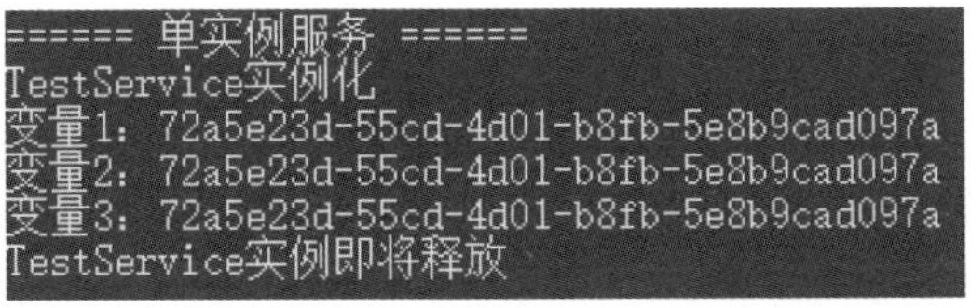

图 3-2　3 个变量的 GUID 相同

图 3-3　4 个变量的 GUID 均不相同

最后测试作用域服务的实例化行为。

```
// 创建服务容器
var container = new ServiceCollection();
```

```
// 注册服务
container.AddScoped<TestService>();
// 构建位于根部的 ServiceProvider
using var rootProvider = container.BuildServiceProvider();

Console.WriteLine("----- 开始 子作用域-1 -----");
// 创建第 1 个子作用域
using (var scoped1 = rootProvider.CreateScope())
{
    // 获取 3 次服务实例
    var v1 = scoped1.ServiceProvider.GetRequiredService<TestService>();
    var v2 = scoped1.ServiceProvider.GetRequiredService<TestService>();
    var v3 = scoped1.ServiceProvider.GetRequiredService<TestService>();
    // 输出 GUID
    Console.WriteLine($"变量 1: {v1.MyID}");
    Console.WriteLine($"变量 2: {v2.MyID}");
    Console.WriteLine($"变量 3: {v3.MyID}");
}
Console.WriteLine("----- 结束 子作用域-1 -----");

Console.WriteLine("----- 开始 子作用域-2 -----");
// 创建第 2 个子作用域
using (var scoped2 = rootProvider.CreateScope())
{
    // 获取 4 次服务实例
    var k1 = scoped2.ServiceProvider.GetRequiredService<TestService>();
    var k2 = scoped2.ServiceProvider.GetRequiredService<TestService>();
    var k3 = scoped2.ServiceProvider.GetRequiredService<TestService>();
    var k4 = scoped2.ServiceProvider.GetRequiredService<TestService>();
    // 输出 GUID
    Console.WriteLine($"变量 1: {k1.MyID}");
    Console.WriteLine($"变量 2: {k2.MyID}");
    Console.WriteLine($"变量 3: {k3.MyID}");
    Console.WriteLine($"变量 4: {k4.MyID}");
}
Console.WriteLine("----- 结束 子作用域-2 -----");
```

上述代码创建了两个子作用域。在第 1 个子作用域中，获取 3 次服务实例并向屏幕输出 GUID；在第 2 个子作用域中，获取了 4 次服务实例并输出 GUID，如图 3-4 所示。

从结果中可以看到，在同一个作用域内，各变量输出的 GUID 相同，表明它引用的是同一个服务实例。

对于作用域服务，需要注意在默认情况下是不能通过位于根部的 ServiceProvider 对象获取服务实例的，只能先调用 CreateScope()（或 CreateAsyncScope()）方法创建子作用域，然后通过子作用域的 ServiceProvider 对象获取服务实例。

```
====== 作用域服务 ======
----- 开始 子作用域-1 -----
TestService实例化
变量1: 8d77bb11-736e-45b6-9424-be103bc1ade3
变量2: 8d77bb11-736e-45b6-9424-be103bc1ade3
变量3: 8d77bb11-736e-45b6-9424-be103bc1ade3
TestService实例即将释放
----- 结束 子作用域-1 -----
----- 开始 子作用域-2 -----
TestService实例化
变量1: 205c56cc-d82c-4762-bc3b-2141c2e91daf
变量2: 205c56cc-d82c-4762-bc3b-2141c2e91daf
变量3: 205c56cc-d82c-4762-bc3b-2141c2e91daf
变量4: 205c56cc-d82c-4762-bc3b-2141c2e91daf
TestService实例即将释放
----- 结束 子作用域-2 -----
```

图 3-4　同一个作用域内 GUID 相同

若确实需要从根 ServiceProvider 处获取作用域服务的实例，可以在调用 BuildServiceProvider()方法时将 validateScopes 参数设置为 false，如

```
// 创建容器
var container = new ServiceCollection();
// 注册服务
container.AddScoped<IDriver, MyDriver>();
// 创建位于容器根部的 ServiceProvider
using var serviceProvider = container.BuildServiceProvider
(validateScopes: false);
// 获取服务
IDriver drv = serviceProvider.GetRequiredService<IDriver>();
```

3.6　GetService()方法与 GetRequiredService()方法的区别

GetService()方法与 GetRequiredService()方法都可以通过 ServiceProvider 类从容器中获取服务实例，其区别如下。

（1）GetService()方法：如果服务已在容器中注册，就返回其实例，否则返回 null，不会引发异常。程序代码需要进行 null 检查。

（2）GetRequiredService()方法：如果服务已在容器中注册，就返回服务类型的实例，否则会引发 InvalidOperationException 异常。

也就是说，GetRequiredService()方法要求服务必须已经注册到容器中。

若使用 GetService()方法获取服务实例，随后需要在代码中验证服务实例是否为 null，如

```
ITest? obj = serviceProvider.GetService<ITest>();
if(obj != null)
{
```

```
    // 使用服务实例
}
```

若使用的是 GetRequiredService()方法，建议将代码写在 try…catch 语句块中，以捕捉可能发生的异常，如

```
try
{
    ITest obj2 = serviceProvider.GetRequiredService<ITest>();
    _ = obj2.GetNum();                          //使用服务
}
catch
{
    // 处理异常
}
```

3.7 注入多个服务实例

假设有一个 IDemo 接口，它有两个实现类 A 和 B。

```
public interface IDemo { }

public class A : IDemo { }

public class B : IDemo { }
```

将 A、B 类添加到服务容器中。

```
container.AddTransient<IDemo, A>();
container.AddTransient<IDemo, B>();
```

当请求一个服务实例时，默认返回最后添加到容器中的服务类的实例。

```
IDemo sv = serviceProvider.GetRequiredService<IDemo>();
```

在该示例中，GetRequiredService()方法返回的是 B 类的实例。

在有些特殊应用场合，开发人员希望同时获取 A、B 类的实例，此时应该调用 GetServices()方法，它将返回一个服务实例列表，其中包含 A、B 类的实例。

```
IEnumerable<IDemo> insts = serviceProvider.GetServices<IDemo>();
foreach(var s in insts)
{
    …
}
```

接下来给出一个比较完整的示例。该示例将实现向 MVC 控制器的构造函数注入多个服务实例。

先定义名为 ICalculator 的接口类型，它表示一个计算器程序的通用模型。该接口包含一个 GetResult()的方法，用于获取计算结果（输入参数和返回结果均为 double 类型）。

```
public interface ICalculator
{
    double GetResult(double x);
}
```

再定义 3 个类，它们都实现 ICalculator 接口。

```
public class CalculatorA : ICalculator
{
    public double GetResult(double x)
    {
        return Math.Pow(x, 2d);
    }
}

public class CalculatorB : ICalculator
{
    public double GetResult(double x)
    {
        return Math.Pow(x, 3d);
    }
}

public class CalculatorC : ICalculator
{
    public double GetResult(double x)
    {
        return Math.Pow(x, 4d);
    }
}
```

CalculatorA 类求 x 的平方（二次方），CalculatorB 类求 x 的立方（三次方），CalculatorC 类求 x 的四次方。

将以上 3 个类添加到服务容器中，其生存期均为瞬时（暂时）服务。

```
builder.Services.AddTransient<ICalculator, CalculatorA>();
builder.Services.AddTransient<ICalculator, CalculatorB>();
builder.Services.AddTransient<ICalculator, CalculatorC>();
```

本示例将使用到 MVC 功能，因此也需要将相关的功能服务添加到容器中。

```
builder.Services.AddControllersWithViews();
```

调用 Build()方法创建 Web 应用程序实例后，还需要向 HTTP 管线添加 MVC 相关的终结点。

```
var app = builder.Build();

// 使用 MVC 终结点
app.MapControllerRoute("mvc", "{controller=Home}/{action=Default}");

app.Run();
```

MapControllerRoute()方法在添加 MVC 终结点时需要指定一条路由规则。本示例中将路由规则命名为 mvc，规则中大括号内的字符串表示路由参数的名称。controller 表示控制器的名字，此处分配一个默认值 Home；action 表示控制器内要调用的操作方法的名称，默认为 Default。当客户端请求的 URL 为根地址（/）时，路由规则中的参数会使用默认值，使请求的 URL 变为/Home/Default，即调用 Home 控制器中的 Default 方法。例如，访问 https://somehost/就会解析为 https://somehost/Home/Default。

创建一个类，派生自 Controller 类（位于 Microsoft.AspNetCore.Mvc 命名空间），该类就会被识别为 MVC 控制器。本示例中控制器名为 Home。

```
public class HomeController : Controller
{
    readonly IEnumerable<ICalculator> _calculators;
    public HomeController(IEnumerable<ICalculator> cals)
    {
        _calculators = cals;
    }

    public IActionResult Default()
    {
        double input = 0.5d;
        // 存入 ViewData 中，以便在视图代码中读取
        ViewData.Add("number", input);
        IList<double> results = new List<double>();
        // 分别计算
        foreach(ICalculator cal in _calculators)
        {
            double r = cal.GetResult(input);
            // 存储计算结果
            results.Add(r);
        }
        return View(results);
    }
}
```

HomeController 类定义了类型为 IEnumerable<ICalculator>的私有字段，该字段用于从类构造函数的参数中接收注入的服务实例。在运行阶段，它将包含 3 个服务实例，分别为 CalculatorA、CalculatorB、CalculatorC 类型。

在 Default 方法中，依次调用这些服务实例的 GetResult()方法，然后把计算结果存放到一个 double 列表中。最后返回与此方法关联的视图（View），同时将 double 列表对象作为模型对象（Model）传递给视图。

新建一个目录，命名为 Views，再在 Views 目录下新建一个子目录 Home。在 Home 目录下添加一个 Razor 视图文件，文件名与 Default()方法相同，即 Default.cshtml。文件内容如下。

```
@model IList<double>
<html>
```

```
<head>
<title>示例</title>
<meta charset="utf-8" />
</head>
<body>
    @{
        // 取出 ViewData 中存放的内容
        if(ViewData.TryGetValue("number", out var v))
        {
            // 显示内容
            <p>输入数字：@v</p>
        }
    }
    @if (Model.Count == 0)                //没有计算结果
    {
<div>无计算结果</div>
    }
    else
    {
        foreach(double val in Model)
        {
            <div>@val</div>
        }
    }
</body>
</html>
```

第 1 行的@model 指令声明此视图的模型类型为 double 列表，而模型的值将从 Model 属性获取（Default()方法末尾调用 View()方法时传递）。

视图中使用 foreach 语句访问 double 列表中的所有计算结果，并呈现在<div>元素中。运行应用程序，结果如图 3-5 所示。

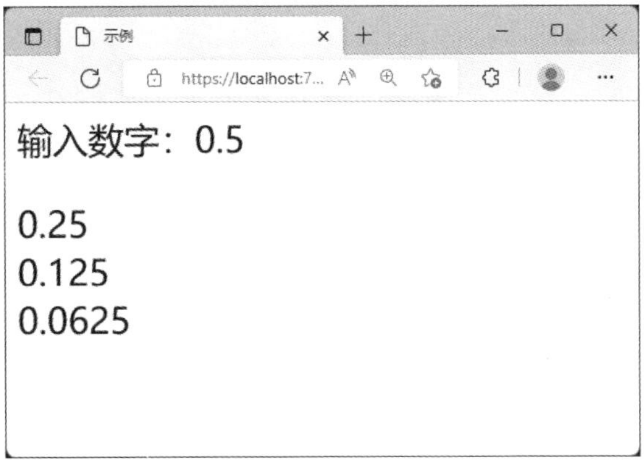

图 3-5　呈现 3 个计算结果

3.8 容易被忽略的问题

由于容器中的服务存在生存期的差异，在设计服务类时需要考虑它们之间的依赖关系是否会导致生存周期的改变。下面的例子将反映该问题。

有两个服务类：Service1 和 Service2。Service2 类需要引用 Service1 类，即在服务容器中，Service1 类的实例会被注入 Service2 类的构造函数中。

```
public class Service1 : IDisposable
{
    readonly Guid _id;
    public Service1()
    {
        _id = Guid.NewGuid();
    }

    // 获取实例标识
    public Guid ID => _id;

    public void Dispose()
    {
        Console.WriteLine("{0}实例即将被清理", GetType().Name);
    }
}

public class Service2 : IDisposable
{
    readonly Guid _id;
    readonly Service1 _service1;
    public Service2(Service1 sv)
    {
        _service1 = sv;
        _id = Guid.NewGuid();
    }

    // 获取实例标识
    public Guid ID => _id;
    // 获取 Service1 实例的引用
    public Service1 OtherService => _service1;

    public void Dispose()
    {
        Console.WriteLine("{0}实例即将被清理", GetType().Name);
    }
}
```

Service1 类在容器中注册为瞬时服务，而 Service2 类则注册为单个实例服务。

```
// 创建服务容器
var services = new ServiceCollection();
// 注册服务
services.AddTransient<Service1>();
services.AddSingleton<Service2>();
// 创建 ServiceProvider 实例
using var svProvd = services.BuildServiceProvider();
```

下面的代码将从容器中连续获取 3 次 Service2 服务实例。

```
for (int i = 1; i < 4; i++)
{
    var svobj = svProvd.GetRequiredService<Service2>();
    Console.WriteLine("第{0}轮: ", i);
    Console.WriteLine($"Service1.ID:  {svobj.OtherService.ID}");
    Console.WriteLine($"Service2.ID:  {svobj.ID}");
    Console.WriteLine("\n");
}
```

运行后的屏幕输出结果如图 3-6 所示。

图 3-6　瞬时服务意外变成了单实例服务

由运行结果可以发现，由于 Service2 是单个实例服务，它内部对 Service1 的引用会导致 Service1 的生存期被延长——从瞬时服务变成了单实例服务（3 次获取到的 Service1 实例的 GUID 相同，属于同一个实例）。

因此，在处理服务类之间的依赖关系时，应避免将生存期短的服务注入生存期较长的服务中。

第 4 章　配置应用程序

本章要点

- 应用程序配置的结构
- 多种配置源：
  - 环境变量与命令行参数
  - JSON 文件与 XML 文件
  - INI 文件
  - 内存中的配置
  - 自定义配置源
- 配置与依赖注入
- 链接多棵配置树

4.1　配置的基本结构

ASP.NET Core（或.NET）应用程序的配置数据为字典结构——以键/值对的方式读写（如 key1=value1，key2=value2）。

以下接口将用于访问配置数据。

（1）IConfigurationRoot：表示配置数据中的根节点。

（2）IConfigurationSection：表示配置中的小节。

（3）IConfiguration：表示配置数据的最原始接口。既可以访问键/值对组成的配置项，也可以访问指定的配置子节点。IConfigurationRoot 和 IConfigurationSection 接口都派生自该接口。

假设有应用配置的层次结构如下。

```
Sect1（节点）
  |-Key1: Val1（配置项）
  |-Key2: Val2（配置项）
Sect2（节点）
```

```
    |-Sect3（节点）
        |-Key3: Val3（配置项）
        |-Key4: Val4（配置项）
```

若要获得配置值 Val2，可以通过给索引器指定完整的路径定位配置项。

```
Config["Sect1:Key2"]
```

同理，若要获得配置值 Val3，应该指定的路径为 Sect2:Sect3:Key3，即

```
Config["Sect2:Sect3:Key3"]
```

配置路径使用冒号（必须是英文符号）分隔每个段（包括节点和键名）。

也可以先获得某个节点的引用，然后再获得与指定键对应的值。例如，要获得配置值 Val1，可以这样处理：

```
section = config["Sect1"];
value = section["Key1"];
```

4.2　在.NET 应用程序中使用配置

在非 ASP.NET Core 应用程序中使用配置功能，需要手动添加以下 Nuget 包的引用。

```
Microsoft.Extensions.Configuration
Microsoft.Extensions.Configuration.Abstractions
Microsoft.Extensions.Configuration.Json
```

由于存在依赖关系的包会自动安装，因此在应用程序项目中只需要添加对 Microsoft.Extensions.Configuration 和 Microsoft.Extensions.Configuration.Json 包的引用即可。

接下来以控制台应用程序为例，读取 JSON 文件中的配置，最后打印到控制台窗口上。

（1）新建应用程序项目后，可以执行以下命令添加 Nuget 包的引用。

```
dotnet add package Microsoft.Extensions.Configuration
dotnet add package Microsoft.Extensions.Configuration.Json
```

（2）在项目中添加一个 JSON 文件，命名为 test.json，然后输入以下内容。

```
{
  "appTitle": "示例应用程序",
  "appIcon": "app.ico",
  "paths": {
  "mainPath": "/bin/appx.exe",
  "tmpPath": "/tmp",
  "infPath": "/my/local/xte.inf"
  }
}
```

将该文件的“复制到输出目录”属性设置为“始终复制”，或者在项目文件中添加以下节点。

```
<ItemGroup>
<None Update="test.json">
<CopyToOutputDirectory>Always</CopyToOutputDirectory>
</None>
```

```
</ItemGroup>
```

（3）打开 Program.cs 文件，在代码中引入以下命名空间。

```
using Microsoft.Extensions.Configuration;
using System.Text;
```

（4）创建 ConfigurationBuilder 实例，稍后用于生成可访问配置数据的对象。

```
IConfigurationBuilder cfgbuilder = new ConfigurationBuilder();
```

（5）将 test.json 文件作为配置数据的来源。

```
cfgbuilder.AddJsonFile("test.json");
```

AddJsonFile()是扩展方法，用于添加 JSON 文件。可添加多个文件，在生成配置信息接口对象时会自动将来自多个文件的数据合并。

（6）调用 Build()方法生成 IConfiguration 类型的对象。

```
IConfiguration config = cfgbuilder.Build();
```

（7）读取各个配置项的值，存放到变量中。

```
string apptitle = config["appTitle"];
string appicon = config["appIcon"];
string mainpath = config["paths:mainPath"];
string tmppath = config["paths:tmpPath"];
string infpath = config["paths:infPath"];
```

mainPath、tmpPath、infPath 这 3 项配置是在 paths 节点下面的，因此在通过索引器访问时要指定完整路径。配置项的路径及键名均不区分大小写，如将 infPath 写成 infpath 也可以。所有配置项都返回字符串类型的值。

（8）将读入的配置信息打印到屏幕上。

```
var strBuilder = new StringBuilder();
strBuilder.AppendLine($"应用标题：{apptitle}")
        .AppendLine($"应用图标：{appicon}")
        .AppendLine($"主程序路径：{mainpath}")
        .AppendLine($"临时目录：{tmppath}")
        .AppendLine($"inf 文件位置：{infpath}");
Console.WriteLine("配置信息：");
Console.WriteLine("-----------------------------------------");
Console.WriteLine(strBuilder.ToString());
```

运行程序，结果如图 4-1 所示。

图 4-1　从 JSON 文件中读取到的配置信息

4.3　在 ASP.NET Core 应用程序中使用配置

ASP.NET Core 项目默认已经引用与配置相关的包，开发者不需要手动添加。而且，用于访问配置的对象都在服务容器中注册，可通过依赖注入使用它。注入的类型一般选用 IConfiguration 接口，使用时可以强制转换为 IConfigurationRoot 接口类型。

4.3.1　配置的数据来源

多数情况下，应用程序可以从以下几个渠道获取配置数据。

（1）命令行：在运行应用程序时通过命令行参数传递配置。

（2）环境变量：不管是系统级别的环境变量，还是用户级别的环境变量都可以。同时包括在运行应用程序前设置的自定义环境变量，如 MY_ROOT=/test/app。

（3）JSON 文件：从 JSON 文档中加载配置。

（4）ini 文件：从 ini 文件（初始化文件）中加载配置。

（5）内存：从内存中加载配置，通常是直接在代码中设置。

（6）XML：从 XML 文档中加载配置。

不管配置来自何种渠道，在生成 IConfiguration 相关对象时会将各个数据源合并，形成一棵完整的配置树。如果相同的配置键名出现在不同数据源，那么后面添加的数据会替换前面的数据。假设环境变量配置了 KeyA="x"，命令行参数配置了 KeyA="y"，两个键名相同，若先添加了环境变量再加载命令行参数，那么配置 KeyA 只留"y"——命令行参数替换了环境变量。

4.3.2　查看所有配置信息

IConfigurationRoot 接口派生自 IConfiguration 接口，既可以读取配置节点信息，也可以直接获取配置项的值。默认的实现类是 ConfigurationRoot。同时，IConfigurationRoot 接口还有一个名为 GetDebugView()的扩展方法，可以获得当前应用程序中已加载的所有配置信息，以字符串形式返回，包含整个配置树结构以及每配置项的数据来源。

下面的代码将演示 GetDebugView()扩展方法的使用。

```
app.MapGet("/", () =>
{
    IConfigurationRoot cfg = (IConfigurationRoot)app.Services.
    GetRequiredService<IConfiguration>();
    return cfg.GetDebugView();
});
```

从服务容器取出配置信息对象时，类型参数要使用 IConfiguration 接口，然后再强制转换为 IConfigurationRoot 接口类型。如果类型参数直接使用 IConfigurationRoot 接口，会发生错误。

运行应用程序后，使用 Web 浏览器打开服务器的根地址，就会看到当前应用程序中已加载的完整配置信息，如图 4-2 所示。

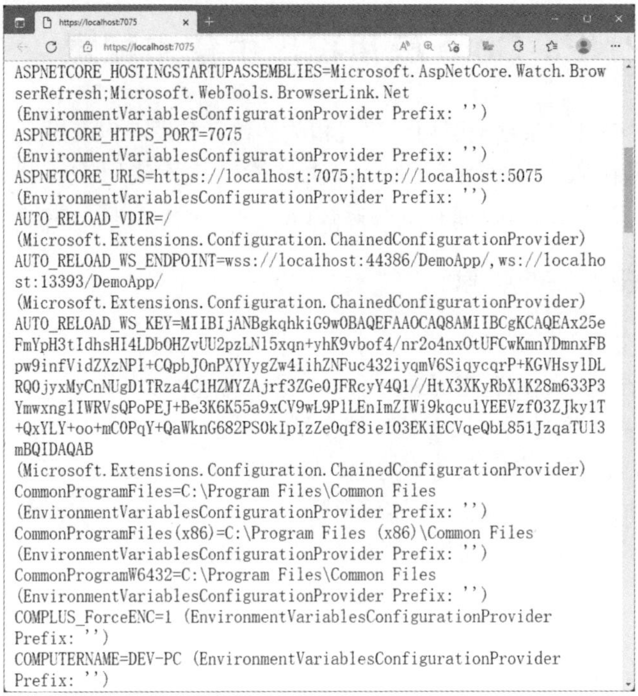

图 4-2　完整的配置信息

4.4　IConfigurationBuilder 接口

IConfigurationBuilder 接口表示一个用于构建应用程序配置的对象，它的默认实现类是 ConfigurationBuilder。其中，Sources 属性表示配置的数据源列表。可以调用 Add()方法添加配置数据源。最后调用 Build()方法将所有配置数据源合并，返回类型是实现 IConfigurationRoot 接口的类型实例（默认为 ConfigurationRoot）。

下面的示例演示 IConfigurationBuilder 接口的使用方法。

（1）声明 IConfigurationBuilder 类型的变量，然后用 ConfigurationBuilder 实例初始化。

```
IConfigurationBuilder cfgbuilder = new ConfigurationBuilder();
```

（2）添加配置源。本示例将使用内存配置源。

```
IDictionary<string, string> initDic = new Dictionary<string, string>();
// 初始化数据
initDic.Add("app_name", "Test");
initDic.Add("app_ver", "1.0.15.1050");
initDic.Add("app_desc", "Demo Production");
initDic.Add("update_url", "ftp://121.16.0.3/lite/vers");
// 添加来自内存的配置源
cfgbuilder.AddInMemoryCollection(initDic);
```

AddInMemoryCollection()扩展方法将 MemoryConfigurationSource 对象添加到 Sources 列表中。

（3）调用 Build()方法构建 IConfigurationRoot 对象。

```
IConfigurationRoot config = cfgbuilder.Build();
```

（4）现在，config 变量中已包含从内存中加载的配置信息。下面的代码将尝试读出这些配置。

```
StringBuilder strb = new();
strb.AppendLine($"应用名称: {config["app_name"]}");
strb.AppendLine($"应用版本: {config["app_ver"]}");
strb.AppendLine($"应用简介: {config["app_desc"]}");
strb.AppendLine($"更新地址: {config["update_url"]}");
Console.WriteLine(strb);
```

（5）运行应用程序，结果如图 4-3 所示。

图 4-3　输出内存中的配置信息

4.5　ConfigurationManager 类

ConfigurationManager 类同时实现 IConfigurationBuilder 和 IConfigurationRoot 接口，因此它兼具 ConfigurationBuilder 类和 ConfigurationRoot 类的功能，既可以用来添加配置源、生成配置信息，也可以直接通过配置的节点名称或键名称获得配置项的值。

ConfigurationManager 类的构造函数将默认添加内存配置源，因此，在初始化 ConfigurationManager 对象后即可直接用于读写配置信息。

下面的代码将使用 ConfigurationManager 类从 JSON 文件中加载配置，然后将配置信息打印到屏幕上。

```
// 创建 ConfigurationManager 实例
ConfigurationManager cfgmgr = new();

// 添加配置源
cfgmgr.AddJsonFile("app.json");

// 直接可访问配置信息
string val1 = cfgmgr["item_1"];
string val2 = cfgmgr["item_2"];
string val3 = cfgmgr["item_3"];
// 输出
```

```
Console.WriteLine("item1 = {0}", val1);
Console.WriteLine("item2 = {0}", val2);
Console.WriteLine("item3 = {0}", val3);
```

app.json 文件的内容如下。

```
{
"item_1": "test_data_1",
"item_2": "test_data_2",
"item_3": "test_data_3"
}
```

运行应用程序，读取并打印的配置信息如图 4-4 所示。

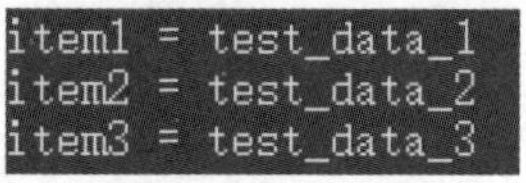

图 4-4　从 JSON 文件中读取的 3 个配置项

ConfigurationManager 类在访问配置前不需要调用 Build()方法。若确实要调用该方法，要先把变量强制转换为 IConfigurationBuilder 接口再调用，这是因为 ConfigurationManager 类是显式实现 IConfigurationBuilder 接口的，只有 IConfigurationBuilder 接口类型的变量才能访问 Build()方法。

4.6 IConfigurationSource 接口与 IConfigurationProvider 接口

IConfigurationSource 接口表示配置的数据来源。实现该接口的类型必须实现 Build()方法，返回一个实现 IConfigurationProvider 接口的类型实例。

IConfigurationProvider 接口要求实现一系列方法，支持应用程序对配置信息的基础访问操作。这些方法如下。

（1）GetChildKeys()：返回指定父节点的直接子级（键名或子节点）列表。如果配置信息来自多个数据源，就会把所有配置源中的子级列表合并。合并子级列表的代码在 GetChildrenImplementation()扩展方法中。该方法仅在 Microsoft.Extensions.Configuration 程序集内部使用，部分源代码如下。

```
internal static IEnumerable<IConfigurationSection>
GetChildrenImplementation(this IConfigurationRoot root, string? path)
{
    using ReferenceCountedProviders? reference = (root as
    ConfigurationManager)?.GetProvidersReference();
    IEnumerable<IConfigurationProvider> providers = reference?
    .Providers ?? root.Providers;

    IEnumerable<IConfigurationSection> children = providers
    .Aggregate(Enumerable.Empty<string>(),
        (seed, source) => source.GetChildKeys(seed, path))
    .Distinct(StringComparer.OrdinalIgnoreCase)
    .Select(key => root.GetSection(path == null ? key : ConfigurationPath.
```

```
    Combine(path, key)));

        if (reference is null)
        {
            return children;
        }
        else
        {
            return children.ToList();
        }
    }
```

（2）TryGet()：根据参数传入的键名，返回与之对应的配置项的值。

（3）Set()：设置指定配置项的值。

（4）GetReloadToken()：返回一个实现了 IChangeToken 接口的对象实例，用于跟踪配置信息的更新。

（5）Load()：从配置源中加载配置信息。

4.6.1　自定义扩展点

开发人员可能需要实现自定义的配置源，以满足应用程序的特殊需求。直接实现 IConfigurationSource 和 IConfigurationProvider 接口显得有些复杂，开发人员可以考虑从现有的抽象类派生出自定义的类型。

ConfigurationProvider 类实现了 IConfigurationProvider 接口。它是一个抽象类，实现时重点是完成 Load()方法的逻辑，从配置源中加载配置信息，并保存到 Data 属性中（结构为字典类型）。

自定义配置源要求实现 IConfigurationSource 接口，并且从 Build()方法返回 ConfigurationProvider 的实现类实例。

另外，还有两组抽象类也可以用于自定义配置源。

（1）StreamConfigurationSource 与 StreamConfigurationProvider 类。实现这两个抽象类，可以从流（Stream）对象中加载配置。

（2）FileConfigurationSource 与 FileConfigurationProvider 类。这两个抽象类支持从文件加载配置信息。

上述类型之间的派生关系如图 4-5 和图 4-6 所示。

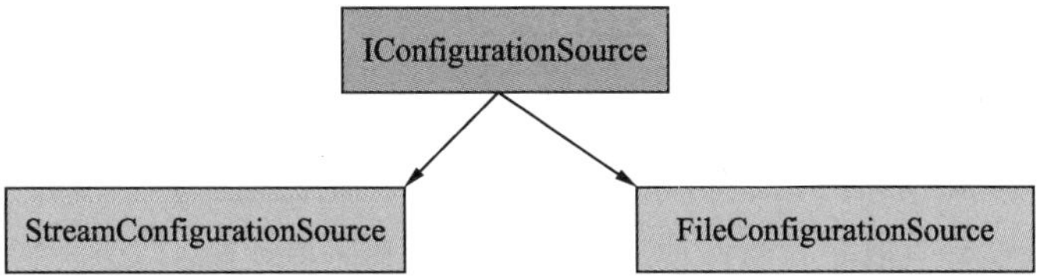

图 4-5　IConfigurationSource 的派生关系

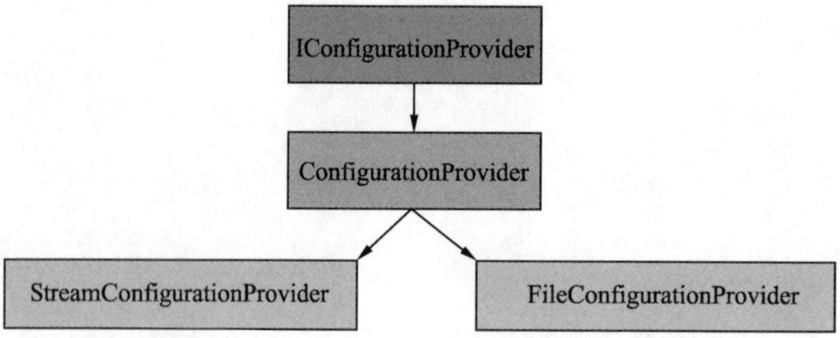

图 4-6　IConfigurationProvider 的派生关系

4.6.2　示例：来自 CSV 文件的配置

本示例将加载 CSV 文件的内容，并从中解析出配置信息。CSV 文件为纯文本文件，一行表示一条数据记录，数据字段用英文的半角逗号（也可以不使用逗号）分隔。示例程序对 CSV 的格式要求如下。

（1）第 1 行中的字段为配置项的键名。

（2）第 2 行中的字段为配置项对应的值。

假设 CSV 文档内容如下。

```
A,B,C
look,book,hook
```

可以解析为以下配置信息。

```
A=look        // config["A"]
B=book        // config["B"]
C=hook        // config["C"]
```

由于本示例是从 CSV 文件中读出数据，表明该配置来源是面向文件操作的，因此，可以选择从 FileConfigurationSource 类派生出自定义类型，命名为 CsvFileConfigurationSource，代码如下。

```
public class CsvFileConfigurationSource : FileConfigurationSource
{
    public override IConfigurationProvider Build(IConfigurationBuilder
    builder)
    {
        EnsureDefaults(builder);
        return new CsvConfigurationProvider(this);
    }
}
```

实现 Build()方法，返回 CsvConfigurationProvider 实例。在 Build()方法返回之前还调用了 EnsureDefaults()方法（在基类 FileConfigurationSource 中定义的方法）。其作用是当代码没有为 FileProvider 属性设置有效值时，默认使用 PhysicalFileProvider（支持访问物理文件）。

以下是 CsvConfigurationProvider 类的完整代码，派生自 FileConfigurationProvider 类。

```
public override void Load(Stream stream)
{
    using StreamReader reader = new(stream);
    // 开始读取配置
    string? line;
    // 第 1 行
    line = reader.ReadLine();
    if(line is null or { Length: 0 })
    {
        throw new FormatException("CSV 文件中没有记录");
    }
    // 读键列表
    string[] keys = line.Split(',');
    if(keys.Length == 0)
    {
        throw new FormatException("未读到键列表");
    }
    // 第 2 行
    line = reader.ReadLine();
    if(line is null or { Length: 0})
    {
        throw new FormatException("CSV 文件中未包含值列表");
    }
    // 读值列表
    string[] values = line.Split(',');
    if(values.Length == 0)
    {
        throw new FormatException("未读到有效的值列表");
    }
    // 比较两个列表的元素个数是否相等
    if(keys.Length != values.Length)
    {
        throw new FormatException("键列表与值列表不匹配");
    }
    // 加载配置信息
    for(int i = 0; i < keys.Length; i++)
    {
        string key = keys[i];
        string val = values[i];
        Data.Add(key, val);
    }
}
```

从 FileConfigurationProvider 类派生后需要实现 Load()抽象方法，在该方法内编写从 CSV 文件读取配置信息的代码，方法参数 stream 是流对象，表示 CSV 文件的内容。本示例所用到的 CSV 文件只需要两行文本即可：第 1 行为配置项的键名列表，第 2 行为配置项的值列表。两行中的字段是一一对应的，因此两行文本的字段数量必须一致。配置信息的加载过程如下。

（1）读出第 1 行，用逗号作为分隔符提取出所有字段，存放到 keys 变量中，它的类型为字符串数组。

（2）读出第 2 行，用逗号分隔所有字段，作为配置项的值列表，并存放到 values 变量中。

（3）验证 keys 和 values 中的元素个数是否相等，如果不等，说明配置格式有误。

（4）以 keys 中的元素为键名，values 中的元素为值，将所有已读出的键/值对添加到 Data 属性中（该属性在 ConfigurationProvider 类中已定义，为字典类型）。

为了在 ASP.NET Core 应用中能够方便添加 CSV 文件的配置，还需要编写一组扩展方法（被扩展的类型为 IConfigurationBuilder），可轻松地将 CsvFileConfigurationSource 实例添加到配置源列表中。

先实现一个 AddCsv()方法。

```
public static IConfigurationBuilder AddCsv(
      this IConfigurationBuilder builder,
      IFileProvider? fileProvider,
      string path,
      bool optional,
      bool reloadOnChanged)
{
   // 验证 path 参数是否有效
   if(string.IsNullOrEmpty(path))
   {
      throw new ArgumentNullException("path");
   }
   return builder.Add<CsvFileConfigurationSource>(s =>
   {
      s.FileProvider = fileProvider;
      s.Path = path;
      s.Optional = optional;
      s.ReloadOnChange = reloadOnChanged;
   });
}
```

第 1 个参数 builder 是被扩展的对象；fileProvider 参数提供一个实现了 IFileProvider 接口的对象，用于访问文件，若为 null，则默认使用 PhysicalFileProvider 类；path 参数指定 CSV 文件的路径（相对路径或绝对路径）；optional 参数表示从该 CSV 文件中加载配置是否为可选；reloadOnChanged 参数表示当配置源更新后是否重新加载。

最后，为 AddCsv()定义 3 个重载方法。

```
public static IConfigurationBuilder AddCsv(this IConfigurationBuilder
builder, string path)
    => builder.AddCsv(null, path, false, false);

public static IConfigurationBuilder AddCsv(this IConfigurationBuilder
builder, string path, bool optional)
    => builder.AddCsv(null, path, optional, false);
```

```
public static IConfigurationBuilder AddCsv(this IConfigurationBuilder
builder, string path, bool optional, bool reloadOnChanged)
    => builder.AddCsv(null, path, optional, reloadOnChanged);
```

在 ASP.NET Core 应用程序的主代码中，可通过 WebApplicationBuilder 类的 Configuration 属性添加 CSV 文件。

```
var builder = WebApplication.CreateBuilder(args);

// 添加 CSV 文件配置
builder.Configuration.AddCsv("test.csv");

var app = builder.Build();
…
```

test.csv 文件的内容如下。

```
item1,item2,item3
Test Data 1,Test Data 2,Test Data 3
```

将调用 app.MapGet()方法的代码修改为

```
app.MapGet("/", () =>
{
    // 读出配置项
    string v1 = app.Configuration["item1"];
    string v2 = app.Configuration["item2"];
    string v3 = app.Configuration["item3"];
    return $"item1 = {v1}\nitem2 = {v2}\nitem3 = {v3}";
});
```

运行应用程序后，打开 Web 浏览器转到本示例的根 URL，就能看到应用程序从 CSV 文件中读取的配置信息了，如图 4-7 所示。

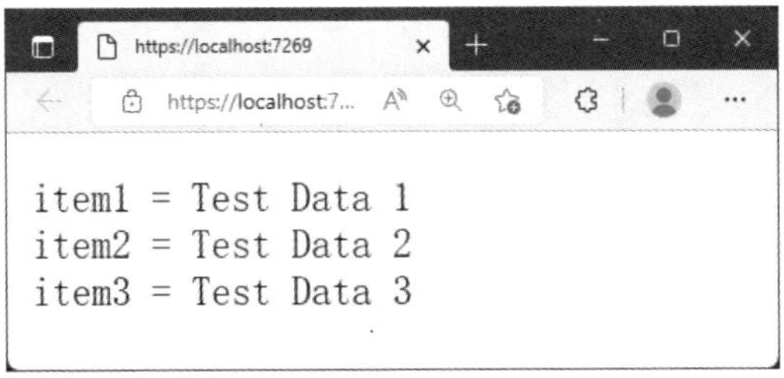

图 4-7　从 CSV 文件中加载的应用配置信息

4.7　JSON 配置

在前面的一些示例中，读者已经接触过 JSON 配置文件。JSON 配置有以下两个数据来源。

（1）JsonConfigurationSource：JSON 配置来自文件，可以指定相对路径或绝对路径。

（2）JsonStreamConfigurationSource：JSON 配置来自流对象，可以是内存中的流，也可以

是文件流或网络流。

调用 AddJsonFile()扩展方法可以轻松地将 JSON 文件添加到配置源列表中，此方法可以多次调用，即添加多个配置文件；调用 AddJsonStream()扩展方法则是把流对象添加到配置数据源中。

下面的示例演示了如何将内存流对象设置为 JSON 配置源。首先定义一个 MakeJsonConfig()方法，生成一个 JSON 文档，并存储在内存流中。

```
Stream MakeJsonConfig()
{
    // 创建内存流实例
    MemoryStream mstream = new MemoryStream();
    JsonWriterOptions options = new JsonWriterOptions();
    options.Indented = true;                              // 包含文本缩进
    Utf8JsonWriter writer = new(mstream, options);
    // 开始标记
    writer.WriteStartObject();
    // 第 1 个属性
    writer.WriteString("item1", "First Object");
    // 第 2 个属性
    writer.WriteNumber("item2", 1860);
    // 写入 1 个小节
    writer.WriteStartObject("sect");
    // 小节下的第 1 个属性
    writer.WriteBoolean("enb", true);
    // 小节下的第 2 个属性
    writer.WriteString("lc", "CN");
    writer.WriteEndObject();                              // 结束标记
    writer.WriteEndObject();                              // 结束标记
    writer.Flush();
    writer.Dispose();

    // 将生成的 JSON 文档打印到屏幕上
    mstream.Position = 0L;
    using StreamReader rd = new StreamReader(mstream, leaveOpen:true);
    Console.WriteLine(rd.ReadToEnd());

    // 重新调整流的当前位置
    mstream.Position = 0L;
    return mstream;
}
```

本示例使用了 Utf8JsonWriter 类生成 JSON 文档。它的构造函数有两个参数：第 1 个参数是要写入数据的流对象，此处用的是内存流；第 2 个参数是一个 JsonWriterOptions 结构体实例，本示例将 Indented 属性设置为 true，表示生成的 JSON 文档中包括文本缩进。实际使用中一般不需要将 Indented 属性设置为 true，本示例中启用文本缩进只是为了方便读者查看打印到屏幕上的 JSON 内容。

WriteStartObject()方法写入一个 JSON 对象的开始标记(即左大括号), 而 WriteEndObject()方法则是写入结束标记（即右大括号）。注意这两个方法必须成对出现，且必须准确配对。也就是说，调用了 WriteStartObject()方法后必须在适当的地方调用 WriteEndObject()方法。

MakeJsonConfig()方法生成的 JSON 文档包括 4 个配置项。其中，item1 和 item2 位于根节点下，enb 和 lc 位于 sect 节点下。最终生成的 JSON 文档如下。

```
{
"item1": "First Object",
"item2": 1860,
"sect": {
  "enb": true,
  "lc": "CN"
  }
}
```

回到 ASP.NET Core 应用的主代码，在创建 WebApplicationBuilder 实例后，可通过 Configuration 属性添加配置源。本示例将调用 AddJsonStream()方法添加内存中的 JSON 文档。

```
var builder = WebApplication.CreateBuilder(args);
builder.Configuration.AddJsonStream(MakeJsonConfig());
var app = builder.Build();
```

修改项目模板生成的 MapGet()方法，读出配置信息，然后以字符串形式返回给客户端。

```
IConfigurationRoot config = (IConfigurationRoot)app.Configuration;
// 读取根节点下的配置项
string val1 = config["item1"];
string val2 = config["item2"];
// 获取小节
IConfigurationSection section = config.GetSection("sect");
// 读取 sect 节点下的配置项
string val3 = section["enb"];
string val4 = section["lc"];
string msg = $"item1 = {val1}\nitem2 = {val2}\n" +
            $"\nsect 节点下的配置信息：\nenb = {val3}\nlc = {val4}";
return msg;
```

运行结果如图 4-8 所示。

图 4-8　从内存流中加载 JSON 配置

4.7.1 示例：访问 JSON 数组对象

JSON 文档中的配置值可能会出现数组类型，如

```
{
    "demo": [
    {
    "width": 505,
    "height": 370
    },
    {
    "width": 1215,
    "height": 860
    }
    ]
}
```

配置项 demo 的值是一个数组对象，它包括两个结构相同的 JSON 对象（均有 width 和 height 属性）。这种情况下，可以通过索引访问需要的元素。例如，要访问第 1 个元素，就使用索引 0，访问第 2 个元素就使用索引 1。使用索引后的配置项路径依然使用英文冒号分隔。下面的代码将读出上述 JSON 配置。

```
app.MapGet("/", () =>
{
    IConfigurationRoot config = (IConfigurationRoot)app.Configuration;
    // 读出第 1 组配置
    string w1 = config["demo:0:width"];
    string h1 = config["demo:0:height"];
    // 读出第 2 组配置
    string w2 = config["demo:1:width"];
    string h2 = config["demo:1:height"];
    // 返回结果
    return "第 1 组：\n" +
          $"width = {w1}, height = {h1}\n\n" +
          "第 2 组：\n" +
          $"width = {w2}, height = {h2}";
});
```

demo:0:...表示访问数组中第 1 个元素中的配置项；同理，demo:1:...表示访问数组中第 2 个元素的配置项。运行结果如图 4-9 所示。

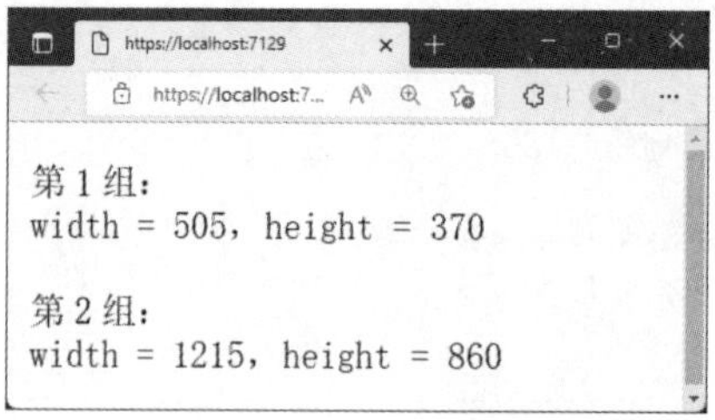

图 4-9　读出两组配置

4.7.2　示例：自动重新加载配置

AddJsonFile()方法有以下重载。

```
public static IConfigurationBuilder AddJsonFile(this IConfigurationBuilder
builder, IFileProvider provider, string path, bool optional, bool
reloadOnChange);
public static IConfigurationBuilder AddJsonFile(this IConfigurationBuilder
builder, string path, bool optional, bool reloadOnChange);
```

以上重载都包含两个 bool 类型的参数。

（1）optional：指定此 JSON 文件是否为可选的。如果此参数为 true，当文件不存在时应用程序不会抛出异常；如果此参数为 false，当文件不存在时会抛出 FileNotFoundException 异常。

（2）reloadOnChange：当文件被修改后，是否重新加载配置信息。

ASP.NET Core 应用程序默认添加的 appsettings.json 和 appsettings.{运行环境}.json 文件都是可选的，并且自动重新加载（optional、reloadOnChange 参数都设置为 true）。源代码如下。

```
config.AddJsonFile("appsettings.json", optional: true, reloadOnChange:
true).AddJsonFile($"appsettings.{env.EnvironmentName}.json", optional:
true, reloadOnChange: true);
```

可通过以下示例验证 appsettings.json 文件是否自动重新加载。

```
var builder = WebApplication.CreateBuilder(args);
var app = builder.Build();
app.MapGet("/", () => $"testItem: {app.Configuration["testItem"]}");
app.Run();
```

在 appsettings.json 文件中，添加名为 testItem 的配置项。

```
{
    …,
    "testItem": "cat"
}
```

运行应用程序后，使用 Web 浏览器访问应用程序的根地址，结果如图 4-10 所示。

此时不要退出应用程序，打开 appsettings.json 文件，将 testItem 配置项的值改为 dog，然后保存。切换回 Web 浏览器，单击“刷新”按钮或按 F5 快捷键，网页刷新后就会看到 testItem 配置项的最新值，如图 4-11 所示。

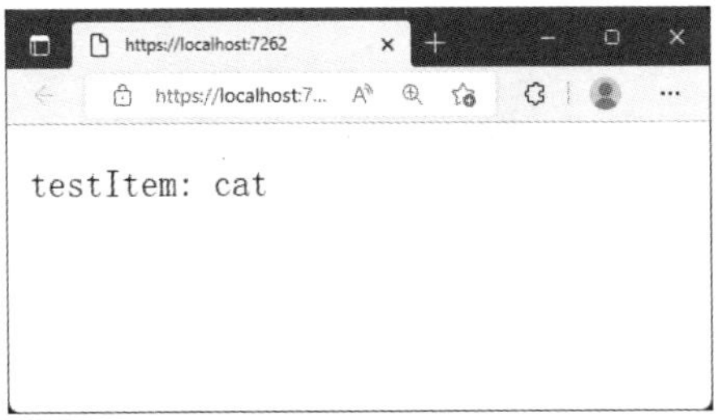

图 4-10　testItem 配置项的初始值

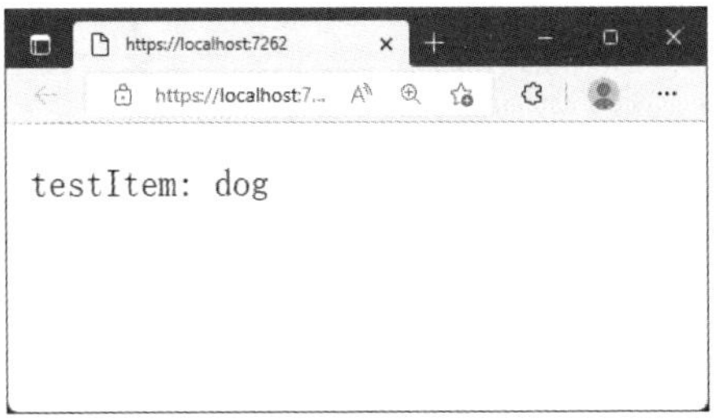

图 4-11　testItem 配置项已重新加载

自动重载选项（通过 reloadOnChange 参数设定）不仅在 JSON 配置文件中可用，对于 XML 文件和 ini 文件的配置源也适用。相应的扩展方法声明如下。

```
public static IConfigurationBuilder AddXmlFile (this IConfigurationBuilder
builder, string path, bool optional, bool reloadOnChange);
public static IConfigurationBuilder AddXmlFile (this IConfigurationBuilder
builder, IFileProvider provider, string path, bool optional, bool
reloadOnChange);
public static IConfigurationBuilder AddIniFile (this IConfigurationBuilder
builder, string path, bool optional, bool reloadOnChange);
public static IConfigurationBuilder AddIniFile (this IConfigurationBuilder
builder, IFileProvider provider, string path, bool optional, bool
reloadOnChange);
```

4.8 XML 配置

与 JSON 配置相似，基于 XML 文档的配置来源既可以是 XML 文件（通过 AddXmlFile() 扩展方法添加），也可以是来自流对象的 XML 文档（通过 AddXmlStream()扩展方法添加）。目前，ASP.NET Core 尚不支持带命名空间前缀的 XML 元素，以下 XML 文档会发生错误。

```
<root>
<data xmlns:d="test/somedata">…</data>
</root>
```

ASP.NET Core 应用程序将按照以下规则从 XML 文档中加载配置。

（1）根元素会被忽略。例如，下面的 XML 文档中，配置路径为 menu:items:open。

```
<root>
    <menu>
        <items>
            <open>打开文件</open>
        </items>
    </menu>
</root>
```

因为根元素始终被忽略，配置键的路径将从 menu 元素开始，所以在编写 XML 配置文档时，根元素可以使用任意有效的命名。例如，下面的 XML 文档中，配置键的路径为 myapp。

```
<config>
    <myapp>test</myapp>
</config>
```

（2）按照 XML 文档的层次以及元素出现的顺序构造路径，如

```
<root>
    <headers>
        <len>120</len>
        <type>png</type>
    </headers>
    <blocks>
```

```
        <tag>v4</tag>
        <layer>3</layer>
    </blocks>
</root>
```

上述 XML 文档将解析出 4 个配置项。

```
headers:len = 120
headers:type = png
blocks:layer = 3
blocks:tag = v4
```

元素的内容将被解析为配置项的值，如上述示例中的 120、png 等。

（3）如果某个 XML 元素含有特性列表（Attributes），那么每个特性也被解析为一个配置项。其中，特性名称为配置项的键名，特性值是配置项的值。例如：

```
<data>
    <center x="176" y="308"/>
    <line>
        <start x="45" y="15" />
        <end x="195" y="416" />
    </line>
</data>
```

上述 XML 文档将解析出 6 个配置项。

```
center:x = 176
center:y = 308
line:start:x = 45
line:start:y = 15
line:end:x = 195
line:end:y = 416
```

若 XML 文档既存在特性列表，也存在元素内容，那么两者都会被解析为配置项的值。例如：

```
<root>
    <dir>/usr/opt/mgon</dir>
    <tools vset="1.0">
        <item label="copy"/>
        <item label="cut" />
        <item label="save" />
    </tools>
</root>
```

上述 XML 文档可以解析出以下配置项。

```
dir = /usr/opt/mgon
tools:vset = 1.0
tools:item:0:label = copy
tools:item:1:label = cut
tools:item:2:label = save
```

<tools>元素下面有 3 个<item>元素，因为元素名称相同，被解析为数组。可通过索引定

位要访问的<item>元素。

（4）如果某个元素上存在名为 name 的特性，就会使用此特性的值充当配置项的键名。例如：

```
<config>
    <content>
        <media name="DVD" isrc="CN-D07-09-236-00/V.K">家园</media>
    </content>
</config>
```

<media>元素的内容“家园”的配置路径为

```
content:media:DVD = 家园
```

即 DVD 将被视为“家园”的键名。

下面的示例将演示 XML 配置文件的简单使用。

在 ASP.NET Core 项目中添加 XML 文件，并将“生成操作”属性设置为“内容”，“复制到输出目录”属性设置为“如果较新则复制”。也可以直接在项目文件中修改：

```
<ItemGroup>
    <Content Include="config.xml">
        <CopyToOutputDirectory>PreserveNewest</CopyToOutputDirectory>
    </Content>
</ItemGroup>
```

在 config.xml 文件中配置应用程序的启动 URL。

```
<config>
    <urls>https://localhost:6777;http://localhost:8426</urls>
</config>
```

在 Program.cs 文件中添加 config.xml 为配置源。

```
var builder = WebApplication.CreateBuilder(args);
builder.Host.ConfigureHostConfiguration(configbuilder =>
{
    configbuilder.AddXmlFile("config.xml", optional:true, reloadOnChange:
    true);
});
var app = builder.Build();
…
```

ConfigureHostConfiguration()方法只能在 builder.Build()方法调用之前调用，否则会发生错误。

运行应用程序后，config.xml 文件中配置的 URL 已被应用。控制台将输出以下内容。

```
info: Microsoft.Hosting.Lifetime[14]
      Now listening on: https://localhost:6777
info: Microsoft.Hosting.Lifetime[14]
      Now listening on: http://localhost:8426
info: Microsoft.Hosting.Lifetime[0]
```

```
    Application started. Press Ctrl+C to shut down.
info: Microsoft.Hosting.Lifetime[0]
      Hosting environment: Development
info: Microsoft.Hosting.Lifetime[0]
      Content root path: C:\Users\…\
```

4.9　环境变量

环境变量是操作系统或用户指定的一组参数，应用程序在运行后可以通过这些参数获取相关信息。如常见的环境变量 PATH，在配置系统时可以把常用的目录路径添加到 PATH 变量中，当在系统中执行某个命令时就不需要输入可执行文件的完整路径，系统会在 PATH 变量中查找要执行的文件。正因为环境变量提供的是环境参数，所以必须在应用程序运行之前设置环境变量。

可以先通过环境变量进行配置，再运行.NET 或 ASP.NET Core 应用程序。EnvironmentVariablesConfigurationSource 类将用于产生 EnvironmentVariablesConfigurationProvider 实例，读取系统环境变量列表，然后以字典形式存储到 ConfigurationManager 对象中。调用 AddEnvironmentVariables()扩展方法可轻松添加环境变量配置源。默认情况下，ASP.NET Core 应用程序已添加来自环境变量的配置，开发者不需要在代码中调用 AddEnvironmentVariables()方法，除非有特殊要求（如添加自定义的前缀）。

下面的代码实现返回当前应用程序中来自环境变量的配置。

```
app.MapGet("/", ()=>
{
    IConfigurationRoot config = (IConfigurationRoot)app.Configuration;
    // 获取来自环境变量的 ConfigurationProvider
    var envProviders = config.Providers.Where(x => x is
    EnvironmentVariablesConfigurationProvider);

    StringBuilder lines = new();
    foreach(var envProvider in envProviders)
    {
        foreach (string key in envProvider.GetChildKeys(Enumerable.Empty
        <string>(), null))
        {
            // 提取环境变量
            if (envProvider.TryGet(key, out string val))
            {
                lines.AppendLine($"{key} = {val}");
            }
        }
    }

    return lines.ToString();
});
```

上述代码首先调用 Where()扩展方法从 config.Providers 列表中筛选出类型为 EnvironmentVariablesConfigurationProvider 的实例集合，再调用 GetChildKeys()方法分别获取每个 EnvironmentVariablesConfigurationProvider 实例的键名列表，然后通过键名就能检索出相应的配置信息。运行结果如图 4-12 所示。

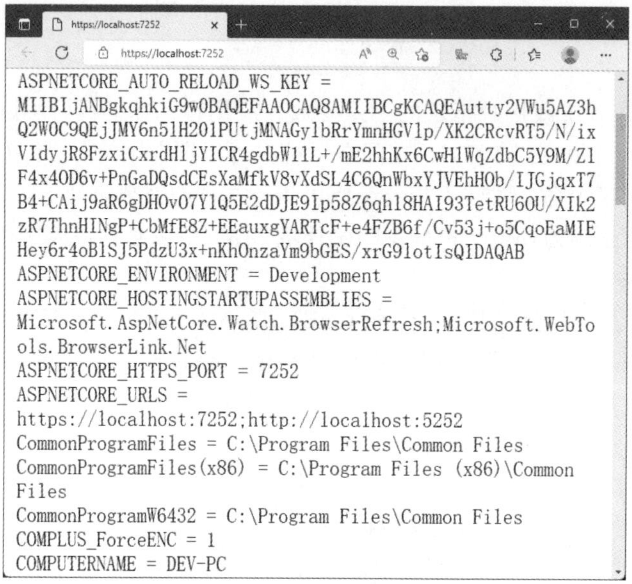

图 4-12　获得环境变量列表

4.9.1　设置环境变量前缀

在添加环境变量配置时，可以设定一个前缀，实际上是一个自定义的字符串。应用程序在加载配置时会进行筛选，提取以指定前缀匹配的环境变量。在将环境变量添加到配置项集合时，会删除前缀字符串。例如，应用程序初始化时默认添加 ASPNETCORE_前缀，若设置环境变量 ASPNETCORE_URLS = http://cooltools.com，那么应用程序加载后的配置键名会变为 URLS。

AddEnvironmentVariables()扩展方法有一个带字符串参数的重载版本。

```
public static IConfigurationBuilder AddEnvironmentVariables(this
IConfigurationBuilder configurationBuilder, string prefix);
```

prefix 参数可以为应用程序要提取的环境变量设置自定义前缀。下面的代码设置自定义前缀为 APP_，即应用程序会筛选出诸如 APP_ICON、APP_ID 的环境变量。

```
var builder = WebApplication.CreateBuilder(args);
builder.Configuration.AddEnvironmentVariables(prefix: "APP_");
...
```

在运行应用程序前，若使用的是本地环境，可以在 launchSettings.json 文件中设置环境变量。

```
"environmentVariables": {
…
"APP_TAG1": "Test-Value-1",
"APP_TAG2": "Test-Value-2",
"APP_TAG3": "Test-Value-3"
}
```

若通过命令行方式运行应用程序，可以在执行应用程序前设置环境变量。

```
// Windows
set APP_TAG1=Test-Value-1
set APP_TAG2=Test-Value-2
set APP_TAG3=Test-Value-3

// Linux
APP_TAG1=Test-Value-1
APP_TAG2=Test-Value-2
APP_TAG3=Test-Value-3
export APP_TAG1 APP_TAG2 APP_TAG3
```

下面的代码将读取上述 3 个配置项。

```
app.MapGet("/", () =>
{
    IConfigurationRoot config = (IConfigurationRoot)app.Configuration;
    // 获取 3 个配置项
    string val1 = config["tag1"];
    string val2 = config["tag2"];
    string val3 = config["tag3"];
    // 构造结果字符串
    StringBuilder lines = new();
    lines.AppendLine($"tag1 = {val1}");
    lines.AppendLine($"tag2 = {val2}");
    lines.AppendLine($"tag3 = {val3}");
    return lines.ToString();
});
```

运行应用程序后，通过 Web 浏览器访问其根地址，服务器将返回如图 4-13 所示的结果。

图 4-13　获取自定义前缀的环境变量

4.9.2 替换默认的 ASPNETCORE_前缀

ASP.NET Core 应用程序在初始化时会在配置来源列表中添加前缀为 ASPNETCORE_的环境变量。Web 主机初始化时使用的一些特殊配置信息（如表示运行环境的键名 environment），在 WebApplicationBuilder 类型实例化后会失效。例如：

```
var builder = WebApplication.CreateBuilder(args);
builder.Configuration.AddEnvironmentVariables(prefix: "MY_");
var app = builder.Build();
```

环境变量的前缀被设置为 MY_，使用 MY_ENVIRONMENT 环境变量配置应用程序的运行环境将不起作用。应用程序会使用 Production 作为默认运行环境。这是因为应用程序在初始化时默认设置的环境变量前缀为 ASPNETCORE_，所以只能识别 ASPNETCORE_ENVIRONMENT 等环境变量。在 WebApplicationBuilder 实例化之后，Web 主机已完成基础配置，不再查找键名为 environment 的配置项，所以应用程序不会读取 MY_ENVIRONMENT 环境变量的值。因此，要加载配置中的运行环境，必须在 WebApplication.CreateBuilder()方法调用之前完成。

4.9.3 示例：替换环境变量前缀

接下来将通过示例阐述如何替换 ASPNETCORE_前缀。

（1）在所有代码开始前创建 ConfigurationManager 类的实例，并添加环境变量配置源，前缀设置为 DEMO_。

```
var config = new ConfigurationManager();
config.AddEnvironmentVariables(prefix: "DEMO_");
```

（2）创建 WebApplicationOptions 实例，并初始化其属性。

```
WebApplicationOptions appOpt = new()
{
    Args = args,
    EnvironmentName = config["environment"]
};
```

Args 属性直接提取的是传递给 Main 方法（程序入口点）的 args 参数；EnvironmentName 属性值为配置中的 environment 键（来自环境变量 DEMO_ENVIRONMENT）。

这里要注意，WebApplicationOptions 类必须在调用构造函数的同时对属性赋值，不能创建实例再通过变量引用来赋值，即下面的代码是错误的。

```
appOpt.Args = args;
appOpt. EnvironmentName = config["environment"];
```

这是因为 WebApplicationOptions 类的属性没有定义 set 访问器，而是 init 访问器。

```
public string[]? Args { get; init; }
public string? EnvironmentName { get; init; }
```

```
public string? ApplicationName { get; init; }
public string? ContentRootPath { get; init; }
public string? WebRootPath { get; init; }
```

init 访问器必须在初始化对象时赋值——在 new 关键字之后。

```
var x = new Sometype()
{
    Prop1 = Val1,
    Prop2 = Val2
};
```

（3）创建 WebApplicationBuilder 实例，并传递 WebApplicationOptions 对象。

```
var builder = WebApplication.CreateBuilder(appOpt);
```

（4）将已创建的配置信息合并到 builder 的配置中，这样可以在应用程序的其他地方访问（如访问其他以 DEMO_开头的环境变量）。

```
builder.Configuration.AddConfiguration(config);
```

（5）设置环境变量。

```
// Windows
set DEMO_ENVIRONMENT=Working
// Linux
export DEMO_ENVIRONMENT=Working
```

运行应用程序后，在控制台屏幕上会看到应用程序的运行环境已变为 Working。

```
…
info: Microsoft.Hosting.Lifetime[0]
      Hosting environment: Working
…
```

4.9.4 分层配置结构

配置路径采用英文的冒号（:）分隔各节点，但在设置环境变量时，出于兼容性考虑，将采用双下画线（__）分隔路径中的节点。

假设要使用环境变配置以下内容。

```
table:
  header:
     width = 270
  body:
     fontSize = 16
     showLinks = yes
```

首先，在 ASP.NET Core 的项目代码中添加环境变量配置源，设置前缀字符串为 TBF_。

```
builder.Configuration.AddEnvironmentVariables(prefix: "TBF_");
```

在 MapGet()方法中读取配置信息，再将结果返回给客户端。

```
app.MapGet("/", () =>
{
```

```
    IConfigurationRoot config = (IConfigurationRoot)app.Configuration;
    StringBuilder strlines = new StringBuilder();
    // 获取 table:header 节点
    var hdsection = config.GetSection("table:header");
    strlines.AppendLine("------ header ------");
    string widthval = hdsection["width"];
    strlines.AppendLine($"width = {widthval}");
    // 获取 table:body 节点
    var bdsection = config.GetSection("table:body");
    strlines.AppendLine("\n------ body ------");
    string fntsizeval = bdsection["fontSize"];
    string showlinksval = bdsection["showLinks"];
    strlines.AppendLine($"fontSize = {fntsizeval}");
    strlines.AppendLine($"showLinks = {showlinksval}");
    return strlines.ToString();
});
```

上述代码读取配置的方式为调用 GetSection()方法分别获得 table:header 和 table:body 两个节点，然后依次获取各节点下的配置项。

在运行应用程序之前，设置环境变量如下。

```
// Windows
set TBF_TABLE__HEADER__WIDTH=270
set TBF_TABLE__BODY__FONTSIZE=16
set TBF_TABLE__BODY__SHOWLINKS=yes

// Linux
TBF_TABLE__HEADER__WIDTH=270
TBF_TABLE__BODY__FONTSIZE=16
TBF_TABLE__BODY__SHOWLINKS=yes
export TBF_TABLE__HEADER__WIDTH TBF_TABLE__BODY__FONTSIZE TBF_TABLE__
BODY__SHOWLINKS
```

应用程序在加载环境变量时会自动将“__”替换为“:”。配置读取结果如下。

```
------ header ------
width = 270

------ body ------
fontSize = 16
showLinks = yes
```

4.10 命令行参数

ASP.NET Core 应用程序默认已添加命令行参数配置源，在调用 WebApplication.CreateBuilder()方法时需要同时传递应用程序的命令行参数。

```
svar builder = WebApplication.CreateBuilder(args);
```

args 参数来自程序的入口点——Main 方法。

以下 3 种命令行格式均可用于应用程序配置。

（1）使用等号（=）连接，即 key=value。

```
dotnet test.dll key1=a key1=b key3=c
```

（2）以斜杠（/）开头，格式为/key value 或/key=value。

```
dotnet test.dll /key1=a /key2=b /key3=c
dotnet test.dll /key1 a /key2 b /key3 c
```

（3）以双横线（--）开头，格式为--key=value 或--key value。

```
dotnet test.dll --key1 year --key2 month --key3 day
dotnet test.dll --key1=year --key2=month --key3=day
```

对于带分层结构的配置数据，可以用英文的冒号分隔。

```
dotnet run root:ad:key1=cat root:key2=cap root:td:key3=cash
dotnet run /root:ad:key1 cat /root:key2 cap /root:td:key3 cash
dotnet run /root:ad:key1=cat /root:key2=cap /root:td:key3=cash
dotnet run --root:ad:key1 cat --root:key2 cap --root:td:key3 cash
dotnet run --root:ad:key1=cat --root:key2=cap --root:td:key3=cash
```

在调用 AddCommandLine()扩展方法添加命令行配置源时，也可以通过以下重载传递一个字典对象，用于设定命令行参数与配置键名之间的映射关系。

```
public static IConfigurationBuilder AddCommandLine(this
IConfigurationBuilder configurationBuilder, string[] args, IDictionary
<string, string> switchMappings);
```

提供命令行参数到配置键的映射关系，使命令行传递参数更加灵活。尤其是在分层的配置结构中，配置项的路径往往比较长，此时可以为每个配置键定义一个简短的命令行参数名，方便用户记忆和使用。

例如，某个应用程序需要以下配置。

```
root:
 logging:
    format = messageOnly
```

format 配置键的路径为 root:logging:format，使用的命令行参数如下。

```
dotnet run --root:logging:format messageOnly
```

在实际应用中，如上所述的命令行参数名称过于冗长，不便于记忆和输入。因此，开发人员可以定义一个映射关系，将 format 参数映射到 root:logging:format 配置键上，命令行参数就比原来更加简洁明了了。

```
dotnet run --format messageOnly
```

接下来给出一个示例，该示例使用的分层配置结构如下。

```
info:
 appkey = …
 subject = …
```

```
runtime:
 workDir = …
 backgroundTasks = …
```

从上述配置结构可知，该示例需要 4 个配置项，即需要提供 4 个命令行参数。下面的代码添加命令行参数配置源，同时定义 4 个参数与配置的映射关系。

```
var builder = WebApplication.CreateBuilder(args);

builder.Configuration.AddCommandLine(args, new Dictionary<string, string>()
{
    ["--key"] = "info:appkey",
    ["--sub"] = "info:subject",
    ["--wd"] = "runtime:workDir",
    ["--bts"] = "runtime:backgroundTasks"
});

var app = builder.Build();
```

修改 app.MapGet()方法中的代码，使其返回配置信息列表。

```
app.MapGet("/", () =>
{
    IConfiguration config = app.Configuration;
    // 显示配置信息
    StringBuilder strs = new();
    // 获取 info 节点
    var sectinfo = config.GetSection("info");
    strs.AppendLine("info:");
    // 获取 info 节点下的配置项
    string appkey = sectinfo["appkey"];
    string subject = sectinfo["subject"];
    strs.AppendLine($"  appkey = {appkey}");
    strs.AppendLine($"  subject = {subject}");
    // 获取 runtime 节点
    var sectruntime = config.GetSection("runtime");
    // 获取 runtime 节点下的配置项
    string workdir = sectruntime["workDir"];
    string bgtasks = sectruntime["backgroundTasks"];
    strs.AppendLine("\nruntime:");
    strs.AppendLine($"  workDir = {workdir}");
    strs.AppendLine($"  backgroundTasks = {bgtasks}");
    return strs.ToString();
});
```

上述代码采取分节读取的方法，即先获取 info 节点的引用，再读出其下面的两个配置项；同理，第 2 个节点 runtime 也使用相同的方法。

运行示例程序，并传递以下命令参数。

```
dotnet run --key=d96ec44ae58f1f76bc6 --sub=Demo --wd=/test/rd --bts=5
```

在测试阶段，也可以在 launchSettings.json 文件中配置命令行参数（在 commandLineArgs 属性中配置）。

```
"profiles": {
"DemoApp": {
"commandName": "Project",
…
"commandLineArgs": "--key=d96ec44ae58f1f76bc6 --sub=Demo --wd=/test/rd
--bts=5"
},
…
```

运行示例程序，结果如图 4-14 所示。

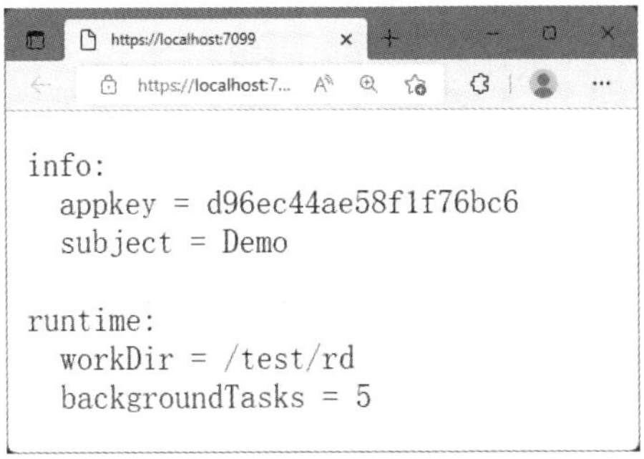

图 4-14　输出命令行配置信息

4.11　ini 配置

ini 文件即初始化文件（Initialization File），常用于存储 Windows 系统中的配置信息；在其他操作系统中也可以用于软件配置。

ini 文件由 3 部分组成：节（section）、键（key）、值（value）。大致结构如下。

```
[section 1]
key1=value1
key2=value2
…

[section 2]
key3=value3
key4=value4
…
```

在 ini 文件中，节点名称被放在一对中括号（[]）中。等号（=）两边允许出现空格，如

```
[data]
fd1 = v1
fd2 = v2
```

与 JSON、XML 等配置文件的使用一样，ini 配置可以使用.ini 文件，也可以包含 ini 配置数据的流对象。对应的扩展方法如下。

（1）AddIniFile()：添加配置源时指定.ini 文件的路径，绝对路径和相对路径都可以。

（2）AddIniStream()：直接通过流对象（类型为 Stream 的派生类）添加 ini 配置源。

下面给出一个使用 ini 配置文件的示例。

（1）创建一个空白的 ASP.NET Core 应用程序项目。

（2）在项目中添加一个 settings.ini 文件，内容如下。

```
[basic options]
cmdName = ffmpeg
audio = yes
sink = Graphics_out

[format options]
vdFormat = MPEG-4
adFormat = AC3

[bitrates]
vr = 9600
ar = 320
```

（3）将 settings.ini 文件的“生成操作”属性设置为“内容”，“复制到输出目录”属性设置为“如果较新则复制”，或直接修改项目文件。

```
<ItemGroup>
<Content Include="settings.ini">
<CopyToOutputDirectory>PreserveNewest</CopyToOutputDirectory>
</Content>
</ItemGroup>
```

（4）在应用程序的初始化代码中，将 settings.ini 文件添加为配置源。

```
var builder = WebApplication.CreateBuilder(args);
builder.Configuration.AddIniFile("settings.ini", optional:false,
reloadOnChange:true);
…
```

（5）修改 MapGet()方法中的代码，读出 settings.ini 文件中配置信息。

```
app.MapGet("/", () => {
    IConfiguration config = app.Configuration;
    StringBuilder msglines = new();

    // 获取"基本配置"节
    var basicSect = config.GetSection("basic options");
    msglines.AppendLine("======= 基本配置 =======");
    // 读取该节下面的配置
    string cmd = basicSect["cmdName"];
    string isaudio = basicSect["audio"];
```

```
        string outsink = basicSect["sink"];
        msglines.AppendLine($"目标命令：{cmd}");
        msglines.AppendLine($"是否包含音频：{isaudio}");
        msglines.AppendLine($"输出通道：{outsink}\n");

        // 获取“格式选项”节
        var fmtSect = config.GetSection("format options");
        msglines.AppendLine("======= 格式配置 =======");
        // 读取该节下的配置
        string vdfmt = fmtSect["vdFormat"];
        string adfmt = fmtSect["adFormat"];
        msglines.AppendLine($"视频格式：{vdfmt}");
        msglines.AppendLine($"音频格式：{adfmt}\n");

        // 获取“比特率”节
        var brSect = config.GetSection("bitrates");
        msglines.AppendLine("======= 比特率配置 =======");
        // 读取该节下的配置
        string vdrate = brSect["vr"];
        string adrate = brSect["ar"];
        msglines.AppendLine($"视频比特率：{vdrate}");
        msglines.AppendLine($"音频比特率：{adrate}");

        return msglines.ToString();
    });
```

（6）运行示例程序，结果如图 4-15 所示。

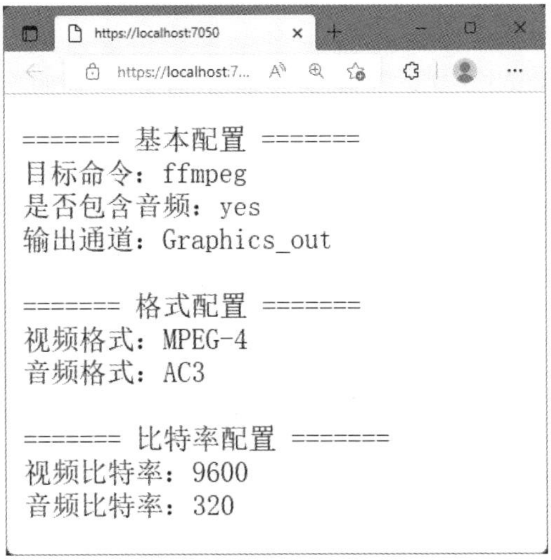

图 4-15　读取 ini 配置

4.12 配置与依赖注入

为了便于在应用程序内访问配置信息，在应用程序加载完配置信息后，会将 ConfigurationManager 对象注册到服务容器中，公开的服务类型为 IConfiguration 接口。所以，在通过依赖注入提取配置信息时要使用 IConfiguration 类型（通过 GetRequiredService()方法或构造函数参数接收对象实例）。

下面给出两个通过依赖注入访问配置信息的示例。

4.12.1 示例：将 IConfiguration 注入 MVC 控制器

本示例将使用 JSON 文件（应用程序项目模板生成的 appsettings.json 文件）配置应用程序。随后创建一个名为 Demo 的 MVC 控制器，控制器类的构造函数存在一个 IConfiguration 类型的参数，通过依赖注入获得服务实例的引用。在 GetConfig()操作方法中，代码将读出配置信息并返回给调用者。

（1）打开 appsettings.json 文件，添加 3 个配置项，然后保存并关闭文件。

```
{
…
"api-ver": "2.3.1",
"api-body": "json",
"api-req": "text"
}
```

（2）在初始化代码中添加与 MVC 功能相关的服务。

```
var builder = WebApplication.CreateBuilder(args);
builder.Services.AddControllers();
…
```

（3）构建 WebApplication 实例后，调用 MapControllerRoute()扩展方法，为 MVC 注册一条路由规则。

```
app.MapControllerRoute("test", "opt/{controller}/{action}");
```

test 是该路由规则的命名，若无特殊用途，可随意命名。第 2 个参数的值 opt/{controller}/{action}是路由规则，其中，{controller}与{action}是该规则的命名参数。路由规则与客户端请求 URL 匹配时，这些命名参数将被替换。例如，要访问 my 控制器中的 run 操作方法，那么客户端应发出的请求 URL 为 https://abc.com/opt/my/run/，即

```
controller = my
action = run
```

（4）在项目中添加一个类，命名为 DemoController，对应的 MVC 控制器名为 Demo（Controller 后缀被去掉）。此名称不区分大小写，使用 demo 也是有效的。

```
public class DemoController : Controller
{
```

```
    IConfiguration _config;

    public DemoController(IConfiguration cfg)
    {
        _config = cfg;
    }

    public IActionResult GetConfig()
    {
        string apiVer = _config["api-ver"];
        string apiBody = _config["api-body"];
        string reqType = _config["api-req"];
        string ct = $"API Version: {apiVer}\n" +
                    $"API Body Type: {apiBody}\n" +
                    $"API Request Type: {reqType}";
        return Content(ct);
    }
}
```

控制器类中定义了 IConfiguration 类型的私有字段_config，作用是引用通过依赖注入获得的 IConfiguration 对象（实为 ConfigurationManager 对象）。在构造函数中对该字段赋值。

GetConfig()方法会被应用程序识别为 MVC 中的 Action（操作方法）。在该方法中，将 3 个配置项读出，最后组成字符串返回给客户端。对于本示例中的路由规则，要访问 GetConfig()方法，需要的 URL 为/opt/demo/getconfig。

（5）运行示例程序，在 Web 浏览器中定位到 https://<本机地址>/opt/demo/getconfig，就可以看到被读取的配置信息，如图 4-16 所示。

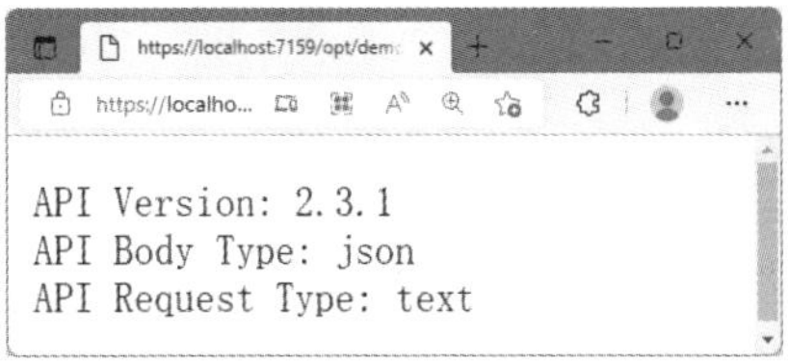

图 4-16　从控制器中读取配置信息

4.12.2　示例：通过配置选择哈希算法

本示例的主要功能是用户输入文本内容后，单击按钮提交到服务器，服务器计算出输入文本的哈希值，最后将结果反馈给用户。计算哈希值的算法将通过应用程序配置来选择，支持的算法有 MD5、SHA1、SHA256、SHA512。配置键名为 hash，如要使用 MD5 算法，其配置为 hash = md5。如果未配置任何算法，则默认使用 SHA1 算法。

本示例定义了 HashComputer 类并注册到服务容器中，用于访问应用程序配置的 IConfiguration 对象将通过构造函数注入 HashComputer 服务中。HashComputer 类读取键名为 hash 的配置值，然后根据该值决定使用哪种哈希算法。

具体实现步骤如下。

（1）在应用项目中添加一个新类，命名为 HashComputer，代码如下。

```
public class HashComputer : IDisposable
{
    readonly HashAlgorithm _hash;
```

```
    readonly string _hashName;

    public HashComputer(IConfiguration config)
    {
        // 获取配置项 hash 的值
        string alg = config["hash"];
        // 根据配置项的值决定使用的哈希算法
        _hash = alg.ToLower() switch
        {
            "md5" => MD5.Create(),
            "sha1" => SHA1.Create(),
            "sha256" => SHA256.Create(),
            "sha512" => SHA512.Create(),
            _ => SHA1.Create()
        };
        _hashName = alg ?? "sha1";
    }

    public void Dispose()
    {
        _hash?.Dispose();
    }

    public string GetHash(string message)
    {
        // 将字符串转换为字节数组
        byte[] data = Encoding.UTF8.GetBytes(message);
        // 计算哈希值
        byte[] hash = _hash.ComputeHash(data);
        // 转换为字符串返回
        return Convert.ToHexString(hash);
    }

    // 获得哈希算法的名称
    public string HashName => _hashName;
}
```

上述代码中用到的与哈希算法有关的类型有 MD5、SHA1、SHA256、SHA512。它们都有一个共同的基类——HashAlgorithm。由于要使用的哈希算法是通过配置来选择的，这使得_hash 字段在运行期间的类型不是固定的，因此使用基类 HashAlgorithm（MD5、SHA1 等类型的实例都可以赋值给_hash 字段），访问时只须调用它的 ComputeHash()方法即可。

（2）将 HashComputer 类添加到服务容器中。

```
builder.Services.AddTransient<HashComputer>();
```

（3）启用 RazorPages 功能。

```
builder.Services.AddRazorPages().WithRazorPagesAtContentRoot();
```

WithRazorPagesAtContentRoot()扩展方法表示当前项目的根目录被设置为查找 Razor 页面文件的路径（默认是根目录下的 Pages 目录，本示例设置为根目录是为了方便添加页面）。

（4）构建 WebApplication 实例后，调用 MapRazorPages()方法，使 HTTP 管线支持对 Razor 页面的处理。

```
app.MapRazorPages();
```

（5）在项目中添加 Index.cshtml 文件。

（6）在 Index.cshtml 文件的第 1 行输入以下指令。

```
@page
```

@page 指令指示该页面专用于 RazorPages 功能（而不是 MVC 的视图）。

（7）在 Index.cshtml 文件的第 2 行输入@inject 指令，表示获取指定类型的依赖注入。

```
@inject HashComputer hasher
```

（8）在 Index.cshtml 文件中输入以下 HTML 代码。

```
<html lang="zh-cn">
<head>
<title>哈希计算器</title>
</head>
<body>
<div>
<form method="post">
              @Html.AntiForgeryToken()
<label for="msg">请输入内容：</label>
<input id="msg" name="msg" type="text" value="@ViewBag.msg"/>
<button type="submit">提交</button>
</form>
</div>
<div>
@if(string.IsNullOrEmpty(hashResult) == false)
{
<span style="background-color:darkblue;color:lightcyan;padding:2px 4px;">
@hasher.HashName</span>
<span style="word-break:break-all">计算结果：@hashResult</span>
}
</div>
</body>
</html>
```

hashResult 是此页面类的字段，当此变量所引用的字符串非空时，会在 HTML 页面上显示哈希值计算结果。<form>元素中的 HTML 代码用于输入文本内容，“提交”按钮触发表单的提交行为，将用户输入的文本发送到 Web 服务器。

（9）在 Index.cshtml 文件最后用@functions 指令定义服务器代码区域。

```
@functions{
    string hashResult;
```

```
    public void OnPost(string msg)
    {
        // 保存视图状态
        ViewBag.msg = msg;
        // 计算哈希值
        hashResult = hasher.GetHash(msg);
    }
}
```

当用户单击“提交”按钮后，Web 浏览器会将 msg 字段发送到服务器，OnPost()方法被调用，并将 msg 字段的值与 msg 参数绑定。所以，通过 msg 参数能获取用户输入的文本。

在计算哈希值前，将 msg 参数的值保存到 ViewBag 中。这是为了当计算哈希值完成后，页面会重新发回给 Web 浏览器，此时<form>元素中的<input>元素可以从 ViewBag 中提取前面输入的文本。

（10）打开 appsettings.json 文件，添加 hash 配置，将其值设置为 sha256。

```
{
…
"hash": "sha256"
}
```

（11）运行应用程序，在页面中随机输入文本，单击“提交”按钮，就可以得到相应的哈希值，如图 4-17 所示。

图 4-17　计算输入文本的哈希值

4.13　链接多棵配置树

ConfigurationBuilder 添加各种配置源后，调用 Build()方法（或者使用 ConfigurationManager 类）后整合所有配置源中的数据，产生一个类似树状结构（包含配置节点和配置项）的配置集合（以 IConfiguration 类型对外公开），以供程序代码访问。

通过阅读本章前面的内容，读者已了解到常见的配置源有命令行参数、环境变量、JSON 文件、XML 文件、ini 文件等（包括自定义的配置源）。另外，运行库还提供一种配置源，名为 ChainedConfigurationSource。这种配置源比较特殊，它的数据来源于其他 IConfiguration 对象——它可以将另一棵配置树链接到当前配置树中，然后合并所有配置项，形成一个新的配置集合。

调用 AddConfiguration()扩展方法可以将已有的 IConfiguration 对象添加（链接）到当前的 IConfigurationBuilder 对象中。这种链接方案一般用于应用程序的初始化过程。应用程序在初始化的不同阶段，会加载相应的配置，而这些被分阶段加载的配置可能来自相同的配置源（有些可能已经存在）。在应用程序完成初始化时，最终产生的配置树已合并了各个阶段所加载的配置数据，以供用户代码后期访问。

假设在构建 Web 主机阶段，应用程序比较关心环境变量、命令行参数、JSON 文件中是否配置了一些必要数据，如 Environment、WebRoot（Web 静态资源的存放目录，默认为 wwwroot）、ContentRoot（应用程序的内容根目录）等基本配置，还会读取以 DOTNET_、ASPNETCORE_ 等开头的环境变量。

到了初始化 Web Application 阶段，程序还会查找所有配置源中的 Urls 配置，以确定 Web 服务器的侦听地址。随后应用程序可能还要查找 Logging 节点，以加载与日志记录有关的配置。

WebApplicationBuilder 实例创建后，用户代码有可能继续添加各种配置源。当代码通过 WebApplication 实例或依赖注入访问 IConfiguration 对象时，该配置树已经将上述所有配置信息合并。

下面的示例将模拟应用程序分 3 个阶段加载配置。第 1 阶段加载来自命令行和环境变量的配置；第 2 阶段加载来自内存的配置；第 3 阶段加载来自 XML 文件的配置。

```
// 第 1 阶段：环境变量和命令行参数
IConfigurationBuilder configbd1 = new ConfigurationBuilder();
configbd1.AddEnvironmentVariables();
configbd1.AddCommandLine(args);
IConfiguration config_s1 = configbd1.Build();
// 第 2 阶段：来自内存的配置
Process p = Process.GetCurrentProcess();
IConfigurationBuilder configbd2 = new ConfigurationBuilder();
configbd2.AddInMemoryCollection(new Dictionary<string, string>
{
    // 当前进程名称
    ["proc"] = p.ProcessName,
    // 当前进程号
    ["pid"] = p.Id.ToString(),
    // 当前进程的启动时间
    ["start_time"]= p.StartTime.ToLongTimeString()
});
IConfiguration config_s2 = configbd2.Build();
// 第 3 阶段：来自 XML 文件
IConfigurationBuilder configbd3=new ConfigurationBuilder();
configbd3.AddXmlFile("test.xml");
// 将前面两个阶段产生的配置信息链接到当前配置中
configbd3.AddConfiguration(config_s1, shouldDisposeConfiguration: true);
configbd3.AddConfiguration(config_s2, shouldDisposeConfiguration: true);
// 产生最终合并后的配置信息树
IConfigurationRoot myConfig = configbd3.Build();
```

（1）第 1 阶段设置的自定义环境变量如下。

```
set ENV_Greeting=Hello App
```

设置的命令行参数如下。

```
--production=试用产品
```

（2）第 2 阶段的配置均来自内存数据，分别是当前进程名称（proc）、当前进程编号（pid）和当前进程的启动时间（start_time）。这些配置都是通过 GetCurrentProcess()方法动态获取。

（3）第 3 阶段使用的 XML 文件如下。

```
<config>
    <binOptions>
        <pathBase>/local/apps</pathBase>
        <ext>.exe;.com</ext>
    </binOptions>
</config>
```

示例程序最终将前两个阶段所生成的配置树都链接到 myConfig 中。下面的代码尝试随机访问部分配置信息。

```
Console.WriteLine($"自定义环境变量:ENV_Greeting = {myConfig["ENV_Greeting"]}");
Console.WriteLine($"命令行参数: Production = {myConfig["production"]}");
Console.WriteLine($"系统环境变量: windir = {myConfig["windir"]}");
Console.WriteLine($"XML 配置: pathBase = {myConfig["binOptions:pathBase"]}");
Console.WriteLine($"当前进程的启动时间: start_time = {myConfig["start_time"]}");
```

运行示例程序，结果如图 4-18 所示。

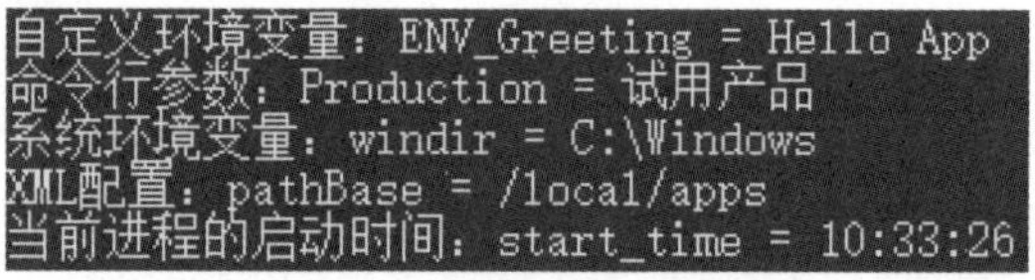
自定义环境变量：ENV_Greeting = Hello App
命令行参数：Production = 试用产品
系统环境变量：windir = C:\Windows
XML配置：pathBase = /local/apps
当前进程的启动时间：start_time = 10:33:26

图 4-18　链接多棵配置树

第 5 章　选 项 模 式

本章要点

- ❑ 了解与选项模式有关的接口
- ❑ 选项类的配置与访问
- ❑ 选项类的验证
- ❑ 实现选项接口

5.1　选项模式概述

选项模式通过类描述应用程序配置，并以强类型的方式访问。

例如，要让应用程序在发生异常后自动导航到处理错误的 URL，可以设定 ExceptionHandlerOptions 选项类（位于 Microsoft.AspNetCore.Builder 命名空间）。代码如下。

```
var builder = WebApplication.CreateBuilder(args);
// 设置选项类的属性
builder.Services.Configure<ExceptionHandlerOptions>(opt =>
{
    // 设置异常处理 URL
    // 当发生异常时自动跳转到此路径
    opt.ExceptionHandlingPath = "/exce";
});
…
// 必须先调用该方法
// 添加可捕捉异常的中间件
app.UseExceptionHandler();
// 发生异常后会访问 /exce
app.MapGet("/exce", () =>"哦，出错了");

// 根 URL
```

```
app.MapGet("/", () =>
{
    Results.StatusCode(StatusCodes.Status500InternalServerError);
    // 抛出异常
    throw new Exception("Error");
});
…
```

builder.Services.Configure()是服务容器的扩展方法，它有一个类型参数 TOptions。方法声明如下。

```
public static IServiceCollection Configure<TOptions>(this
IServiceCollection services, Action<TOptions> configureOptions)
where TOptions : class;
```

类型参数 TOptions 的约束条件比较简单，只要该类型是类（class）即可，此类型便是选项类。选项类与普通类没有区别，但其命名常以 Options 结尾。这并不是语法要求，只是一种使用习惯，以明确其用途。

TOptions 类型的实例由服务容器负责创建，调用 Configure()方法时会通过一个 Action 类型的委托设置选项类的属性，因此，选项类如何配置可以由代码编写者来掌控。

选项类配置完毕后，其实例也会被添加到服务容器中，使它支持依赖注入。不过，在请求选项类实例时不能直接使用 TOptions 类，而应该使用 IOptions<TOptions>接口。例如，将上述代码中/exce 终结点的 MapGet()方法代码做以下修改。

```
app.MapGet("/exce", () =>
{
    IOptions<ExceptionHandlerOptions> opt = app.Services.
    GetRequiredService<IOptions<ExceptionHandlerOptions>>();
    return "哦，出错了\n 异常处理路径：" + opt.Value.ExceptionHandlingPath;
});
```

上述代码中，先从服务容器中取出 IOptions<ExceptionHandlerOptions>类型的实例，再通过它的 Value 属性获得 ExceptionHandlerOptions 实例的引用，随后访问它的 ExceptionHandlingPath 属性。

与项目模式相关的 API 封装在以下程序集中。

（1）Microsoft.Extensions.Options。

（2）Microsoft.Extensions.Options.ConfigurationExtensions。

（3）Microsoft.Extensions.Options.DataAnnotations。

ASP.NET Core 应用程序项目会自动引用以上程序集，.NET 应用程序项目需要通过 Nuget 手动安装。命令如下。

```
dotnet add package Microsoft.Extensions.Options
dotnet add package Microsoft.Extensions.Options.ConfigurationExtensions
dotnet add package Microsoft.Extensions.Options.DataAnnotations
```

5.2　服务容器的扩展方法

为了方便开发人员配置选项类，OptionsServiceCollectionExtensions 类提供了一组扩展方法。

（1）AddOptions()：该方法向服务容器添加选项模式功能。以下类型会被添加到容器中：IOptions<T>、IOptionsSnapshot<T>、IOptionsMonitor<T>、IOptionsFactory<T>、IOptionsMonitorCache<T>。其中，类型参数 T（或者 TOptions）表示选项类。这些类型将被用于依赖注入。AddOptions()方法一般不需要开发人员去调用，应用程序在初始化时（或与 Configure<TOptions>类似的方法被调用时）将自动调用此方法。

（2）Configure<TOptions>：该方法最常用，可以直接通过委托配置选项类实例。

（3）PostConfigure<TOptions>：在完成选项类配置后调用（即在 Configure<TOptions>方法之后），可以对选项类进行后期配置。例如，为未赋值的属性设置默认值。

（4）ConfigureAll<TOptions>与 PostConfigureAll<TOptions>：如果选项类使用多个配置组，那么这两个方法可以同时配置所有分组中的选项类。

（5）ConfigureOptions()：要注意，这个方法不是用来配置选项类的，而是用来注册实现 IConfigureOptions<TOptions>、IPostConfigureOptions<TOptions>、IValidateOptions<TOptions> 接口的服务类型。内置的实现类型有 ConfigureOptions<TOptions>、PostConfigureOptions<TOptions>、ValidateOptions<TOptions>等。ConfigureOptions()方法主要用于自定义扩展。

5.3　各接口之间的关系

选项功能的运作都依靠服务容器来完成。根据用途实现各种接口类型，然后注册到服务容器中，再通过依赖注入将这些对象组合起来，形成一个小型系统。

5.3.1　IConfigureOptions<TOptions>接口与 IConfigureNamedOptions<TOptions>接口

IConfigureOptions<TOptions>接口用来配置选项类，它有一个 Configure()方法，实现代码可以在该方法中设置选项类的属性。此接口的默认实现类是 ConfigureOptions<TOptions>。

为了增加灵活性，将设置选项类属性的代码交给开发者去实现——通过传递给构造函数的委托实现。

```
public ConfigureOptions(Action<TOptions>? action)
```

在 Configure 方法的实现代码中，只是调用该委托而已。

```
public virtual void Configure(TOptions options)
{
    …
```

```
    Action?.Invoke(options);
}
```

IConfigureNamedOptions<TOptions>接口的默认实现类是ConfigureNamedOptions<TOptions>。其逻辑与ConfigureOptions<TOptions>类相同，只是多了一个name参数，表示命名选项组的名称。

在调用服务容器的Configure<TOptions>扩展方法时，实际上应用程序只是将Configure-NamedOptions<TOptions>对象添加到服务容器，并没有真正给选项类的属性赋值（关联的委托并未调用）。多次调用Configure<TOptions>就会添加多个实例。

5.3.2 IPostConfigureOptions<TOptions>接口

IPostConfigureOptions<TOptions>接口包含一个PostConfigure()方法。在选项类进行后期配置时调用。PostConfigure()方法声明如下。

```
void PostConfigure(string name, TOptions options)
```

默认实现类为PostConfigureOptions<TOptions>，思路与ConfigureNamedOptions<TOptions>相似，也是通过外部的委托实现后期配置。源代码如下。

```
public virtual void PostConfigure(string? name, TOptions options)
{
    …
    // Null name is used to initialize all named options.
    if (Name == null || name == Name)
    {
        Action?.Invoke(options);
    }
}
```

5.3.3 IValidateOptions<TOptions>接口

实现IValidateOptions<TOptions>接口用于验证选项类，主要实现Validate()方法，声明如下。

```
ValidateOptionsResult Validate(string name, TOptions options)
```

默认的实现类是ValidateOptions<TOptions>。具体的验证逻辑将交由外部的委托实现，最后通过构造函数传递。源代码如下。

```
public ValidateOptions(string? name, Func<TOptions, bool> validation,
string failureMessage)
```

在调用OptionsBuilder<TOptions>类的Validate()方法时，会向服务容器注册ValidateOptions<TOptions>对象。源代码如下。

```
public virtual OptionsBuilder<TOptions> Validate(Func<TOptions, bool>
validation, string failureMessage)
{
```

```
    ...
    Services.AddSingleton<IValidateOptions<TOptions>>(new ValidateOptions
    <TOptions>(Name, validation, failureMessage));
    return this;
}
```

5.3.4　IOptionsFactory<TOptions>接口

实现 IOptionsFactory<TOptions>接口，负责创建选项类的实例。实现类必须包含 Create()方法的实现，声明如下。

```
TOptions Create(string name)
```

默认的实现类是 OptionsFactory<TOptions>。前面所介绍的 IConfigureOptions<TOptions>、IConfigureNamedOptions<TOptions>、IValidateOptions<TOptions>接口所关联的服务实例都会注入 OptionsFactory<TOptions>类的构造函数中。该构造函数的声明如下。

```
public OptionsFactory(IEnumerable<IConfigureOptions<TOptions>> setups,
IEnumerable<IPostConfigureOptions<TOptions>> postConfigures, IEnumerable
<IValidateOptions<TOptions>> validations)
```

使用 IEnumerable<T>是因为同一接口类型可能被多次注册，如多次调用 Configure<TOptions>就会注册多个 ConfigureNamedOptions<TOption>。

实现 Create()方法，创建选项类实例，并通过从构造函数参数中获取的各种功能对象配置和验证选项类。源代码如下。

```
public TOptions Create(string name)
{
    TOptions options = CreateInstance(name);
    // 配置选项类
    foreach (IConfigureOptions<TOptions> setup in _setups)
    {
        if (setup is IConfigureNamedOptions<TOptions> namedSetup)
        {
            namedSetup.Configure(name, options);
        }
        else if (name == Options.DefaultName)
        {
            setup.Configure(options);
        }
    }
    // 后期配置
    foreach (IPostConfigureOptions<TOptions> post in _postConfigures)
    {
        post.PostConfigure(name, options);
    }

    // 验证选项类
```

```
        if (_validations.Length > 0)
        {
            var failures = new List<string>();
            foreach (IValidateOptions<TOptions> validate in _validations)
            {
                ValidateOptionsResult result = validate.Validate(name, options);
                if (result is not null && result.Failed)
                {
                    failures.AddRange(result.Failures);
                }
            }
            if (failures.Count > 0)
            {
                throw new OptionsValidationException(name, typeof(TOptions),
                failures);
            }
        }

        return options;
    }

    protected virtual TOptions CreateInstance(string name)
    {
        return Activator.CreateInstance<TOptions>();
    }
```

5.3.5 完整的流程图

选项接口的工作流程如图 5-1 所示。

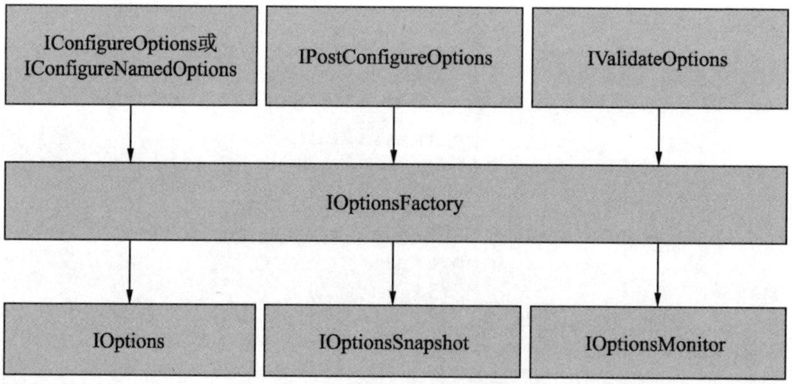

图 5-1 选项接口的工作流程

默认情况下，IOptions<TOptions>接口对应的实现类型是 UnnamedOptionsManager<TOptions>，IOptionsSnapshot<TOptions>接口对应的实现类型是 OptionsManager<TOptions>，IOptionsMonitor<TOptions>接口对应的实现类型是 OptionsMonitor<TOptions>。当要从服务容

器获取这些服务时，IOptionsFactory<TOptions>会注入它们的构造函数中。

例如，使用 IOptions<X>接口获取 X 选项类的实例，服务容器会返回 UnnamedOptionsManager<X>实例；同时，OptionsFactory<X>实例被注入 UnnamedOptionsManager<X>类的构造函数中；最后调用 OptionsFactory<X>实例的 Create()方法得到 X 选项类的实例。

5.4 选项类的封装接口

选项类虽然支持依赖注入，但应用程序并不是直接将类实例添加到服务容器中的，而是通过一组特定的对象将其封装，然后再以服务的形式注册到容器中。

在依赖注入中常用到的封装接口如下。

（1）IOptions<TOptions>：这是最基础的接口类型，通常使用该接口作为服务类型就可以从服务容器中提取服务实例。该接口定义了 Value 属性，用于获得 TOptions 类（即选项类）的实例引用。该接口的实现类型有 UnnamedOptionsManager<TOptions>、OptionsManager<TOptions>、OptionsWrapper<TOptions>，其中 UnnamedOptionsManager<TOptions>是内部类型，没有对外公开。

（2）IOptionsSnapshot<TOptions>：该接口继承了 IOptions<TOptions>接口，新增了 Get()方法，可以根据名称获取对应的选项类实例。

（3）IOptionsMonitor<TOptions>：实现该接口的类型支持更改通知。当选项类被更新后调用指定的委托。

以上几个接口在注册到服务容器中时所给定的服务生存期是不同的，通过 AddOptions()方法（IServiceCollection 的扩展方法）的源代码了解它们在容器中的生存周期。

```
public static IServiceCollection AddOptions(this IServiceCollection
services)
{
    …
    services.TryAdd(ServiceDescriptor.Singleton(typeof(IOptions<>),
    typeof(UnnamedOptionsManager<>)));
    services.TryAdd(ServiceDescriptor.Scoped(typeof(IOptionsSnapshot<>),
    typeof(OptionsManager<>)));
    services.TryAdd(ServiceDescriptor.Singleton(typeof(IOptionsMonitor<>),
    typeof(OptionsMonitor<>)));
    …
    services.TryAdd(ServiceDescriptor.Singleton(typeof
    (IOptionsMonitorCache<>), typeof(OptionsCache<>)));
    return services;
}
```

选项类封装接口的服务生存期如表 5-1 所示。

表 5-1 选项类封装接口的服务生存期

接　口	生 存 期	备　注
IOptions<TOptions>	单实例	只创建一个实例，可以注入任意生存期的服务中
IOptionsSnapshot<TOptions>	作用域	生存期仅限于特定作用域，不能注入单实例服务中
IOptionsMonitor<TOptions>	单实例	可以注入任意生存期的服务中
IOptionsMonitorCache<TOptions>	单实例	可以注入任意生存期的服务中

5.4.1 示例：在 MVC 控制器中访问选项类

本示例所使用的自定义选项类如下。

```
public class DemoOptions
{
    public long MaxMemSize { get; set; }
    public long MinMemSize { get; set; }
}
```

在 Program.cs 文件中，向服务容器注册与 MVC 相关的服务类型。

```
builder.Services.AddControllers();
```

调用 Configure()扩展方法配置 DemoOptions 选项类。

```
builder.Services.Configure<DemoOptions>(opt =>
{
    opt.MaxMemSize = 1024 * 1024 * 16;
    opt.MinMemSize = 1024 * 1024 * 48;
});
```

在创建 WebApplication 后，调用 MapControllerRoute()方法向 HTTP 管道插入 MVC 终结点。

```
app.MapControllerRoute("test", "{controller=home}/{action=default}");
```

下面是 Home 控制器的完整代码。

```
public class HomeController : Controller
{
    readonly DemoOptions _myOptions;

    public HomeController(IOptions<DemoOptions> options)
    {
        // 从注入的 IOptions<DemoOptions>对象中
        // 获得选项类的引用
        _myOptions = options.Value;
    }

    public ActionResult Default()
    {
        // 获取选项类的属性值
```

```
        long maxmen = _myOptions.MaxMemSize;
        long minmen = _myOptions.MinMemSize;
        string ct = $"最大内存：{maxmen}\n" +
                    $"最小内存：{minmen}";
        return Content(ct);
    }
}
```

控制器类的构造函数定义类型为 IOptions<DemoOptions>的参数，在控制器被激活时会通过依赖注入自动获取到封装 DemoOptions 选项类实例的对象。随后在 Default()方法中就可以访问 MaxMemSize 和 MinMemSize 属性了。

运行示例程序，结果如图 5-2 所示。

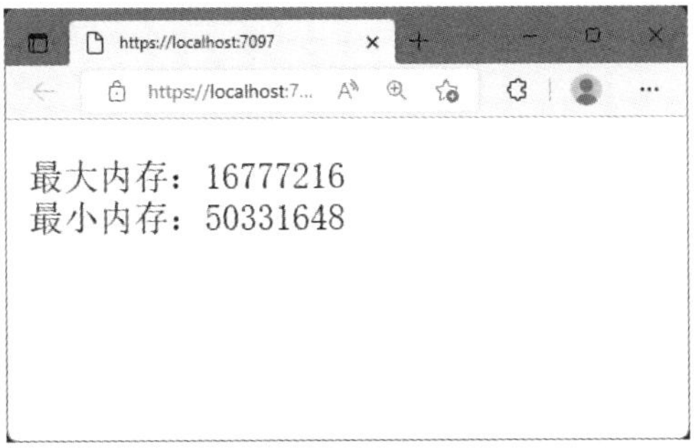

图 5-2　访问选项类的属性

5.4.2　示例：自动更新选项类

在依赖注入时，如果使用 IOptionsMonitor<TOptions>接口类型，那么当选项类的数据源更新后会自动重新载入选项类实例。

在本示例中，选项类的属性值来自 appsettings.json 配置文件。应用程序启动后会在 Web 浏览器上显示选项类各属性的值，然后修改 appsettings.json 文件中的选项配置。应用程序不需要重新启动，只要在 Web 浏览器刷新一下即可读取到选项的最新数据。

具体实现步骤如下。

（1）定义 ColorOptions 类，在本示例中将作为选项类使用。

```
public class ColorOptions
{
    public string? TitleColor { get; set; }
    public string? BodyColor { get; set; }
    public string? CommentColor { get; set; }
}
```

（2）在服务容器中注册以上选项类，并将其与 appsettings.json 文件中的 colors 节点绑定。

```
builder.Services.Configure<ColorOptions>(builder.Configuration.
GetSection("colors"));
```

（3）在 Home 控制器类中，使用 IOptionsMonitor<TOptions>接口接收依赖注入的对象。

```
readonly ColorOptions _colorOptions;
public HomeController(IOptionsMonitor<ColorOptions> monitor)
{
    colorOptions = monitor.CurrentValue;
}
```

CurrentValue 属性获取未命名的 ColorOptions 类的实例。如果选项类在注册时是自定义命名的，则需要调用 Get()方法来获取。本示例使用的是未命名的选项类。

（4）在 Index()操作方法中读出 ColorOptions 对象各个属性的值，并返回给客户端。

```
public IActionResult Index()
{
    StringBuilder msgs = new();
    msgs.AppendLine($"标题颜色：{_colorOptions.TitleColor}");
    msgs.AppendLine($"正文颜色：{_colorOptions.BodyColor}");
    msgs.AppendLine($"注释颜色：{_colorOptions.CommentColor}");
    return Content(msgs.ToString());
}
```

（5）打开 appsettings.json 文件，添加 colors 节点配置并保存。

```
{
  …
  "colors": {
  "titleColor": "blue",
  "bodyColor": "black",
  "commentColor": "green"}
}
```

（6）运行示例程序，此时浏览器中显示上述 JSON 文件所配置的内容，如图 5-3 所示。

（7）回到 appsettings.json 文件，将配置内容修改为以下内容并保存。

```
"colors": {
"titleColor": "gray",
"bodyColor": "red",
"commentColor": "pink"
}
```

（8）不需要重新启动应用程序，只要在浏览器中刷新就能看到最新配置的选项内容，如图 5-4 所示。

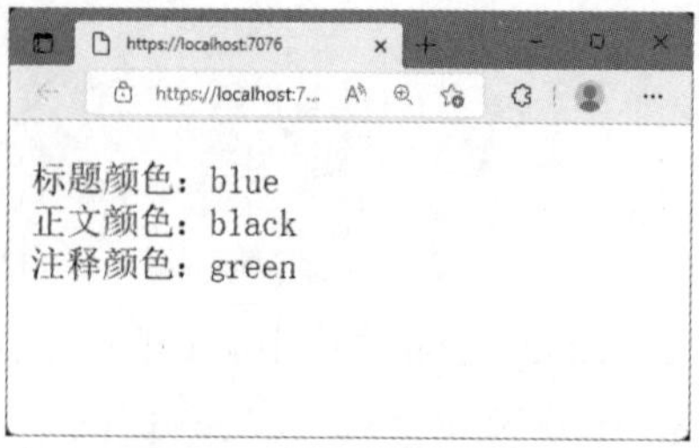

图 5-3　已配置的选项

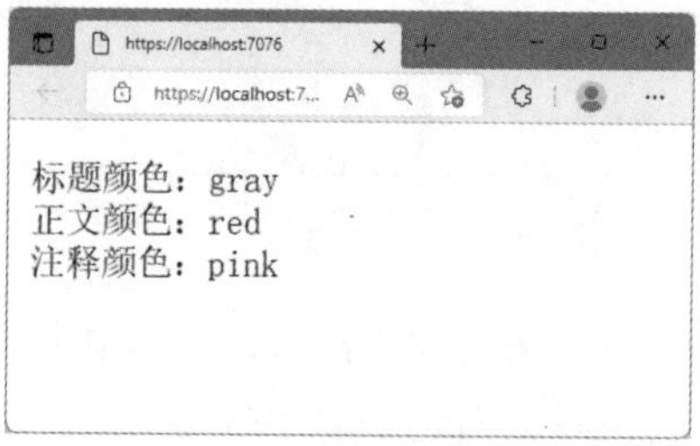

图 5-4　选项类的属性已自动更新

5.5 带名称的选项组

选项类在服务容器中注册时可以为其指定一个名称，这使得相同类型的选项存在多个分组，每个命名分组的选项类实例都具有独立的数据。在获取选项类的实例引用时，依赖注入类型可选择 IOptionsSnapshot<TOptions>或 IOptionsMonitor<TOptions>。它们都有一个 Get()方法，通过选项组名称获取对应的选项类实例。

选项配置的默认命名由 Options.DefaultName 字段定义，其值为空白字符串。可以认为是未命名的选项组。前文示例中使用过的 IOptions<TOptions>接口就是基于默认选项名称的。

下面的示例演示了通过选项分组设置 HTML 页面主题颜色的方法。

（1）定义选项类。

```
public class ThemeOptions
{
    /// <summary>
    /// 标题文本颜色
    /// </summary>
    public string? HeaderColor { get; set; }
    /// <summary>
    /// 段落正文颜色
    /// </summary>
    public string? ContentColor { get; set; }
    /// <summary>
    /// 列表项颜色
    /// </summary>
    public string? ListColor { get; set; }
    /// <summary>
    /// 元素背景颜色
    /// </summary>
    public string? BackColor { get; set; }
}
```

上述选项类定义了 4 个选项：标题文本的颜色（HeaderColor）、正文段落的文本颜色（ContentColor）、列表项的文本颜色（ListColor）、HTML 容器元素的背景色（BackColor）。

（2）添加两组配置，分别命名为 theme-1 和 theme-2。

```
builder.Services.Configure<ThemeOptions>("theme-1", topt =>
{
    topt.HeaderColor = "red";
    topt.ContentColor = "black";
    topt.ListColor = "purple";
    topt.BackColor = "snow";
});
builder.Services.Configure<ThemeOptions>("theme-2", topt =>
{
    topt.HeaderColor = "yellow";
```

```
        topt.ContentColor = "white";
        topt.ListColor = "lightgreen";
        topt.BackColor = "black";
    });
```

上述代码调用的是 Configure<TOptions>方法的一个重载版本，它带有 string 类型的 name 参数，用来指定选项的分组名称。本示例注册了两组选项：theme-1 和 theme-2。

（3）在 Home 控制器类中，通过 IOptionsSnapshot<ThemeOptions>类型从服务容器中获取选项类相关的封装对象。

```
IOptionsSnapshot<ThemeOptions> _opt;
public HomeController(IOptionsSnapshot<ThemeOptions> opt)
{
    _opt = opt;
}
```

（4）调用 Get()方法并提供分组的名称，即可获取相应的 ThemeOptions 实例。

```
public IActionResult Index()
{
    // 初始主题
    ThemeOptions t = _opt.Get("theme-1");
    ViewBag.header = t.HeaderColor;
    ViewBag.content = t.ContentColor;
    ViewBag.list = t.ListColor;
    ViewBag.back = t.BackColor;
    return View();
}

public IActionResult SetTheme(string theme)
{
    ThemeOptions tho = _opt.Get(theme);
    ViewBag.header = tho.HeaderColor;
    ViewBag.content = tho.ContentColor;
    ViewBag.list = tho.ListColor;
    ViewBag.back = tho.BackColor;
    // 保存当前选择项
    ViewBag.currentSelect = theme;
    return View("Index");
}
```

在 SetTheme()方法中，选项分组名称由 theme 参数的值来决定。该参数的值通过视图中的<select>元素提交。

```
<form asp-action="SetTheme">
    @{
        string sel = ViewBag.currentSelect;
    }
<select name="theme">
```

```
        @if (string.IsNullOrEmpty(sel))
        {
            <option value="theme-1" selected>主题 1</option>
            <option value="theme-2">主题 2</option>
        }
        else
        {
            if (sel == "theme-2")
            {
                <option value="theme-1">主题 1</option>
                <option value="theme-2" selected>主题 2</option>
            }
            else
            {
                <option value="theme-1" selected>主题 1</option>
                <option value="theme-2">主题 2</option>
            }
        }
</select>
<button type="submit">提交</button>
</form>
```

当用户从下拉列表（<select>元素）中选择一个主题后，会向 Web 服务器提交被选的值。服务器调用 SetTheme()方法。处理完毕后，将选择的主题名称存储到 ViewBag.currentSelect 动态属性中。当重新加载 HTML 视图时，会根据 ViewBag 中所存放的内容控制<select>元素中默认选中的子项。如果用户提交时选择了 theme-2，那么重新加载视图后默认会选中“主题 2”选项。如果是第 1 次加载视图（ViewBag.currentSelect 动态属性无值），则默认选中“主题 1”选项。

（5）运行示例程序，结果如图 5-5 和图 5-6 所示。

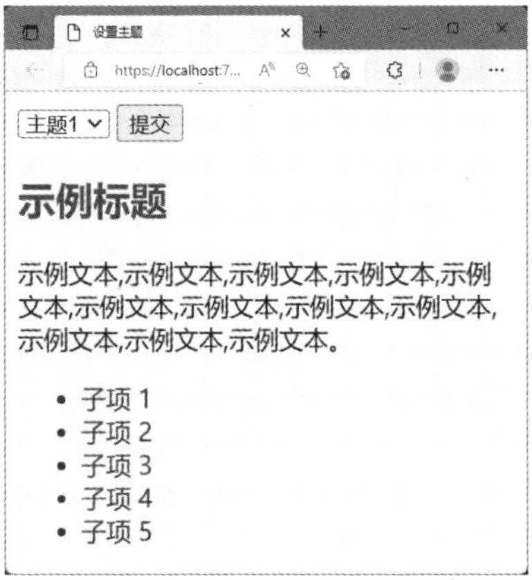

图 5-5　theme-1 颜色主题

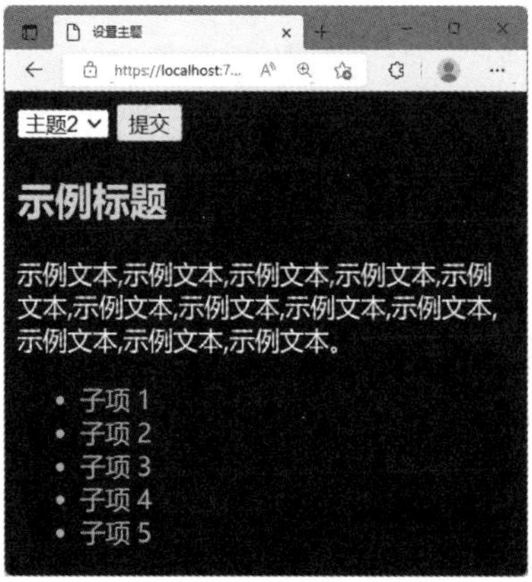

图 5-6　theme-2 颜色主题

5.6 后期配置

后期选项配置通过调用 PostConfigure()扩展方法，向服务容器添加 IPostConfigureOptions<TOptions>服务，其配置过程与 Configure<TOptions>方法一样。后期配置可以在选项类完成配置后对其进行再次修改。典型的作用如下。

（1）在前期配置中没有为选项类的个别属性赋值，在后期配置中可以为这些属性分配默认值。

（2）因其他选项类的配置导致当前选项类的某些属性值不合适，需要修改。例如，选项类 A 有一个 CurrVal 属性，而这个属性与选项类 B 的 Max 属性相关，若 A.CurrVal＞B.Max，就把 A.CurrVal 修改为 B.Max。

接下来给出一个示例。

（1）定义选项类。

```
public class TestOptions
{
    public string? FrameworkName { get; set; }
    public Version? FrameworkVersion { get; set; }
    public string? Platform { get; set; }
}
```

（2）先配置选项类。

```
builder.Services.Configure<TestOptions>(opt =>
{
    opt.FrameworkVersion = new Version(2, 0, 1292);
    opt.FrameworkName = "DBLand";
});
```

（3）由于上面的配置中未设置 Platform 属性的值，且 FrameworkVersion 属性所表示的版本号可能与实际需求不符合，于是可以通过后期配置进行调整。

```
builder.Services.PostConfigure<TestOptions>(opt =>
{
    if(opt.FrameworkVersion == null || opt.FrameworkVersion.Major < 3)
    {
        opt.FrameworkVersion = new Version(3, 0, 100);
    }
    if(opt.Platform == null)
    {
        opt.Platform = "All OS";
    }
});
```

如果 FrameworkVersion 属性指定的主版本号小于 3，就将属性改为版本号 3.0.100；如果 Platform 属性未设置，就设置为 All OS。

（4）修改 MapGet()方法中的代码，输出该选项类实例的属性值。

```
    app.MapGet("/", () =>
    {
        IOptions<TestOptions> optWrapper = app.Services.GetRequiredService
        <IOptions<TestOptions>>();
        TestOptions theOpt = optWrapper.Value;
        string s = $"Framework Name: {theOpt.FrameworkName}\n" +
                   $"Framework Version: {theOpt.FrameworkVersion}\n" +
                   $"Platform: {theOpt.Platform}";
        return s;
    });
```

（5）运行示例程序，结果如图 5-7 所示。

图 5-7　选项类的后期配置

5.7　选项类的验证

IValidateOptions<TOptions>接口用于实现选项类的验证功能。该接口只需要实现以下方法。

```
    ValidateOptionsResult Validate(string? name, TOptions options);
```

name 参数表示选项组的命名，未命名选项可使用 Options.DefaultName（空白字符串）字段的值；options 参数代表被验证的选项类的实例。

Validate()方法的内部实现选项类的验证过程，返回结果为 ValidateOptionsResult 类的实例。为了方便使用，该类定义了一些静态成员，具体如下。

（1）Success 字段：如果选项类验证通过，可直接从 Validate()方法返回 Success 字段。

（2）Skip 字段：表示跳过此次验证，可从 Validate()方法直接返回该字段。一般来说，当调用 Validate()方法时给 name 参数传递的选项组命名不匹配时可以跳过验证。

（3）Fail()方法：当验证失败时调用此方法并从 Validate()方法返回。可以指定自定义的错误信息。

要使用选项的验证功能，需要先调用服务容器的 AddOptions<TOptions>扩展方法，从返回值中获得一个 OptionsBuilder<TOptions>实例；然后通过这个 OptionsBuilder<TOptions>实例的方法（或扩展方法）添加验证功能。

5.7.1 内置的验证方式

ValidateOptions<TOptions>类（位于 Microsoft.Extensions.Options 命名空间）实现 IValidateOptions<TOptions>接口，提供通过的选项类验证逻辑。不过，ValidateOptions<TOptions>类在实现 Validate()方法时并没有提供实际的验证逻辑。具体的验证逻辑交给来自外部的委托对象。这使得 ValidateOptions<TOptions>类与选项类的具体验证过程分离，提高了通用性。

用于验证的委托对象从构造函数传入，由 Validation 属性引用。部分源代码如下。

```
public ValidateOptions(string? name, Func<TOptions, bool> validation,
string failureMessage)
{
    …
    Validation = validation;
    …
}

public Func<TOptions, bool> Validation { get; }
```

Validation 属性使用的委托将以选项类实例为输入参数，如果验证成功则返回 true，否则返回 false。

开发人员不需要直接使用 ValidateOptions<TOptions>类，而是通过调用 OptionsBuilder<TOptions>类的 Validate()方法来完成。调用方法时传递带验证逻辑的委托即可。

下面给出一个示例。

（1）定义选项类。

```
public class KeyOptions
{
    public string? Key1 { get; set; }
    public string? Key2 { get; set; }
    public int KeySize { get; set; }
}
```

（2）由于本示例要对选项类进行验证，因此需要调用服务容器的 AddOptions()扩展方法，然后依次调用 Configure()和 Validate()方法。

```
builder.Services.AddOptions<KeyOptions>()
  .Configure(opt =>                                      // 配置选项类
  {
     opt.Key1 = "T+f59y2j6enX29m6";
     opt.Key2 = "45163390";
     opt.KeySize = 2048;
  })
  .Validate(opt =>                                       // 验证选项类
  {
     if (opt.Key1 == null || opt.Key2 == null)
         return false;
```

```
        // Key1 必须以一个字母开头，后跟+符号
        // 接着是 12～15 个非空字符
        var res = Regex.Match(opt.Key1, "^[a-zA-Z]{1}\\+\\S{12,15}");
        if (!res.Success)
        {
            return false;
        }
        // Key2 必须是 8 个数字字符
        res = Regex.Match(opt.Key2, "^\\d{8}$");
        if (!res.Success)
        {
            return false;
        }
        // KeySize 是大于 0 的整数
        // KeySize 必须能被 1024 整除
        if(opt.KeySize <= 0 || (opt.KeySize % 1024) != 0)
        {
            return false;
        }
        return true;
    },
    "选项类未通过验证");
```

在 Configure()方法中设置选项类各属性的值；在 Validate()方法中处理验证逻辑。验证规则：Key1 属性必须以“<单个字母>+”开头，后接 12～15 个非空字符；Key2 属性必须是 8 个数字字符；KeySize 属性必须能被 1024 整除。Key1 和 Key2 属性是通过正则表达式进行验证的。Validate()方法的最后一个参数指定验证失败时 OptionsValidationException 异常所包含的错误信息。

（3）修改项目模板生成的 MapGet()方法，读取选项类实例的属性值并返回给客户端。

```
app.MapGet("/", () =>
{
    string msg = string.Empty;
    // 尝试访问选项类
    try
    {
        var iopt = app.Services.GetRequiredService<IOptions<KeyOptions>>();
        msg += $"Key1: {iopt.Value.Key1}\n";
        msg += $"Key2: {iopt.Value.Key2}\n";
        msg += $"Key Size: {iopt.Value.KeySize}";
    }
    catch(OptionsValidationException ve)
    {
        // 验证未通过时引发异常
        msg = ve.Message;
    }
    return msg;
});
```

若选项类验证失败，会抛出 OptionsValidationException 异常，因此访问选项类的代码应写在 try...catch 语句块中。

（4）运行示例程序，结果如图 5-8 所示。

（5）如果验证失败，会返回自定义的错误信息，如图 5-9 所示。

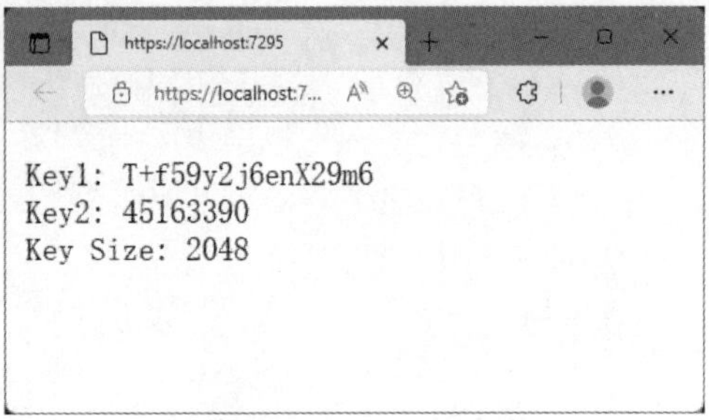

图 5-8 选项类验证成功

图 5-9 选项类验证失败

5.7.2 使用数据批注

数据批注是位于 System.ComponentModel.DataAnnotations 命名空间的一组特性类。将这些特性应用于选项类的属性（或字段）成员上，如果该成员的值不满足数据批注所设定的条件，则表示验证失败。

DataAnnotationValidateOptions<TOptions> 类实现 IValidateOptions<TOptions> 接口，在 Validate()方法的实现代码中，调用 Validator.TryValidateObject()方法验证选项类实例的数据批注。关键源代码如下。

```
public ValidateOptionsResult Validate(string? name, TOptions options)
{
    // Null name is used to configure all named options.
    if (Name != null && Name != name)
    {
        // Ignored if not validating this instance.
        return ValidateOptionsResult.Skip;
    }

    …

    var validationResults = new List<ValidationResult>();
    if (Validator.TryValidateObject(options, new ValidationContext
    (options), validationResults, validateAllProperties: true))
    {
        return ValidateOptionsResult.Success;
    }

    …
```

```
        var errors = new List<string>();
        foreach (ValidationResult result in validationResults)
        {
         // 添加错误信息
        }

        return ValidateOptionsResult.Fail(errors);
    }
```

OptionsBuilderDataAnnotationsExtensions 类中为 OptionsBuilder<TOptions> 类定义了 ValidateDataAnnotations()扩展方法。调用方法时，会将 DataAnnotationValidateOptions<TOptions> 实例添加到服务容器中。

```
public static OptionsBuilder<TOptions> ValidateDataAnnotations<TOptions>
(this OptionsBuilder<TOptions> optionsBuilder) where TOptions : class
{
    optionsBuilder.Services.AddSingleton<IValidateOptions<TOptions>>
    (new DataAnnotationValidateOptions<TOptions>(optionsBuilder.Name));
    return optionsBuilder;
}
```

常用的批注特性如下。

（1）Required：表示该属性（或字段）必须有值。

（2）MaxLength 和 MinLength：指定数组和字符串的最大和最小长度。

（3）Range：指定属性（或字段）值的范围。MaxLength 和 MinLength 特性是针对数组类型的元素个数而言，而 Range 特性是针对单个值的，如 int、double 类型的值。Range 特性支持实现了 IComparable 接口的类型，因此 DateTime、Byte 等类型也可以使用 Range 属性指定范围。

（4）EmailAddress：指定属性（或字段）的值必须符合电子邮箱的格式。

（5）Phone：属性（或字段）值必须符合手机号码格式。

（6）Url：指定属性（或字段）必须为有效的 URL 格式。

（7）RegularExpression：通过正则表达式约束属性（或字段）的值。

（8）FileExtensionsAttribute：表示属性（或字段）必须包含特性提供的文件扩展名。其格式为用逗号分隔的扩展名，如“docx,jpg,exe”。

（9）StringLength：对于字符串类型的属性（或字符），限制其字符个数。

下面的代码定义了一个选项类，并应用数据批注特性。

```
public class MyOptions
{
    [Range(3, 10, ErrorMessage = "{0}属性的值无效，有效范围：[{1},{2}]")]
    public int ThreadCount { get; set; }

    [Required]
    [RegularExpression("^\\d{2,4}:[a-z0-9]{5,7}", ErrorMessage = "{0}属性
    的格式不正确")]
```

```
    public string? Identity { get; set; }

    [MaxLength(3, ErrorMessage = "{0}属性的最大元素个数为{1}")]
    public uint[]? Segs { get; set; }

    [StringLength(16, ErrorMessage = "{0}属性的字符串最大长度：{1}")]
    public string? Tag { get; set; }
}
```

ThreadCount 属性要求的值在[3,10]范围内，错误消息（由 ErrorMessage 属性指定，下同）的格式化参数：{0}表示当前属性名（此处为 ThreadCount）；{1}表示最小值；{2}表示最大值。

Identity 属性为必须成员，并且用正则表达式设置了格式：以 2～4 个数字字符开头，之后是冒号（:）；冒号之后是 5～7 个字符，可以使用小写字母和数字字符。错误消息的格式化参数：{0}表示当前属性名；{1}表示指定的正则表达式。

Segs 属性是 uint 值数组，最大元素个数为 3。错误消息的格式化参数：{0}表示当前属性名；{1}为指定的最大元素个数。

Tag 属性为字符串类型，最大长度是 16 个字符。错误消息的格式化参数：{0}表示当前属性名；{1}表示指定的最大长度。

下面的代码配置选项类，并启用验证功能。

```
// 配置选项类
optBuilder.Configure(o =>
{
    o.ThreadCount = 4;
    o.Identity = "27:xe6r80";
    o.Segs = new uint[] { 113, 42};
    o.Tag = "test_test";
});
// 开启数据批注验证
optBuilder.ValidateDataAnnotations();
optBuilder.ValidateOnStart();
```

选项类在默认情况下是被用户代码请求时才会进行验证。若调用了 ValidateOnStart()方法，那么应用程序在调用 app.Run()方法的过程中就进行验证。一旦验证失败，就会抛出 OptionsValidationException 异常。

```
try
{
    app.Run();
}
catch(OptionsValidationException ex)
{
    app.Logger.Log(LogLevel.Error, ex.Message);
}
```

5.8　处理带参数的构造函数

一般情况下，推荐保留选项类的默认构造函数（即无参数的构造函数）。但在实际开发中，难免会遇到特殊情况——选项类的构造函数带有参数。

请读者回顾一下 OptionsFactory 类的实现代码（5.3.4 节），官方源代码位于 https://github.com/dotnet/runtime 仓库中，文件路径为 src/libraries/Microsoft.Extensions.Options/src/OptionsFactory.cs。创建选项类实例的关键代码如下。

```
public TOptions Create(string name)
{
    TOptions options = CreateInstance(name);
    …
    return options;
}

protected virtual TOptions CreateInstance(string name)
{
    return Activator.CreateInstance<TOptions>();
}
```

重点关注 CreateInstance()方法，也请读者记住此方法，本节后面的示例中会用到。此方法声明为虚方法，意味着在派生类中可以重写。在这个方法中，创建选项类实例只有一行代码——调用 Activator 类的 CreateInstance()静态方法。由于 TOptions 只是类型参数，并没有明确特定的类型（约束条件是必须为类，存在默认构造函数），所以无法直接使用 new 运算符调用构造函数，只能用 Activator 类创建实例。

从上述源代码中不难发现，它只能调用无参数的构造函数。若选项类的构造函数带有参数，开发人员就得自己实现 IOptionsFactory<TOptions>接口。更简单的方法是从 OptionsFactory<TOptions>类派生，然后重写 CreateInstance()方法即可。

接下来给出一个示例。假设有以下选项类。

```
public class TestOptions
{
    // 构造函数
    public TestOptions(string? k1, string? k2, ushort num = 1000)
    {
        KeyItem1 = k1;
        KeyItem2 = k2;
        U16Num = num;
    }

    public string? KeyItem1 { get; set; }
    public string? KeyItem2 { get; set; }
    public ushort U16Num { get; set; }
}
```

上述选项类存在有参数的构造函数，并在构造函数中为 3 个属性赋值（其中 num 参数默认值为 1000，调用构造函数时可以不赋值）。

然后从 OptionsFactory 类派生出 TestOptionsFactory 类，类型参数 TOptions 直接用 TestOptions 类填充。

```
public class TestOptionsFactory : OptionsFactory<TestOptions>
{
    public TestOptionsFactory(
            IEnumerable<IConfigureOptions<TestOptions>> optCfgs,
            IEnumerable<IPostConfigureOptions<TestOptions>> postCfgs
            ): base(optCfgs, postCfgs)
    { }

    public TestOptionsFactory(
            IEnumerable<IConfigureOptions<TestOptions>> optConfigs,
            IEnumerable<IPostConfigureOptions<TestOptions>> optPostConfigs,
            IEnumerable<IValidateOptions<TestOptions>> validaters
            ): base(optConfigs,optPostConfigs,validaters)
    { }

    // 覆写方法
    protected override TestOptions CreateInstance(string name)
    {
        return (TestOptions)Activator.CreateInstance(typeof(TestOptions),
        "defaultItem", string.Empty, (ushort)0)!;
    }
}
```

要注意以下两点。

（1）构造函数要声明 IEnumerable<IConfigureOptions<TestOptions>>、IEnumerable<IPostConfigureOptions<TestOptions>>、IEnumerable<IValidateOptions<TestOptions>>类型的参数，并传递给基类的构造函数。在配置选项类时需要这些类型。

（2）上述代码在重写 CreateInstance()方法时，使用的是 Activator 类的成员方法 CreateInstance()创建 TestOptions 实例。第 1 个参数是要创建实例的 Type，从第 2 个参数起就是要传给 TestOptions 构造函数的参数值。

其实，在本示例中可以不使用 Activator 类，因为 TestOptionsFactory 类是专门为 TestOptions 选项类而编写的，即类型参数 TOptions 是明确的，因此可以直接调用 TestOptions 类的构造函数，这样做性能会更好。于是，CreateInstance()方法中的代码也可以这样写：

```
protected override TestOptions CreateInstance(string name)
{
    return new TestOptions("defaultItem", string.Empty, (ushort)0);
}
```

在 Program.cs 文件中，需要将 TestOptionsFactory 类注册到服务容器中。

```
var builder = WebApplication.CreateBuilder(args);
builder.Services.AddTransient<IOptionsFactory<TestOptions>,
```

```
    TestOptionsFactory>();
```

还可以配置一下选项类。

```
builder.Services.Configure<TestOptions>(opt =>
{
    opt.U16Num = 45500;
    opt.KeyItem2 = "newItem";
});
```

尝试使用选项类。

```
app.MapGet("/", () =>
{
    IOptions<TestOptions> iopt = app.Services.GetRequiredService
    <IOptions<TestOptions>>();
    TestOptions theOpt = iopt.Value;
    string msg = $"Key Item 1: {theOpt.KeyItem1}\n";
    msg += $"Key Item 2: {theOpt.KeyItem2}\n";
    msg += $"U16 Num: {theOpt.U16Num}";
    return msg;
});
```

运行示例程序，结果如图 5-10 所示。

在服务容器中添加 TestOptionsFactory 类后，并不影响选项模式的默认行为。为了验证这一点，可以尝试在服务容器中配置一下 ConsoleLoggerOptions 选项类。

```
builder.Services.Configure<ConsoleLoggerOptions>(csopt =>
{
    csopt.FormatterName = ConsoleFormatterNames.Json;
});
```

上述配置是将控制台窗口的日志输出格式修改为 JSON。运行示例程序后，会看到控制台窗口上的记录消息变为 JSON 结构，如图 5-11 所示。这说明默认的选项模式能正常工作，示例中自定义的 TestOptionsFactory 类并没有破坏原有的选项模式。

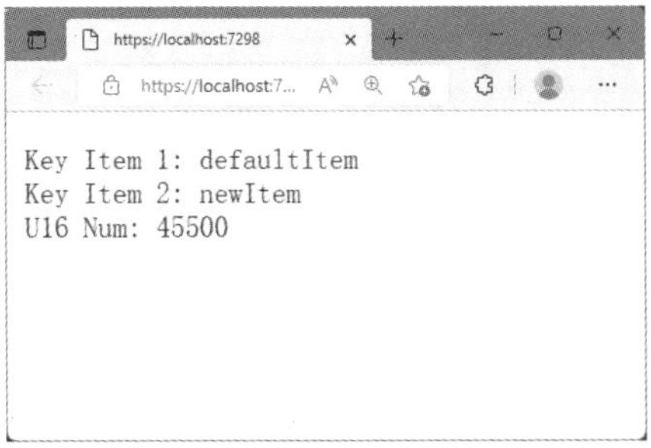

图 5-10 TestOptions 各属性的值

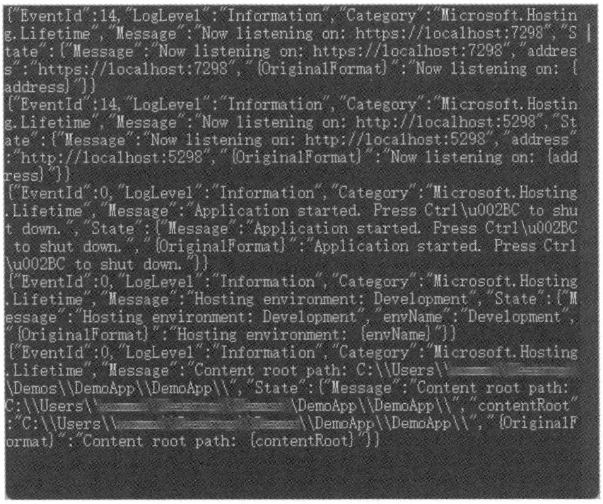

图 5-11 日志记录输出为 JSON 格式

5.9 直接实现 IOptions 接口

如果某个类直接实现 IOptions<TOptions>接口，是否能用于选项模式呢？接下来将围绕这个问题做一个尝试。

定义一个名为 AppOptions 的类，并让它显式实现 IOptions<TOptions>接口。类型参数 TOptions 替换为 AppOptions 类。

```
public class AppOptions : IOptions<AppOptions>
{
    public DateTime Key1 { get; set; }
    public double Key2 { get; set; }
    public int Key3 { get; set; }

    // 显式实现接口
    AppOptions IOptions<AppOptions>.Value => this;
}
```

此时将该类注册到服务容器中就可以正常使用了。

```
builder.Services.AddSingleton<IOptions<AppOptions>, AppOptions>();
// 配置选项
builder.Services.Configure<AppOptions>(opt =>
{
    opt.Key1 = new DateTime(2002, 11, 8, 14, 55, 13);
    opt.Key2 = 107.0212d;
    opt.Key3 = -5303;
});

/*-------------------------------------------------------------*/
// MVC 控制器
public class HomeController : Controller
{
    IOptions<AppOptions> _option;
    // 构造函数
    public HomeController(IOptions<AppOptions> opt)
    {
        _option = opt;
    }

    public ActionResult Default()
    {
        // 访问选项类
        AppOptions appopt = _option.Value;
        string returnStr = string.Format("Key1: {0}", appopt.Key1);
        returnStr += string.Format("\nKey2: {0}", appopt.Key2);
        returnStr += $"\nKey3: {appopt.Key3}";
        return Content(returnStr);
```

```
        }
    }
```

运行代码后，通过 IOptions<AppOptions>接口类型能获得 AppOptions 实例，但 Services.Configure<AppOptions>方法所做的配置没有生效，如图 5-12 所示。

原因是 AppOptions 类直接实现了 IOptions<TOptions>接口，在依赖注入时未使用 UnnamedOptionsManager<TOptions>等类型，这使得 IOptionsFactory<TOptions>、IConfigureOptions<TOptions>等类型也不会被使用。AppOptions 实例由服务容器创建，而不是 IOptionsFactory 服务。

解决方法是为 AppOptions 类定义一个可以接收 IConfigureOptions<TOptions>对象的构造函数，然后对 AppOptions 实例进行配置，这样 AppOptions 自身的属性才会被赋值。

```
public class AppOptions : IOptions<AppOptions>
{
    // 构造函数
    public AppOptions(IEnumerable<IConfigureOptions<AppOptions>> configs)
    {
        foreach(var cfg in configs)
        {
            // 配置属性
            cfg.Configure(this);
        }
    }
    …
}
```

再次运行程序，选项类的配置就生效了，如图 5-13 所示。

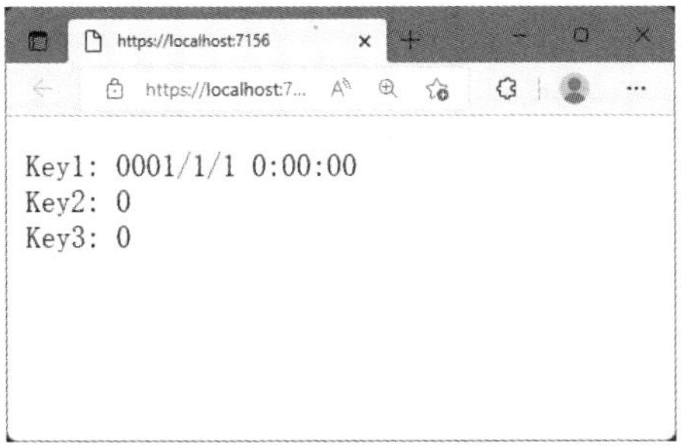

图 5-12　选项类配置未生效

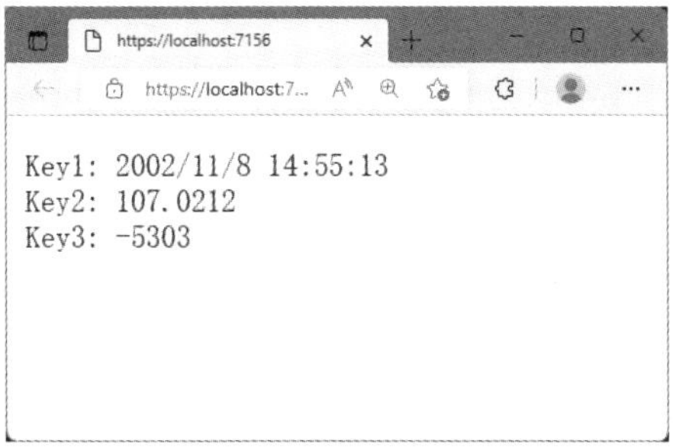

图 5-13　选项类配置已生效

第 6 章　HTTP 管道

本章要点

- ❑ 中间件的实现方法
- ❑ 映射终结点
- ❑ 终结点的命名与元数据管理
- ❑ 有条件地执行中间件

6.1　HTTP 管道与中间件

Web 服务器接收到客户端的请求后会读取请求内容，经处理后把响应内容返回给客户端。处理 HTTP 请求的过程如同工厂车间里的流水线，由一道道工序（组件）链接而成。这些组成 HTTP 管道的最小组件也称为中间件（Middleware）。

HTTP 管道上每个中间件都可在调用下一个中间件前后执行相关代码，也可以选择不调用下一个中间件。中间件之间的逻辑关系类似于栈结构，调用时，从第 1 个中间件开始，直到最后一个；而返回时则从最后一个中间件开始，“原路返回”到第 1 个中间件。中间件的执行过程如图 6-1 所示。

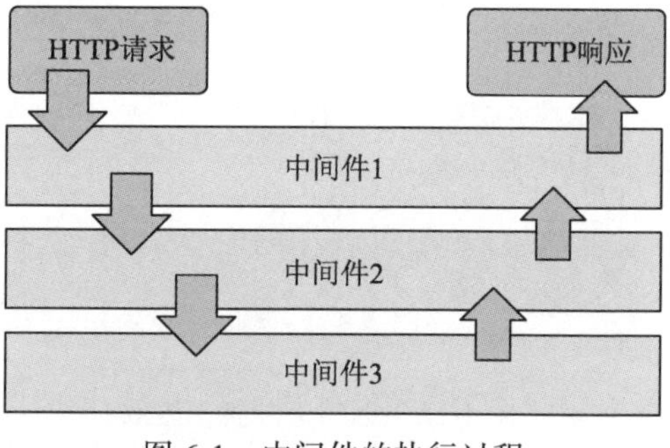

图 6-1　中间件的执行过程

6.2　中间件的实现方法

以下方法都可以实现中间件，并且它们可以混合使用。

（1）通过委托实现。调用 Use()扩展方法（由 UseExtensions 类公开的成员，被扩展类型为 IApplicationBuilder），然后通过委托构建代码。

（2）定义中间件类实现，实现约定的方法成员，通过 UseMiddleware()扩展方法（UseMiddlewareExtensions 类的成员）调用。

（3）通过 IMiddleware 接口实现，由 IMiddlewareFactory 接口负责创建和释放实例。使用此方法实现的中间件必须注册到服务容器中。

6.3 通过委托实现中间件

Use()方法有以下两个常用的重载。

```
public static IApplicationBuilder Use(this IApplicationBuilder app,
Func<HttpContext, RequestDelegate, Task> middleware);
public static IApplicationBuilder Use(this IApplicationBuilder app,
Func<HttpContext, Func<Task>, Task> middleware);
```

尽管上述两个重载方法中 middleware 委托在输入参数上有类型差异，但其含义是相同的。该委托有两个输入参数：第 1 个参数是 HttpContext 实例，它表示一个上下文对象，该对象在整个 HTTP 管道的各个环节之间传递，使得 HTTP 请求相关的状态信息保持一致（如请求的 URL 路径、携带的 Cookie 等）；第 2 个参数可以是 RequestDelegate 委托或 Func<Task>委托，表示 HTTP 管道中下一个中间件。middleware 委托返回 Task 类型，表示当前中间件的执行结果。

为了简化代码的调用，WebApplication 类对早期 ASP.NET Core 版本的代码调用进行了二次封装。WebApplication 类实现了 IApplicationBuilder 接口，因此可以在 WebApplication 对象上调用 Use()方法。实际被调用的是 ApplicationBuilder 类的 Use()方法（WebApplication 类的内部引用了 ApplicationBuilder 实例）。

ApplicationBuilder.Use()方法的源代码如下。

```
public IApplicationBuilder Use(Func<RequestDelegate, RequestDelegate>
middleware)
{
    _components.Add(middleware);
    return this;
}
```

middleware 参数是委托类型，输入和输出参数的类型都是 RequestDelegate 委托，输入参数表示下一个中间件的代码，输出参数表示当前中间件要运行的代码。Use()方法的实现很简单——把 middleware 引用的委托放入_components 列表。_components 是一个委托列表，其声明如下。

```
private readonly List<Func<RequestDelegate, RequestDelegate>> _components
= new();
```

这表明 Use()方法的调用只是把中间件代码放到列表对象中，并没有把它们链接起来。真正完成中间件链接操作的是在 Build()方法中。源代码如下。

```
public RequestDelegate Build()
{
    RequestDelegate app = context =>
    {
       ...
       // 设置 HTTP 响应代码为 404
        context.Response.StatusCode = StatusCodes.Status404NotFound;
        return Task.CompletedTask;
    };

   // 这里是关键
    for (var c = _components.Count - 1; c >= 0; c--)
    {
        app = _components[c](app);
    }

    return app;
}
```

在 Build()方法中，首先定义了一个特殊的中间件 app。这个中间件表示 HTTP 管道的最后一个中间件，它会向客户端返回 404 状态码。把中间件链接起来的关键是随后的 for 循环。要注意，该循环是从_components 列表的最后一个元素开始链接的。

假设最后返回 404 状态码的中间件名为 last，而通过 Use()方法添加的中间件依次为 1，2，…。中间件的链接过程如下。

（1）中间件 last 作为输入参数传给中间件 *n*，在中间件 *n* 中调用，返回中间件 *n*。

（2）返回的中间件 *n* 作为输入参数传给中间件 *n*−1，返回中间件 *n*−1。

（3）中间件 *n*−1 作为输入参数传给中间件 *n*−2，返回中间件 *n*−2。

（4）……

（5）中间件 3 作为输入参数传给中间件 2，返回中间件 2。

（6）中间件 2 作为输入参数传给中间件 1，返回中间件 1。

（7）中间件 1 最终被 Build()方法返回。

链接时是反向的，但调用时是正向的。

（1）中间件 1 被调用，在中间件 1 中会调用中间件 2。

（2）中间件 2 被调用，在中间件 2 中会调用中间件 3。

（3）中间件 3 被调用，在中间件 3 中会调用中间件 4。

（4）……

（5）中间件 *n*−2 被调用，在中间件 *n*−2 中调用中间件 *n*−1。

（6）中间件 *n*−1 被调用，在中间件 *n*−1 中调用中间件 *n*。

（7）中间件 *n* 被调用，最后调用中间件 last。

中间件的链接和调用过程如图 6-2 所示。

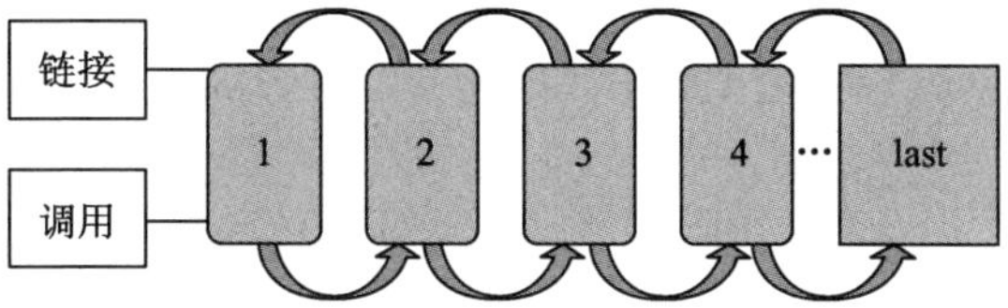

图 6-2　中间件的链接和调用过程

也可以用下面的示例模拟这个过程。

```
List<Func<Action, Action>> actions = new();

// 添加中间件
int i = 1;
for (int n = 1; n < 7; n++)
{
    Func<Action, Action> fun = next =>
    {
        return () =>
        {
            Console.WriteLine($"中间件-{i++}");
            // 调用下一个中间件
            next();
        };
    };
    actions.Add(fun);
}

// 链接中间件
Action last = () => Console.WriteLine("结束标记-404");

Action app = last;
// 从最后一个元素开始链接
for(int c = actions.Count - 1; c >= 0; c--)
{
    app = actions[c](app);
}

// 尝试调用一下，看会发生什么
app();
```

上述代码中创建的列表实例使用Func<Action, Action>委托作为元素类型，即输入和输出参数都是 Action 委托对象。last 表示中间件列表的结束标志，在 for 循环中从列表的最后开始链接。循环执行完成后，尝试调用 app 变量所引用的委托实例。最后得到如图 6-3 所示的结果。

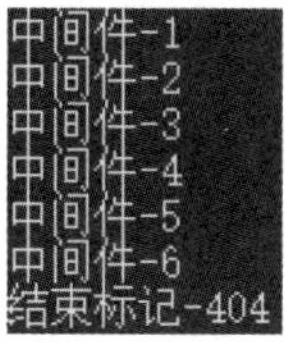

图 6-3　模拟中间件的链接

6.3.1 示例：Use()方法的简单用法

下面的示例将调用扩展方法 Use()实现 3 个中间件，这些中间件构成 HTTP 管道。

```
var builder = WebApplication.CreateBuilder(args);
var app = builder.Build();

app.Use(async (context, next) =>
{
    Console.WriteLine("第 1 个中间件");
    // 调用下一个中间件
    await next();
});

app.Use(async (context, next) =>
{
    Console.WriteLine("第 2 个中间件");
    // 调用下一个中间件
    await next(context);
});

app.Use(async (context, next) =>
{
    Console.WriteLine("第 3 个中间件");
    // 调用最后一个中间件（返回 404 状态码）
    await next();
});

app.Run();
```

上述代码中，第 1 个和第 3 个中间件的实现代码调用了 Use()方法的以下重载。

```
public static IApplicationBuilder Use(this IApplicationBuilder app,
Func<HttpContext, Func<Task>, Task> middleware);
```

而第 2 个中间件的实现代码则是调用了 Use()方法的以下重载。

```
public static IApplicationBuilder Use(this IApplicationBuilder app,
Func<HttpContext, RequestDelegate, Task> middleware);
```

RequestDelegate 是委托类型，其声明如下。

```
delegate Task RequestDelegate(HttpContext context);
```

它与 Func<Task>的区别在于是否带有 HttpContext 类型的输入参数。因此，Use()方法的两个重载在调用上是很相似的。

运行示例程序，用 Web 浏览器访问服务器会得到 HTTP 404 错误，如图 6-4 所示。这是因为 HTTP 管道上最后一个中间件会返回 404 状态码。

图 6-4 返回 404 状态码

6.3.2　HTTP 管道的“短路”

HTTP 管道在默认情况下会执行已链接的所有中间件，这使得应用程序总是返回 404 状态码。此状态码仅在给定的 URL 找不到资源时才返回。正常情况下，Web 服务器应当返回 200 状态码，表示 HTTP 请求已成功处理，并返回客户端所需的数据（如要浏览的网页的 HTML 文档、被请求的图像资源等）。可见，在实际开发中，应用程序不能总返回 404 状态码。这意味着 HTTP 管道上的最后一个中间件可以不调用。

当下一个中间件不被调用时，HTTP 管道将终止。这种行为称为“短路”。短路并不一定在 HTTP 管道的最后发生，HTTP 管道上任意位置都可能发生短路。

下面的示例实现了两个中间件：第 1 个中间件向 HTTP 响应消息中加入名为 X-Tag 的标头（Header）；第 2 个中间件要判断请求的 URL 中是否包含名为 flag 的查询参数。如果 flag 参数存在且它的值为 1（如 https://localhost:7185?flag=1），就不再调用下一个中间件；如果 flag 参数不为 1 或不存在，那么就继续调用下一个中间件。

```
app.Use(async (context, next) =>
{
    // 设置 HTTP 标头
    context.Response.Headers.Add("X-Tag", "Demo");
    // 调用下一个中间件
    await next();
});

app.Use(async (context, next) =>
{
    // 检查 URL 查询中是否有 flag 字段
    // 如果存在，且值为 1，就短路 HTTP 管道
    var request = context.Request;
    var response = context.Response;
    if (request.Query["flag"] == "1")
    {
        await response.WriteAsync("Flag = 1");
    }
    else
    {
        // 否则调用下一个中间件
        await next();
    }
});
```

运行示例程序，若访问时不加 flag 参数或参数的值不为 1，那么服务器将返回 404 状态码，如图 6-5 所示；否则返回字符串 Flag = 1，如图 6-6 所示。

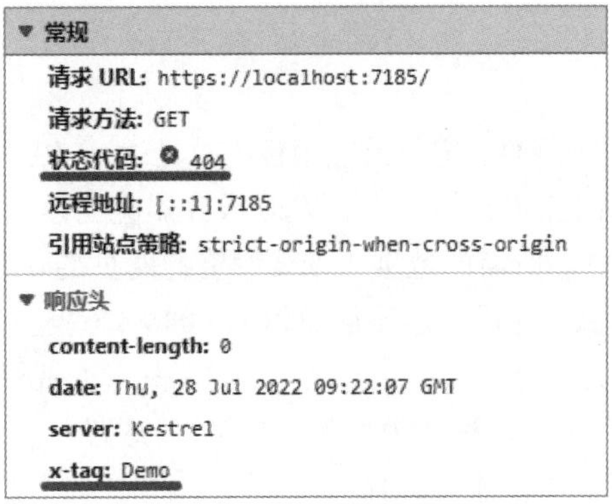

图 6-5　URL 不带 flag 参数

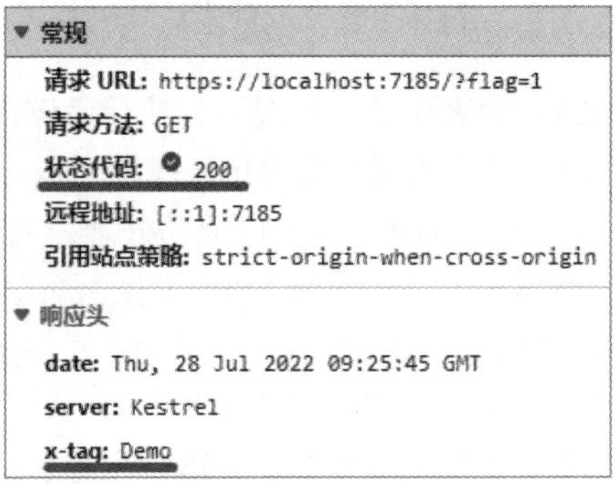

图 6-6　URL 带 flag 参数且值为 1

6.3.3　Run()方法

这里提到的 Run()方法是 RunExtensions 类中定义的扩展方法，虽然名称相同，但并非用于启动应用程序的 Run()方法。

调用 Run()方法使 HTTP 管道"短路"是一种更简单的方法，它不需要判断是否调用下一个中间件（因为在 Run()方法中直接忽略掉下一个中间件）。Run()方法声明如下。

```
public static void Run(this IApplicationBuilder app, RequestDelegate
handler);
```

它只需要一个 RequestDelegate 委托作为参数即可，通过 HttpContext 写入响应消息。因此，Run()方法的使用非常简单，示例如下。

```
app.Run(async context =>
{
    // 设置编码为 UTF-8
    context.Response.ContentType = "text/plain; charset=UTF-8";
    // 写入响应消息
    await context.Response.WriteAsync("你好，世界");
});
```

由于写入响应流中的文本包含中文字符，所以要将 ContentType 属性设置为 UTF-8 编码，防止出现乱码。在 HTTP 响应时会生成 content-type 标头，如图 6-7 所示。

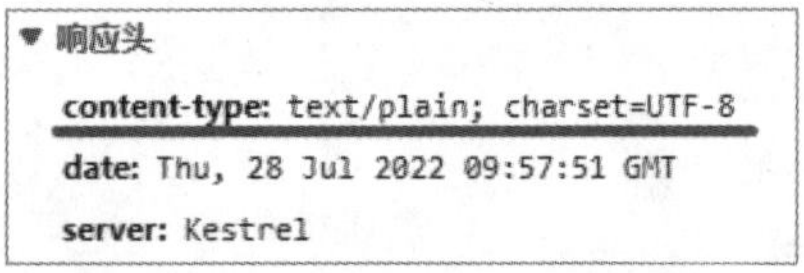

图 6-7　返回内容为 UTF-8 编码

由于 Run()方法会直接使 HTTP 管道终止，所以在 Run()方法调用之后的中间件将不会被调用。例如，下面的代码中后两个中间件都不会被调用。

```
app.Run(async context =>
{
    Console.WriteLine("第 1 个中间件");
});
```

```
// 不会被调用
app.Use(async (context, next) =>
{
    Console.WriteLine("第 2 个中间件");
    await next();
});

// 不会被调用
app.Use(async (context, next) =>
{
    Console.WriteLine("第 3 个中间件");
    await next();
});
```

6.4　通过类实现中间件

如果中间件的代码比较多，可以考虑将其转移到一个自定义的类中。通过定义新类实现中间件，则要求符合以下约定。

（1）类中必须包含 Invoke()或 InvokeAsync()方法。

（2）方法返回的类型为 Task。

（3）方法至少存在一个参数，第 1 个参数必须是 HttpContext 类型。

要将中间件类添加到 HTTP 管道中，需要调用 app.UseMiddleware()扩展方法。该方法的两个重载如下。

```
IApplicationBuilder UseMiddleware(this IApplicationBuilder app, Type
middleware, params object?[] args);

IApplicationBuilder UseMiddleware<TMiddleware>(this IApplicationBuilder
app, params object?[] args);
```

第 1 个重载需要通过 middleware 参数指定中间件类的 Type。第 2 个重载是泛型方法，直接使用类型参数 TMiddleware 指定中间件类。args 是可选参数，它表示要传递给中间件类构造函数的参数列表，如果不需要，可忽略此参数。

下面的代码定义了一个名为 TestMiddleware 的类，用于实现中间件。命名以 Middleware 结尾是方便人们识别其用途。

```
public class TestMiddleware
{
    readonly RequestDelegate _next;
    public TestMiddleware(RequestDelegate next)
    {
        _next = next;
    }
```

```
    public async Task InvokeAsync(HttpContext context)
    {
        Console.WriteLine($"中间件被调用");
        // 调用下一个中间件
        await _next(context);
    }
}
```

需要注意的是，下一个中间件的引用（上述代码中的_next 字段）是通过类构造函数的参数传递的，类型为 RequestDelegate 委托。

中间件类包含 Invoke()或 InvokeAsync()方法（二选一即可）。上述代码使用的是 InvokeAsync()方法，返回类型为 Task，参数类型为 HttpContext 类型。

中间件的逻辑代码就写在 InvokeAsync()方法中，本示例仅仅向控制台输出文本“中间件被调用”。如果需要下一个中间件，则调用_next 字段所引用的委托对象即可。

在 Program.cs 文件中，需要通过 UseMiddleware()方法使用 TestMiddleware 中间件。

```
var builder = WebApplication.CreateBuilder(args);
var app = builder.Build();

app.UseMiddleware<TestMiddleware>();
…
app.Run();
```

TestMiddleware 中间件也可以与基于委托实现的中间件一起使用。

```
app.UseMiddleware<TestMiddleware>();
app.Use(async (context, next) =>
{
    // …
    await next();
});
```

6.4.1 带参数的中间件

如果中间件类需要参数，则可以在调用 UseMiddleware()扩展方法时进行传递，再由中间件类的构造函数接收参数值。

例如，下面的代码定义了名为 TestMiddleware 的中间件类。

```
public class TestMiddleware
{
    // 字段
    readonly RequestDelegate _next;
    readonly string _arg1;
    readonly int _arg2;
    readonly string _arg3;

    // 除了 next，其他均为该中间件的参数
    public TestMiddleware(string arg1, RequestDelegate next, int arg2,
```

```
string arg3)
    {
        _arg1 = arg1;
        _arg2 = arg2;
        _arg3 = arg3;
        _next = next;
    }

    public Task Invoke(HttpContext context)
    {
        Console.WriteLine($"arg1 = {_arg1}\n" +
                          $"arg2 = {_arg2}\n" +
                          $"arg3 = {_arg3}");
        return _next(context);
    }
}
```

在定义 TestMiddleware 类的构造函数时，类型为 RequestDelegate 委托的参数（即 next 参数）代表 HTTP 管道中的下一个中间件。它在参数列表中的位置没有严格的要求，如本示例中 next 为第 2 个参数。

在向 TestMiddleware 中间件传递参数时，应用忽略 next 参数，它由框架运行时自动赋值。

在应用程序创建 WebApplication 实例后，可以调用 UseMiddleware()扩展方法应用 TestMiddleware 中间件。代码如下。

```
var builder = WebApplication.CreateBuilder(args);
var app = builder.Build();

app.UseMiddleware<TestMiddleware>(6699, "val-1", "val-2");

app.Run(async context =>
{
    context.Response.ContentType = "text/plain; charset=UTF-8";
    await context.Response.WriteAsync("示例程序");
});
...
```

TestMiddleware 中间件的构造函数需要赋值 3 个参数：arg1、arg2、arg3。其中，由于 arg2 是唯一一个 int 类型参数，因此在参数值列表中，无论 int 类型的值出现在哪个位置，都会传递给 arg2；而 arg1 和 arg3 都是 string 类型，就按参数的声明顺序赋值，即 arg1 参数的值为 val-1，arg3 参数的值为 val-2，剩下的 6699 是 int 类型数值，直接赋值给 arg2。next 参数不需要提供，应用程序会自动赋值（引用 HTTP 管道中下一个中间件）。

不过，为了避免赋值顺序的错误，推荐的方法是按参数的声明顺序来赋值，即

```
app.UseMiddleware<TestMiddleware>("val-1", 6699,"val-2");
```

6.4.2 中间件类与依赖注入

由于中间件类的构造函数用于传递调用参数，因此依赖注入将通过 Invoke()或 InvokeAsync()

方法实现。Invoke()或 InvokeAsync()方法的第 1 个参数必须是 HttpContext 类型，从第 2 个参数起，均可用于获取来自服务容器的实例引用，即支持依赖注入。

下面的代码将定义一个名为 TestMiddleware 的中间件类。

```
public class TestMiddleware
{
    // 字段
    readonly RequestDelegate _next;

    // 构造函数
    public TestMiddleware(RequestDelegate next)
    {
        _next = next;
    }

    public async Task Invoke(HttpContext context, ILogger<TestMiddleware>
    logger, IWebHostEnvironment env)
    {
        // 写入日志
        logger.LogInformation("当前应用程序名称：{0}", env.ApplicationName);
        // 调用下一个中间件
        await _next(context);
    }
}
```

在 Invoke()方法的参数列表中：第 1 个参数必须引用 HttpContext 对象；第 2 个参数为 ILogger<T>实例，它用于写入日志；第 3 个参数为 IWebHostEnvironment 实例，上述代码将通过 ApplicationName 属性获得当前应用程序名称，并用 ILogger<T>对象输出日志。

在 HTTP 管道中添加 TestMiddleware 中间件，代码如下。

```
app.UseMiddleware<TestMiddleware>();
```

运行示例程序，只要有客户端向服务器发出请求，TestMiddleware 中间件就会执行，随后向日志写入当前应用程序的名称，如图 6-8 所示。

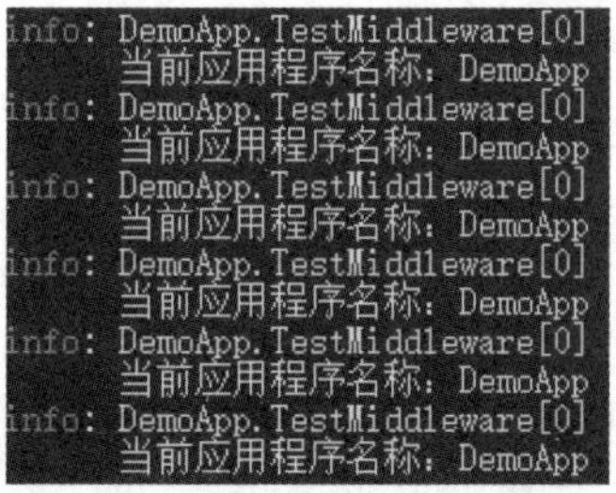

图 6-8　应用程序名称已写入日志

6.5 通过 IMiddleware 接口实现中间件

IMiddleware 接口（位于 Microsoft.AspNetCore.Http 命名空间）的定义如下。

```
public interface IMiddleware
{
    Task InvokeAsync(HttpContext context, RequestDelegate next);
}
```

实现此接口的类型要求包含 InvokeAsync()方法。第 1 个参数是 HttpContext 对象；第 2 个参数是下一个中间件的引用。这说明通过实现 IMiddleware 接口创建的中间件是在 InvokeAsync()方法中传递下一个中间件的。

下面给出一个示例。

```
public class TestMiddleware : IMiddleware
{
    public async Task InvokeAsync(HttpContext context, RequestDelegate
    next)
    {
        // 获取 response 对象
        var response = context.Response;
        // 添加 HTTP 头
        response.Headers.Add("test_header", "Web Project");
        // 调用下一个中间件
        await next(context);
    }
}
```

该中间件在使用前必须注册到服务容器中。一般情况下，可以考虑注册为瞬时服务或作用域服务。下面的代码将 TestMiddleware 中间件类注册为作用域服务。

```
builder.Services.AddScoped<TestMiddleware>();
```

在构建 WebApplication 实例后，需要通过 UseMiddleware()扩展方法调用中间件。

```
app.UseMiddleware<TestMiddleware>();
```

运行示例程序，并在 Web 浏览器中访问。如果示例应用的响应 HTTP 消息中带有 test_header 标头，表明 TestMiddleware 中间件已运行，如图 6-9 所示。

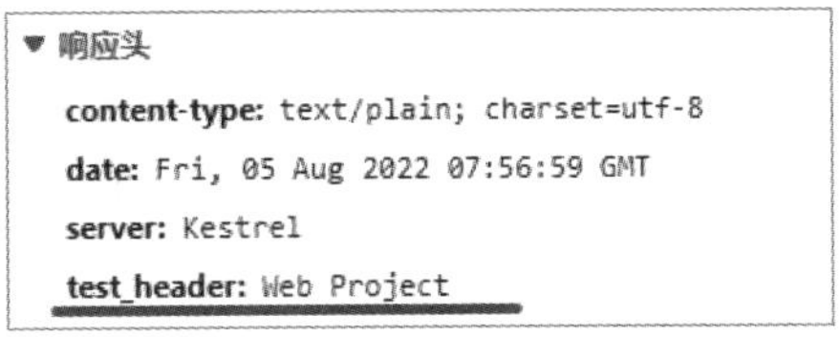

图 6-9　通过接口实现中间件

6.6　终结点

终结点会建立一系列 URL 路由规则与 HTTP 处理程序的映射关系（通过 Map*()扩展方法），仅当路由规则匹配成功时，HTTP 处理程序才会执行。要在 HTTP 管道中启用终结点，需要先调用 UseRouting()方法，然后再调用 UseEndpoints()方法。两个方法之间可以插入其他中间件，如

```
app.UseRouting();
app.Use(…);
app.Use(…);
app.UseEndpoints();
```

WebApplication 类在实例化过程中已调用 UseRouting()和 UseEndpoints()方法，通常开发人员无须调用。不过，如果开发人员有特殊需求，也可以在适当时再次调用它们。

调用 UseRouting()方法会向 HTTP 管道插入 EndpointRoutingMiddleware 中间件。此中间件

会用当前客户端请求的 URL 匹配路由规则。调用 UseEndpoints()方法会向 HTTP 管道添加 EndpointMiddleware 中间件，已匹配的终结点将在此中间件内执行。

如果路由规则匹配成功，应用程序将为当前 HttpContext 对象分配一个 Endpoint 实例（可以通过 EndpointHttpContextExtensions.GetEndpoint()扩展方法获取）；如果路由规则匹配失败，HTTP 管道返回 404 状态码。

在默认情况下，开发人员所定义的中间件会插入 UseRouting 与 UseEndpoints 之间。这样做的目的是保证开发人员定义的中间件能够被调用（不管路由规则是否成功匹配）。

终结点通常会与特定功能关联，如 MVC、Web API、Blazor 等。为了方便开发人员调用，ASP.NET Core 框架提供一系列 Map*()扩展方法，可以快速地添加终结点，如表 6-1 所示。

表 6-1　常用的Map*()扩展方法

扩 展 方 法	用　　途	相 关 功 能
MapGet()	当客户端以HTTP GET方式请求时，若URL与路由规则匹配，将执行该终结点	
MapPost()	当客户端以HTTP POST方式请求时，若URL与路由规则匹配，将执行该终结点	
MapPut()	当客户端以HTTP PUT方法请求时，若路由规则匹配成功，将执行该终结点	
MapDelete()	当客户端以HTTP DELETE方式发出请求时，且路由规则匹配，将执行该终结点	
MapMethods()	当路由规则匹配时，该方法可以指定多个HTTP请求方式。若客户端发出的请求方式存在于指定的请求方式列表中，将执行该终结点	
Map()	只要HTTP请求与路由规则匹配，就会执行此终结点，不考虑请求方式	
MapControllers()、MapDefaultControllerRoute()、MapControllerRoute()、MapAreaControllerRoute()	当默认或自定义的路由规则匹配成功后，执行相应的MVC控制器中的操作方法	MVC/Web API
MapHub()	添加SignalR终结点	SignalR
MapRazorPages()	添加用于定位Razor页面的终结点	Razor Pages
MapBlazorHub()	当使用服务器端的Blazor时，添加相关终结点	Blazor
MapFallback()、MapFallbackToPage()、MapFallbackToController()、MapFallbackToAreaController()	当其他终结点都无法与路由规则匹配时，将使用这些方法提供的默认终结点	

Map*()扩展方法仅添加路由规则与 HTTP 处理程序的映射关系，并未在 HTTP 管道上添加中间件，也不会改变中间件的执行顺序。

6.6.1　示例：常见的 HTTP 请求方式

本示例会为应用程序定义多条 URL 路由规则，以响应几种常用的 HTTP 请求方式，如 HTTP GET、HTTP POST、HTTP PUT 等。

由于 ASP.NET Core 应用程序默认已调用 UseRouting()和 UseEndpoints()方法，因此本示例代码中并不需要调用它们。在 WebApplication 对象初始化完成后，直接调用 Map*()扩展方法添加 URL 路由规则即可。关键代码如下。

```
// 1. HTTP GET
app.MapGet("/", ()=>"以 HTTP-GET 方式访问根 URL");

// 2. HTTP POST，路径：/send
app.MapPost("/send", ()=>"以 HTTP-POST 方式访问 /send");

// 3. HTTP DELETE，路径：/remove
app.MapDelete("/remove", ()=>"以 HTTP-DELETE 方式访问 /remove");

// 4. HTTP PUT，路径：/newdata
app.MapPut("/newdata", ()=>"以 HTTP-PUT 方式访问 /newdata");

// 5. 未指明请求方式，兼容各种方案
app.Map("/anymeth", (HttpContext context)=>
{
    // 获取访问的 URL
    var url = context.Request.GetDisplayUrl();
    // 获取请求方式
    var reqMth = context.Request.Method;

    return $"以 HTTP-{reqMth}方式访问 {url}";
});
```

上述代码添加了 5 条路由规则。

（1）以 HTTP GET 方式访问，URL 为服务器的根地址。

（2）以 HTTP POST 方式访问，URL 为/send（相对于根地址）。

（3）以 HTTP DELETE 方式访问，URL 为/remove（相对地址）。

（4）以 HTTP PUT 方式访问，URL 为/newdata（相对地址）。

（5）URL 为/anymeth（相对地址），此规则允许以任意 HTTP 方式请求。可通过 HttpContext.Request.Method 属性获取请求方式。

Map*()扩展方法一般有两种调用方案：第 1 种是指定一个 RequestDelegate 委托，它绑定的方法需要一个 HttpContext 类型的输入参数，并返回 Task；第 2 种是使用一个自定义委托，此委托可以有参数，也可以无参数。参数和返回值可以使用基础类型（string、int、byte 等），也可以使用自定义的类型（如类），应用程序会使用 JSON 格式序列化对象实例并返回给客户端。

为了方便对本示例进行测试，可以编写一个控制台应用程序，使用 HttpClient 类（位于 System.Net.Http 命名空间）模拟发送 HTTP 请求。

具体代码如下。

```
using System.Net.Http;
using System;

// 服务器的基地址
Uri baseUrl = new Uri("https://localhost:7273");

HttpClient client = new();
client.BaseAddress = baseUrl;

string responseBack;
// HTTP GET
responseBack = await client.GetStringAsync("/");
Console.WriteLine("1. HTTP-GET:   /");
Console.WriteLine($"服务器响应: {responseBack}\n");
// HTTP POST  /send
responseBack = await (await client.PostAsync("/send", null)).Content.
ReadAsStringAsync();
Console.WriteLine("2. HTTP-POST:  /send");
Console.WriteLine($"服务器响应: {responseBack}\n");
// HTTP DELETE  /remove
responseBack = await (await client.DeleteAsync("/remove")).Content.
ReadAsStringAsync();
Console.WriteLine("3. HTTP-DELETE:  /remove");
Console.WriteLine($"服务器响应: {responseBack}\n");
// HTTP PUT  /newdata
responseBack = await (await client.PutAsync("/newdata", null)).Content.
ReadAsStringAsync();
Console.WriteLine("HTTP-PUT:  /newdata");
Console.WriteLine($"服务器响应: {responseBack}");

Console.WriteLine("--------------------------------------------------");
// 同一 URL 多种请求方式
string relUrl = "/anymeth";
Console.WriteLine(relUrl);
responseBack = await (await client.PostAsync(relUrl, null)).Content.
ReadAsStringAsync();
Console.WriteLine("POST: {0}", responseBack);
responseBack = await(await client.GetAsync(relUrl)).Content.
ReadAsStringAsync();
Console.WriteLine("GET: {0}", responseBack);
responseBack = await(await client.PatchAsync(relUrl, null)).Content.
ReadAsStringAsync();
Console.WriteLine("PATCH: {0}", responseBack);
```

```
// 释放对象及其资源
client.Dispose();
Console.Read();
```

注意 baseUrl 变量的值，它表示 ASP.NET Core 应用程序运行后的 URL，请读者根据配置信息（如 launchSettings.json 文件中的配置）将 https://localhost:7273 替换为实际的地址。

HttpClient 类封装了一组方法，可以简单地向服务器发出 HTTP 请求，并返回服务器的响应消息。例如，调用 GetAsync()方法向服务器发出 HTTP GET 请求并返回 HttpResponseMessage 实例。HttpResponseMessage 类封装了与 HTTP 消息相关的内容，其中，Headers 属性表示 HTTP 标头的集合，Content 属性表示 HTTP 消息的正文部分。也可以调用 GetByteArrayAsync()、GetStringAsync()等方法，可以直接返回 byte 数组、string 等数据类型。

设置 HttpClient 实例的 BaseAddress 属性，将提供一个基础地址，随后发出 HTTP 请求时所指定的地址将作为 BaseAddress 的相对地址。例如，上述代码中，BaseAddress 属性设置为 https://localhost:7273，调用 DeleteAsync()方法时指定地址为/remove，最终组成的完整地址为 https://localhost:7273/remove。

上述代码最后访问的是/anymeth，此地址支持任意有效的 HTTP 请求方式。此处只演示了 GET、POST、PATCH 这 3 种请求方式。

在测试本示例时，先运行 ASP.NET Core 项目，再运行控制台项目，结果如图 6-10 所示。

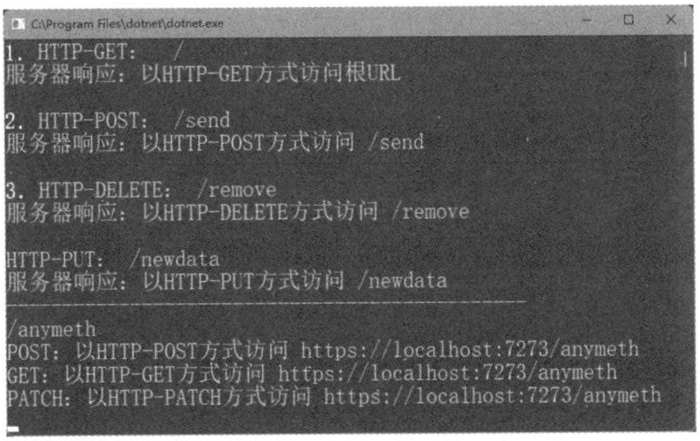

图 6-10　终结点的访问测试

6.6.2　示例：同时使用 Razor Pages 和 MVC

ASP.NET Core 应用程序允许同时启用多项功能，并可以在 HTTP 管道中通过 URL 路由规则匹配终结点的方式共存（路由规则不能有冲突）。

例如，/account/reguser 指向 reguser.cshtml 页面，使用的是 Razor Pages 功能；/chat 指向名为 chat 的 SignalR 服务中心。

本示例将在同一个 ASP.NET Core 应用程序中同时使用 MVC 和 Razor Pages 两项功能。

（1）新建一个类，命名为 DemoController，派生自 Controller 类（位于 Microsoft.AspNetCore.

Mvc 命名空间）。该类表示一个 MVC 控制器，控制器名称为 Demo（不区分大小写）。该类包含一个名为 Home()的方法，即 MVC 控制器的操作方法。

```
[Route("[controller]/[action]")]
public class DemoController : Controller
{
    public ActionResult Home()
    {
        return View();
    }
}
```

Route 特性指定该控制器的 URL 路由规则。其中，controller 表示控制器的名称占位符，action 是操作方法名称的占位符。要访问 Demo 控制器中的 Home()操作方法，则需要的 URL 为/demo/home。

（2）在项目中添加名为 Views 的目录，再在 Views 目录下新建子目录 Demo。

（3）在 Demo 目录下新建 Home.cshtml 文件，此为 MVC 的视图文件，内容如下。

```
<html>
<head>
<meta charset="UTF-8"/>
<title>示例程序</title>
</head>
<body>
<div>
<h3>欢迎来到本站</h3>
</div>
<div>
<a href="/about" target="_blank">关于...</a>
</div>
</body>
</html>
```

（4）在项目根目录下创建 Pages 目录，再在 Pages 目录下新建 About.cshtml 文件，内容如下。

```
@page

<html>
<head>
<meta charset="UTF-8"/>
<title>关于本站</title>
</head>
<body>
<h5>我的小站</h5>
<p>版本：1.0.0</p>
<p>邮箱：sdt050@abc.com</p>
<p>支持热线：55855622</p>
</body>
</html>
```

About.cshtml 文件用于 Razor Pages，其 URL 为/about。

（5）回到 Program.cs 文件，向服务容器添加 MVC 和 Razor Pages 相关服务。

```
var builder = WebApplication.CreateBuilder(args);

// 注册相关服务
builder.Services.AddControllersWithViews();
builder.Services.AddRazorPages();
…
```

AddControllersWithViews()扩展方法表示添加的 MVC 功能将包含视图。如果只使用控制器相关功能而不需要视图，则调用 AddControllers()扩展方法。

（6）在 HTTP 管道上同时使用 Razor Pages 和 MVC 两个终结点。

```
// Razor Pages 终结点
app.MapRazorPages();
// MVC 终结点
app.MapControllers();
…
```

（7）运行程序，在 Web 浏览器中定位到 https://<启动 URL>/demo/home，可以浏览与 Demo 控制器的 Home()操作方法匹配的视图，如图 6-11 所示。

（8）单击页面上的“关于…”链接，将打开 About.cshtml 页面，如图 6-12 所示。

图 6-11　执行 Demo 控制器

图 6-12　About.cshtml 页面

6.6.3　为终结点分配名称

有些情况下，应用程序可能需要获取终结点的信息，以便实现特定的逻辑。例如，在记录应用程序日志时，希望记录各个终结点被访问的情况，为了区分各个终结点，可以为每个终结点分配一个友好的名称，以此作为标识。

在调用 Map*()方法后可以直接通过 WithDisplayName()扩展方法为当前终结点设置一个友好的名称。终结点的名称是允许重复出现的，如下面的代码所示，访问/foo 和访问/bar，尽管两者匹配到不同的终结点，但它们都有相同的名称——test。

```
app.MapGet("/foo", (HttpContext context) =>
{
```

```
        return context.GetEndpoint()?.DisplayName;
    }).WithDisplayName("test");

    app.MapGet("/bar", (HttpContext context) =>
    {
        return context.GetEndpoint()?.DisplayName;
    }).WithDisplayName("test");
```

下面的代码将为与根 URL 匹配的终结点分配名称 Root_Endpoint，为路径/data 分配名称 Fetch_Data。

```
    app.MapGet("/", (HttpContext context) =>
    {
        // 获取终结点信息
        var myEndp = context.GetEndpoint();
        if(myEndp == null)
        {
            return "未获取到终结点引用";
        }
        return string.Format("友好名称：{0}", myEndp.DisplayName);
    }).WithDisplayName("Root_Endpoint");

    app.MapGet("/data", (HttpContext context) =>
    {
        // 获取终结点信息
        Endpoint? ep = context.GetEndpoint();
        if(ep == null)
        {
            return "未找到终结点";
        }
        return $"友好名称：{ep.DisplayName}";
    }).WithDisplayName("Fetch_Data");
```

在 MapGet()方法引用的委托内部，通过 HttpContext 对象的 GetEndpoint()方法获取当前请求关联的终结点信息（类型为 Endpoint），并通过它的 DisplayName 属性得到终结点的名称。

运行上述代码后，若访问根 URL，则返回名称 Root_Endpoint，如图 6-13 所示；若访问/data，则返回名称 Fetch_Data，如图 6-14 所示。

图 6-13　访问根 URL

图 6-14　访问/data

6.6.4　元数据

在调用 Map*()方法之后，不仅可以为终结点分配名称，还可以添加元数据。元数据是一个对象集合，可以添加任意类型的对象。被添加到元数据集合中的对象将作为终结点的附加信息，用于对终结点进行补充说明。

下面的代码调用 MapGet()方法为根 URL 定义处理代码，并为其终结点添加一个 Version 实例作为元数据。

```
app.MapGet("/", (HttpContext context) =>
{
    return "Hello Web App";
})
.WithMetadata(new Version(1,5,220867));
```

为终结点添加元数据，可以调用 WithMetadata()扩展方法，其参数个数是可变的。上述代码只传递了一个 Version 实例。

在 MapGet()方法之后，添加一个中间件，读出终结点的元数据。然后检查主版本号，若主版本号小于 2，则返回错误信息，否则执行下一个中间件。此处，下一个中间件正是待执行的终结点。所以，如果 next()方法不被调用，那么终结点就不会执行。

```
app.Use(async (context, next) =>
{
    // 设置响应内容的类型为普通文本
    context.Response.ContentType = "text/plain; charset=UTF-8";
    // 获取终结点
    var endPoint = context.GetEndpoint();
    if(endPoint != null)
    {
        // 获取版本信息
        object? obj = endPoint.Metadata.FirstOrDefault(x => x is Version);
        if(obj != null)
        {
            Version ver = (Version)obj;
            if(ver.Major < 2)
            {
                // 设置响应状态码
                context.Response.StatusCode = StatusCodes.
                Status500InternalServerError;
                await context.Response.WriteAsync("由于版本过低，暂时无法使用");
                return;
            }
        }
    }
    // 调用下一个中间件
    await next();
});
```

Endpoint 类的 Metadata 属性为 EndpointMetadataCollection，表示一个 object 类型的集合。在本示例中，由于只添加了一个 Version 对象作为元数据，所以调用 FirstOrDefault()方法读取类型为 Version 的第 1 个元素即可。

由于版本号被指定为 1.5.220867，主版本号为 1，小于 2，所以服务器会返回如图 6-15 所示的版本错误信息。

现在，把 MapGet()方法的调用代码改一下，将主版本号改为 5。

```
app.MapGet("/", (HttpContext context) =>
{
   ...
})
.WithMetadata(new Version(5,5,220867));
```

再次运行程序，由于版本验证通过，终结点正常执行，如图 6-16 所示。

图 6-15　版本错误信息

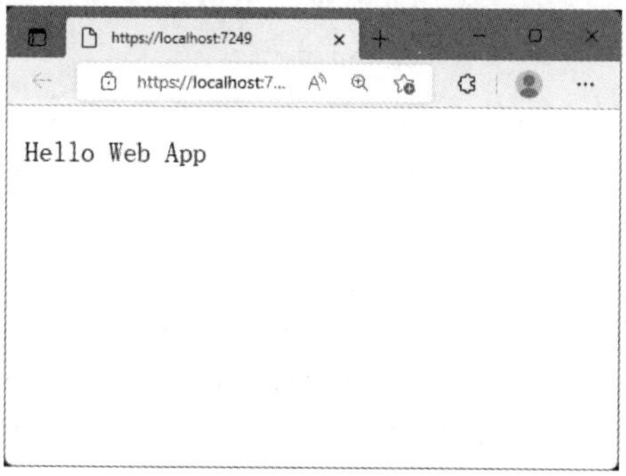

图 6-16　终结点正常执行

6.7　有条件地执行中间件

调用 UseWhen()扩展方法（在 UseWhenExtensions 类中公开）可以指定一个委托（predicate 参数）——以 HttpContext 对象作为输入参数，返回值为布尔类型。该扩展方法声明如下。

```
public static IApplicationBuilder UseWhen (this IApplicationBuilder app,
Func< HttpContext,bool>predicate, Action< IApplicationBuilder>configuration);
```

如果 predicate 参数的调用结果为 true，那么 configuration 参数所指定的中间件会被调用，否则不调用。也就是说，UseWhen()方法为 HTTP 管道创建了一个分支，但这个分支是否运行是有条件的。

另外，还有一个 MapWhen()方法（在 MapWhenExtensions 类中公开），它的声明与 UseWhen()方法相同。

```
public static IApplicationBuilder MapWhen (this IApplicationBuilder app,
Func< HttpContext,bool> predicate, Action< IApplicationBuilder>
configuration);
```

6.7.1　示例：调用包含 user_id 字段的中间件

本示例演示了 UseWhen()方法的使用。应用程序会读取请求 URL 中的查询字段（“?”后面的内容），当存在名为 user_id 的字段时才会执行后面的 Run()方法。

```
app.UseWhen(context =>
{
    // 查找 user_id 字段
    var request = context.Request;
    if(request.Query.TryGetValue("user_id",out StringValues fd))
    {
        if(fd.Count > 0)
        {
            // 只取出第 1 个值
            string first = fd[0];
            // 如果值不为空，就表示条件成立
            if(first is not null or {Length: 0})
            {
                return true;
            }
        }
    }
    return false;
},
appbuilder =>
{
    appbuilder.Run(async context =>
    {
        // 写入响应消息
        string html =
                        "<html>" +
                        "<head>" +
                        "<meta charset=\"UTF-8\">" +
                        "</head>" +
                        "<body>" +
                        "<h3>欢迎使用本站</h3>" +
                        "<p>示例 Web 应用程序</p>" +
                        "<p>用户 ID: {0}</p>" +
                        "</body>" +
                        "</html>";
        // 获取 user_id 字段
        string userid = context.Request.Query["user_id"];
        // 格式化字符串
        html = string.Format(html, userid);
        // 消息内容类型
        context.Response.ContentType = "text/html; charset=UTF-8";
        await context.Response.WriteAsync(html);
```

```
    });
});
```

UseWhen()方法的第 1 个参数是返回布尔值的委托。从 HttpContext.Request.Query 集合中查找是否存在 user_id 字段。如果存在 user_id 字段，还要进一步分析它的值是否为空。如果存在 user_id 字段且不为空，表示中间件的调用条件成立，返回 true，否则返回 false。

第 2 个参数也是委托类型，输入参数为 IApplicationBuilder 类型，用于添加中间件（如调用 Use()方法）。在本示例中，直接调用 Run()方法，执行代码逻辑后终止 HTTP 管道。变量 html 是字符串类型，包含一个完整的 HTML 文档。同时还存在一处格式化占位符（“用户 ID：{0}”中的“{0}”是占位符），随后代码会从 URL 的查询字符串中读出 user_id 字段的值，保存到变量 userid 中。最后调用 string.Format()方法进行格式化——变量 userid 的值会替换 html 变量中的“{0}”。

运行示例程序后，假设服务器的根 URL 为 https://localhost:7023，在发出请求时需要加上 user_id 字段，最终格式为 https://localhost:7023/? user_id=1506，结果如图 6-17 所示。

图 6-17　访问时加上 user_id 字段

6.7.2　示例：只允许以 POST 方式调用 Web API

本示例将通过 MapPost()方法编写一个简易版的 Web API，输入参数为字符串对象，然后使用 SHA256 算法计算该字符串的哈希值（Hash），最后以十六进制字符串的格式返回给客户端。同时，本示例也结合 MapWhen()方法，限制只能以 POST 方式访问此 API。

示例的关键代码如下。

```
app.MapWhen(context =>
{
    // 验证请求方式是否为 POST
    if (context.Request.Method == "POST")
    {
        return true;
    }
    return false;
```

```
    },
    appBuilder =>
    {
        // 必须先调用以下方法
        appBuilder.UseRouting();

        appBuilder.UseEndpoints(endPoints =>
        {
            endPoints.Map("/api/demo", ([FromBody]string msg) =>
            {
                if (string.IsNullOrEmpty(msg))
                {
                    return "输入内容为空";
                }
                // 转换为字节数组
                byte[] data = Encoding.UTF8.GetBytes(msg);
                // 计算哈希值 - SHA256
                SHA256 sha = SHA256.Create();
                byte[] result = sha.ComputeHash(data);
                // 返回计算结果，用十六进制字符串表示
                return Convert.ToHexString(result).ToLower();
            });
        });
    });
```

在传递给 Map()方法的匿名委托中，msg 参数应用了 FromBody 特性，这是为了告诉应用程序要从 HTTP 消息的正文部分读取数据。这是因为字符串属于基础类型，ASP.NET Core 应用程序在进行类型推断时会在请求 URL 的查询字段中读取 msg 参数的值，而不是从正文部分读取。否则会导致无法找到 msg 参数的值。

需要注意的是，MapWhen()方法的第 2 个参数用于构建新的 HTTP 管道分支，必须先调用 UseRouting()方法，再调用 UseEndpoints()方法。

调用上述 Web API 时，提交给服务器的数据要使用 JSON 格式。由于 msg 参数是字符串类型，因此，提交的 JSON 内容无须加上大括号，直接以 JavaScript 字符串表达式提交即可，如

```
"some data"
```

文本内容可以用双引号（英文）包装，也可以用单引号（英文）。

本示例将使用 Python 脚本测试 Web API 的调用，代码如下。

```
from http.client import HTTPSConnection, HTTPResponse
import ssl

# 服务器名
host = 'localhost'
# 服务器端口
port = 7092
# 取消服务器证书验证
```

```
sslcontext = ssl._create_unverified_context()

conn = HTTPSConnection(host, port, context=sslcontext)

# 要提交给服务器的数据
postData = '"测试内容"'.encode('utf-8')
# HTTP 标头
headers = {
"Content-Type": "application/json",
"Accept": "application/json"
}

# 向服务器发出请求
conn.request("POST", "/api/demo", postData, headers)
# 获取服务器的响应消息
response = conn.getresponse()
if response.status == 200:          # 成功
    print('服务器响应: {}'.format(response.read(512).decode('utf-8')))
else:                               # 失败
    print(f'操作未成功完成，状态码: {response.status}')
```

上述代码采用 HTTPS 进行调用测试，用到了 HTTPSConnection 类。该类在 http.client 模块中定义，使用前需要用 import 语句导入。调用 ssl._create_unverified_context()函数是为了取消对服务器证书的验证，避免发生错误。

在测试本示例时，先运行 ASP.NET Core 应用程序。待服务器启动后，调用 Python 脚本。

```
python test.py
```

假设测试脚本的文件名为 test.py。执行命令后，将看到以下输出。

```
服务器响应: 05a17114410b01aaefead38464050aa4be82ff685b17b91206fb631915b4646b
```

这说明 Web API 已成功调用，服务器返回测试数据的哈希值。

第 7 章　HTTP 状态存储

本章要点

- ❑ 了解 HTTP 上下文对象
- ❑ HTTP 消息头
- ❑ URL 查询字符串与表单数据
- ❑ Cookie 与会话

7.1　HTTP 上下文

HTTP 是无状态通信，即客户端与服务器的交互是“没有记忆”的。当客户端请求某个 URL 时，服务器仅发回响应消息，如访问 http://test/index，服务器接收到消息后，将“Hello World”发回给客户端。无论客户端访问多少次 http://test/index，服务器都会返回“Hello World”。如果用户使用 Web 浏览器访问服务器，然后在服务器响应的 HTML 页面上输入了用户名和密码，最后提交给服务器进行验证登录。虽然用户的身份通过了验证，服务器跳转并返回另一个 HTML 页面（通常是主页或用户登录前试图访问的页面），但是，服务器不会记忆身份验证结果，当用户再次访问同一资源时，服务器不知道用户是否之前登录过，于是，服务仍然跳转到登录页面让用户输入用户名和密码。

对于相互独立的网页，有无状态是不会影响使用的。例如，浏览某条新闻的具体内容，只要客户端请求的是该条新闻的 URL，服务器就会返回该新闻的内容，不需要考虑状态信息。不过，如果用户想要对此新闻发表评论，就会出现问题。通常登录网站、浏览新闻和发表评论是相互独立的，在提交评论内容时，服务器并没有记录是谁登录了网站，也不知道评论是谁发表的。

HTTP 通信就是简单的“请求–答复”机制，同一个请求中的各个阶段，以及不同的请求之间不会保留任何数据，也无法共享任何数据。为了让 HTTP 通信能够携带状态数据，需要构建一个“上下文”对象负责保持状态数据。在同一个 HTTP 请求的各个阶段传递状态数据，或者在多个 HTTP 请求之间共享状态数据。

HttpContext类（位于Microsoft.AspNetCore.Http命名空间）是一个抽象类，它描绘了HTTP上下文的基本雏形，没有任何实现代码。如果有特定需求，开发人员可以实现此抽象类自定义HTTP上下文。但一般不需要这样做，因为ASP.NET Core内部已经提供了一个默认实现类——DefaultHttpContext。内部实现的类型是符合HTTP交互的通用模型的，可满足绝大多数的应用场景。

DefaultHttpContext类是由一系列Feature对象组合而成的，每个Feature对象负责一个小功能。例如，HttpConnectionFeature提供与HTTP基础连接相关的信息，QueryFeature提供与URL查询参数有关的数据，FormFeature提供用户提交的表单数据（通常以POST方式提交），等等。Feature对象可以是任意数据类型，也可以实现IHttpRequestFeature、IHttpResponseFeature、IQueryFeature、ISessionFeature、IEndpointFeature等接口。

为了使HTTP上下文对象能够集中管理各种Feature对象，ASP.NET Core公开了IFeatureCollection接口，用于规范实现管理Feature对象的集合对象。默认的实现类是FeatureCollection，可以访问HttpContext类的Features属性获得该集合的实例引用。

在编写代码时，要从Feature对象集合中获取状态数据并不方便，因此，HttpContext类通过属性对外公开了以下对象。

（1）HttpRequest：默认的实现类为DefaultHttpRequest，该类未对外公开。HttpRequest封装了与HTTP请求消息相关的数据，即客户端发送到服务器的消息，包括URL查询字符串（Query、QueryString）、请求消息的HTTP标头集合（Headers）、请求正文的内容数据（Body）、客户端发送的Cookie、表单数据（Form、ReadFormAsync）等。

（2）HttpResponse：默认实现类为DefaultHttpResponse，该类也未对外公开。HttpResponse封装与响应消息相关的数据，即服务器发回给客户端的消息，包括HTTP响应标头（Headers）、响应正文（Body）、发给客户端的Cookie、HTTP状态码（StatusCode）等。

（3）Session和Cookie：用于保持状态数据。HTTP标头、正文等数据在请求完成后不会被保留，而Session和Cookie会被保留和存储。但它们都有过期时间。Session（HTTP会话）一般存放于服务器的内存中，也可以自定义Session的存储；Cookie由客户端维护，一般由Web浏览器负责存储。Cookie的内容为简单的文本（明文或密文），并且以域名标识。例如，来自www.bac.com的Cookie只能用于该域名下的请求消息，即在域名www.xyz.com下面的URL不能使用www.abc.com的Cookie。Session和Cookie通常会同时使用，可以在用户登录验证通过后，将用户名存入Session中，并把Session的当前会话ID写到Cookie中发回给客户端。当客户端再次向服务器发出请求时，服务器可以检查Session中是否存在已登录的用户名，如果存在说明用户已登录过，这时可以跳过登录页面直接进入主页。

7.1.1 示例：在中间件中设置响应标头

本示例分别通过Use()方法和类定义创建了两个中间件。其中，Use()方法实现的中间件将修改server响应头；而通过类实现的中间件会向响应消息中添加名为invoke-time的标头。

不管是Use()方法还是中间件类中的Invoke()或InvokeAsync()方法，都会自动获取HttpContext

对象的引用。设置 HTTP 响应标头可通过 HttpContext.Response 属性访问 HttpResponse 对象，然后读写 HttpResponse.Headers 属性。

下面的代码将定义 DemoMiddleware 类，它是一个中间件类，在 InvokeAsync()方法中添加 invoke-time 响应标头。

```
public class DemoMiddleware
{
    private readonly RequestDelegate _next;

    public DemoMiddleware(RequestDelegate next)
    {
        _next = next;
    }

    public Task InvokeAsync(HttpContext context)
    {
        // 添加响应标头
        context.Response.Headers.Add("invoke-time", DateTime.Now.
        ToLongTimeString());
        return _next(context);
    }
}
```

Headers 为字典结构，可以调用 Add()方法添加新的标头，也可以直接通过索引器设置，即

```
context.Response.Headers["invoke-time"] = DateTime.Now.ToLongTimeString();
```

接着，在应用程序初始化代码中通过 UseMiddleware()扩展方法使用上述中间件。

```
app.UseMiddleware<DemoMiddleware>();
```

以下代码通过 Use()方法构建中间件，在此中间件中会修改 server 响应标头。

```
app.Use((context, next) =>
{
    // 修改响应标头
    context.Response.Headers["server"] =
    "Test Web Server";
    return next();
});
```

运行程序，打开 Web 浏览器的“开发人员工具”，然后刷新页面，随即会抓取到 HTTP 消息。在“响应头”节点下会看到应用程序设置的 HTTP 标头，如图 7-1 所示。

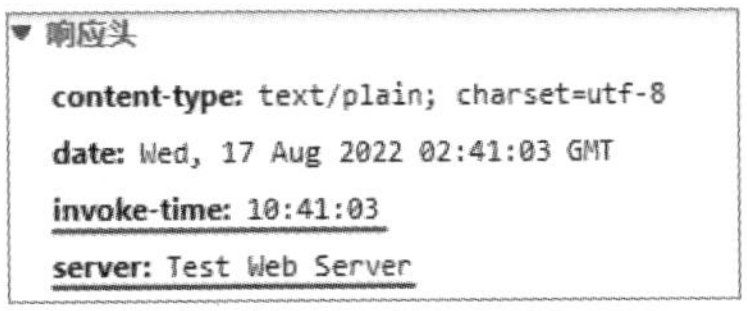

图 7-1　查看 HTTP 响应头

7.1.2　示例：在 Map*()方法中访问 HTTP 上下文

Map*()方法有两类重载。

（1）使用 RequestDelegate 委托作为参数，该委托在定义上就带有 HttpContext 类型的输入参数。因此，调用此类重载方法会自动获取对 HttpContext 对象的引用。

（2）使用通用委托类型作为参数，即 Delegate 类型。它是所有委托类型的公共基类，所以此参数可以接受任意类型的委托实例。若想获得对 HttpContext 对象的引用，可以使用带 HttpContext 类型输入参数的委托实例。

本示例将演示调用以上两类重载方法时获取 HttpContext 引用的过程。

```
// 参数类型为 RequestDelegate 委托
app.MapGet("/", async context =>
{
    context.Response.ContentType = "text/plain; charset=UTF-8";
    await context.Response.WriteAsync("示例程序");
});

// 当参数是任意类型的委托时，所使用的委托需提供 HttpContext 类型的参数
app.MapGet("/test", (HttpContext context) =>
{
    return "示例程序：/test";
});
```

7.1.3 示例：使用 Razor 标记呈现 HTTP 请求标头

Razor 标记将 C#代码混合到 HTML 标记中，当代码在服务器上执行时，Razor 标记中混合的 C#代码能够控制 HTML 标记的生成，使 Web 页面具备动态性。例如，A 页面用于呈现某城市的天气预报信息，当 Web 浏览器请求指定的 URL 时，应用程序先提取天气数据（数据可能来自数据库,可能来自第三方的服务接口),然后根据天气数据生成 HTML 标签。由于 HTML 是根据天气数据生成的，所以浏览器每次访问 A 页面都会呈现不同的结果（最新的气温值、相对湿度值等）。

本示例将使用 Razor Pages 呈现所有 HTTP 请求标头。具体步骤如下。

（1）在应用程序初始化代码中，依次添加 Razor Pages 相关的服务，以及映射终结点。

```
var builder = WebApplication.CreateBuilder(args);
// 添加相关的服务
builder.Services.AddRazorPages();
var app = builder.Build();
// 添加终结点映射
app.MapRazorPages();
// 开始运行应用
app.Run();
```

（2）在项目中新建目录，命名为 Pages。然后在 Pages 目录下新建 Razor 代码文件，命名为 Index.cshtml。

（3）在 Index.cshtml 文件中输入 HTML 和 C#代码。

```
@page

<html>
```

```
<head>
<meta charset="utf-8" />
<title>示例程序</title>
</head>
<body>
<style>
…
</style>
<div class="tb">
@foreach (var ent in HttpContext.Request.Headers)
{
<div class="tbline">
<div>@ent.Key</div><div>@ent.Value</div>
</div>
}
</div>
</body>
</html>
```

扩展名为.cshtml 的文件表示使用了 Razor 标记的 HTML 文档。在文档首行加上@page 指令，表示此页面为 Razor Pages 专用（非 MVC 视图）。Index.cshtml 文档隐式继承了 Page 类（位于 Microsoft.AspNetCore.Mvc.RazorPages 命名空间），而 Page 类继承了 PageBase 类。因此，Index.cshtml 文档继承了 PageBase 类的 HttpContext 属性。通过该属性就可以获得与当前 HTTP 请求关联的 HttpContext 实例引用，然后再由 Request.Headers 属性获得所有 HTTP 请求标头。foreach 循环读出标头集合中的元素，再混合 HTML 标记呈现到 Web 页面上。

运行示例程序，结果如图 7-2 所示。

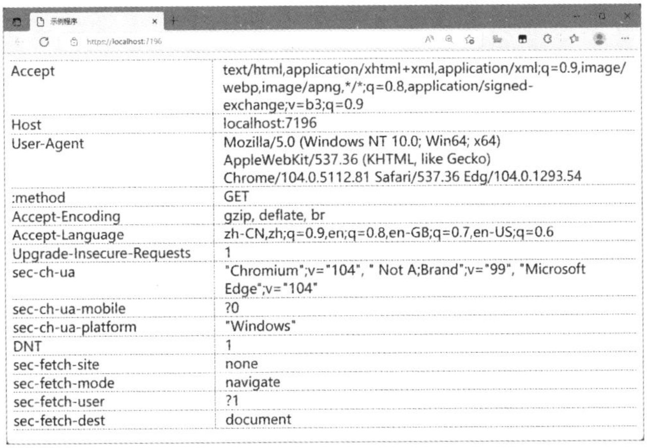

Accept	text/html,application/xhtml+xml,application/xml;q=0.9,image/webp,image/apng,*/*;q=0.8,application/signed-exchange;v=b3;q=0.9
Host	localhost:7196
User-Agent	Mozilla/5.0 (Windows NT 10.0; Win64; x64) AppleWebKit/537.36 (KHTML, like Gecko) Chrome/104.0.5112.81 Safari/537.36 Edg/104.0.1293.54
:method	GET
Accept-Encoding	gzip, deflate, br
Accept-Language	zh-CN,zh;q=0.9,en;q=0.8,en-GB;q=0.7,en-US;q=0.6
Upgrade-Insecure-Requests	1
sec-ch-ua	"Chromium";v="104", " Not A;Brand";v="99", "Microsoft Edge";v="104"
sec-ch-ua-mobile	?0
sec-ch-ua-platform	"Windows"
DNT	1
sec-fetch-site	none
sec-fetch-mode	navigate
sec-fetch-user	?1
sec-fetch-dest	document

图 7-2　显示所有请求标头

如果 Razor Pages 采用视图和代码分离的方案，会使用@model 指令让 Razor 文档与某个类关联。这个类从 PageModel 类（位于 Microsoft.AspNetCore.Mvc.RazorPages 命名空间）派生，表示 Razor 页面专用的模型类。PageModel 类也公开了 HttpContext 属性，因此，它也可以直接获取到 HTTP 上下文对象的引用。

7.1.4 示例：在 MVC 中访问 HTTP 上下文

MVC（包括 Web API）的控制器均以 ControllerBase 为基类。该类中存在公共属性 HttpContext，可以获得 HTTP 上下文对象的引用。而 MVC 视图文档隐式继承了 RazorPage<TModel>类（TModel 为 dynamic，即动态类型），RazorPage<TModel>派生自 RazorPage 类，继承了 Context 属性。所以，在 MVC 视图文档中，可以通过 Context 属性获取 HttpContext 实例的引用。

本示例将实现在 MVC 视图中呈现客户端 IP 地址和端口。具体步骤如下。

（1）在应用程序初始化代码中，添加 MVC 相关的服务，以及 HTTP 管线中的终结点映射。

```
var builder = WebApplication.CreateBuilder(args);
// 添加服务
builder.Services.AddControllersWithViews();
var app = builder.Build();

// 映射终结点
app.MapControllerRoute("test",
"{controller=demo}/{action=home}");

// 开始运行应用程序
app.Run();
```

（2）定义 DemoController 类，继承 Controller 类，它表示一个 MVC 控制器。

```
public class DemoController : Controller
{
    public IActionResult Home() => View();
}
```

控制器中仅包含一个 Home()操作方法，执行后返回同名视图文件（即 Home.cshtml）。

（3）在项目中新建目录，命名为 Views。然后在 Views 目录下创建子目录，命名为 Demo（与控制器名称相同）。

（4）在 Demo 目录下添加 Razor 标记文档，命名为 Home.cshtml。

```
<html>
<head>
<title>示例程序</title>
<meta charset="utf-8" />
</head>
<body>
<div>
客户端 IP：@Context.Connection.RemoteIpAddress?.ToString()
</div>
<div>
客户端端口：@Context.Connection.RemotePort
```

```
</div>
</body>
</html>
```

访问 Context 属性可以获得 HttpContext 实例的引用。再通过 Connection 属性获得 ConnectionInfo 实例，它封装了 HTTP 基础连接相关的信息。RemoteIpAddress 属性表示远程主机的 IP 地址，即客户端的 IP 地址；同理，RemotePort 属性表示远程主机的端口号，即客户端发出 HTTP 请求时所使用的端口。

（5）打开 appsettings.json 文件，为应用程序配置 URL。

```
{
"Logging": {
…
},
"AllowedHosts": "*",
"urls": "https://*:443;http://*:80"
}
```

443 是 HTTPS 的默认端口，80 是 HTTP 的默认端口。*表示侦听服务器上的所有地址。配置默认端口后，Web 浏览器在访问上述 URL 时不需要指定端口。例如，直接访问 http://192.168. 1.100 或 https://192.168.1.100 即可。

（6）运行示例程序，然后尝试在本地或远程机器上访问 Web 服务器。Web 页面上会显示客户端的 IP 地址和端口，如图 7-3 所示。

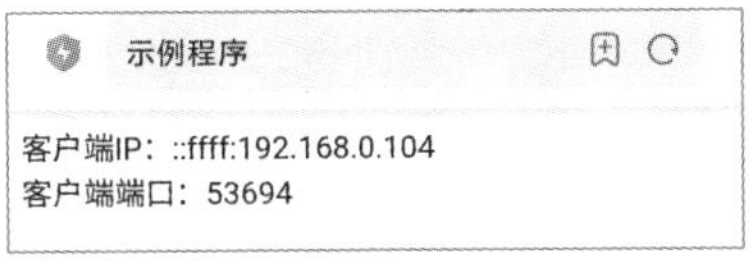

图 7-3 呈现客户端的 IP 地址和端口

7.2 HTTP 消息头

HTTP 消息头（简称 HTTP 头）出现在 HTTP 消息正文的前面，由一系列字段组成。字段名称不区分大小写，名称与值之间用英文冒号（:）分隔。消息头是 HTTP 正文的附加信息，既可以描述与正文内容相关的信息，也可以描述 HTTP 消息本身。标准的 HTTP 消息头也简称为 HTTP 标头。

每个 HTTP 消息头占用一行，与正文之间保留一个空白行，如

```
name1: value1
name2: value2
name3: value3

<消息正文>
abcdabcdabcdabcd
```

请求消息与响应消息的结构一样，因此，请求消息和响应消息都有各自的消息头区域。请求消息的头部由 HttpRequest 对象的 Headers 属性提供；响应消息的头部则由 HttpResponse 对象的 Headers 属性提供。

下面的示例将接收客户端发送的 HTTP-GET 请求，然后找出自定义的消息头（本示例假设自定义消息头的字段名均以 dk_开头）。

```
app.MapGet("/", (HttpContext context) =>
{
    var request = context.Request;
    // 查找以 dk_开头的 HTTP 头
    var selres = request.Headers.Where(x => x.Key.StartsWith("dk_"));
    // 返回结果
    StringBuilder strbd = new();
    strbd.AppendLine("你提供的自定义消息头：");
    foreach(var item in selres)
    {
        strbd.AppendLine($"{item.Key} = {item.Value}");
    }
    return strbd.ToString();
});
```

.NET 工具库提供了一个用于 Web API 测试的命令行工具——httprepl。该工具可以通过交互方式发送 HTTP 请求。在使用前需要安装该工具，方法是打开任意可用的命令行窗口，输入以下命令。

```
dotnet tool install -g Microsoft.dotnet-httprepl
```

以上命令将 httprepl 安装为全局工具，即在任意路径下都可以运行命令。若希望安装为本地工具（只对当前目录及子目录有效），需要先创建清单文件。命令如下。

```
dotnet new tool-manifest
```

然后执行以下安装命令。

```
dotnet tool install --local Microsoft.dotnet-httprepl
```

本书建议安装为全局工具，调用起来比较方便。

运行示例程序，打开任意命令行窗口，输入 httprepl 命令，按 Enter 键确认后就会启动 httprepl 工具并且进入交互模式。

使用 connect 命令连接服务器，本示例中服务器的 URL 为 https://localhost:10230，命令如下。

```
connect https://localhost:10230
```

使用 get 命令发送 HTTP-GET 请求，并通过-h 参数添加 3 个自定义消息头。

```
get -h dk_name=test -h dk_ver=3.0 -h
dk_publisher=Jack
```

消息头的字段名与字段值之间也可以用英文冒号分隔，如

```
get -h dk_name:test …
```

按 Enter 键确认后，工具会向服务器发送 HTTP 请求，并显示服务器的响应消息，如图 7-4 所示。

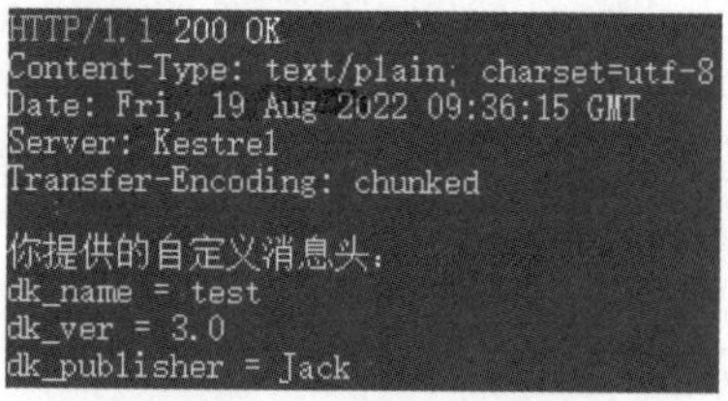
```
HTTP/1.1 200 OK
Content-Type: text/plain; charset=utf-8
Date: Fri, 19 Aug 2022 09:36:15 GMT
Server: Kestrel
Transfer-Encoding: chunked

你提供的自定义消息头：
dk_name = test
dk_ver = 3.0
dk_publisher = Jack
```

图 7-4　自定义的消息头

7.2.1　HeaderNames 类

HeaderNames 类（位于 Microsoft.Net.Http.Headers 命名空间）是一个静态类，它通过只读字段公开标准 HTTP 头的名字。使用 HeaderNames 类可以轻松地设置或读取常见的 HTTP 头。

下面的示例将使用 HeaderNames 类的字段，从 HTTP 请求消息中获取 Accept-Language、Accept-Encoding、Host、User-Agent 等标准头的值。

```
app.Map("/", (HttpContext context) =>
{
    var request = context.Request;

    // Accept-Language
    var acceptLan = request.Headers[HeaderNames.AcceptLanguage];

    // Accept-Encoding
    var acceptEnc = request.Headers[HeaderNames.AcceptEncoding];

    // Host
    var host = request.Headers[HeaderNames.Host];
    // 或者
    //var host = request.Headers.Host;

    // User-Agent
    var uagent = request.Headers[HeaderNames.UserAgent];
    // 或者
    //var uagent = request.Headers.UserAgent;

    StringBuilder bd = new();
    bd.AppendLine($"{HeaderNames.AcceptLanguage}: {acceptLan}")
      .AppendLine($"{HeaderNames.AcceptEncoding}: {acceptEnc}")
      .AppendLine($"{HeaderNames.Host}: {host}")
      .AppendLine($"{HeaderNames.UserAgent}: {uagent}");
    return bd.ToString();
});
```

request.Headers 属性（定义的类型为 IHeaderDictionary）自身也公开了一些属性，可以访问常用的 HTTP 消息头，如 UserAgent、Accept、Host 等。因此，对于这些常用的消息头，除了使用 HeaderNames 类的字段检索外，还可以直接访问这些属性，如上述代码中的

```
var uagent = request.Headers.UserAgent;
var host = request.Headers.Host;
```

运行示例程序，结果如图 7-5 所示。

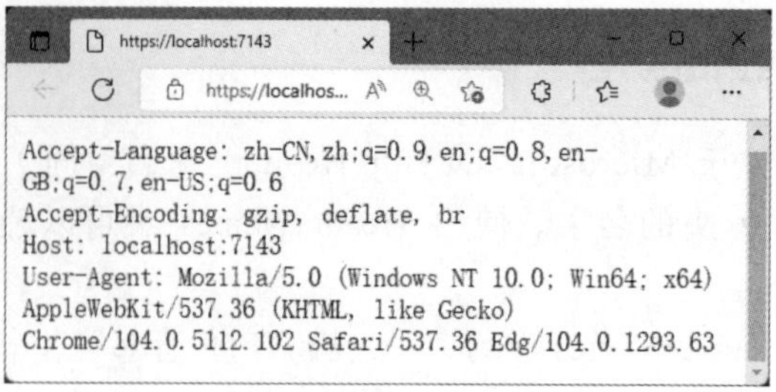

图 7-5　呈现常用的 HTTP 消息头

7.2.2　消息头的分类

根据使用的场景和上下文，HTTP 消息头一般可分为以下 4 种类型。

（1）请求头：客户端在发送 HTTP 消息时所添加的头部。这些消息头通常会包含客户端的相关信息，如 Host、Accept 等。这些消息头由客户端定义。

（2）响应头：服务器发送的响应消息中包含的头部信息，如 Server、Location 等。这些消息头由服务器定义。

（3）通用头：同时适用于请求消息和响应消息的头部，如 Connection、Date 等。

（4）实体头：用于描述 HTTP 消息正文部分的头部。常用的有 Content-Length、Content-Type、Content-Encoding 等。

有些消息头可以在客户端与服务器之间搭配使用。例如，客户端在发出 HTTP 请求时，可以添加 Accept 消息头，告诉服务器“我能接收视频流”；服务器在发送响应消息时可以添加 Content-Type 消息头，告诉客户端“我刚刚发的是 MP4 格式的视频”。

下面的示例将演示客户端（Web 浏览器）向根 URL 发送 GET 请求后，服务器将发送一张 JPEG 格式的图片作为回应。服务器在返回图像文件时会设置 Content-Type 消息头为 image/jpeg，即 JPEG 图像。

```
app.MapGet("/", () =>
{
    // 获取项目根目录
    string rootDir = app.Environment.ContentRootPath;
    // 文件名
    string fileName = Path.Combine(rootDir, "test.jpg");
    return Results.File(fileName, "image/jpeg");
});
```

首先通过 Environment.ContentRootPath 属性获得当前应用程序所在的目录，接着调用 Path.Combine()静态方法拼接路径（应用程序根目录下有一个名为 test.jpg 的文件），最后直接通过 Results.File()方法将文件内容返回。

在上述示例中，开发人员不需要手动设置 Content-Type 消息头，而是在调用 Results.File() 方法时通过第 2 个参数（contentType）指定。

运行示例程序，使用 Web 浏览器访问应用程序的根 URL，服务器所返回的图像文件将显示在浏览器窗口中，如图 7-6 所示。

图 7-6　服务器返回图像文件

7.2.3　分析复杂消息头

尽管 HTTP 消息头的值都是字符串，但对于一些特定的消息头，它们有专用的格式。例如 Accept-Charset 请求消息头，它用于告知服务器客户端能够处理的字符集。它可以使用多个字符集名称，每个名称用英文逗号（,）分隔。如果需要，可以为字符集指定权重比例。权重值以英文分号（;）开头，使用如“q=0.8”的格式表示。权重取值为 0 ~ 1，可精确到小数点后 3 位。权重越大，优先级越高；反之，优先级越小。

下面的 Accept-Charset 消息头表示客户端可接收 UTF-8、ISO-8859-1 和 Unicode-1-1 字符集。

```
Accept-Charset: utf-8, iso-8859-1;q=0.5, unicode-1-1;q=0.8
```

其中，UTF-8 省略了权重值，默认权重为 1.0，即它的优先级最高；其次是 Unicode-1-1，优先级最低的是 ISO-8859-1。

在 Microsoft.AspNetCore.Http.Headers 命名空间中，有两个类可用于处理特殊消息头的值：RequestHeaders 类处理的是请求消息的头部信息，ResponseHeaders 类则用于处理响应消息的头部信息。这两个类中包含通用头（如 CacheControl）和实体头（如 ContentType、ContentLength）。

为了能够分析各种消息头信息的值，还需要一组专门的封装类型。

（1）MediaTypeHeaderValue：表示多媒体类型。格式由主体类型和子类型组成，用/连接。例如，image/jpeg 表示主体类型是图像数据，子类型表明该内容为.jpg 图像文件（或数据流）。多个类型用英文的逗号（,）分隔。如果需要参数，每组参数都由 name 和 value 组成，以英文的分号（;）开头。例如：

```
text/plain;p1=v1;p2=v2
```

（2）NameValueHeaderValue：由 name 和 value 组成的头部值，每组之间以英文的逗号（,）分隔。

（3）SetCookieHeaderValue：用于 Set-Cookie 响应头的值。格式由 name 和 value 组成，可以添加参数（特性）。参数以英文的分号（;）开头。例如：

```
cookieName=cookieValue;path=/
```

（4）ContentRangeHeaderValue：表示 Content-Range 响应头。返回消息正文的部分数据，常用于断点续传。其格式为“<计量单位><开始索引>-<结束索引>/<总大小>”。其中，“计量单位”一般为字节（bytes）。例如：

```
    bytes 0-105/200
```

（5）RangeHeaderValue：表示 Range 请求头。该消息头会请求服务器返回某个资源（或文档）中的部分内容。Range 头可以指定多个分段。例如：

```
bytes=0-100,300-499,1000-1500
```

上述指令表示向服务器请求了 3 段数据，分别是内容开头的 101 字节、从索引 300 处读取 200 字节、从索引 1000 处读取 501 字节。注意索引是从 0 开始的，即第 1 个字节的索引为 0，最后一个字节是 n−1。Range 消息头所指定的范围是起始值和终止值的索引，如 1000-1500 表示起始索引为 1000，终止索引为 1500，共 501 字节。

如果服务器能正确响应，返回 206 状态码。如果请求的只有一个片段的数据，那么响应消息中的 Content-Type 头表示数据的内容类型（如 image/png），Content-Range 头描述返回的数据片段位于整个文档（或文件）的位置。如果请求的是多个片段，那么响应头 Content-Type 为 multipart/byteranges，并使用多组 Content-Type 和 Content-Range 头描述每个数据片段。例如：

```
HTTP/1.1 206 Partial Content
...
Content-Length: 1741
Content-Type: multipart/byteranges; boundary=<分隔符>

--<分隔符>
Content-Type: application/pdf
Content-Range: bytes 234-639/8000

65c2b6e538f4d28
--<分隔符>
Content-Type: application/pdf
Content-Range: bytes 4590-7999/8000

x5ck6g8fs4e64s8eb
--<分隔符>--
```

由于消息正文由多部分组成，为了让客户端能够正确处理，在 Content-Type 消息头中通过 boundary 参数指定一个分隔符。服务器返回的各个数据片段都用这个分隔符来分隔，避免属于不同片段的数据被错误地混合在一起。

（6）ContentDispositionHeaderValue：表示 Content-Disposition 消息头。当它用于响应消息时，用于告诉客户端如何处理从服务器返回的数据。inline 表示返回的内容将和 Web 页面一起呈现在浏览器；attachment 表示以附件的形式处理服务器返回的内容，可以用 filename 参数指定默认保存的文件名。文件下载功能正是通过 Content-Disposition 消息头实现的。例如：

```
Content-Type: text/plain; charset=utf-8
Content-Disposition: attachment; filename="recs.txt"
```

服务器返回的内容将以文本文件的形式下载到本地，文件名为 recs.txt。

当 Content-Disposition 消息头用于多部分消息主体中的某个片段时，Content-Type 可以指定为 multipart/form-data，每个片段都可以用 Content-Disposition 来定义。例如：

```
POST /test/update HTTP/1.1
Host: example.com
Content-Type: multipart/form-data;boundary="<分隔符>"

--<分隔符>
Content-Disposition: form-data; name="name 1"

value1
--<分隔符>
Content-Disposition: form-data; name="name 2"; filename="upload.dat"

value2
--<分隔符>--
```

（7）StringWithQualityHeaderValue：表示带有权重值的 HTTP 消息头的值——带有参数 q，如 q=0.5。

下面的示例将分析客户端发送的自定义消息头。此消息头由一组 name-value 格式的值构成。

```
var builder = WebApplication.CreateBuilder(args);
builder.Services.AddHttpContextAccessor();
var app = builder.Build();

app.MapGet("/", () =>
{
    // 获取 IHttpContextAccessor 服务
    IHttpContextAccessor accessor = app.Services.GetRequiredService
    <IHttpContextAccessor>();
    // 获得 HttpContext 实例
    HttpContext context = accessor.HttpContext!;

    RequestHeaders reqHeaders = new(context.Request.Headers);
```

```
        // 获得 demo-header 消息头的值列表
        IList<NameValueHeaderValue> values = reqHeaders.GetList
        <NameValueHeaderValue>("demo-header");
        if(values.Count > 0)
        {
            string s = "自定义消息头的值列表：\n";
            // 将分析得到的值追加到字符串中
            foreach (var val in values)
            {
                s += val.ToString() + '\n';
            }
            return s;
        }

        return "未找到自定义消息头";
    });

    // 开始运行应用程序
    app.Run();
```

本示例是通过 IHttpContextAccessor 服务获取 HttpContext 实例的，因此在构建应用程序阶段，需要调用 AddHttpContextAccessor()扩展方法。

由于 demo-header 消息头的值中存在多组 name-value 值，所以应当调用 RequestHeaders 对象的 GetList<T>方法，其中类型参数 T 为 NameValueHeaderValue。最后把这些分析出来的值组装为字符串实例返回给客户端。

运行示例程序，在命令行窗口中运行 httprepl 工具。执行 connect 命令开始与应用程序通信。

```
connect https://localhost:9102
```

向服务器发送 GET 请求，并提供名为 demo-header 的消息头。

```
get -h demo-header=key1=val1,key2=val2,
key3=val3
```

注意消息头有多个值时要用英文的逗号分隔，并且整个-h 参数值不能有空格。执行以上命令后，命令行窗口会打印服务器响应的结果，如图 7-7 所示。

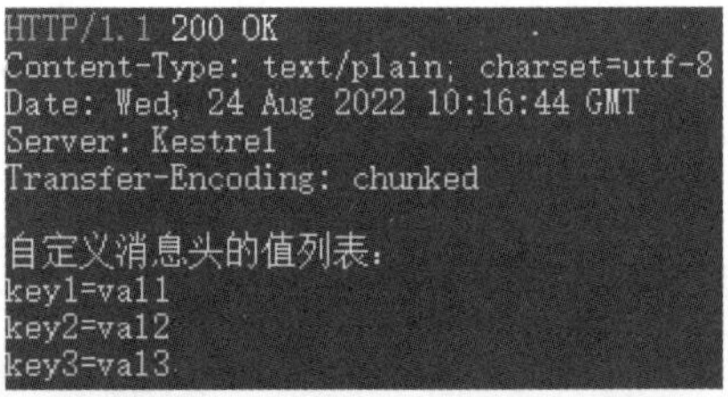

图 7-7 服务器响应 demo-header 消息头的分析结果

7.3 查询字符串

查询字符串出现在 URL 的末尾，以问号（?）开头。查询参数由 name 和 value 组成，其中，name 表示字段名称，value 表示字段的值。参数之间使用字符&连接。例如：

```
http://abc.com/index?a=1&b=2&c=3
```

上述 URL 传递了 3 个参数：a=1，b=1，c=3。

7.3.1　读取查询参数

HttpRequest 类的 Query 属性用于访问查询字符串。它类似于字典结构，可以通过 Key 检索查询参数的值，也可以枚举所有参数。另外，QueryString 属性可以获取完整的查询字符串——URL 中从问号（?）开始的所有字符（包括问号），如

```
?type=1&bit=0。
```

以下示例将通过 QueryString 属性提取出查询字符串，然后从 Query 属性中枚举所有查询参数。

```
app.MapGet("/", async context =>
{
    // 获取 HttpRequest 对象
    var request = context.Request;
    string msg = $"查询字符串：{request.QueryString}\n\n";
    msg += "查询参数：\n";
    foreach (var p in request.Query)
    {
        msg += $"{p.Key}={p.Value}\n";
    }
    // 获取 HttpResponse 对象
    var response = context.Response;
    // 设置响应头
    response.Headers.Add(HeaderNames.ContentType, "text/plain;charset=
    UTF-8");
    // 写入响应消息
    await response.WriteAsync(msg);
});
```

运行示例程序，在 Web 浏览器中查看以下 URL。

```
https://localhost:7216/?name=myApp&desc=Demo
```

假设服务器的域名为 localhost:7216，上述 URL 提供了两个查询参数——name 和 desc。服务器分析查询参数后，会发送如图 7-8 所示的响应消息。

图 7-8　提取查询参数

7.3.2 多值参数

客户端在提供 URL 查询字符串时，如果同一名称的参数多次出现，服务器在处理时会将名称相同的参数值合并为一个值。例如：

```
http://test.org/api/cars?id=15&id=17&flag=on
```

上述查询字符串中，id 参数出现了两次。当 HTTP 消息到达服务器后，应用程序会把 15 和 17 合并为一个值，由 StringValues 结构体表示。

以下示例演示读取查询参数 items 的所有值。

```
app.MapGet("/", (HttpContext context) =>
{
    // 获取 HttpRequest 对象
    var request = context.Request;
    // 判断是否存在 items 参数
    if(request.Query.ContainsKey("items"))
    {
        string values = "items 的参数值：\n";
        foreach(var val in request.Query["items"])
        {
            values += $"{val}\n";
        }
        return values;
    }
    // 未找到 items 参数，返回
    return "未提供 items 参数";
});
```

Query 属性中的参数值是 StringValues 结构体，它由多个字符组成，实现了 ICollection<string>、IEnumerable<string>、IList<string>等接口，可以用 foreach 语句枚举所有字符串实例。

运行示例程序，在客户端发出 HTTP-GET 请求时添加 3 个 items 查询参数。

```
https://localhost:7005/?items=桃子
&items=香蕉&items=李子
```

经过 URL 编码后，变为

```
https://localhost:7005/?items=
%E6%A1%83%E5%AD%90&items=%E9%A6%99
%E8%95%89&items=%E6%9D%8E%E5%AD%90
```

应用程序在读取参数后会返回如图 7-9 所示的内容。

图 7-9 items 参数的多个值

7.4 表单数据

表单数据（Form Data）通常以 HTTP-POST 方式发送到服务器。对于由简单的 name 和 value 组成的表单，Content-Type 消息头一般为 application/x-www-form-urlencoded。数据正文格式与 URL 查询字符串相似，name 和 value 之间用符号=连接，字段之间用符号&连接。例如：

```
data1=value1&data2=value2&data3=value3
```

对于复杂的表单数据，可能包含文本、二进制数据（如文件、音频、图像等），通常会设置 Content-Type 消息头为 multipart/form-data，通过 boundary 参数设置分隔符，用于分隔消息正文中的子内容。

每个子内容都包含 Content-Disposition 头，并指定为 form-data 类型。name 参数设置字段名称。

```
Content-Disposition: form-data; name="<字段名>"

<字段值>
```

完整的 HTTP 消息形如

```
POST /demo HTTP/1.1
Host: hello.com
Content-Type: multipart/form-data;boundary="<分隔符>"

--<分隔符>
Content-Disposition: form-data; name="field1"

value1
--<分隔符>
Content-Disposition: form-data; name="field2"; filename="abc.docx"

value2
--<分隔符>--
```

7.4.1 读取简单的表单数据

IFormCollection 接口提供类似字典的数据结构，可以通过 Key 检索数据。其默认实现类型是 FormCollection，由 FormFeature 类提供，通过 HttpRequest 对象的 Form 属性公开。

下面的代码将演示从客户端提交的消息正文中读出所有表单数据。

```
app.MapPost("/addnews", async context =>
{
    var request = context.Request;
    var response = context.Response;
    response.Headers.ContentType = "text/plain;charset=UTF-8";
```

```
    if(request.HasFormContentType)
    {
        // 读出所有字段
        var postData = from p in request.Form
                       select $"{p.Key} = {p.Value}";
        await response.WriteAsync(string.Join('\n', postData));
        return;
    }

    await response.WriteAsync("未发现消息正文");
});
```

在读取表单数据前，可以访问 HasFormContentType 属性，检查一下客户端发送的 HTTP 消息中的 Content-Type 标头是否为 application/x-www-form-urlencoded 或 multipart/form-data。

上述示例仅仅将表单数据全部读出，然后作为响应消息发回给客户端，并未作其他处理。下面的代码将用于访问服务器测试。

```
HttpClient client = new();

// 要发送的数据
IDictionary<string, string> data = new Dictionary<string, string>();
data["title"] = "测试标题";
data["content"] = "测试内容";
data["author"] = "小明";

// 封装数据
FormUrlEncodedContent form = new(data);

// HTTP 消息
HttpRequestMessage reqmsg = new HttpRequestMessage(HttpMethod.Post,
"http://localhost:6982/addnews");
reqmsg.Content = form;

// 发送请求
var response = await client.SendAsync(reqmsg);
// 读取响应消息
if(response.IsSuccessStatusCode)
{
    string respmsg = await response.Content.ReadAsStringAsync();
    Console.WriteLine("服务器响应：");
    Console.WriteLine(respmsg);
}
```

上述代码首先通过一个字典对象构建要提交的表单，然后作为参数传递给 FormUrl-EncodedContent 类的构造函数。该类会自动封装并处理表单数据，开发人员不需要手动拼接 HTTP 消息正文。

发送 POST 请求后，从服务器的响应消息中读出正文（ReadAsStringAsync()方法可将正文内容以字符串形式返回），最后显示在控制台窗口上。

同时运行 ASP.NET Core 应用程序和测试客户端程序，当服务器成功运行后，切换到客户端程序的控制台窗口，按下键盘上的任意键，客户端程序就会将测试表单数据发送给服务器，并输出服务器的响应消息，如图 7-10 所示。

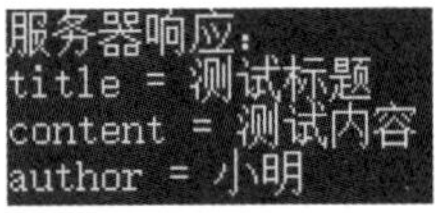

图 7-10　服务器发回表单数据

7.4.2　文件上传

文件属于表单中的一种特殊字段，一般长度较大，而且是二进制数据。在向服务器提交包含文件的表单时，Content-Type 标准头应使用 multipart/form-data。文件字段可以出现多次，即可以上传多个文件。例如，下面的 HTTP 消息将上传一个图像文件。

```
POST /test HTTP/1.1
Host: localhost:6500
Content-Type: multipart/form-data; boundary=abcdefghijk
Accept-Encoding: gzip, deflate

--abcdefghijk
Content-Disposition: form-data; name="file"; filename="123.jpg"

xxxxxxxxxxxxxxxxxxxxxx 文件内容 xxxxxxxxxxxxxxxxxxxxxxx
--abcdefghijk--
```

每个字段中都用 Content-Disposition 头部定义字段信息，如 name 指定字段的名称，filename 设置文件名。

在 ASP.NET Core 应用程序中，HttpRequest 对象的 Form 属性为 IFormCollection 集合，该集合公开了 Files 属性，类型为 IFormFileCollection 集合。客户端通过表单上传的文件都会被添加到 IFormFileCollection 集合中，其中，每个文件会使用 IFormFile 类来封装。调用 OpenReadStream()方法可以打开文件流，方便进一步处理（如把文件加密后再保存到服务器）。也可以直接调用 Copy()方法将文件内容复制到其他流中，如内存流。

下面给出一个示例。该示例只读取一个文件并保存在服务器上（只获取 IFormFileCollection 集合中的第 1 个文件）。

（1）在应用程序初始化时，注册 MVC 控制器功能。

```
builder.Services.AddControllers();
```

（2）修改 Web 服务器（内置的 Kestrel 服务器）对请求正文的大小限制。

```
builder.Services.Configure<KestrelServerOptions>(opt =>
{
    // 不限制
    opt.Limits.MaxRequestBodySize = null;
});
```

文件内容通常比较大，因此必须修改 MaxRequestBodySize 属性，以允许接受体积较大的

HTTP 消息正文。本示例将该属性设置为 null，表示大小不受限制。

（3）应用程序构建完成后，需要为 MVC 功能映射终结点。

```
var app = builder.Build();
...
app.MapControllers();
```

（4）在项目中添加新类，派生自 Controller 类，它是一个 MVC 控制器类。类名为 DemoController，去掉 Controller 后缀后，控制器的名称为 Demo。

```
[Route("demo/[action]")]
public class DemoController : Controller
{
    // 限制文件大小为 500MB
    [RequestFormLimits(MultipartBodyLengthLimit = 500 * 1024 * 1024)]
    public IActionResult Upload(IFormFileCollection my_files,
    [FromServices] IHostEnvironment env)
    {
        if (my_files.Count == 0)
        {
            return Content("未发现文件");
        }

        IFormFile theFile = my_files[0];
        // 当前目录
        string dir = env.ContentRootPath;
        // 文件保存路径
        string newFilePath = Path.Combine(dir, theFile.FileName);
        // 写入文件
        using(FileStream fsout = System.IO.File.OpenWrite(newFilePath))
        {
            theFile.CopyTo(fsout);
        }
        return Content("文件上传完成");
    }
}
```

Upload()方法用于接收并处理客户端上传的文件。RequestFormLimits 特性设置多部分正文中大小限制，此处设置大小为 500MB，即可以上传 500MB 以内的文件。虽然在应用程序初始化时设置了 Kestrel 服务器不限制请求正文的大小，但针对表单数据正文需要再次进行配置，否则体积较大的文件仍然无法上传。

由于 MVC 有内置的模型绑定机制，本示例中不需要手动访问 HttpContext 对象及其相关对象，而是在定义 Upload()方法时指定一个 IFormFileCollection 类型的参数。需要注意的是，参数的名称要与客户端提交的表单字段名称相同。例如，客户端在提交时设定字段名为 my_files，那么 IFormFileCollection 类型的参数名称也必须为 my_files，否则无法激活自动绑定。

Upload()方法的第 2 个参数加了 FromServices 特性，表示 env 参数的值将从服务容器中获取（由依赖注入获得）。

获得上传的文件后，在服务器的本地磁盘中新建文件流（由 File.OpenWrite()方法返回），然后直接调用 CopyTo()方法把文件内容复制到新文件流中，就可以将文件保存到服务器上了。

（5）下面的代码实现了一个简单的测试客户端（控制台应用程序）。

```
// 输入文件路径
Console.Write("请输入文件路径：");
// 行尾的 Trim()方法用于删除路径首尾可能出现的双引号或单引号
string? fileToUpload = Console.ReadLine()?.Trim("", '\");
if(!File.Exists(fileToUpload))
{
    Console.WriteLine("请输入有效的文件路径");
    return;
}

// 服务器根地址
Uri baseUrl = new Uri("https://localhost:7104");

// 文件内容
StreamContent streamContent = new(File.OpenRead(fileToUpload));
// Multipart 内容
MultipartFormDataContent form = new();
// 添加字段
form.Add(streamContent, "my_files", Path.GetFileName(fileToUpload));

HttpClient client = new();
client.BaseAddress = baseUrl;
// 上传
var response = await client.PostAsync("/demo/upload", form);
// 显示结果
if(response.IsSuccessStatusCode)
{
    Console.WriteLine("服务器响应：");
    Console.WriteLine(await response.Content.ReadAsStringAsync());
}

Console.Read();
```

在使用 HttpClient 对象发送 HTTP 请求之前，先用 StreamContent 对象读取文件流，然后添加到 MultipartFormDataContent 对象中，最后才调用 HttpClient 对象的 PostAsync()方法。

（6）同时运行 ASP.NET Core 应用程序（服务器）和测试客户端。在客户端控制台中输入要上传文件的路径，然后按 Enter 键。

（7）文件上传结束，会显示服务器的响应消息，如图 7-11 所示。

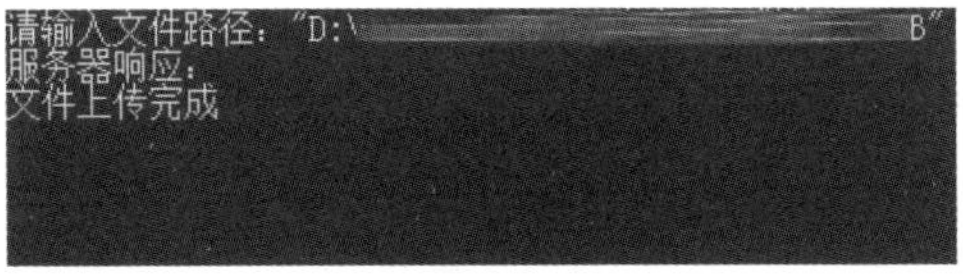

图 7-11　上传单个文件

还有一种更简单的方法，直接把文件的内容作为 HTTP 消息正文来提交。请看下面的示例。

```
app.MapPost("/upload", async context =>
{
    var request = context.Request;
    var response = context.Response;
    response.ContentType = "text/plain;charset=UTF-8";
    // 创建目录
    string dir = Path.Combine(app.Environment.ContentRootPath, "Uploads");
    if(Directory.Exists(dir) == false)
    {
        Directory.CreateDirectory(dir);
    }

    // 通过自定义的消息头获得文件名
    if(!request.Headers.TryGetValue("file-name", out var fileName))
    {
        fileName = "test.bin";
    }
    // 组合路径，文件保存到 Uploads 目录下
    fileName = Path.Combine(dir, fileName);

    // 直接复制文件内容
    bool isOk = false;
    try
    {
        using (FileStream fs = File.OpenWrite(fileName))
        {
            await request.Body.CopyToAsync(fs);
        }
        isOk = true;
    }
    catch
    {
        isOk = false;
    }
    // 返回结果
    if (isOk)
    {
        await response.WriteAsync("文件上传完成");
    }
    else
    {
        await response.WriteAsync("文件上传失败");
    }
});
```

由于 HTTP 消息的正文部分填充了文件的内容，所以文件名只能通过自定义的消息头来

指定。在上述示例中，指定文件名的消息头为 file-name。文件最终被存放到 Uploads 目录下。

下面是客户端上传文件的代码。

```
Uri rootUrl = new("https://localhost:7082");

// 要上传的文件
string filePath = "demo.mp3";
StreamContent content = new(File.OpenRead(filePath));

HttpClient client = new HttpClient();
client.BaseAddress = rootUrl;
// 用自定义的消息头指定文件名
client.DefaultRequestHeaders.Add("file-name", filePath);
// 发送请求
var response = await client.PostAsync("/upload", content);
if(response.IsSuccessStatusCode)
{
    Console.WriteLine("服务器响应消息: ");
    Console.WriteLine(await response.Content.ReadAsStringAsync());
}
```

在调用 HttpClient 对象的 PostAsync()方法时，将 StreamContent 对象作为内容提交即可，文件的内容会以流的形式读取，且完全填充 HTTP 消息的正文部分。

7.5 Cookie

Cookie（或复数形式 Cookies）指小型文本文件，用于在客户端（通常是 Web 浏览器）上存储简单的用户信息（如会话标识、用户加入购物车的商品编号等）。它的结构与 HTTP 消息头类似，也是 name = value 格式。

Web 服务器在响应消息上使用 Set-Cookie 标准头设置要发送的 Cookie。如果有多项数据要发送，可以使用多个 Set-Cookie 头。客户端负责存储 Cookie，在下一次向服务器发送请求消息时，使用 Cookie 标准头设置 Cookie 数据，这些数据会与请求消息一同发送到服务器。例如：

```
// 服务器发送响应消息
Set-Cookie: id=12345;Expires=…

// 客户端发送请求消息
Cookie: id=12345
```

对于 Set-Cookie 头，每次只能设置一个数据项。数据后面如果使用参数（如上述示例中的 Expires 参数），可以加上英文的分号（;），然后再追加参数（也是 name=value 格式）。而对于客户端发送的 Cookie 头，可以将多个数据项写在一起。例如：

```
Cookie: name1=value1; name2=value2; name3=value3
```

下面的代码演示在响应消息中添加两项 Cookie 数据。

```
app.MapGet("/", (HttpContext context) =>
{
    // 获取 HttpResponse 对象
    var response = context.Response;
    // 设置 Cookie
    response.Cookies.Append("data-1", "Test-Data-1");
    response.Cookies.Append("data-2", "Test-Data-2");

    return "Hello ASP.NET Core";
});
```

运行示例程序，用 Web 浏览器打开应用程序的根 URL，通过开发人员工具可以查看到服务器的响应消息中包含了两个 Set-Cookie 头，如图 7-12 所示。

当浏览器向服务器发出第 2 次请求时，请求消息中就会带上刚刚获得的 Cookie，如图 7-13 所示。

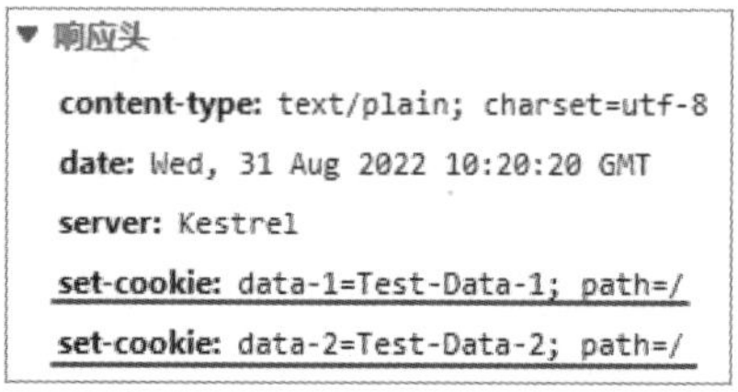

图 7-12　响应标头中的 Cookie

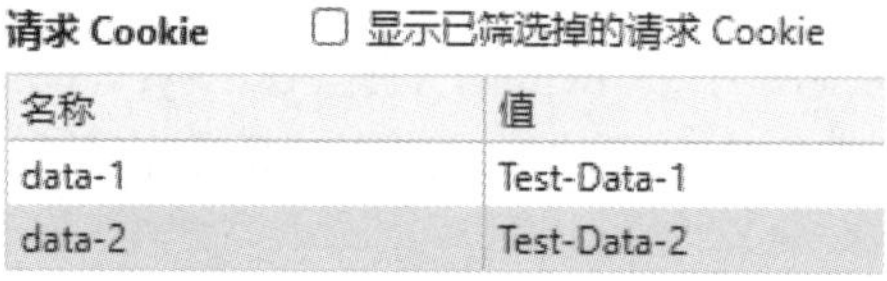

图 7-13　发送请求时携带 Cookie

若要修改 Cookie 的过期时间，需要用到 CookieOptions 类。该类的 Expires 属性用于设置 Cookie 的过期时间，类型为 DateTimeOffset，表示日期时间的偏移量。它设置的是一个时间点，如 3 天后、1 小时后、30 分钟后等。

下面的代码设置 Cookie 将在 15 秒后过期。

```
app.MapGet("/", (HttpContext context) =>
{
    // 设置 Cookie
    CookieOptions options = new()
    {
        // 修改过期时间
        Expires = DateTime.Now.AddSeconds(15)
    };
    context.Response.Cookies.Append("Test", "abcd", options);
    return "Hello World!";
});
```

Cookies.Append()方法有一个可以传递 CookieOptions 实例的重载。

```
void Append(string key, string value, CookieOptions options);
```

运行上述代码，第 1 次访问应用程序时，服务器将设置名为 Test，值为 abcd 的 Cookie。

在 Cookie 的有效期内，再次访问服务器时，Web 浏览器在发送 HTTP 请求时会自动携带 Cookie，如图 7-14 所示；当 Cookie 过期后，Web 浏览器在请求时不会携带 Cookie，如图 7-15 所示。

请求 Cookie　☐ 显示已筛选掉的请求 Cookie

名称	值	Domain	Path	Expires / Max-Age
Test	abcd	localhost	/	2022-09-02T03:23:11.377Z

响应 Cookie

名称	值	Domain	Path	Expires / Max-Age
Test	abcd	localhost	/	2022-09-02T03:23:11.000Z

图 7-14　Cookie 未过期

响应 Cookie

名称	值	Domain	Path	Expires / Max-Age
Test	abcd	localhost	/	2022-09-02T03:24:00.000Z

图 7-15　Cookie 已过期

7.6　HttpContext 类的 Items 属性

HttpContext 类的 Items 属性为字典数据结构，Key 和 Value 都是 object 类型，即应用程序代码可以将任意对象实例存放在 HttpContext.Items 属性中。

Items 属性实现 HTTP 管线上的数据共享，常用于在中间件之间传递数据。下面将通过一个示例阐述 Items 属性的使用。该示例在第 1 个中间件中读取客户端提交的消息正文，然后计算其哈希值（算法为 SHA256），将计算结果暂存于 Items 属性中。第 2 个中间件从 Items 属性中取出计算结果，然后转换为 Base64 字符串返回给客户端。

```
app.Use(async (context, next) =>
{
    var request = context.Request;
    if(request.Method == HttpMethods.Post)
    {
        // 读取正文部分并计算哈希值
        SHA256 sha = SHA256.Create();
        byte[] data = await sha.ComputeHashAsync(request.Body);
        // 暂存计算结果，以供下一个中间件使用
        context.Items.Add("result", data);
    }
    await next();
});

app.Run(async (context) =>
{
    var response = context.Response;
    // 设置响应头
```

```
        response.Headers.ContentType = "text/html;charset=UTF-8";
        // 取出上一个中间件产生的哈希值
        byte[]? data = (byte[]?)context.Items["result"];
        string msg = string.Empty;
        if(data != null)
        {
            msg = $"Shs256: {Convert.ToBase64String(data)}";
        }
        else
        {
            msg = "未获得哈希值";
        }
        // 获取消息正文的字节序列
        byte[] output = Encoding.UTF8.GetBytes(msg);
        // 设置响应消息的正文长度
        response.ContentLength = output.Length;
        // 写入正文数据
        await response.Body.WriteAsync(output);
    });
```

运行示例程序，打开命令行窗口，执行 httprepl 命令启动 HTTP 测试工具。接着执行 connect 命令连接服务器（假设应用程序的 URL 为 https://localhost:7266）。

```
connect https://localhost:7266
```

用 HTTP-POST 方式向服务提交字符串“你好，明天”。

```
post -c 你好，明天
```

服务器将返回哈希结果。

```
HTTP/1.1 200 OK
Content-Length: 53
Content-Type: text/html; charset=UTF-8
Server: Kestrel

Shs256: eHz44X/ykqMx+MCk8+q1cAK+9UIT1ZgKgmLhJlhZip8=
```

7.7 会话

会话（Session）可在服务器上存储（通常存储在内存中，可以自定义实现其他存储方式，如文件）用户信息，然后通过响应消息的 Cookie 把会话标识（Session ID）返回给客户端。客户端在下一次发送请求时可将包含会话标识的 Cookie 一起发送。服务器将根据会话标识检索与用户关联的会话数据。尽管 Cookie 中只包含会话标识，但出于安全考虑，ASP.NET Core 应用程序默认会将会话标识加密（使用 IDataProtector 服务）。

与 Cookie 类似，会话也是字典数据结构，其中，Key 为字符串类型（String），Value 是字节数组（byte[]）。当会话过期后，将无法读取用户信息，服务器会为 HTTP 请求分配新的会话标识。

会话仅用于临时存放用户信息，不要在会话中存放用户的敏感数据（如密码）。对于要永久保存的数据，应该写入文件或数据库中。

7.7.1　ISession 接口

ISession 接口规范了访问会话对象的类型成员，具体如下。

（1）Id 属性：用于返回会话的标识，此标识是唯一的。

（2）IsAvailable 属性：检查当前会话是否已成功加载。

（3）Keys 属性：该会话中包含的所有键（Key）。

（4）TryGetValue()方法：检索会话中与键（Key）对应的值。如果该项存在，则返回 true，否则返回 false。

（5）Set()方法：向会话添加新值。

（6）LoadAsync()方法：从会话的存储位置（如内存）加载会话数据。

（7）Remove()方法：如果键（Key）存在，就将与它匹配的数据删除。

（8）Clear()方法：删除当前会话中所有数据。

（9）CommitAsync()方法：将会话数据保存到存储区中（如内存）。

ISession 接口的默认实现类型是 DistributedSession，即数据存储在分布式内存中。访问 HttpContext 实例的 Session 属性可以引用 ISession 接口的实现类型实例。随后可以调用 TryGetValue()和 Set()方法读写会话数据。

下面的示例使用了两个 Razor 页面，Index 表示应用主页，Login 表示用户登录页。当客户端访问 Index 页面时，检查会话数据中是否存在名为 log_in 的项。如果存在，表明用户已经登录过，可以直接访问 Index 页面，否则跳转到 Login 页面进行登录。

具体实现步骤如下。

（1）在项目中新建目录，命名为 Pages。

（2）在 Pages 目录下添加 Razor 页面，命名为 Index.cshtml，关联的代码文件名为 Index.cshtml.cs。

```
// =============== Index.cshtml 文件 ===============
@page
@model DemoApp.Pages.IndexModel
<h2>主页</h2>
<p>欢迎访问本站</p>

// =============== Index.cshtml.cs 文件 ===============
public class IndexModel : PageModel
{
    public IActionResult OnGet()
    {
        // 检查会话数据
```

```
        ISession session = HttpContext.Session;
        if (session.TryGetValue("log_in", out byte[]? data))
        {
            if (data == null || data[0] != 0x01)
            {
                // 如果值不为 1，跳转到登录页
                return RedirectToPage("/login");
            }
            else
            {
                // 返回当前页面
                return Page();
            }
        }
        // 如果 log_in 数据项不存在，跳转到登录页
        return RedirectToPage("/login");
    }
}
```

通过 HttpContext.Session 属性获得一个 ISession 对象，随后调用 TryGetValue()方法尝试读取键名为 log_in 的值。如果数据项存在，还要检查一个值是否等于 1（0x01）。如果不等，就跳转到 Login 页面，否则返回当前页面（Index）。如果 log_in 数据项不存在，直接跳转到 Login 页面。

（3）在 Pages 目录下新建 Razor 页面，命名为 Login.cshtml，关联的代码文件名为 Login.cshtml.cs。

```
// =============== Login.cshtml 文件 ===============
@page
@model DemoApp.Pages.LoginModel

<p>
    本页面仅作演示
</p>

<div>
    单击下面按钮登录
</div>
<form method="post">
<button type="submit">登录</button>
</form>

// =============== Login.cshtml.cs 文件 ===============
…
public class LoginModel : PageModel
{
    public void OnGet()
    {
    }
```

```
        public IActionResult OnPost()
        {
            // 设置会话
            HttpContext.Session.Set("log_in", new byte[] { 0x01 });
            // 完成登录，跳回主页
            return RedirectToPage("/index");
        }
    }
```

这里的重点是编写 OnPost()方法，单击 HTML 页面上的按钮后，<form>标记会触发以 HTTP-POST 方式提交，OnPost()方法会被调用。在此方法中，先调用 Set()方法设置键名为 log_in 的会话数据项，值为 1（0x01）。最后跳转回主页（Index 页面）。

（4）打开 Program.cs 文件，在应用程序初始化时向服务容器注册以下服务。

```
builder.Services.AddRazorPages();
builder.Services.AddDataProtection();
builder.Services.AddDistributedMemoryCache();
builder.Services.AddSession(opt =>
{
    opt.IdleTimeout = TimeSpan.FromSeconds(5);
});
```

要在 ASP.NET Core 应用程序中使用会话，必须调用 AddSession()方法。本示例中将会话选项的 IdleTimeout 属性设置为 5s，表示会话会在 5s 后过期。注意会话依赖数据保护和分布式缓存技术，因此还要调用 AddDataProtection()和 AddDistributedMemoryCache()方法。

（5）在 WebApplication 对象实例化之后，需要在 HTTP 管线上调用 UseSession()方法添加会话中间件，将 ISession 对象添加到 HttpContext 对象中。

```
app.UseSession();
app.MapRazorPages();
```

UseSession()方法要在 MapRazorPages()方法之前调用，这样才能保证 ISession 对象被顺利加载，使其能在 Razor 页面代码中访问。

（6）运行示例程序，由于用户尚未登录，应用程序自动跳转到登录页，如图 7-16 所示。

（7）单击“登录“按钮，成功登录后应用程序跳回主页，如图 7-17 所示。只要会话未过期，重新访问主页时就不需要再登录了。

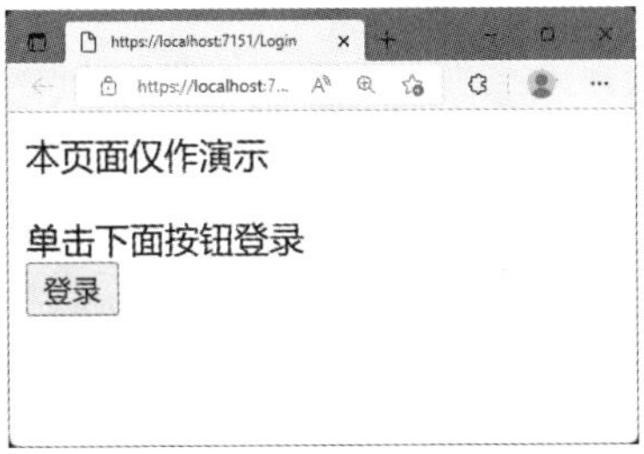

图 7-16　登录页面

图 7-17　进入主页

本示例仅用于演示，所以在登录时并不需要输入用户名和密码。

7.7.2 设置会话 Cookie 的名称

默认情况下，与会话关联的 Cookie 名称为.AspNetCore.Session，如图 7-18 所示。在服务容器上调用 AddSession()扩展方法时可以通过 SessionOptions 选项类的 Cookie 属性访问 CookieBuilder 对象，再通过 CookieBuilder 对象的 Name 属性修改默认的名称。

图 7-18 会话 Cookie 的默认名称

下面的代码将 Cookie 的名称设置为 appSession。

```
// 添加会话功能
builder.Services.AddSession(options =>
{
    options.Cookie.Name = "appSession";
});
// 必须：添加分布式缓存功能
builder.Services.AddDistributedMemoryCache();
// 必须：添加数据保护功能
builder.Services.AddDataProtection();
```

客户端访问后获得包含会话标识的 Cookie，如图 7-19 所示。

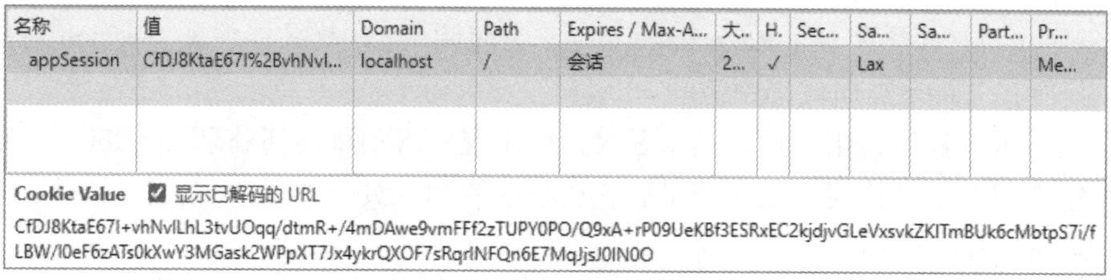

图 7-19 Cookie 的名称已变为 appSession

7.7.3 示例：将会话数据存储到 JSON 文件中

自定义会话的存储方式需要实现两个接口。

一个是 ISessionStore 接口。该接口要求实现 Create()方法返回已实现 ISession 接口的类型实例。

另一个是 ISession 接口。实现此接口的类型实例由 ISessionStore.Create()方法返回。ISession

接口要求实现以下成员。

（1）IsAvailable：只读属性，表示当前会话是否已成功加载到数据。

（2）Id：只读属性，表示当前会话的唯一标识。

（3）Keys：只读属性，获取会话数据的所有键名。

（4）TryGetValue()：该方法根据指定的键名返回对应的会话数据。

（5）LoadAsync()：该方法用于加载会话数据。

（6）Set()：该方法用于设置会话数据。

（7）Remove()：调用该方法删除指定数据项。

（8）Clear()：调用该方法清空当前会话中的所有数据项。

（9）CommitAsync()：调用该方法保存会话数据。

本示例将实现将会话数据保存到 JSON 文件中（使用 JSON 序列化），具体步骤如下。

（1）定义 SessionData 类，用于封装会话数据以及相关属性（创建时间、过期时间）。

```
internal class SessionData
{
    /// <summary>
    /// 创建时间
    /// </summary>
    public DateTime CreateTime { get; set; }
    /// <summary>
    /// 过期时间
    /// </summary>
    public DateTime Expires { get; set; }
    /// <summary>
    /// 会话数据
    /// </summary>
    public IDictionary<string, byte[]>? Data { get; set; }
}
```

Data 属性为字典类型，Key 为字符串类型，Value 是字节数组。

（2）定义 JsonSessionStore 类，用于返回 JsonSession 对象。该类实现 ISessionStore 接口。

```
public class JsonSessionStore : ISessionStore
{
    private readonly IWebHostEnvironment _env;

    public JsonSessionStore(IWebHostEnvironment env)
    {
        // IWebHostEnvironment 用于获取应用程序所在的路径
        _env = env;
    }

    public ISession Create(string sessionKey, TimeSpan idleTimeout,
    TimeSpan ioTimeout, Func<bool> tryEstablishSession,
    bool isNewSessionKey)
        {
```

```
            // 会话数据文件的存放目录
            string dir = Path.Combine(_env.ContentRootPath, "Sessions");
            if(!Directory.Exists(dir))
            {
                Directory.CreateDirectory(dir);
            }
            return new JsonSession(sessionKey, idleTimeout, ioTimeout,
            tryEstablishSession, isNewSessionKey, dir);
        }
    }
```

JsonSessionStore 类的构造函数通过依赖注入获取 IWebHostEnvironment 类型的对象引用，在 Create()方法中将用于获得当前应用程序所在目录路径。

（3）JsonSession 类的代码如下，它实现了 ISession 接口。JsonSession 类是核心，负责会话数据的读写操作。

```
public class JsonSession : ISession
{
    private readonly string sessionKey;
    private readonly TimeSpan idleTimeout, ioTimeout;
    private readonly Func<bool> tryEstablishSession;
    private readonly bool isNewSessionKey;
    private readonly string saveDir;
    private bool isAvailable;
    private SessionData sData;

    // 类构造函数
    internal JsonSession(string sessionKey, TimeSpan idleTimeout,
    TimeSpan ioTimeout, Func<bool> tryEstablishSession,
    bool isNewSessionKey, string saveDir)
    {
        // 初始化各字段
        // 会话标识
        this.sessionKey = sessionKey;
        // 表示对话数据是否可用的字段
        this.isAvailable = false;
        // 封装会话数据的对象
        this.sData = new SessionData();
        // 应用程序配置的会话的超时时限
        this.idleTimeout = idleTimeout;
        // 读写数据的超时时限
        this.ioTimeout = ioTimeout;
        // 此委托指示当前上下文能否建立会话
        this.tryEstablishSession = tryEstablishSession;
        // 是否为刚建立的新会话
        this.isNewSessionKey = isNewSessionKey;
        // 存放会话文件的目录
        this.saveDir = saveDir;
    }
```

```
// 表示当前会话是否包含可用数据
public bool IsAvailable
{
    get
    {
        // 尝试加载一次数据，以确定会话是否有可用数据
        LoadAsync().Wait();
        return isAvailable;
    }
}

// 返回当前会话的标识
public string Id => sessionKey;

// 获取所有会话数据项的键名
public IEnumerable<string> Keys
{
    get
    {
        // 尝试加载一次数据
        LoadAsync().Wait();
        return sData.Data?.Keys.AsEnumerable() ?? Enumerable.Empty
        <string>();
    }
}

// 清空会话数据
public void Clear()
{
    sData.Data?.Clear();
}

// 确认保存会话数据
public async Task CommitAsync(CancellationToken cancellationToken
= default)
{
    string fileName = Path.Combine(saveDir, $"{Id}.json");
    // 设置创建时间
    sData.CreateTime = DateTime.Now;
    // 设置过期时间
    sData.Expires = sData.CreateTime + idleTimeout;
    // 序列化并写入文件
    using (var stream = File.Create(fileName))
    {
        await JsonSerializer.SerializeAsync(stream, sData);
    }
}

// 加载会话数据
```

```
public async Task LoadAsync(CancellationToken cancellationToken =
default)
{
    // 新建的会话
    if (isNewSessionKey)
    {
        isAvailable = true;
        return;
    }

    isAvailable = false;
    string fileName = Path.Combine(saveDir, $"{Id}.json");
    if (!File.Exists(fileName))
    {
        return;
    }

    // 操作超时
    // cancellationToken 参数所指定的超时与 ioTimeout 比较
    // 其中任意一个 Token 超时，操作就会取消
    CancellationTokenSource tksrc2 = new CancellationTokenSource
    (ioTimeout);
    CancellationTokenSource csrc = CancellationTokenSource.
    CreateLinkedTokenSource(cancellationToken, tksrc2.Token);

    // 尝试反序列化
    SessionData? temp;
    using (var stream = File.OpenRead(fileName))
    {
        // 如果已取消就抛出异常
        csrc.Token.ThrowIfCancellationRequested();
        temp = await JsonSerializer.DeserializeAsync<SessionData>
        (stream, (JsonSerializerOptions?)null, csrc.Token);
    }
    if (temp is null)
    {
        // 反序列化未成功
        return;
    }
    // 仅在会话未过期时可用
    if (temp.Data != null && (temp.Expires > DateTime.Now))
    {
        isAvailable = true;
        sData = temp;
    }
}

// 删除指定的数据项
public void Remove(string key)
{
```

```
            sData?.Data?.Remove(key);
        }

        // 设置会话数据项
        public void Set(string key, byte[] value)
        {
            // tryEstablishSession 如果返回 false
            // 表示当前上下文不允许建立会话
            if (!tryEstablishSession())
            {
                throw new InvalidOperationException();
            }
            if (sData.Data == null)
            {
                sData.Data = new Dictionary<string, byte[]>();
            }
            sData.Data.Add(key, value);
        }

        // 获取会话数据项
        public bool TryGetValue(string key, [NotNullWhen(true)] out byte[]?
        value)
        {
            value = null;
            // 加载一次数据
            LoadAsync().Wait();
            if (sData.Data == null)
            {
                return false;
            }
            return sData.Data.TryGetValue(key, out value);
        }
    }
```

开发者通过 HttpContext 类的 Session 属性获取到 JsonSession 实例后，一般不会直接调用 LoadAsync()方法加载数据，而是调用 TryGetValue()、Set()方法读写会话数据。因此，在 TryGetValue()方法中调用一次 LoadAsync()方法，确保能返回最新的数据。

（4）实现 ISession 接口的类型不需要添加到服务容器中，但实现了 ISessionStore 接口的类型需要添加到服务容器中。为了方便调用，可以编写一个 AddJsonSession()扩展方法，在 Program.cs 文件中直接调用即可。

```
public static class JsonSessionServiceExtensions
{
    public static IServiceCollection AddJsonSession(this
    IServiceCollection services)
    {
        return services.AddJsonSession(_ => { });
    }

    public static IServiceCollection AddJsonSession(this IServiceCollection
    services, Action<SessionOptions> options)
```

```
    {
        services.AddDataProtection();
        services.Configure(options);
        services.AddTransient<ISessionStore, JsonSessionStore>();
        return services;
    }
}
```

（5）打开 Program.cs 代码文件，在应用程序初始化过程中注册自定义会话存储功能。

```
builder.Services.AddJsonSession(opt =>
{
    opt.IdleTimeout = TimeSpan.FromSeconds(5);
});
```

上述代码通过 SessionOptions 选项类设置会话的过期时间为 5s。

（6）当访问服务器的客户端多了时，会产生很多 JSON 文件。为了能及时删除已过期的会话文件，可以考虑编写一个服务每隔 30s（时间间隔可根据实际应用情形设定，此处仅作演示）清理一次文件。

```
public class CustTimerService : IHostedService, IAsyncDisposable
{
    private Timer? myTimer;
    private readonly string fileDir;

    public CustTimerService(IWebHostEnvironment env)
    {
        // 获取会话文件的存储目录
        fileDir = Path.Combine(env.ContentRootPath, "Sessions");
    }

    public Task StartAsync(CancellationToken cancellationToken)
    {
        // 启动计时器
        myTimer = new(OnTimer, null, 0, 30_000);
        return Task.CompletedTask;
    }

    public Task StopAsync(CancellationToken cancellationToken)
    {
        // 停止计时器
        myTimer?.Change(Timeout.Infinite, 0);
        return Task.CompletedTask;
    }

    private void OnTimer(object? state)
    {
        // 删除过期的文件
        foreach (string file in Directory.GetFiles(fileDir, "*.json"))
        {
            bool expired = false;
            using (var stream = File.OpenRead(file))
```

```
            {
                SessionData? data = JsonSerializer.Deserialize<SessionData>
                (stream);
                if(data != null)
                {
                    // 测试是否过期
                    expired = data.Expires < DateTime.Now;
                }
            }
            if(expired)
            {
                // 已过期，删除文件
                File.Delete(file);
            }
        }
    }

    public async ValueTask DisposeAsync()
    {
        if (myTimer is IAsyncDisposable disp)
        {
            // 释放计时器实例
            await myTimer.DisposeAsync();
        }
        myTimer = null;
    }
}
```

控制每隔一段时间执行一次代码，可以使用计时器——Timer 类（位于 System.Threading 命名空间）。该类实例化后就自动运行，要使它停止运行，可以调用 Change()方法将执行延时设置为无限大，循环周期间隔为 0。

（7）回到 Program.cs 文件，将 CustTimerService 类也添加到服务容器中。

```
builder.Services.AddHostedService<CustTimerService>();
```

（8）运行示例程序。第 1 次访问时刚建立会话，在 Web 浏览器中刷新一下，再次访问服务器，就会显示上次设置的会话数据，如图 7-20 所示。

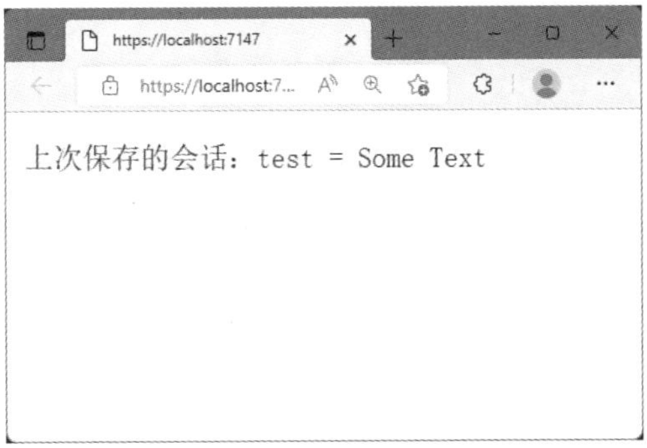

图 7-20　显示已设置的会话数据

第 8 章　Razor　页　面

本章要点

- ❑ Razor 标记基础
- ❑ Razor 页面文件的结构
- ❑ Razor 页面模型类
- ❑ 页面路由
- ❑ 页面处理程序（Handler）

8.1　Razor 页面的特点

Razor 页面（Razor　Pages）也是 MVC 框架的分支，因此它有许多地方与 MVC 是既相似又通用的。Razor 页面具有以下特点。

（1）以页面为基本单位，侧重点是网页。对于功能逻辑不复杂的应用场景，Razor 页面易使用且开发效率高。

（2）与 MVC 相比，Razor 页面不需要创建控制器，也可以不带模型类。只要有一个 Razor 文件（扩展名为.cshtml）即可。

（3）页面文件的相对路径（默认是 Pages 目录下的文件）可以直接作为 URL 使用，规则简单。

（4）使用特定的 Handler 方法处理请求。

（5）与 MVC 共享部分功能，如标记帮助器（Tag Helper）、模型绑定（Model Binding）等。

Razor 页面适用于小型应用程序项目，或者对数据模型没有严格要求的应用程序。基于页面构建应用功能，有较高的自由度和灵活性。对于 ASP.NET Core 的初学者，通过 Razor 页面了解 Razor 语法会变得更轻松。

本章讲述 Razor 页面相关的内容，MVC 将在第 9 章介绍。

8.2　Razor 语法

Razor 是一种语言标记，其作用是把.NET 代码插入 HTML 标记中。当在服务器上执行时，Razor 文档中的 HTML 会原封不动地呈现到 HTML 文档中，而.NET 代码（如 C#代码）会先执行，最后把生成的 HTML 按照其出现的位置插入 HTML 文档中。

Razor 标记以字符@开头，其后紧跟代码表达式或代码块。例如，下面的代码通过 Razor 标记插入一段 for 循环，在循环体内部生成<p>元素，元素内的文本为变量 i 的值。

```
<div>
    @for (int i = 1; i <= 10; i++)
    {
        <p>@i</p>
    }
</div>
```

变量 i 的初始值为 1，循环条件是 i 小于或等于 10，因此上述 for 语句将执行 10 次，i 的值依次是 1，2，…，10。循环执行会生成以下 HTML。

```
<div>
<p>1</p>
<p>2</p>
<p>3</p>
<p>4</p>
<p>5</p>
<p>6</p>
<p>7</p>
<p>8</p>
<p>9</p>
<p>10</p>
</div>
```

再举一例，下面的代码先声明变量 a、b，然后在生成 HTML 时分别呈现两个变量的值，以及它们的和。

```
@{
    double a = 5.65d;
    double b = 12.003d;
}
<span>@a + @b = @(a + b)</span>
```

@a 表示呈现变量 a 的值，@b 表示呈现变量 b 的值，@(a+b)则表示呈现它们相加后的结果。最终生成的 HTML 如下。

```
<span>5.65 + 12.003 = 17.653</span>
```

8.2.1　两种表达式

Razor 表达式有隐式和显式之分。

隐式表达式在符号@后直接跟 C#表达式，但整个表达式中不能有空格（await 表达式除外），也无法使用算术运算符（如+、*等）和泛型。例如，下面的代码将在<div>元素内呈现当前时间。

```
<div>@DateTime.Now.ToString()</div>
```

显式表达式在符号@后紧跟一对圆括号（小括号），C#表达式写在括号内。此表达式可以解决不能出现空格、算术运算符、泛型等的问题。例如，下面的代码呈现 3 个变量的积。

```
@{
    int a = 5, b = 4, c = 6;
}
<p>计算结果：@(a * b * c)</p>
```

@(a * b * c)为显式表达式，不仅出现了空格，还出现了乘法运算符（*）。有了一对圆括号，编译器就能分析出 a * b * c 是一个整体，属于同一表达式。

下面的代码调用了泛型方法，只能使用显式表达式。

```
<div>@(item.GetValue<int>().ToString())</div>
```

隐式表达式适用于对变量的引用，或简单的对象成员访问（如呈现某个属性的值，调用非泛型方法）；显式表达式适用于各种复杂的表达式（如调用泛型方法、带运算符的表达式等）。

Razor 表达式的结尾不需要分号（;），也可以写在 Razor 代码块内部，如

```
@{
<p>6 的立方是：@Math.Pow(6, 3)</p>
<p>当前系统平台：@Environment.OSVersion.Platform</p>
}
```

生成的 HTML 如下。

```
<p>6 的立方是：216</p>
<p>当前系统平台：Win32NT</p>
```

8.2.2 代码块

代码块包含在一对大括号中，紧跟在符号@之后，即

```
@{

}
```

在代码块中，完整的 C#语句必须以分号（;）结尾，与 HTML 混合的 C#表达式不需要分号结尾。例如：

```
@{
<span>@DateTime.Today</span>
    int x = 145;
    if(x % 2 == 0)
```

```
    {

    }
    else
    {

    }
}
```

注意符号@之后不能出现换行符，下面的代码会发生错误。

```
@
{
    string? path = Environment.GetEnvironmentVariable("PATH");
}
```

符号@与大括号之间也不能有空格，下面的代码也会发生错误。

```
@  {
    string? path = Environment.GetEnvironmentVariable("PATH");
}
```

在代码块的内部呈现不包含 HTML 的纯文本内容会出现编译错误。例如：

```
@{
    string curDir = Directory.GetCurrentDirectory();
    当前目录：@curDir                          // 此处将引发错误
}
```

解决方法是在行首加上“@:”，即符号@后紧跟一个冒号。把上述代码修改为

```
@{
    string curDir = Directory.GetCurrentDirectory();
    @:当前目录：@curDir
}
```

8.2.3 注释

Razor 标记支持使用 C#代码注释和 HTML 注释。被注释后的 C#代码不会执行，由代码生成的 HTML 不会输出。但 HTML 注释仍然会输出到 HTML 文档中。请思考下面的代码。

```
@{
    //int x = 5, y = 10;
    //<div>@(x + y)</div>
<!-- HTML 注释 1 -->
}
<!-- HTML 注释 2 -->
```

由于代码被注释，因此<div>元素不会输出，但两处 HTML 注释依旧会呈现在最终的 HTML 文档中，如图 8-1 所示。

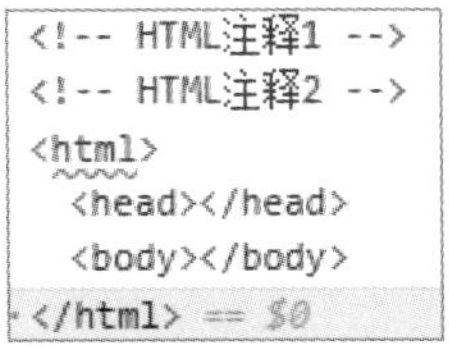

图 8-1　HTML 注释正常输出

若希望阻止 Razor 引擎生成 HTML 注释，可以将整段代码都放在@*和*@之间。这样处理后，Razor 引擎在生成时会删除

@*和*@之间的所有内容。以下代码在运行后不会生成任何内容。

```
@*
<div>...</div>
<span>@DateTime.Today</span>
@{
    uint m = 8000;
    uint n = 5590;
    uint r = m - n;
}
<p>@m - @n = @r</p>
*@
```

8.2.4 流程控制

流程控制语句即 if、for、foreach、switch、while 等语句。在 Razor 标记中控制语句有两种格式。

第 1 种格式是写在@{...}内，请看下面的示例。

```
@{
    Random rand = new((int)DateTime.Now.Ticks);
    // 生成随机整数
    int num = rand.Next(1, 10);

    <h4>抽奖结束，您将获得以下幸运奖品：</h4>
    switch(num)
    {
        case 1:
            <p>洗衣机</p>
            break;
        case 2:
            <p>电热水壶</p>
            break;
        case 3:
            <p>木制筷子一套</p>
            break;
        case 4:
            <p>吸尘器</p>
            break;
        case 5:
            <p>电风扇</p>
            break;
        case 6:
            <p>家用投影仪</p>
            break;
        case 7:
            <p>电吹风</p>
            break;
        default:
            <p>下次再试试手气吧</p>
```

```
            break;
    }
}
```

上述代码先生成一个随机整数，然后使用 switch 语句根据整数值输出不同的 HTML 内容。

第 2 种格式是省略@{...}，直接将 switch 语句写在符号@之后。

```
@{
    Random rand = new((int)DateTime.Now.Ticks);
    // 生成随机整数
    int num = rand.Next(1, 10);
}

<h4>抽奖结束，您将获得以下幸运奖品：</h4>
@switch (num)
{
    case 1:
        <p>洗衣机</p>
        break;
    case 2:
        <p>电热水壶</p>
        break;
    case 3:
        <p>木制筷子一套</p>
        break;
    case 4:
        <p>吸尘器</p>
        break;
    case 5:
        <p>电风扇</p>
        break;
    case 6:
        <p>家用投影仪</p>
        break;
    case 7:
        <p>电吹风</p>
        break;
    default:
        <p>下次再试试手气吧</p>
        break;
}
```

运行上述代码，如图 8-2 所示。

图 8-2　switch 语句示例

8.3　开启 Razor 页面功能

在 ASP.NET Core 项目中使用 Razor 页面，需要完成以下两步。

（1）在服务容器上调用 AddRazorPages()扩展方法，注册相关服务类型。

（2）在 HTTP 管线上调用 MapRazorPages()扩展方法，添加终结点映射，使 URL 路由能够定位 Razor 页面文件。

详细代码示例如下。

```
var builder = WebApplication.CreateBuilder(args);
// 向服务容器注册相关服务类型
builder.Services.AddRazorPages();
// 构建应用程序
var app = builder.Build();

// 映射终结点
app.MapRazorPages();

// 开始运行应用程序
app.Run();
```

8.4 Razor 页面文件

Razor 页面文件扩展名为.cshtml，即 C#代码和 HTML 的混合。Razor 标记语法不仅可用于 Razor 页面，还可以用于 MVC 的视图文件、Razor 组件（如 Blazor 组件）等。因此，如果 Razor 文档是明确用于 Razor 页面的，必须在代码的第 1 行加上@page 指令。例如：

```
@page

<h2>示例代码</h2>
```

下面给出一个完整的示例。

（1）在 Program.cs 文件中注册 Razor 页面相关功能，并映射终结点。

```
var builder = WebApplication.CreateBuilder(args);
// 注册服务
builder.Services.AddRazorPages();
var app = builder.Build();

// 映射终结点
app.MapRazorPages();

app.Run();
```

（2）在项目中创建文件夹，命名为 Pages。

（3）在 Pages 目录下添加新页面文件，命名为 Musics.cshtml。然后输入以下代码。

```
@page

<h3>我的音乐收藏</h3>
<hr/>
<ol>
```

```
<li>金</li>
<li>木</li>
<li>水</li>
<li>火</li>
<li>土</li>
</ol>
```

（4）在 Pages 目录下创建 Movies.cshtml 文件，代码如下。

```
@page

<h3>我的电影</h3>
<hr/>
<ul>
<li>电影 1</li>
<li>电影 2</li>
<li>电影 3</li>
</ul>
```

（5）运行应用程序。假设 Web 服务器的根 URL 为 https://localhost:7340，那么访问 Musics 页面的 URL 为 https://localhost:7340/musics（URL 不区分大小写），如图 8-3 所示。

（6）Movies 页面的访问 URL 为 https://localhost:7340/movies，如图 8-4 所示。

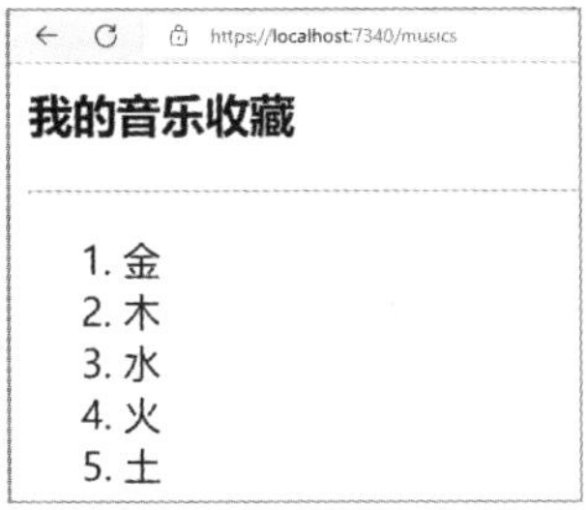

图 8-3　Musics 页面

图 8-4　Movies 页面

可以使用 webapp 模板直接创建包含 Razor 页面基本代码的应用程序项目。执行下面的命令将创建 TestApp 项目。

```
dotnet new webapp -n TestApp -o .
```

8.5　页面文件的搜索路径

ASP.NET Core 应用程序默认会在项目的 Pages 目录下查找 Razor 页面文件。页面的 URL 将由 Pages 目录的子级的相对路径组成。例如，Pages 目录下有 Products 目录，Products 目录下有 Show.cshtml 页面文件，那么 Show 页面的相对根 URL 的地址为/products/show。

本节将介绍如何修改 Razor 页面文件的搜索路径。

8.5.1 配置 RazorPagesOptions 选项类

RazorPagesOptions 类公开了 RootDirectory 属性，可用于配置 Razor 页面文件的根目录（即文件搜索目录），其默认值为/Pages。

下面的示例将页面文件的根目录修改为/Blogs。

```
builder.Services.AddRazorPages(options =>
{
    options.RootDirectory = "/Blogs";
});
```

在项目中新建名为 Blogs 的文件夹。在 Blogs 目录下添加名为 List.cshtml 的页面文件，代码如下。

```
@page

<h5>最近热门博客推荐</h5>

<ol>
<li><a href="#">小高的博客</a></li>
<li><a href="#">小王的博客</a></li>
<li><a href="#">小红的博客</a></li>
<li><a href="#">小田的博客</a></li>
<li><a href="#">小章的博客</a></li>
<li><a href="#">小雷的博客</a></li>
</ol>
```

图 8-5　Blogs 目录中的页面

List.cshtml 页面相对于根 URL 的地址为/list，如图 8-5 所示。

8.5.2 便捷的扩展方法

有两个扩展方法可以很方便地设置 Razor 页面的搜索路径，分别如下。

（1）WithRazorPagesRoot()方法：直接通过 rootDirectory 参数设置页面文件的根目录。

（2）WithRazorPagesAtContentRoot()方法：将当前应用程序的内容根目录作为 Razor 页面的根目录，即 Razor 页面文件将位于项目目录下。

下面的示例将设置 Razor 页面的根目录与应用程序的内容根目录相同。

```
builder.Services.AddRazorPages().WithRazorPagesAtContentRoot();
```

此时，可以直接在项目的根目录下添加 Razor 页面文件。例如，添加一个名为 News.cshtml 的文件，代码如下。

```
@page

<h3>新闻资讯</h3>

<div style="margin-top:7px; background-color:lightyellow">
```

```
<ul>
<li>示例新闻 1</li>
<li>示例新闻 2</li>
<li>示例新闻 3</li>
</ul>
</div>
```

访问 News.cshtml 页面的相对 URL 为/news，如图 8-6 所示。

图 8-6　位于根目录下的 News 页面

8.6　页面路由

默认情况下，Razor 页面的 URL 是基于应用程序配置的根目录（如/Pages）的相对路径。若有特殊需要，可以通过配置路由规则修改页面的 URL。

8.6.1　通过@page 指令设置路由规则

设置 URL 路由规则最简单的方法就是使用@page 指令，在指令后可直接指定路由。例如：

```
@page "/abc"
```

下面的示例将实现缩短页面文件路径的功能，将路径/pictures/items 映射为/pictures。

（1）在 Program.cs 文件中注册相关服务，然后在 HTTP 管线上配置终结点映射。

```
var builder = WebApplication.CreateBuilder(args);
// 注册服务
builder.Services.AddRazorPages();
var app = builder.Build();
// 终结点映射
app.MapRazorPages();

app.Run();
```

（2）在项目目录下新建 Pages 目录。

（3）在 Pages 目录下新建 Pictures 子目录。

（4）在 Pictures 目录下新建 Items.cshtml 文件。代码如下。

```
@page "/pictures"

<h3>个人图册</h3>

<div>
    当前路径：@Path
</div>
```

在@page 指令后（有空格）使用路由规则将当前页面的 URL 定义为/pictures。同时，上述代码输出 Path 属性（当前页面类的属性，编译时自动继承 Microsoft.AspNetCore.Mvc.RazorPages.Page 类）的值，程序运行后就能看该页面的原始路径。

（5）运行应用程序，在浏览器中输入 https://<根 URL>/pictures，结果如图 8-7 所示。

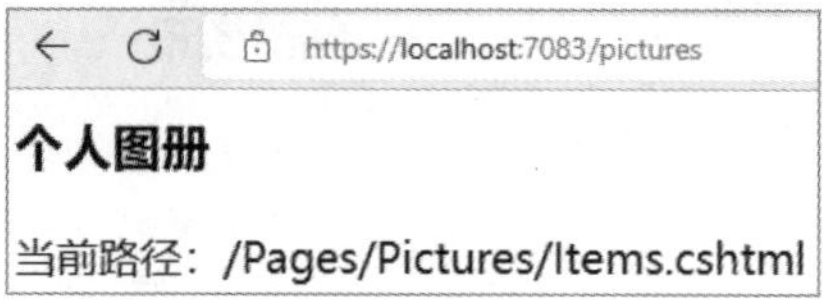

图 8-7　通过路径映射的新路径访问 Items 页面

8.6.2　通过约定模型定义路由规则

RazorPagesOptions 选项类公开 Conventions 属性，表示一个 PageConvention 集合，用于添加内置或自定义的页面约定。

页面约定的公共接口为 IPageConvention。此接口未定义任何成员，一般在自定义方案中不会使用该接口，而是实现从 IPageConvention 接口派生的 3 个接口。

（1）IPageApplicationModelConvention 接口：用于自定义 PageApplicationModel 对象。

（2）IPageHandlerModelConvention 接口：用于自定义 PageHandlerModel 对象。

（3）IPageRouteModelConvention 接口：用于自定义 PageRouteModel 对象。

为了方便开发人员调用，PageConventionCollection 类针对不同的自定义方案公开了对应的方法成员。

（1）AddPageApplicationModelConvention()方法：为指定页面添加可自定义的 PageApplicationModel 对象。

（2）AddPageRouteModelConvention()方法：为指定页面添加自定义的 PageRouteModel 对象。

（3）AddFolderRouteModelConvention()方法：将自定义的 PageRouteModel 对象应用到指定目录的所有页面。

（4）AddFolderApplicationModelConvention()方法：将自定义的 PageApplicationModel 对象应用到指定目录下的所有页面。

AddAreaFolderApplicationModelConvention()、AddAreaFolderRouteModelConvention()、AddAreaPageApplicationModelConvention()、AddAreaPageRouteModelConvention()这 4 个方法与上述 4 个方法含义相同，只是多了一个区域名称（Area Name）。

要通过页面约定自定义路由规则，需要用到 AddPageRouteModelConvention()方法。下面的示例将通过约定将根 URL(/)指向 Default 页面(/Default)，即访问根地址时默认呈现 Default 页面。具体步骤如下。

（1）在项目中创建 Pages 目录。

（2）在 Pages 目录下新建页面文件，命名为 Default.cshtml。然后输入以下代码。

```
@page

<h3>我的主页</h3>

@* 如果存在路由参数 ver，就将其输出 *@
@if(RouteData.Values.ContainsKey("ver"))
{
<div>当前版本：@RouteData.Values["ver"]</div>
}
```

RouteData 属性用于获取当前页面关联的路由参数（当页面文件被执行时，表明针对该页面的路由规则已成功匹配），本示例将获取名为 ver 的路由参数。如果路由规则中存在此参数，就将它的值呈现到页面上。

（3）在 Program.cs 文件中，通过 RazorPagesOptions 选项类的 Conventions 属性为 Default 页面配置路由规则。

```
builder.Services.AddRazorPages(opt =>
{
    opt.Conventions.AddPageRouteModelConvention("/default", model =>
    {
        // 清空原有 Selector
        model.Selectors.Clear();
        // 创建新的 Selector
        SelectorModel newSelector = new()
        {
            AttributeRouteModel = new AttributeRouteModel()
            {
                // 路由规则
                Template = "/{ver?}"
            }
        };
        model.Selectors.Add(newSelector);
    });
});
```

AddPageRouteModelConvention()方法的第 1 个参数是要设置路由规则的页面（本例为 /default，不区分大小写）；第 2 个参数是一个委托对象，其输入参数为 PageRouteModel 实例。

此处的关键是修改 Selectors 属性中的 SelectorModel 对象。AttributeRouteModel 对象的 Template 属性表示路由规则。本示例将设置为/{ver?}，大括号（{}）里面是路由参数，名为 ver。参数名后面的问号（?）表示该参数是可选的，在匹配 URL 时如果没有提供 ver 参数也能匹配成功。

（4）运行示例程序，假设应用程序的 URL 为 https://localhost:7092/，直接以该 URL 访问时表示定义应用程序的根路径（/），根据路由规则可成功匹配页面/default，因此呈现 Default 页面的内容（此时不存在路由参数 ver），如图 8-8 所示。

（5）将访问 URL 改为 https://localhost:7092/6.2，这时 URL 中的 6.2 成功能匹配路由参数 ver，Default 页面会呈现 ver 参数的值，如图 8-9 所示。

图 8-8　未提供路由参数 ver

图 8-9　呈现 ver 参数的值

另外，使用 AddPageRoute()扩展方法向页面约定集合添加页面路由会更简单。因此，本示例在 AddRazorPages()方法中的代码可以进行以下修改。

```
builder.Services.AddRazorPages(opt =>
{
    opt.Conventions.AddPageRoute("/default", "/{ver?}");
});
```

可见，调用 AddPageRoute()扩展方法非常简单，pageName 参数表示要设置路由的页面，此处是/default；route 参数（/{ver?}）指定路由规则。

8.7　页面模型类

Razor 页面允许使用@model 指令关联一个类，以实现用户界面与逻辑代码分离。用于页面模型的类型只要满足以下其中一个条件即可。

（1）继承 PageModel 类（位于 Microsoft.AspNetCore.Mvc.RazorPages 命名空间）。

（2）在类上应用 PageModelAttribute（位于 Microsoft.AspNetCore.Mvc.RazorPages.Infrastructure 命名空间）。

把页面的逻辑处理代码放到模型类中，方便后期维护。当要对代码逻辑进行修改时，只需要修改模型类中的代码，用于呈现用户界面的 Razor 标记文档不需要修改。Razor 标记可以与模型类的属性进行绑定，也可以关联特定 HTTP 请求方式的处理方法（Handler）。例如，当以 HTTP-GET 方式请求页面时，会执行默认的 OnGet 方法（方法在模型类中定义，若存在则执行）。或者当页面中的<form>元素以 HTTP-POST 方式向服务器提交数据时，会调用默认的

OnPost 方法（也是在模型类中定义的方法）。

如果没有为 Razor 页面指定模型类，那么它会把页面自身作为模型类。这是因为在编译时，Razor 页面文件（扩展名为.cshtml）会自动生成一个类，类名由页面文件所处的相对路径组成。例如，对于页面文件/Pages/Index.cshtml，编译后生成的类名就是 Pages_Index。也就是说，如果 Index 页面没有显式指定模型类，那么它的默认模型类就是 Pages_Index。

8.7.1　页面自身作为模型类

当 Razor 页面以其自身所生成的类作为模型类时，逻辑代码不能写在一般的 Razor 代码块（@{…}）中，必须写在@functions 代码块中。请看下面的例子。

```
@page

<div>@SomeMessage</div>

@functions{
    // 公共属性
    public string? SomeMessage{ get; set; }

    public void OnGet()
    {
        // 设置属性值
        SomeMessage = "Some text data";
    }
}
```

上述代码为页面模型定义了一个公共属性 SomeMessage，在 OnGet()方法中为该属性赋值。OnGet()方法就是该页面处理 HTTP-GET 请求的 Handler。当客户端用 GET 方式请求页面时，会先执行 OnGet()方法，然后再初始化 Razor 页面上的 HTML 文档。当<div>元素要获取 SomeMessage 属性的值时，OnGet()方法已执行过，所以 SomeMessage 属性的值能够呈现在网页上，如图 8-10 所示。

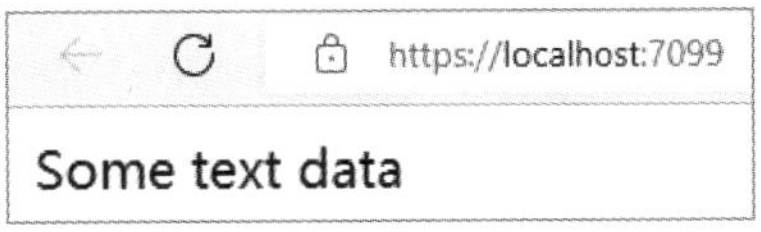

图 8-10　呈现 SomeMessage 属性的值

8.7.2　从 PageModel 派生类

为了实现用户界面设计与代码分离，便于维护和管理，模型类最好单独实现。最常见的做法是定义一个从 PageModel 派生的类。在 Razor 页面文件中，使用@model 指令关联模型类。

在文件的命名上，模型类所在的代码文件名一般与 Razor 页面相同，并加上扩展名.cs（C# 代码文件）。例如，Razor 页面文件命名为 About.cshtml，那么与之关联的模型类通常命名为 AboutModel，所在的代码文件名为 About.cshtml.cs。

当然，以上所述的命名方式仅仅是一种习惯，并非强制性规则。以下示例并没有遵循以上命名方式，但不影响项目的运行。如图 8-11 所示，Index 和 About 页面位于 Pages 目录下（默认），而与它们关联的模型类 IndexModel、AboutModel 位于 PageModels.cs 文件中。

Index.cshtml 文件内容如下。

```
@page
@model MyModels.IndexModel

<h3>@Model.HeaderText</h3>

<a href="/about">关于本站</a>
```

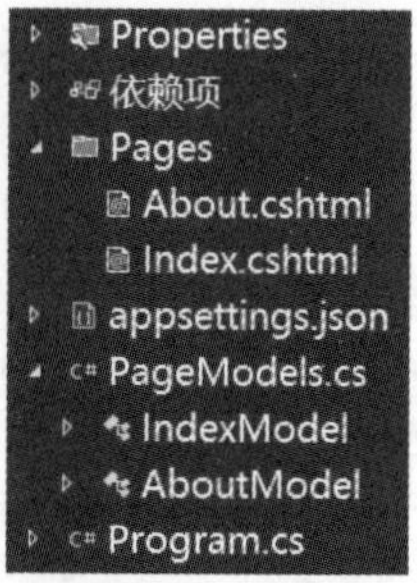

图 8-11　项目结构

通过@model 指令关联 IndexModel 类，并且在页面中输出 IndexModel 实例的 HeaderText 属性值。IndexModel 类的代码如下。

```
public class IndexModel : PageModel
{
    public string? HeaderText { get; set; }

    public void OnGet()
    {
        HeaderText = "欢迎光临";
    }
}
```

About.cshtml 页面也类似，对应的模型类是 AboutModel。详细代码如下。

```
// About.cshtml
@page
@model MyModels.AboutModel

<h3>@Model.PageTitle</h3>

// AboutModel 类
public class AboutModel : PageModel
{
    public string? PageTitle { get; set; }

    public void OnGet()
    {
        PageTitle = "关于本项目";
    }
}
```

8.7.3　通过特性类实现页面模型类

如果某个类应用了 PageModel 特性（类名为 PageModelAttribute），即使它没有从 PageModel 类派生，也可以用作 Razor 页面的模型类。

下面的代码演示了名为 Home.cshtml 的页面文件，它关联的模型类是 HomeModel。

```
@page
@model DemoApp.HomeModel

<p>随机数：@Model.RandVal</p>
```

RandVal 属性来自 HomeModel 类，它的定义如下。

```
[PageModel]
public class HomeModel
{
    public int RandVal { get; set; }

    public void OnGet()
    {
        // 产生随机数
        RandVal = RandomNumberGenerator.GetInt32(10, 90000);
    }
}
```

由于 HomeModel 类并非派生自 PageModel 类，如果要让其被识别为页面模型类，就必须在类上应用 PageModelAttribute。

在 OnGet()方法中，使用了 RandomNumberGenerator 类（位于 System.Security.Cryptography 命名空间）生成随机整数。

页面的呈现效果如图 8-12 所示。

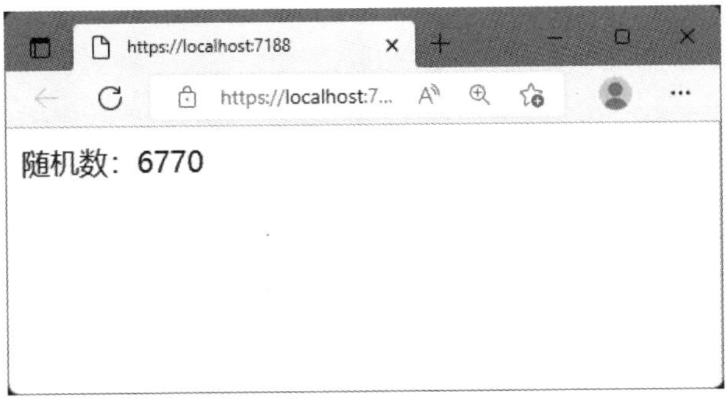

图 8-12　输出随机整数

8.8　页面处理程序

页面处理程序（称为 Handler）是一个或多个在页面模型类中定义的方法，用于处理由用户操作而触发的 HTTP 请求。例如，用户在登录页面中输入了用户名和密码，单击“登录”按钮后会触发一个以 HTTP-POST 方式发送的请求。服务器会将此请求转发给特定的处理程序，即调用相关的方法。

用于页面处理程序的方法成员有特殊的命名约定，格式如下。

```
On<HTTP 请求方式>[Handler 名称][Async]
```

（1）方法必须是公共方法，不能是构造函数，不能是静态方法。

（2）方法的名称以 On 开头。

（3）HTTP 请求方式要求至少出现一个字符，通常首字母大写，如 Get、Post、Put 等。

（4）处理程序名称（即 Handler 名称部分）是可选的，如果存在，则必须首字母大写。

（5）末尾的 Async 是可选的，若方法返回 Task 类型，可以加上此后缀，表明它是异步方法，如 OnGetMusicsAsync。

以下命名都可以被识别为 Handler（页面处理程序）。

```
OnGet
OnDelete
OnPost
OnPostAddUser
OnPostUpdateAsync
```

以下命名不会被识别为页面的 Handler。

```
GetStudents      // 没有以 On 开头
OnWork           // 此命名会被识别成 HTTP 请求方式为 Work，无 Handler 名称。如果客户
                 // 端发出的请求是常见的 GET、POST 等，那么此方法不会被调用。除非客
                 // 户端在发出请求时设置 HTTP 请求方式为 Work
OnAsync          // 只有开头的 On 和结尾的 Async，相当于 HTTP 请求方式和 Handler 名
                 // 称都不存在
```

8.8.1 通用的处理程序

没有指定 Handler 名称（仅给定 HTTP 请求方式）的方法成员为通用处理程序，ASP.NET Core 应用程序在匹配到有效的 HTTP 请求方式后会自动调用，如 OnGet()（当请求方式为 GET 时自动调用）、OnPost()（当请求方式为 POST 时自动调用）。

下面的代码演示 Index 页面的模型类，其中包含通用的处理程序——OnGet()方法。当客户端（Web 浏览器）以 HTTP-GET 方式访问时会自动调用该方法。

```
public class IndexModel : PageModel
{
    public void OnGet()
    {
        ViewData["data1"] = "测试文本";
        ViewData["data2"] = 15563.21d;
    }
}
```

在 OnGet()方法中，向 ViewData 属性写入两个元素，它们的 Key 分别为 data1 和 data2。ViewData 称为“视图数据”，可用于缓存在视图/页面呈现过程中要用到的对象（数据）。它是字典结构，Key 是字符串类型，Value 可以是任意类型（Object）。

OnGet()方法在 Index 页面呈现之前被调用，因此在 Index 页面中可以读取 ViewData 中的对象。

```
@{
    // 获取数据
    string x1 = Convert.ToString(ViewData["data1"]) ?? "<no data>";
```

```
    double x2 = Convert.ToDouble(ViewData
["data2"]);
}

<p>数据项 1：@x1</p>
<p>数据项 2：@x2</p>
```

呈现效果如图 8-13 所示。

图 8-13　从 ViewData 中读到的内容

8.8.2　解决 POST 请求时出现的错误

先看下面的示例，假设页面上有<form>元素，以 POST 方式提交。用户输入城市名称后，单击按钮进行提交。

```
<div>请输入要查询的城市：</div>
<form method="post">
<input type="text" name="city"/>
<button type="submit">提交</button>
</form>
```

如果上述页面上的<form>元素触发提交操作，关联的页面模型类中会自动调用 OnPost() 或 OnPostAsync()方法。

```
public ActionResult OnPost(string city)
{
    return Content($"你输入的城市是：{city}");
}
```

由于要通过 Content()方法返回 HTML 文本，OnPost()方法的返回值类型就不能是 void 了，需要改为 IActionResult 或 ActionResult。此示例在逻辑处理上比较简单，直接将用户提交的城市名称（city 字段）组成新的字符串实例返回。OnPost()方法要接收<input>元素的值，只需要令参数名与<input>元素的 name 特性值相同即可，本例中是 city。

此时运行示例，会发现表单提交后服务器返回 400 状态码。在 ASP.NET Core 应用中，像 MVC、Razor 页面等功能默认都使用 Anti-Forgery Token 验证，以防止跨站请求伪造攻击（Cross-Site Request Forgery，CSRF）。而<form>元素在发送请求时并没有携带有效的防伪令牌，在服务器上无法通过验证，便返回 400 状态码。

第 1 种解决方案是在模型类上应用 IgnoreAntiforgeryToken 特性类。例如：

```
[IgnoreAntiforgeryToken]
public class HomeModel : PageModel
{
    …

    public ActionResult OnPost(string city)
    {
        return Content($"你输入的城市是：{city}");
```

```
    }
}
```

不过，此方案会降低安全性，本书推荐第 2 种方案——使用 Tag Helper（标记帮助器）。在 Razor 页面文件中通过@addTagHelper 指令将 FormTagHelper 标记帮助器导入。

```
@addTagHelper Microsoft.AspNetCore.Mvc.TagHelpers.FormTagHelper,
Microsoft.AspNetCore.Mvc.TagHelpers
```

@addTagHelp 指令有两个值（用英文逗号隔开）：第 1 个值是 FormTagHelper 类的完整命名（包括命名空间）；第 2 个值是该类所在程序集的名称，上述示例中是 Microsoft.AspNetCore.Mvc.TagHelpers。如果想把该程序集下面的所有帮助器类都导入，可以将第 1 个值设置为*。

```
@addTagHelper *, Microsoft.AspNetCore.Mvc.TagHelpers
```

FormTagHelper 标记帮助器会自动与<form>元素关联，而且会自动处理 Anti-Forgery 令牌的验证。

当 Razor 页面呈现到 Web 浏览器后，<form>元素下会自动添加一个隐藏字段（默认名称为__RequestVerificationToken），其中包含加密后的防伪令牌。

```
<form method="post">
<input type="text" name="city">
<button type="submit">提交</button>
<input name="__RequestVerificationToken" type="hidden" value="<加密内容>">
</form>
```

8.8.3 使用多个处理程序

在一个页面模型类中，除了通用的 OnGet()、OnPost()、OnPut()等方法外，开发人员可以根据需要自定义处理程序。Razor 页面文件可以针对不同应用目的选择对应的处理程序。

下面将通过一个示例阐述具体做法。

（1）添加 MainPort 页面。它的模型类定义如下。

```
public class MainPortModel : PageModel
{
    // 0：管理员
    // 1：普通用户
    // 2：游客
    public int RoleType { get; set; }

    public void OnGetAdmin()
    {
        RoleType = 0;
    }

    public void OnGetUser()
    {
        RoleType = 1;
```

```
    }

    public void OnGetGuest()
    {
        RoleType = 2;
    }
}
```

RoleType 属性是一个整数值，0 表示进入该页面的是管理员；1 表示普通用户；2 表示游客（未注册的用户）。

3 个处理程序都是面向 HTTP-GET 请求的，方法成员分别为 OnGetAdmin()、OnGetUser()、OnGetGuest()，对应的 Handler 名称分别为 Admin、User、Guest。

在 MainPort.cshtml 文件中，读取 RoleType 属性的值，并根据此值呈现不同的信息。

```
@if(Model.RoleType == 0)
{
<p>欢迎回来，管理员</p>
}
else if(Model.RoleType == 1)
{
<p>欢迎回来，普通用户</p>
}
else
{
<p>欢迎，您是游客</p>
}
```

（2）添加 Index 页面。在页面上声明 3 个<a>元素，它们都指向 MainPort 页面，但使用不同的处理程序。

```
@page
@addTagHelper *, Microsoft.AspNetCore.Mvc.TagHelpers

<div>请选择角色：</div>
<p>
<a asp-page="MainPort" asp-page-handler="admin">我是管理员</a>
</p>
<p>
<a asp-page="MainPort" asp-page-handler="user">我是普通用户</a>
</p>
<p>
<a asp-page="MainPort" asp-page-handler="guest">我是游客</a>
</p>
```

asp-page 和 asp-page-handler 两个特性来自 AnchorTagHelper 标记帮助器，因此在文件中需要使用@addTagHelper 指令将 ASP.NET Core 框架自带的所有标记帮助器全部导入。也可以只导入 AnchorTagHelper。

```
@addTagHelper Microsoft.AspNetCore.Mvc.TagHelpers.AnchorTagHelper,
Microsoft.AspNetCore.Mvc.TagHelpers
```

asp-handler 特性用于指明要使用的处理程序名称，其值不区分大小写。

（3）运行示例程序。

（4）在页面上单击“我是管理员”链接，如图 8-14 所示。随后应用程序跳转到 MainPort 页面，并输出“欢迎回来，管理员”，如图 8-15 所示。

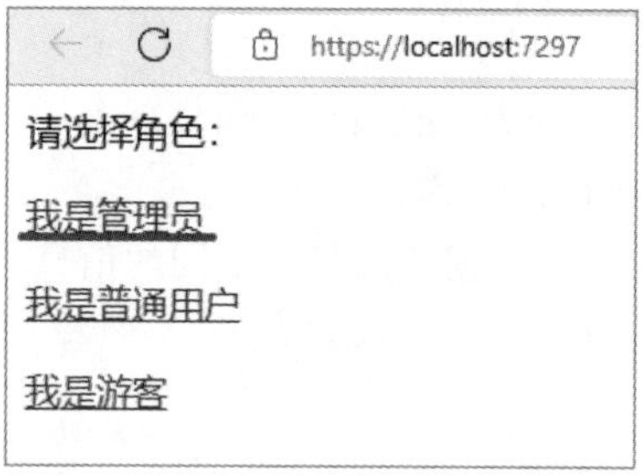

图 8-14　单击“我是管理员”链接

图 8-15　MainPort 页面

（5）在 Web 浏览器单击后退按钮回到 Index 页面，单击“我是游客”链接，跳转后页面输出“欢迎，您是游客”。

读者可能已经发现，在跳转到 MainPort 页面时，ASP.NET Core 是通过 URL 传递名为 handler 的参数选择处理程序的。例如，要执行 OnGetUser()方法，那么指向 MainPort 页面的 URL 就是/MainPort?handler=user。

8.8.4　通过路由参数选择处理程序

在 8.8.3 节的示例中，读者已经发现，可以在 URL 中使用 handler 查询参数选择处理程序。如果希望在 URL 中隐藏 handler 关键字，可以改用路由参数。路由参数的名称依然是 handler，但它不会出现在 URL 中。

请看下面的示例（Test.cshtml 页面文件）。

```
@page "/{handler?}"

<h3>@Branch</h3>

@functions
{
    public string? Branch { get; set; }

    public void OnGet()
    {
        Branch = "欢迎访问总站";
    }

    public void OnGetGz()
```

```
        {
            Branch = "欢迎访问广州分站";
        }

        public void OnGetSh()
        {
            Branch = "欢迎访问上海分站";
        }

        public void OnGetCd()
        {
            Branch = "欢迎访问成都分站";
        }
    }
```

文件的第 1 行，在@page 指令中指定 Test 页面的 URL 路由，即其访问 URL 为根路径（/），大括号（{}）中的 handler 是路由参数，问号（?）表示该参数可选，不提供 handler 参数也能匹配路由规则。

随后，<h3>标签中输出 Branch 属性的值。该属性在@functions 代码块中定义。另外还有 4 个 Handler 方法。

有了路由参数，只需要在 URL 的末尾加上处理程序的名称即可调用对应的方法。例如，访问 http://localhost:7081/sh，就会调用 OnGetSh()方法，如图 8-16 所示。如果 URL 中未提供 handler 参数值，即访问 http://localhost:7081，就会调用 OnGet()方法，如图 8-17 所示。

图 8-16　调用了 OnGetSh()方法

图 8-17　未提供 handler 参数值

8.8.5　自定义的处理程序模型

多数情况下，只需要将页面模型类中的方法按照约定命名，ASP.NET Core 应用程序会自动识别为页面处理程序，如 OnGet()、OnPostWork()等。然而，在有些特殊应用场景中，有些公共方法成员的命名不符合默认的约定，就无法被识别为页面处理程序。

此时就要考虑对应用程序进行扩展了，具体方案是自定义一个类，实现 IPageApplicationModelConvention 接口，再将此类型的实例添加到 RazorPagesOptions 选项类的 Conventions 集合中。IPageApplicationModelConvention 接口要求实现以下方法。

```
void Apply(PageApplicationModel model);
```

此方法只有一个 model 参数，类型为 PageApplicationModel，表示 Razor 页面模型相关的信息。其中，它有一个列表类型的属性——HandlerMethods。开发人员可以通过编写代码，将不符合默认命名约定但又需要在程序中使用的方法成员添加到 HandlerMethods 列表中。

每个处理程序模型都用 PageHandlerModel 类表示。先通过反射技术从页面模型类中查找出符合要求的方法成员(用 MethodInfo 表示),再用符合要求的方法成员创建 PageHandlerModel 实例。另外，还需要设置以下几个属性。

（1）Name：处理程序的描述性名称，一般与方法成员同名。

（2）HandlerName：处理程序的名字。可以直接用方法的名字，也可以使用处理过的名称。如 OnGetCars()方法，HandlerName 可以使用 Cars，也可以用 OnGetCars。这个值要和 URL 路由中的 handler 参数的值相同，即页面在选择处理程序时所使用的名称。

（3）HttpMethod：HTTP 的请求方式，如 GET、POST。

（4）Parameters：如果方法成员有参数，则需要向此列表中添加 PageParameterModel 对象。一个方法参数对应一个 PageParameterModel 对象实例。

本节所实现的示例中，将从页面模型类中找出符合条件的方法成员，并作为页面的处理程序。方法成员需要满足以下条件。

（1）实例方法，非静态成员。

（2）公共方法。

（3）方法上应用了 PostHandlerAttribute 特性类。

示例的具体实现步骤如下。

（1）定义 PostHandlerAttribute 类。它是特性类，应用于方法成员上。只有应用此特性类的方法才会被识别为页面处理程序。

```
[AttributeUsage(AttributeTargets.Method, AllowMultiple = false, Inherited
= true)]
public sealed class PostHandlerAttribute : Attribute
{
    // 如果此属性为 true，表明目标方法可作为页面处理程序使用
    // 如果此属性为 false，表明目标方法将不作为页面处理程序使用
    public bool IsEnabled { get; set; } = true;

    public PostHandlerAttribute(bool enabled) => IsEnabled = enabled;
}
```

IsEnabled 属性可以方便开发人员控制方法成员是否应作为页面处理程序。当不希望目标方法被用作处理程序时，只需要将 IsEnabled 属性改为 false 即可。

（2）定义 CustPageApplicationModelConvention 类，实现 IPageApplicationModelConvention 接口。在 Apply()方法中实现筛选页面模型类中的方法成员，并让合适的方法作为页面处理程序。

```
public class CustPageApplicationModelConvention :
IPageApplicationModelConvention
{
    public void Apply(PageApplicationModel model)
```

```
        {
            var handlerType = model.HandlerType;
            // 找出应用了 PostHandlerAttribute 的公共方法成员
            MethodInfo[] methods = handlerType.GetMethods(BindingFlags.Public
            | BindingFlags.Instance);
            foreach(var mth in methods)
            {
                PostHandlerAttribute? att = mth.GetCustomAttribute
                <PostHandlerAttribute>();
                if(att == null)
                {
                    // 方法不包含 PostHandlerAttribute
                    continue;
                }
                if(att.IsEnabled == false)
                {
                    // 此方法不作为页面处理程序使用
                    continue;
                }
                PageHandlerModel handlerModel = new(mth, new List<object> { att });
                // 请求方式为 POST
                handlerModel.HttpMethod = HttpMethods.Post;
                handlerModel.Name = mth.Name;
                // 去掉 Async 后缀
                handlerModel.HandlerName = mth.Name.Replace("Async", "");
                // 方法是否包含参数
                ParameterInfo[] parms = mth.GetParameters();
                foreach (var p in parms)
                {
                    PageParameterModel pmmodel = new(p, new List<object>());
                    pmmodel.ParameterName = p.Name!;
                    handlerModel.Parameters.Add(pmmodel);
                }
                model.HandlerMethods.Add(handlerModel);
            }
        }
    }
```

访问 model.HandlerType 属性可以获得定义方法的类型信息，一般是页面的模型类。在对每个 MethodInfo 对象进行判断时，不仅要判定方法是否应用了 PostHandlerAttribute 特性类，还要保证 IsEnabled 属性的值为 true。

在分析带参数的方法成员时，每个参数都要有对应的 PageParameterModel 实例。这里还必须为 ParameterName 属性赋值，否则在对方法参数进行数据绑定时会发生错误。ParameterName 通常使用与方法参数相同的名字即可。

（3）在 Program.cs 文件中，注册 Razor 页面相关服务时，要将 CustPageApplicationModelConvention 实例添加到 Conventions 集合中。

```
builder.Services.AddRazorPages(opt =>
{
    opt.Conventions.Add(new CustPageApplicationModelConvention());
});
```

（4）在项目中添加 Index 页面。页面上有<form>元素，首先包括两个输入框，可用于输入两个整数；接着是 4 个按钮，依次代表加、减、乘、除四则运算。

```
@page "{handler?}"
@model DemoApp.Pages.IndexModel
@addTagHelper *, Microsoft.AspNetCore.Mvc.TagHelpers

<form method="post">
<div>
        第 1 个整数:
<input type="number" name="x"/>
</div>
<div>
        第 2 个整数:
<input type="number" name="y"/>
</div>
<div style="margin-top:6px">
<button type="submit" asp-page-handler="add">加</button>
<button type="submit" asp-page-handler="sub">减</button>
<button type="submit" asp-page-handler="mult">乘</button>
<button type="submit" asp-page-handler="div">除</button>
</div>
</form>

@if(Model.Result != null)
{
<p>计算结果: @Model.Result.Value</p>
}
```

4 个按钮都选用各自的处理程序。相关的方法成员在 IndexModel 类中定义。

```
public class IndexModel : PageModel
{
    public int? Result { get; set; }

    public void OnGet()
    {
        Result = null;
    }

    [PostHandler(true)]
    public void Add(int x, int y)
    {
        Result = x + y;
    }
```

```
    [PostHandler(true)]
    public void Sub(int x, int y)
    {
        Result = x - y;
    }

    [PostHandler(true)]
    public void Mult(int x, int y)
    {
        Result = x * y;
    }

    [PostHandler(false)]
    public void Div(int x, int y)
    {
        if(y == 0)
        {
            return;
        }
        Result = x / y;
    }
}
```

Add()、Sub()、Mult()、Div()这 4 个方法都应用了 PostHandlerAttribute，可作为页面处理程序使用。但 Div()方法上的 PostHandlerAttribute 特性除外，它的 IsEnabled 属性为 false。

（5）运行示例程序，依次在两个输入框中输入整数值，再单击页面上的按钮。随后页面上会呈现计算结果。例如，输入 4 和 28，单击“乘”按钮，将得到结果 112，如图 8-18 所示。

图 8-18　呈现计算结果

在 Web 浏览器与服务器交互中是不会保留输入框的内容的，如果希望保留，可以将这些值存放在 ViewData 字典中，然后在页面中读取。请看下面的 SetView()方法。

```
// 页面模型类
[PostHandler(true)]
public void Add(int x, int y)
{
    Result = x + y;
    SetView(x, y);
}

[PostHandler(true)]
public void Sub(int x, int y)
{
    Result = x - y;
    SetView(x, y);
}
```

```
[PostHandler(true)]
public void Mult(int x, int y)
{
    Result = x * y;
    SetView(x, y);
}

[PostHandler(false)]
public void Div(int x, int y)
{
    if(y == 0)
    {
        return;
    }
    Result = x / y;
    SetView(x, y);
}

private void SetView(int x, int y)
{
    ViewData["x"] = x;
    ViewData["y"] = y;
}

// Razor 页面
<form method="post">
<div>
        第 1 个整数：
<input type="number" name="x" value="@ViewData["x"]"/>
</div>
<div>
        第 2 个整数：
<input type="number" name="y" value="@ViewData["y"]"/>
</div>
…
</form>
```

修改后，在提交到服务器进行计算后，输入框中仍显示输入的整数值，如图 8-19 所示。

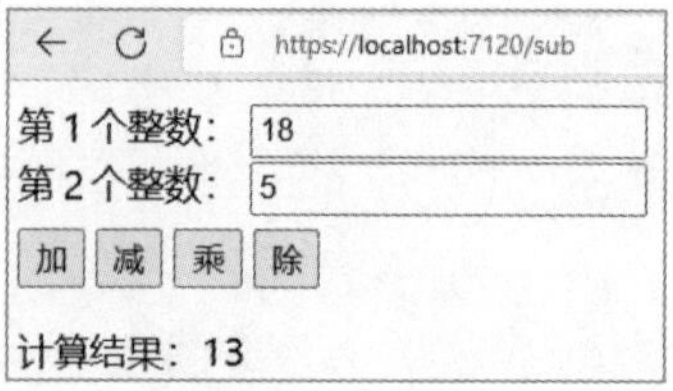

图 8-19　输入框的内容被保留

第 9 章 MVC 框架

本章要点

- ❑ MVC 的基本使用
- ❑ MVC 的 URL 路由规则
- ❑ 区域
- ❑ MVC 视图
- ❑ 局部视图与视图组件
- ❑ 发现其他程序集中的 MVC

9.1 MVC 基本概念

MVC（即 Model-View-Controller，译为模型、视图和控制器）实现业务逻辑与用户界面的分离，使应用程序拥有多种呈现形式。

模型（Model）封装了业务流程所需要的各种数据对象，具备与各种数据源之间的交互能力。模型仅描述数据的构成与特征，并不考虑数据以何种格式呈现（可能是纯文本格式，或者是 JSON、XML 等格式），兼容性好，可以为各种视图提供数据。

视图（View）即用户界面（在 Web 项目中常以 HTML 形式输出），它将模型数据以友好的方式呈现给用户，如表格、饼状图、图像等。在 ASP.NET Core 项目中，视图一般使用 Razor 标记生成 HTML。

控制器（Controller）一般不进行太复杂的逻辑处理（复杂的业务逻辑应在模型中完成），它主要负责接收输入（如 HTTP 请求），然后从模型中选取出相关的数据，最后用合适的视图显示数据（即输出）。控制器负责模型与视图之间的适配。

9.2 启用 MVC 功能

要在 ASP.NET Core 应用程序项目中使用 MVC 框架，需要完成以下两个步骤。

（1）在构建应用程序前，需要在服务容器中注册与 MVC 功能相关的服务类。

（2）构建应用程序后，需要在终结点上添加 MVC 专用的路由规则映射。

在调用 WebApplicationBuilder 对象的 Build()方法前，需要使用 AddControllers()或 AddControllersWithViews()扩展方法向服务容器注册相关的服务。这两个扩展方法都会注册 MVC 相关的服务，其区别在于：AddControllers()方法不注册与 Razor 视图相关的服务，控制器不能返回 Razor 标记；而 AddControllersWithViews()方法则允许控制器返回 Razor 视图。

无论使用的是 AddControllers()方法还是 AddControllersWithViews()方法，开发人员都可以用手动写入响应消息的方式构建 HTML 页面，也可以返回诸如 XML、JSON 等格式的数据。以上方式虽然没有返回可呈现的用户界面，但可作为中间数据使用——传递给其他应用程序进一步处理。假设客户端不是 Web 浏览器，而是一个桌面应用程序，那么控制器可以将数据以 JSON 格式返回。桌面程序接收到 JSON 格式的数据后再打开对应的窗口，通过专门的控件显示数据。因此，当控制器返回无视图数据时，可将其作为 Web 应用程序接口（Application Program Interface，API）调用。

在 WebApplicationBuilder.Build()方法调用后，需要调用 WebApplication 对象的 MapControllers()或 MapControllerRoute()方法进行终结点映射。这样，符合路由规则的 HTTP 请求才能激活控制器。MapControllers()方法添加终结点映射时不添加全局的路由规则，每个控制器都需要定义自己的路由规则；而 MapControllerRoute()方法会添加一个全局的路由规则，应用程序中所有控制器都会应用此规则，当然，控制器也可以使用自己的路由规则。

下面的代码简单演示了在 ASP.NET Core 项目中启用 MVC 功能的过程。

```
var builder = WebApplication.CreateBuilder(args);
// 添加 MVC 相关的服务
builder.Services.AddControllersWithViews();
var app = builder.Build();

// 添加终结点映射
app.MapControllers();

app.Run();
```

9.3 控制器

先来看看 ASP.NET Core 应用程序是如何发现控制器类的。下面的代码片段来自 ASP.NET Core 源代码。

```
protected virtual bool IsController(TypeInfo typeInfo)
{
    if (!typeInfo.IsClass)
    {
        return false;
    }
```

```
    if (typeInfo.IsAbstract)
    {
        return false;
    }

    if (!typeInfo.IsPublic)
    {
        return false;
    }

    if (typeInfo.ContainsGenericParameters)
    {
        return false;
    }

    if (typeInfo.IsDefined(typeof(NonControllerAttribute)))
    {
        return false;
    }

    if (!typeInfo.Name.EndsWith(ControllerTypeNameSuffix,
    StringComparison.OrdinalIgnoreCase) &&
    !typeInfo.IsDefined(typeof(ControllerAttribute)))
    {
        return false;
    }

    return true;
}
```

上述代码通过一系列 if 语句进行排除，只要其中有一处 if 语句的判断条件成立，IsController()方法就返回 false，表示此类型不是控制器；若所有if语句均未执行，则IsController()方法返回 true，表示此类型可认定为控制器。

具体的分析过程如下。

（1）如果目标类型不是类（class），则不是控制器。

（2）如果目标类型是抽象类，则不是控制器。

（3）如果不是公共类型，则不是控制器。

（4）如果目标类型存在泛型参数，则不是控制器。

（5）如果类型上应用了 NonControllerAttribute 特性，则不是控制器。

（6）如果目标类型没有应用 ControllerAttribute 特性，并且类名不是以 Controller 结尾的，则不是控制器。

ASP.NET Core 内部提供了两个基础类：ControllerBase 和 Controller，帮助开发人员快速

构建自定义的控制器类。ControllerBase 是抽象类，并且应用了 ControllerAttribute 特性。因此，所有派生自 ControllerBase 的类都会被识别为控制器。在编写控制器时，既可以选择从 ControllerBase 类派生，也可以从 Controller 类派生。通常，如果控制器不使用 Razor 视图功能，直接从 ControllerBase 类派生即可；如果控制器使用 Razor 视图功能，应从 Controller 类派生。

控制器中可包含若干数量的操作方法（Action）——控制器类中的公共方法成员。以下代码同样来自 ASP.NET Core 源代码，用于排除不合适的方法成员。

```
internal static bool IsAction(TypeInfo typeInfo, MethodInfo methodInfo)
{
    if (typeInfo == null)
    {
        throw new ArgumentNullException(nameof(typeInfo));
    }

    if (methodInfo == null)
    {
        throw new ArgumentNullException(nameof(methodInfo));
    }

    // The SpecialName bit is set to flag members that are treated in a special
    // way by some compilers,
    // such as property accessors and operator overloading methods
    if (methodInfo.IsSpecialName)
    {
        return false;
    }

    if (methodInfo.IsDefined(typeof(NonActionAttribute)))
    {
        return false;
    }

    // Overridden methods from Object class, e.g. Equals(Object),
    // GetHashCode(), etc., are not valid
    if (methodInfo.GetBaseDefinition().DeclaringType == typeof(object))
    {
        return false;
    }

    // Dispose method implemented from IDisposable is not valid
    if (IsIDisposableMethod(methodInfo))
    {
        return false;
    }

    if (methodInfo.IsStatic)
    {
        return false;
    }
```

```
        if (methodInfo.IsAbstract)
        {
            return false;
        }

        if (methodInfo.IsConstructor)
        {
            return false;
        }

        if (methodInfo.IsGenericMethod)
        {
            return false;
        }

        return methodInfo.IsPublic;
    }
```

IsAction()方法的处理过程如下。

（1）如果方法成员是特殊成员（如属性中的 get 和 set 访问器），则不是控制器的操作方法。

（2）如果方法成员应用了 NonActionAttribute 特性，则不是控制器的操作方法。

（3）如果该方法重写了 Object 类的方法（如 GetHashCode()等），则不是控制器的操作方法。

（4）如果该方法是 IDisposable 接口的实现方法（如 Dispose()方法），则不是控制器的操作方法。

（5）如果该方法是静态方法，则不是控制器的操作方法。

（6）如果该方法是抽象方法，则不能作为控制器的操作方法。

（7）如果该方法是构造函数，则不能作为控制器的操作方法。

（8）如果该方法是泛型方法，则不是控制器的操作方法。

控制器通过公开操作方法实现用户输入与模型之间的交互。例如，控制器用于管理员工信息，可能需要以下操作。

（1）ListAll()：查看所有员工信息。

（2）AddNew()：有新员工入职，添加新的档案信息。

（3）Delete()：删除员工信息（员工已离职）。

（4）Update()：修改员工信息。

9.3.1　示例：从 ControllerBase 类派生

对于不使用 Razor 视图的控制器，可以直接从 ControllerBase 类派生，不需要调用 View()方法。下面的代码定义了 TestController 类。

```
public class TestController : ControllerBase
{
```

```
    public IActionResult GoodMorning()
    {
        return Content("早上好");
    }

    public IActionResult GoodAfternoon()
    {
        return Content("下午好");
    }

    public IActionResult GoodEvening()
    {
        return Content("晚上好");
    }
}
```

控制器的类名一般以 Controller 结尾，因此，上述代码定义的控制器的实际名称为 Test。其中 3 个公共方法均被视为控制器的操作方法。操作方法的返回值通常是实现 IActionResult 接口的类型。上述代码调用了 Content()方法，返回 ContentResult 类型的对象，表示将指定的字符串写入 HTTP 响应消息中发回给客户端。

操作方法也可以返回非 IActionResult 类型的值。例如，可以将上述代码中的 GoodEvening()方法修改为

```
public string GoodEvening()
{
    return "晚上好";
}
```

下面的代码为应用程序开启 MVC 功能，并为终结点映射全局路由规则。

```
var builder = WebApplication.CreateBuilder(args);
builder.Services.AddControllers();
var app = builder.Build();

app.MapControllerRoute("default", "{controller}/{action}");

app.Run();
```

全局的 MVC 路由规则为{controller}/{action}。其中，大括号内的是路由规则的参数，将作为 URL 的占位符使用。假设请求的 URL 为 https://myweb.com/home/about，用于匹配的 URL 段为 home/about。将此 URL 段与路由规则对照，会得到 controller 参数的值为 home，action 参数的值为 about。因此，该 URL 将访问的是 home 控制器中的 about()操作方法。

对于上述 Test 控制器，若要调用 GoodMorning()操作方法，其 URL 应为 https://myweb.com/test/goodmorning。路由规则在匹配时不区分大小写，即 test/goodmorning 与 Test/GoodMorning 都能成功匹配。

9.3.2 示例：从 Controller 类派生

当 MVC 应用程序需要使用 Razor 视图时，控制器应从 Controller 类派生。Controller 类提供了 View()方法，可以返回指定的 Razor 文档。若调用 View()方法时不提供参数，那么返回的 Razor 文件名与当前操作方法的名称相同。例如：

```
public IActionResult LoadBooks()
{
    return View();
}
```

操作方法名称为 LoadBooks()，由于调用 View()方法时未提供参数，所以默认返回的视图文件名为 LoadBooks.cshtml。

下面的示例定义了 TestController 类，即控制器名为 Test，其中包含操作方法 Info()。该操作方法返回项目根目录下的 Test.cshtml 视图。

```
public class TestController : Controller
{
    public IActionResult Info() => View("~/Test.cshtml");
}
```

Test.cshtml 文件内容如下。

```
<div>
<h3>示例视图</h3>
</div>
```

在构建应用程序时，应使用 AddControllersWithViews()方法注册与 MVC 相关的服务（AddControllers()方法所注册的服务不能使用视图功能）。

```
var builder = WebApplication.CreateBuilder
(args);
builder.Services.AddControllersWithViews();
var app = builder.Build();

app.MapControllerRoute("app",
"{controller}/{action}");
app.Run();
```

运行示例程序后，在 Web 浏览器中定位到 https://<根URL>/test/info，即可看到 Test.cshtml 视图输出的 HTML 内容，如图 9-1 所示。

图 9-1 返回 Test.cshtml 视图

9.3.3 示例：使用 ControllerAttribute

在类上应用 ControllerAttribute 可使其被判定为 MVC 控制器。例如，下面的代码所定义的 Demo 类，控制器名称为 Demo。

```
[Controller]
public class Demo
{
    [NonAction]
    IActionResult TextContent(string text)
    {
        return new ContentResult
        {
            // 要返回的文本
            Content = text,
            // 内容类型
            ContentType = "text/plain;charset=UTF-8",
            // 状态码为 200
            StatusCode = StatusCodes.Status200OK
        };
    }

    // 此方法才是 MVC 的操作方法
    public IActionResult SayHello()
    {
        return TextContent("你好，小明");
    }
}
```

MVC 应用程序会将控制器类的公共方法视为操作方法。在 Demo 类中，TextContent()方法的功能是创建并返回 ContentResult 对象，并不用于 MVC 的操作方法。为避免应用程序将其视为操作方法，需要在方法上应用 NonActionAttribute。SayHello()方法的请求 URL 为 https://<主机名>/demo/sayhello。响应结果如图 9-2 所示。

图 9-2　Demo 控制器的响应结果

9.3.4　示例：使用 Controller 后缀

如果某个类的名称以 Controller 结尾，同样会被 MVC 应用程序判定为控制器。例如：

```
public class GameController
{
    // game/start
    public string Start() =>"游戏开始";

    // game/stop
    public string Stop() =>"游戏结束";
}
```

控制器的名称为 Game，并且 Start()和 Stop()方法也会被判定为控制器的操作方法。调用 Start()方法的 URL 为 http://localhost/game/start；调用 Stop()方法的 URL 则是 http://localhost/game/stop。

若不希望 GameController 类被应用程序判定为控制器，可以在类上应用 NonController-Attribute 特性类。

```
[NonController]
public class GameController
{
    …
}
```

此时访问 http://localhost/game/start 将找不到匹配的资源——GameController 类不再被判定为控制器。

9.3.5　自定义控制器的名称

默认情况下，MVC 应用程序会将控制器类的名称用于控制器名（不包括 Controller 后缀），如

```
public class HomeController …
```

那么控制器的名称就是 Home（不区分大小写，访问时可以用 home）。如果希望控制器的名称与类名不同，那么开发人员需要实现 IControllerModelConvention 接口（位于 Microsoft.AspNetCore.Mvc.ApplicationModels 命名空间）。该接口定义如下。

```
public interface IControllerModelConvention
{
    void Apply(ControllerModel controller);
}
```

实现该接口，只要实现 Apply()方法即可。在 Apply()方法的实现代码中，修改 ControllerModel 类的 ControllerName 属性，就可以自定义控制器的名称了。

为了让 IControllerModelConvention 接口的实现类能作用到控制器上，该类还要从 Attribute 类派生——使其变成一个特性类，最后将该特性类应用于控制器类上。

9.3.6　示例：ControllerNameAttribute 类

本示例将定义名为 ControllerNameAttribute 的特性类，通过构造函数可指定控制器的名称，并通过只读属性 Name 公开。

```
[AttributeUsage(AttributeTargets.Class, AllowMultiple = false, Inherited
= true)]
public class ControllerNameAttribute : Attribute, IControllerModelConvention
{
    // 获取控制器的名称
    public string Name { get; }

    public ControllerNameAttribute(string controllerName)
    {
        // 设置控制器的名称
```

```
        Name = controllerName;
    }

    public void Apply(ControllerModel controller)
    {
        // 修改控制器的名称
        controller.ControllerName = Name;
    }
}
```

将上述特性类应用于控制器类就可以自定义控制器的名称了，如

```
[ControllerName("BookStore")]
public class BookController : Controller
{
    public string GetStock()
    {
        return "库存：60";
    }
}
```

如果没有应用 ControllerNameAttribute，上述控制器的默认名称为 Book。通过 ControllerNameAttribute 类将控制器的名称重命名为 BookStore。因此，GetStock()操作方法的访问 URL 为 https://<主机名>/bookstore/getstock，结果如图 9-3 所示。

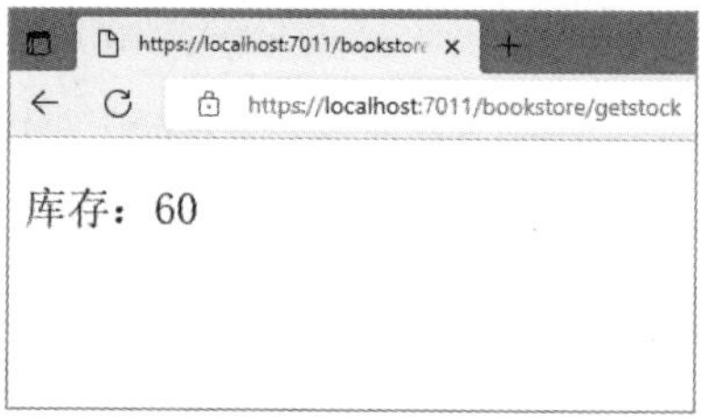

图 9-3　从 BookStore 控制器返回的内容

9.3.7　自定义操作方法的名称

自定义操作方法的名称有以下两种方案。

（1）实现 IActionModelConvention 接口，在 Apply()方法中修改 ActionModel 对象的 ActionName 属性。

（2）在操作方法上应用内置的 ActionNameAttribute 特性类（该类位于 Microsoft.AspNetCore.Mvc 命名空间）。

9.3.8　示例：CustActionNameAttribute 类

下面的代码定义 CustActionNameAttribute 类，除了实现 IActionModelConvention 接口，还派生自 Attribute 类，使 CustActionNameAttribute 成为特性类，可以直接应用到操作方法上。通过类的构造函数传递自定义的操作方法名称。

```
using Microsoft.AspNetCore.Mvc.ApplicationModels;

[AttributeUsage(AttributeTargets.Method, AllowMultiple = false, Inherited
= true)]
public class CustActionNameAttribute : Attribute, IActionModelConvention
{
```

```
    // 构造函数
    public CustActionNameAttribute(string customName)
    {
        CustName = customName;
    }

    // 属性，获取操作方法的名称
    public string CustName { get; }

    public void Apply(ActionModel action)
    {
        // 修改操作方法的名称
        action.ActionName = CustName;
    }
}
```

随后就可以使用该特性类了。

```
public class MessageController : ControllerBase
{
    [CustActionName("send")]
    public string SendMessage() =>"发送消息";

    [CustActionName("recv")]
    public string ReceiveMessage() =>"接收消息";
}
```

此时，Message 控制器中的两个操作方法分别被重命名为 send 和 recv。访问的 URL 分别为 https://<主机名>/message/send 和 https://<主机名>/message/recv，结果如图 9-4 所示。

9.3.9 示例：ActionNameAttribute 类

下面的代码定义了名为 Home 的控制器，其中包含名为 me 和 index 的操作方法。

```
public class HomeController
{
    [ActionName("me")]
    public string AboutMe()
    {
        return "关于本站...";
    }

    [ActionName("index")]
    public string MainPage()
    {
        return "欢迎来到官方主页";
    }
}
```

应用了 ActionNameAttribute 类后，两个操作方法的访问路径分别变为/home/me 和

/home/index，结果如图 9-5 所示。

图 9-4　访问 Message 控制器

图 9-5　访问 Home 控制器

9.4　MVC 路由规则

在 MVC 应用程序中，可以定义以下两种路由规则。

（1）全局路由规则：应用于所有控制器。

（2）局部路由规则：通过在控制器类或操作方法上应用 RouteAttribute 特性类指定路由规则。此规则仅对当前控制器有效，且优先级高于全局路由规则（会覆盖全局路由规则）。

MVC 的路由规则中会用到两个特殊的字段（或称为占位符）：controller 代表控制器名称的占位符，action 代表操作方法的名称。在全局规则中，这两个字段要写在大括号（{ }）中；而在局部规则中，要写在方括号（[ ]）中。

例如，最常用的全局路由规则为

```
{controller}/{action}
```

路由规则将从 URL 的根路径（/）后开始匹配，在指定时不能以/开头（写成/{controller}/{action}会报错）。假设请求的 URL 为 http://demo.cn/abc/play，那么，abc 将传递给 controller 字段，play 将传递给 action 字段。最后确定要调用的是 abc 控制器中的 play()操作方法。也可以为这些字段指定默认值，如

```
{controller=students}/{action=index}
```

如果请求的 URL 为 http://test.org/，此 URL 并未提供 controller 和 action 字段的值，于是将设置默认值——访问的控制器为 students，操作方法是 index()。

除了上述两个特定的字段外，路由规则中还可以映射操作方法的参数，如

```
{controller}/{action}/{name?}
```

假设访问 URL 为/demo/sayhello/Jack，那么，经过路由规则匹配后，确定调用的对象是 demo 控制器，操作方法是 sayhello()，并且此方法有一个名为 name 的参数，字符串 Jack 将赋值给 name 参数。路由规则中 name 字段后面的问号（?）表示该字段的值是可选的。如果 URL 不提供 name 的值，该规则也能匹配成功。

9.4.1　全局路由规则

注册全局路由规则的方法是在 WebApplication 实例上调用 MapControllerRoute()扩展方法。MVC 支持使用多条路由规则进行匹配，因此需要为每条路由规则指定一个唯一的名称，如

```
app.MapControllerRoute("pattern1", …);
app.MapControllerRoute("pattern2", …);
app.MapControllerRoute("pattern3", …);
```

路由规则的优先级取决于它们的注册顺序。当服务器接收到客户端请求时，先将 URL 与 pattern1 进行匹配测试，若成功就执行相应的控制器；若不成功，则与 pattern2 进行匹配；若仍不成功，再与 pattern3 进行匹配，直到成功为止。如果所有规则都匹配失败，服务器返回 HTTP-404 状态码。

9.4.2　示例：注册两条全局路由规则

下面的代码将定义名为 Test 的控制器，其中包含名为 GetData()的操作方法。

```
public class TestController : ControllerBase
{
    public string GetData()
    {
        // 获取终结点信息
        var ep = HttpContext.GetEndpoint();
        string s = string.Empty;
        // 获取终结点元数据
        IRouteNameMetadata? rn = ep.Metadata.GetMetadata
        <IRouteNameMetadata>();
        if (rn != null)
        {
            s += $"匹配的规则名称：{rn.RouteName}";
        }
        // 获取已匹配的路由规则
        if(ep is RouteEndpoint routeep)
        {
            s += $"\n 匹配的路由规则：{routeep.RoutePattern.RawText}";
            // 获取路由规则中必备字段的值
            s += "\n 匹配的字段：";
            foreach (var item in routeep.RoutePattern.RequiredValues)
            {
                s += $"\n\t{item.Key} = {item.Value}";
            }
        }
        return s;
    }
}
```

在 GetData()方法中，先通过 HttpContext 对象的 GetEndpoint()方法获取当前已执行的终结点实例。接着在 Metadata 集合中查找 IRouteNameMetadata 接口类型（或实现此接口的类型）的终结点元数据。该元数据对象有一个 RouteName 属性，可以获取当前成功匹配的路由规则的名称。最后，判断一下当前终结点对象是否为 RouteEndpoint，如果是，就可以通过 RoutePattern 属性得到当前路由规则的相关信息（如规则内容、成功匹配后各字段的值等）。

在应用程序初始化代码中注册两条路由规则。

```
var builder = WebApplication.CreateBuilder(args);
…
app.MapControllerRoute("A", "{controller}/{action}");
app.MapControllerRoute("B", "other/data", new
{
    controller = "test",
    action = "getdata"
});
…
```

注意，在添加规则 B 时由于没有提供 controller、action 字段的值（两者皆为必需字段），因此需要通过匿名对象初始化 controller 字段的值为 test，初始化 action 字段的值为 getdata，即指向 TestController.GetData()方法。

运行示例程序后，在 Web 浏览器中访问 https://<主机名>/test/getdata，将匹配路由规则 A，服务器返回内容如图 9-6 所示。

访问 https://<主机名>/other/data，该 URL 将与规则 B 匹配，服务器返回内容如图 9-7 所示。

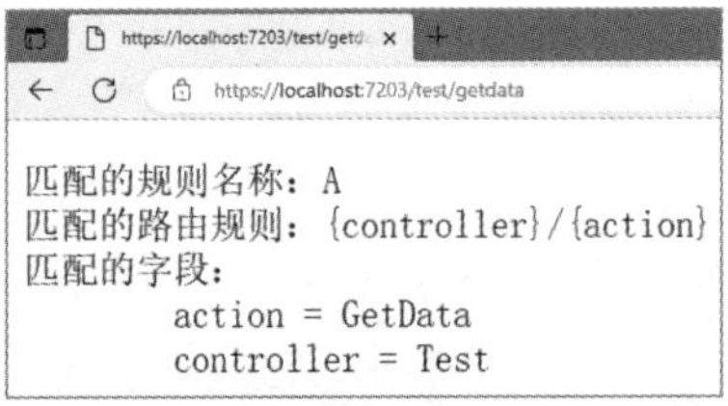

图 9-6　匹配路由规则 A

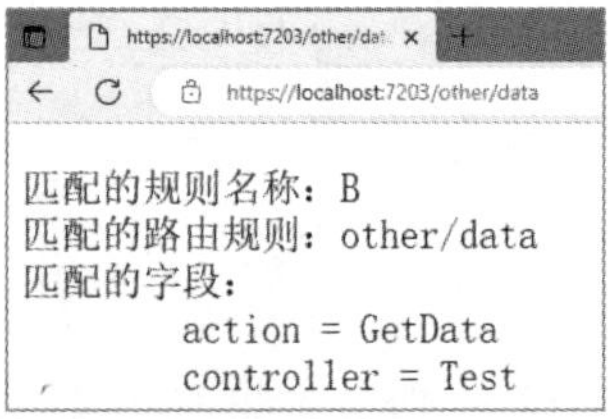

图 9-7　匹配路由规则 B

9.4.3　局部路由规则

RouteAttribute 特性类既可以应用于类，也可以应用于方法成员。使用该特性类可以为控制器（包括控制器中的操作方法）注册局部路由规则。一个控制器（或操作方法）可以指定多条路由规则，这些规则仅对当前控制器有效。

在局部路由规则中，controller、action 等必备字段要放在方括号内，如

```
[Route("[controller]/[action]")]
public class StockController : ControllerBase
{
```

```
    …
}
```

如果控制器和操作方法上都应用了 Route 特性，最终的路由规则就是二者的组合，如

```
[Route("news")]
public class NewsCentreController : ControllerBase
{
    [Route("all")]
    public IList<string> GetTitleList()
    {
        List<string> list = new List<string>();
        list.Add("Sample Text-1");
        list.Add("Sample Text-2");
        list.Add("Sample Text-3");
        return list;
    }
}
```

要访问 NewsCentre 控制器中的 GetTitleList()方法，请求的 URL 为/news/all。

9.4.4 IRouteTemplateProvider 接口

IRouteTemplateProvider 接口为基于特性的路由规则提供数据。注册局部路由规则时所使用的 RouteAttribute 特性类正是实现了该接口。

IRouteTemplateProvider 接口定义了以下属性。

（1）Name：表示路由规则的名称。

（2）Order：表示路由规则的执行顺序。此属性的值越小，则执行优先级越高。如果返回 null，表示顺序未设置，应用程序将使用默认值 0。

（3）Template：路由模板，即路由规则字符串，如[controller]/[action]。

下面的示例将定义一个实现 IRouteTemplateProvider 接口的特性类，使用通用的[controller]/[action]路由规则作为基础，添加一个后缀，如[controller]/[action]/abc。

```
[AttributeUsage(AttributeTargets.Class | AttributeTargets.Method,
AllowMultiple = true, Inherited = true)]
public class SuffixRouteAttribute : Attribute, IRouteTemplateProvider
{
    // 私有字段
    private readonly string _suffix;

    // 构造函数
    public SuffixRouteAttribute(string suffix)
    {
        _suffix = suffix;
    }

    // 路由规则末尾加上后缀，分隔符依然是/
```

```
    // 此属性为只读，不允许修改
    public string? Template =>"[controller]/[action]/" + _suffix;

    // 执行顺序
    public int? Order { get; set; } = null;
    // 路由规则名称
    public string? Name { get; set; }
}
```

下面的代码定义一个名为 Home 的控制器，在类定义上应用 SuffixRouteAttribute，指定后缀为 test。

```
[SuffixRoute("test")]
public class HomeController : ControllerBase
{
    public IActionResult DoSomething()
    {
        return Content("测试应用程序");
    }
}
```

在初始化应用程序时，由于本示例不需要指定全局路由规则，因此调用 MapControllers() 扩展方法即可。

```
var builder = WebApplication.CreateBuilder(args);
builder.Services.AddControllers();
var app = builder.Build();

app.MapControllers();
app.Run();
```

运行示例程序后，访问控制器时要加上后缀，即 https://<主机名>/home/dosomething/test，结果如图 9-8 所示。

图 9-8　带后缀的路由规则

9.4.5　通过实现约定接口定义路由规则

ASP.NET Core MVC 应用程序允许开发人员通过实现约定接口完成各种自定义方案。MVC 框架中不同层面的组件都有对应的约定接口。这些接口都包含一个 Apply()方法，开发者必须实现该方法，而该方法的参数则对应着不同的模型对象（应用程序模型、控制器模型、控制器操作方法模型、操作方法的参数模型）。

（1）应用程序层：对应的约定接口为 IApplicationModelConvention，Apply()方法使用的参数类型为 ApplicationModel。该约定接口可以修改应用程序相关的属性，以及各种控制器的相关信息（如控制器的名称）。

（2）控制器层：对应的约定接口为 IControllerModelConvention，Apply()方法的参数为 ControllerModel 类。该约定接口可用于修改控制器及其操作方法相关的信息。

（3）操作方法层：对应的约定接口为 IActionModelConvention，Apply()方法的参数类型为

ActionModel。该约定接口可用于自定义操作方法的相关信息。

（4）操作方法的参数：对应的约定接口为 IParameterModelConvention，参数类型为 ParameterModel。该约定接口用于自定义操作方法的参数。

要通过约定接口自定义路由规则，用到的是 IControllerModelConvention 和 IActionModel-Convention 接口。有以下两种思路可供选择。

（1）仅实现 IControllerModelConvention 接口。如果使用如[controller]/[action]的路由规则，那么只需要为控制器设置路由规则即可，操作方法的路由规则可以忽略；如果控制器与操作方法都要设置路由规则，那么可以先设置控制器的规则，再通过 Actions 属性获取各操作方法相关的 ActionModel 实例，再逐个设置路由规则。

（2）先实现 IControllerModelConvention 接口，为控制器设置路由规则；再实现 IActionModelConvention 接口，为操作方法独立设置路由规则。

设置路由规则需要用到 AttributeRouteModel 类，它包含 Name、Template、Order 等属性，其含义及用法与 IRouteTemplateProvider 接口相同。

9.4.6 示例：CustPrefixRouteConvention 类

CustPrefixRouteConvention 类实现 IControllerModelConvention 接口，为路由规则的 URL 模板添加自定义前缀。

```
public class CustPrefixRouteConvention : IControllerModelConvention
{
    private readonly string _prefix;
    public CustPrefixRouteConvention(string prefix)
    {
        _prefix = prefix;
    }

    public void Apply(ControllerModel controller)
    {
        // 如果已设置过路由规则，就跳过
        if(controller.Selectors.Any(x => x.AttributeRouteModel != null))
        {
            return;
        }
        // 前缀 + / + [controller]/[action]
        string template = $"{_prefix}/[controller]/[action]";
        // 设置路由规则
        foreach(var selector in controller.Selectors)
        {
            selector.AttributeRouteModel = new()
            {
                Template = template
            };
        }
```

```
    }
}
```

表示 URL 前缀的字符串保存在私有字段_prefix 中。URL 的合并方法是在[controller]/[action]前面加上_prefix 字段的内容。

把 CustPrefixRouteConvention 添加到 MvcOptions 选项类的 Conventions 列表中，这使得该约定会应用到所有控制器上。

```
var builder = WebApplication.CreateBuilder(args);
builder.Services.AddControllers(options =>
{
    options.Conventions.Add(new CustPrefixRouteConvention("demo"));
});
var app = builder.Build();

app.MapControllers();

app.Run();
```

本示例为路由 URL 加上了 demo 前缀。由于已通过约定生成路由规则，所以在映射终结点时调用 MapControllers()扩展方法即可，不用指定全局路由规则。

接下来定义两个控制器，验证加了前缀的路由规则是否生效。

```
public class PlayerListController : ControllerBase
{
    public int Nums() => 100;
    public string Info() =>"Demo Data";
}

public class MiniPlayerController : ControllerBase
{
    public string RunNow() =>"正在播放";
    public float Speed() => 1.05f;
}
```

第 1 个控制器名为 PlayerList，访问它的操作方法时应使用以下 URL。

```
/demo/playerlist/nums
/demo/playerlist/info
```

第 2 个控制器名为 MiniPlayer，其中两个操作方法的访问 URL 为

```
/demo/miniplayer/runnow
/demo/miniplayer/speed
```

访问上述其中一个 URL 后服务器返回的结果如图 9-9 所示。

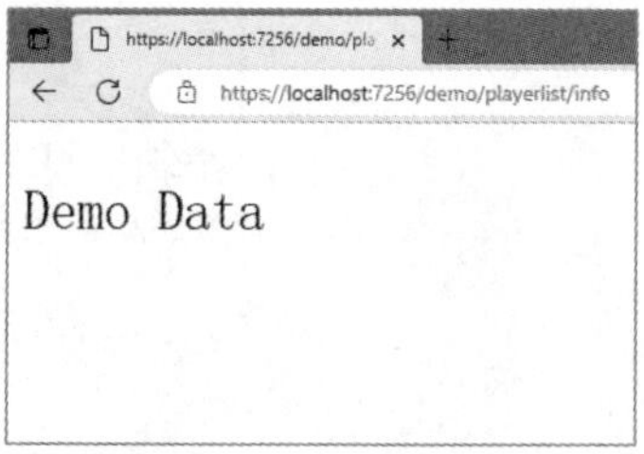

图 9-9 带 demo 前缀的 URL

9.5　限制操作方法所支持的 HTTP 请求

实现 IActionHttpMethodProvider 接口时，可以限制操作方法所支持的 HTTP 请求。例如，只允许客户端发出 HTTP-POST 请求。实现该接口的类型必须包含 HttpMethods 属性，它是一个字符串列表，其中列出所支持的 HTTP 请求方法。

9.5.1　示例：只支持 HTTP-PUT 请求的操作方法

本示例将实现一个特性类，将此特性类应用到某个操作方法上，使操作方法只接受 HTTP-PUT 方法的请求。

```
[AttributeUsage(AttributeTargets.Method, AllowMultiple = true, Inherited
= true)]
public class HttpPutOnlyAttribute : Attribute, IActionHttpMethodProvider
{
    // 私有字段
    private readonly string[] _supportedMethods = { "PUT" };
    // 属性：返回所支持的 HTTP 请求方法
    public IEnumerable<string> HttpMethods => _supportedMethods;
}
```

HttpMethods 属性可以返回多个值，本示例中只允许接受 HTTP-PUT 请求，因此只返回 PUT。

定义 Home 控制器，并把 HttpPutOnlyAttribute 应用到 SendMessage()操作方法上。

```
[Route("test/[action]")]
public class HomeController : Controller
{
    [HttpPutOnly]
    public IActionResult SendMessage(string data)
    {
        return Content($"你发送的数据是：{data}");
    }
}
```

SendMessage()方法带有一个 data 参数，客户端发送请求时，可以通过 URL 查询字符串提供参数值，如/test/sendmessage?data=abcd。

下面的代码用于测试发送 HTTP-PUT 请求。

```
// 服务器的基础地址
Uri baseUrl = new Uri("http://localhost:5004");

HttpClient client = new HttpClient();
client.BaseAddress = baseUrl;
// 发出 HTTP-PUT 请求
var resp = await client.PutAsync("/test/sendmessage?data=Mike", null);
```

```
if(resp.IsSuccessStatusCode)
{
    // 打印服务器响应消息
    string res = await resp.Content.ReadAsStringAsync();
    Console.WriteLine($"服务器响应：{res}");
}
else
{
    // 打印错误信息
    Console.Write("请求未成功，返回状态码：");
    Console.Write(resp.StatusCode.ToString());
    Console.WriteLine("({0})", (int)resp.StatusCode);
}
```

执行上述代码，服务器将返回以下内容。

```
你发送的数据是：Mike
```

若改为 HTTP-GET 请求，由于不受支持，会返回 405 状态码。

9.5.2 内置特性类

ASP.NET Core 内置了一批用于限制 HTTP 请求方法的特性类。大多数应用场景中，开发人员不需要自己实现 IActionHttpMethodProvider 接口，直接使用 ASP.NET Core 自身提供的特性类即可。

这些类都实现抽象类 HttpMethodAttribute。该类同时实现了 IActionHttpMethodProvider 和 IRouteTemplateProvider 接口，不仅可用于限制操作方法的 HTTP 请求方法，还可以设置 URL 路由规则（此功能与 RouteAttribute 相同，但 RouteAttribute 不能限制 HTTP 请求方法）。

HttpMethodAttribute 类派生出以下类。

（1）HttpGetAttribute：操作方法支持 HTTP-GET 请求。

（2）HttpPostAttribute：操作方法支持 HTTP-POST 请求。

（3）HttpPutAttribute：操作方法支持 HTTP-PUT 请求。

（4）HttpDeleteAttribute：操作方法支持 HTTP-DELETE 请求。

（5）HttpHeadAttribute：操作方法支持 HTTP-HEAD 请求。

（6）HttpPatchAttribute：操作方法支持 HTTP-PATCH 请求。

（7）HttpOptionsAttribute：操作方法支持 HTTP-OPTIONS 请求。

下面的代码定义 Demo 控制器，它包含 3 个操作方法，分别支持 POST、DELETE、PATCH 请求。

```
public class DemoController : ControllerBase
{
    [HttpPost("demo/act1")]
    public string TestMethod1(int x, double y)
    {
```

```
        return string.Format("你提交的数据：x={0}, y={1}", x, y);
    }

    [HttpDelete("demo/act2")]
    public string TestMethod2()
    {
        return "你发起了 DELETE 请求";
    }

    [HttpPatch("demo/act3")]
    public string TestMethod3()
    {
        return "你发起了 PATCH 请求";
    }
}
```

随后可以添加一个控制台应用程序项目，使用以下代码进行 HTTP 请求测试。

```
// 服务器基础地址
Uri baseAddr = new("http://localhost:5235");

HttpClient client = new HttpClient();
client.BaseAddress = baseAddr;

// 第 1 次请求：POST
IDictionary<string, string> fds = new Dictionary<string, string>();
fds.Add("x", "55000");
fds.Add("y", "249.015");
FormUrlEncodedContent forms = new(fds);
var response1 = await client.PostAsync("/demo/act1", forms);
if(response1.IsSuccessStatusCode)
{
    Console.Write("HTTP POST，服务器响应：");
    Console.WriteLine(await response1.Content.ReadAsStringAsync());
}

// 第 2 次请求：DELETE
var response2 = await client.DeleteAsync("/demo/act2");
if(response2.IsSuccessStatusCode)
{
    Console.Write("HTTP DELETE，服务器响应：");
    Console.WriteLine(await response2.Content.ReadAsStringAsync());
}

// 第 3 次请求：PATCH
var response3 = await client.PatchAsync("/demo/act3", null);
if(response3.IsSuccessStatusCode)
{
    Console.Write("HTTP PATCH，服务器响应：");
```

```
    Console.WriteLine(await response3.Content.ReadAsStringAsync());
}
```

9.6 区域

如果项目规模较大，或者业务逻辑上需要模块化管理，就可以考虑将 MVC 框架拆分为多个独立的功能单元，即区域（Area）。每个区域都可以拥有自己的 MVC 框架。例如，Stock 区域用于管理公司的产品库存信息，它包含 Product 控制器，支持查询各个产品的入库数量与出库数量，或更新产品的库存状态；而 Sale 区域则管理产品的销售信息，该区域也包含 Product 控制器，支持查询各种产品的销售量、售出时间等。

在指定全局路由规则时，区域也有专用的字段名称——area，即可以使用以下路由规则。

```
{area}/{controller}/{action}
```

在控制器的类定义上，可通过 AreaAttribute 特性类指定控制器所在的区域名称。例如：

```
[Area("Users")]
public class AdminController : Controller
{
    public IActionResult ListAll()
    {
        string[] userlist =
        {
            "Lily", "Tom", "Dick", "Bob", "Jim"
        };
        return Ok(userlist);
    }
}
```

上述代码定义了 Admin 控制器，它位于 Users 区域内。在添加终结点映射时，可以使用以下全局路由规则。

```
app.MapControllerRoute("app", "{area}/{controller}/{action}");
```

访问 ListAll()操作方法时，URL 需要加上区域的名称，如 http://<主机名>/users/admin/listall。

如果 area 字段加上 exists 约束，那么要求 URL 必须包含有效的区域名称才能匹配路由规则。

```
app.MapControllerRoute("app", "{area:exists}/{controller}/{action}");
```

也可以调用 MapAreaControllerRoute()扩展方法。路由模板中只需包含 controller 和 action 字段即可。

```
app.MapAreaControllerRoute("app", "Users", "{controller}/{action}");
```

此时，访问 http://<主机名>/admin/listall，就能调用 ListAll()操作方法。MapAreaController-Route()扩展方法将从 areaName 参数获得区域名称，而不是通过 URL 匹配。即使 URL 中未指

明 Users 区域，也能够找到 Admin 控制器。不过，许多时候为了避免多个全局路由规则之间有冲突，可以为路由模板加上前缀。例如：

```
app.MapAreaControllerRoute("app", "Users", "siteusers/{controller}/
{action}");
```

此时，正确的访问 URL 为 https://<主机名>/siteusers/admin/listall。

9.7　视图

视图负责将模型对象以不同的格式向用户展示。在 MVC 项目中，视图通过 Razor 标记生成 HTML 文档。与 Razor Pages 文件相同，视图文件的扩展名为.cshtml，但视图文件中代码文档的首行不使用@page 指令。

当控制器中的操作方法返回的结果为 ViewResult 类型时，应用程序会搜索对应的视图并执行其中的代码，最后将 HTML 文档返回给客户端。控制器类应当派生自 Controller 类，以便调用 View()方法轻松返回视图。View()方法有多个重载，在调用时存在以下几种情况。

（1）在调用 View()方法时若未指定视图名称，那么默认的视图名称与操作方法名称相同。

（2）若指定的是视图文件的路径（相对于应用程序的根目录），那么应用程序将按照路径直接查找视图文件。

（3）若指定的是视图名称而非文件名，那么应用程序将按照视图搜索引擎（View Engines）将依据所配置的路径查找包含视图名称的文件。

9.7.1　视图文件的默认存放路径

在默认配置中，MVC 视图文件存放于/Views 目录下。在 Views 目录下需要创建与控制器同名的子目录，最后将视图文件放置于此子目录内。

可以通过一个示例进一步说明，实现过程可参考以下步骤。

（1）创建名为 Demo 的控制器，其中包含 Pictures()、Movies()、Songs()操作方法。

```
public class DemoController : Controller
{
    // 视图文件：Pictures.cshtml
    public IActionResult Pictures() => View();

    // 视图文件：Movies.cshtml
    public IActionResult Movies() => View();

    // 视图文件：Musics.cshtml
    public IActionResult Songs() => View("Musics");
}
```

前两个操作方法在调用 View()方法时都未指定视图名称或路径，因此视图文件名与操作方法名称相同。第 3 个操作方法（即 Songs()方法）指定了视图的名称，对应的文件名应为

Musics.cshtml。

（2）如果项目的根目录下不存在 Views 目录，则创建它。

（3）在 Views 目录下创建子目录，命名与控制器名称相同，即 Demo。

（4）在 Demo 目录下添加 3 个视图文件，文件名依次为 Movies.cshtml、Musics.cshtml 和 Pictures.cshtml，目录结构如图 9-10 所示。

图 9-10　视图目录的结构

（5）3 个视图文件的内容如下。

```
// Pictures.cshtml
<h3>图集</h3>

// Movies.cshtml
<h3>影集</h3>

// Musics.cshtml
<h3>歌集</h3>
```

每个视图文件中都只有一个<h3>元素，文档结构比较简单，仅用于演示。

（6）在向服务容器注册服务时，应调用 AddControllersWithViews()方法，这样才能保证视图功能相关的服务都被添加到容器中。

```
var builder = WebApplication.CreateBuilder(args);
builder.Services.AddControllersWithViews();
var app = builder.Build();
// 路由模板中已为字段设置了默认值
app.MapControllerRoute("app", "{controller=Demo}/{action=Movies}");

app.Run();
```

（7）运行示例程序，如果请求的 URL 指向根路径，即默认执行 Demo 控制器中的 Movies()操作方法，返回 Movies.cshtml 视图文件，如图 9-11 所示。

（8）访问/demo/songs，服务器将返回 Musics 视图，如图 9-12 所示。

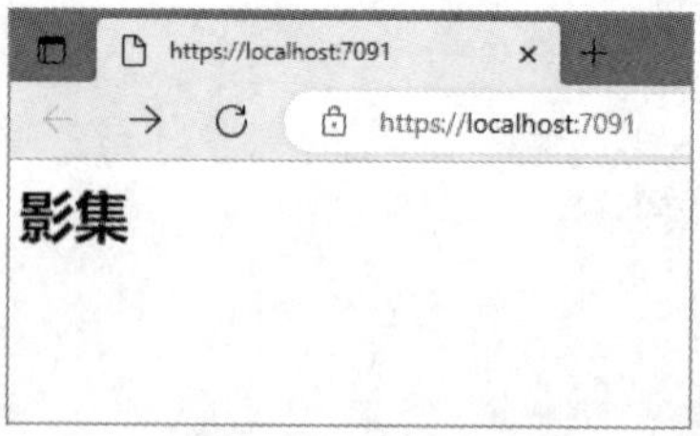

图 9-11　呈现 Movies 视图

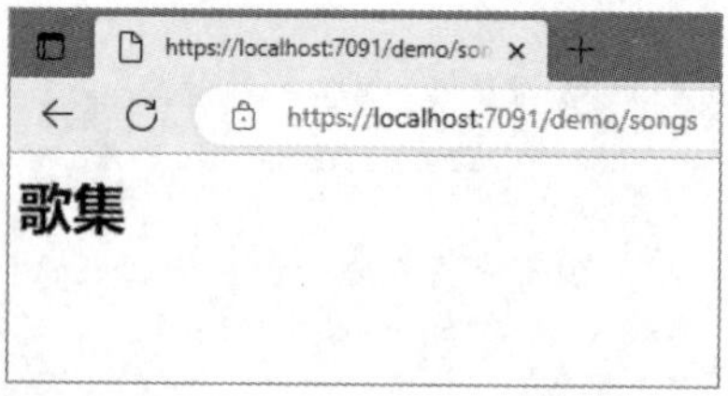

图 9-12　返回 Musics 视图

9.7.2　自定义视图的路径格式

MVC 的视图文件在存储路径上有特殊格式，默认的配置如下。

```
/Views/{1}/{0}.cshtml
/Views/Shared/{0}.cshtml
```

如果 MVC 应用程序包含区域，则视图文件的路径格式如下。

```
/Areas/{2}/Views/{1}/{0}.cshtml
/Areas/{2}/Views/Shared/{0}.cshtml
/Views/Shared/{0}.cshtml
```

其中有 3 处占位符，具体如下。

{0}：视图（或操作方法）名称。如果控制器代码在调用 View()方法时未指定视图名称，那么操作方法名称与视图名称相同；如果二者不相同，则该值为视图名称。

{1}：控制器名称。默认控制器名称与其类名相同，若 ControllerModel 类的 ControllerName 属性被修改（如通过实现 IControllerModelConvention 接口），那么此处{1}应为 ControllerName 属性的值。

{2}：区域名称。

假设有以下控制器。

```
[Area("Checker")]
public class ManageController : Controller
{
    public IActionResult RegNew()
    {
        return View();
    }
}
```

根据上述代码可知，区域名称为 Checker，控制器名称为 Manage，控制器中包含 RegNew()操作方法。由于 View()方法在调用时未指定视图名称，因此视图名称也是 RegNew。文件路径应为 /Areas/Checker/Views/Manage/RegNew.cshtml，其目录层次结构如图 9-13 所示。

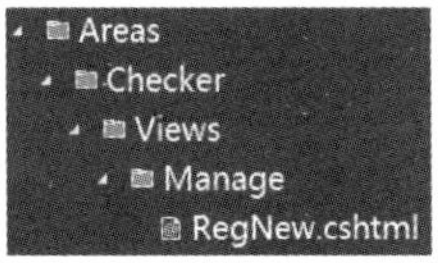

图 9-13　RegNew 视图的目录层次结构

请思考下面的代码。

```
[AttributeUsage(AttributeTargets.Class, AllowMultiple = false, Inherited
= true)]
public class CustControllerNameAttribute : Attribute,
IControllerModelConvention
{
    // 私有字段
    private readonly string _ctrlName;

    public CustControllerNameAttribute(string name)
    {
        _ctrlName = name;
    }

    public void Apply(ControllerModel controller)
```

```
        {
            // 修改控制器名称
            controller.ControllerName= _ctrlName;
        }
    }

    [CustControllerName("Abc")]
    public class TestController : Controller
    {
        public IActionResult ShowOne()
        {
            return View("one");
        }
    }
```

CustControllerNameAttribute 特性类实现了 IControllerModelConvention 接口，可以自定义控制器名称。在定义 Test 控制器时应用了此特性类，将控制器名称设置为 Abc。而 ShowOne()操作方法在调用 View()方法时指定了视图名称为 one。

根据以上信息可知，one 视图的存放路径为/Views/Abc/one.cshtml，其目录层次结构如图 9-14 所示。

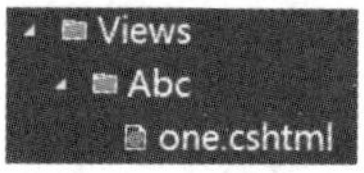

图 9-14 one 视图的目录层次结构

上述两个示例所使用的均为默认的视图路径格式，如果 MVC 项目的视图文件并非位于/Views 目录下，那就必须根据实际的存放路径进行自定义了。

下面的示例将配置自定义的视图查找路径。

```
var builder = WebApplication.CreateBuilder(args);
builder.Services.AddControllersWithViews();
builder.Services.Configure<RazorViewEngineOptions>(opt =>
{
    // 先清空默认的路径格式
    opt.ViewLocationFormats.Clear();
    opt.AreaViewLocationFormats.Clear();
    // 添加自定义的路径格式
    opt.ViewLocationFormats.Add("/{1}Views/{0}" + RazorViewEngine.
    ViewExtension);
    opt.ViewLocationFormats.Add("/SharedViews/{0}" + RazorViewEngine.
    ViewExtension);
    opt.AreaViewLocationFormats.Add("/{2}/{1}Views/{0}" + RazorViewEngine.
    ViewExtension);
    opt.AreaViewLocationFormats.Add("/SharedViews/{0}" + RazorViewEngine.
    ViewExtension);
});
var app = builder.Build();
…
```

配置视图的格式化路径，需要使用 RazorViewEngineOptions 选项类。ViewLocationFormats 属性中的字符串列表为不带区域的 MVC 视图文件的查找路径；而 AreaViewLocationFormats

属性中的字符串列表则是带区域的视图文件路径。上述代码先将两个列表中的内容清空，然后再添加自定义的路径格式。

本示例所定义的视图路径，目录名称由控制器名称（占位符{1}）加 Views 后缀组成。目录内为视图文件（{0}.cshtml）。若包含区域字段，先创建以区域名称命名的目录（占位符{2}），然后在此目录下再创建由控制器名称加 Views 后缀命名的子目录（{1}Views）。子目录内存放视图文件（{0}.cshtml）。另外，还包含一个 SharedViews 目录，用于存放共享的视图，如布局文件_Layout.cshtml 等。

注意路径的大小写问题，如果运行应用程序的操作系统对大小写敏感，那么/Views/{1}/{0}.cshtml 与/views/{1}/{0}.cshtml 表示不同的路径。

下面的代码将定义新的控制器 Nodes，包含操作方法 Elements()。

```
[Route("test/[action]")]
public class NodesController : Controller
{
    public IActionResult Elements() { return View(); }
}
```

控制器名称为 Nodes。调用 View()方法时未指定视图名称，默认与操作方法相同，即视图名称为 Elements。根据上文所配置的路径，应在应用程序项目的根目录下创建名为 NodesViews 的目录，再在 NodesViews 目录下添加视图文件 Elements.cshtml。视图文件中的代码如下。

```
@{
    var list = new string[]
    {
    "Item-A",
    "Item-B",
    "Item-C",
    "Item-D",
    "Item-E"
    };
}

@foreach(string x in list)
{
    <div style="margin-top:2px; margin-bottom:
    2px;background-color: skyblue;color:red">
        @x
    </div>
}
```

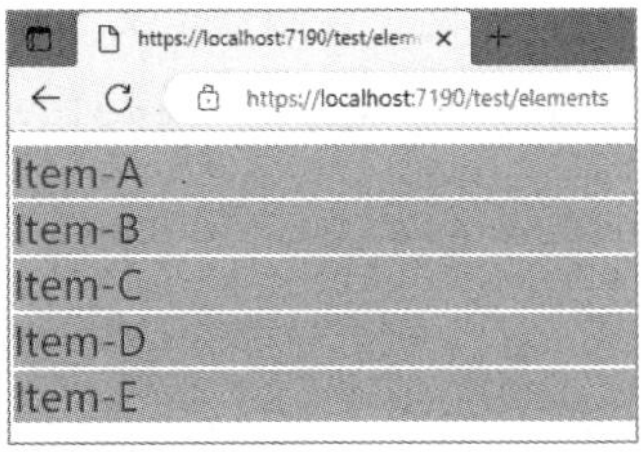

图 9-15　Elements 视图返回的内容

运行示例程序，访问/test/elements，结果如图 9-15 所示。

9.7.3　布局视图

布局视图（Layout View）可用于构建应用程序内各个视图基本结构，其功能类似于 PowerPoint 中的“母版”。例如，某公司的网站，不管是首页，还是产品分类页，或是公司信息页，都会使用相同的背景图片、导航栏、广告条、版权信息等。也就是说，在一个项目中，不同的视图中总会有一部分内容是不变的。这些内容没有必要重复制作，而应该将其放在布局视图中，每个视图只负责填充自身的内容即可。

可以把布局视图比作“入职员工登记表”，而把常规视图比作员工。表格的格式是固定的，可批量打印，新员工将自己的个人信息填上即可。

布局文件（默认名称为_Layout.cshtml）也使用 Razor 标记，其查找的路径与普通视图文件一样，也遵循 ViewLocationFormats 或 AreaViewLocationFormats 中的配置，如

```
/Views/{1}/{0}.cshtml
/Views/Shared/{0}.cshtml
```

如果多个路径中都能找到指定的布局视图，那么其优先级取决于路径添加的顺序。也就是说，如果在第 1 个路径中找到布局文件，那么就不再查找第 2 个、第 3 个路径。

9.7.4 示例：布局视图的查找顺序

本示例将在不同层次的目录中放置布局视图，看看应用程序是如何查找这些文件的。

（1）基本的应用程序配置。

```
…
builder.Services.AddControllersWithViews();
builder.Services.Configure<RazorViewEngineOptions>(opt =>
{
    opt.ViewLocationFormats.Clear();
    opt.ViewLocationFormats.Add("/Views/Shared/{0}.cshtml");
    opt.ViewLocationFormats.Add("/Views/{1}/{0}.cshtml");
});
…
```

上述代码为 MVC 视图的查找配置了自定义路径，其中位于/Views/Shared 目录下的文件优先被查找。

（2）两个控制器定义如下。

```
public class Demo1Controller : Controller
{
    public IActionResult Main() { return View(); }
}

public class Demo2Controller : Controller
{
    public IActionResult Main() { return View();}
}
```

（3）在项目中创建 Views 目录。在 Views 目录下依次创建 Demo1、Demo2 目录（对应上述两个控制器的名称）。

（4）在 Demo1 目录下创建_Layout.cshtml 文件，将作为布局视图使用。内容如下。

```
<!DOCTYPE html>

<html>
<head>
<meta name="viewport" content="width=device-width" />
<title>@ViewBag.Title</title>
</head>
<body>
<h3>Demo1 中的布局</h3>
<hr/>
<div>
@RenderBody()
</div>
</body>
</html>
```

在布局视图中，调用 RenderBody()方法的地方将会呈现普通视图，即 RenderBody()方法仅仅是一个占位符，在程序运行阶段会用指定视图的内容将其替换掉。

（5）在 Demo1 目录下创建 Main.cshtml 文件，它是 Main()操作方法的视图。

```
@{
    // 设置布局
    Layout = "_Layout";
}

<div>
    Demo1 -- Main 视图
</div>
```

Layout 属性用于设置该视图将要使用的布局，这里指定_Layout（即_Layout.cshtml 文件）。

（6）Demo2 目录下的内容与 Demo1 目录相似。

```
// _Layout.cshtml 文件
<!DOCTYPE html>

<html>
<head>
<meta name="viewport" content="width=device-width" />
<title>@ViewBag.Title</title>
</head>
<body>
<h3>Demo2 中的布局</h3>
<hr/>
<div>
@RenderBody()
</div>
</body>
</html>

// Main.cshtml 文件
```

```
@{
    // 设置布局
    Layout = "_Layout";
}

<div>
    Demo2 -- Main 视图
</div>
```

（7）在 Views 目录下创建 Shared 目录，在 Shared 目录下添加布局视图文件_Layout.cshtml。

```
<!DOCTYPE html>

<html>
<head>
<meta name="viewport" content="width=device-width" />
<title>@ViewBag.Title</title>
</head>
<body>
<h3>共享的布局</h3>
<hr/>
<div>
@RenderBody()
</div>
</body>
</html>
```

此时，示例项目中 Views 目录结构如图 9-16 所示。

运行应用程序，不管是访问/demo1/main，还是/demo2/main，MVC 视图使用的布局文件都来自 Shared 目录中的_Layout.cshtml，分别如图 9-17 和图 9-18 所示。

这是因为应用程序在/Views/Shared 目录找到共享的布局视图，子目录中的布局视图被忽略。现在，修改一下示例代码，将/Views/Shared 目录放在控制器目录之后。

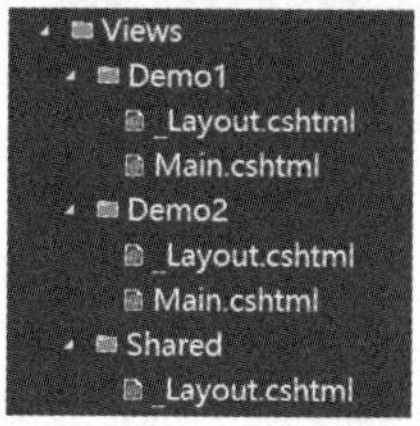

图 9-16　Views 目录结构

图 9-17　Demo1 控制器套用的布局文件

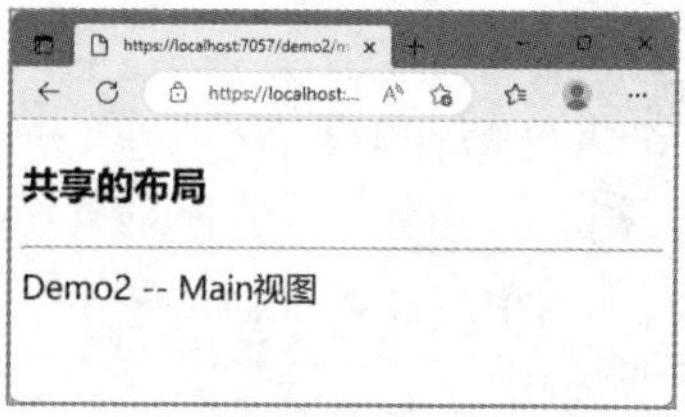

图 9-18　Demo2 控制器套用的布局文件

```
…
builder.Services.Configure<RazorViewEngineOptions>(opt =>
{
    opt.ViewLocationFormats.Clear();
```

```
    opt.ViewLocationFormats.Add("/Views/{1}/{0}.cshtml");
    opt.ViewLocationFormats.Add("/Views/Shared/{0}.cshtml");
});
...
```

再次运行示例程序。如果访问的是/demo1/main，套用的是 Demo1 目录下的布局视图；如果访问的是/demo2/main，套用的就是 Demo2 目录下的布局视图，分别如图 9-19 和图 9-20 所示。

图 9-19　套用 Demo1 目录中的布局文件

图 9-20　套用 Demo2 目录中的布局文件

9.7.5　示例：配置 Razor Pages 布局视图的查找路径

与 MVC 项目一样，Razor Pages 项目也能使用布局视图。默认的布局视图查找路径如下。

```
/Pages/{1}/{0}.cshtml
/Pages/Shared/{0}.cshtml
/Views/Shared/{0}.cshtml
```

其中，有两个格式占位符，具体如下。

{0}：视图名称。Razor Pages 项目的用户接口以页面为单位，因此这里所说的视图是指除页面以外的其他视图，如布局视图和局部视图（Partial View，本书后面会介绍）。

{1}：页面名称。通常，存放布局视图的目录不需要考虑页面名称（因为布局视图可通用于多个页面），所以页面名称的占位符多用于局部视图（针对某个页面可能会用到多个局部视图）。

本示例将布局视图文件的存放路径自定义为/Pages/Others 目录。具体实现步骤如下。

（1）在项目根目录下新建名为 Pages 的目录。这是 Razor 页面文件的默认存储目录。

（2）在 Pages 目录下新建目录并命名为 Others。

（3）在 Others 目录下新建布局文件，命名为_Layout.cshtml（这是默认文件名）。文件内容如下。

```
<!DOCTYPE html>

<html>
<head>
<meta name="viewport" content="width=device-width" />
<title>@ViewBag.Title</title>
</head>
<body>
```

```
<div>
<h3>布局视图</h3>
</div>
<hr />
<div>
@RenderBody()
</div>
</body>
</html>
```

当某个页面应用布局视图后，会将页面内容输出到@RenderBody()所在的位置。

（4）在 Pages 目录下新建一个 Home 页面，文件名为 Home.cshtml。

```
@page
@{
    // 设置要使用的布局视图
    Layout = "_Layout";
    // 设置页面标题
    ViewBag.Title = "Demo App";
}

<p>@(Path)页面</p>
```

（5）在应用程序的初始化代码中，注册 Razor Pages 相关的服务，并为终结点添加映射。

```
var builder = WebApplication.CreateBuilder(args);
builder.Services.AddRazorPages();
builder.Services.Configure<RazorViewEngineOptions>(opt =>
{
    // 清空路径列表
    opt.PageViewLocationFormats.Clear();
    // 添加自定义路径
    opt.PageViewLocationFormats.Add("/Pages/Others/{0}.cshtml");
});
var app = builder.Build();

app.MapRazorPages();

app.Run();
```

调用 AddRazorPages()方法后，对 RazorViewEngineOptions 选项类进行配置。先清空 PageViewLocationFormats 列表中的所有路径，然后添加自定义路径/Pages/Others/{0}.cshtml。在本示例中，格式占位符{0}的值为_Layout。

（6）运行示例程序，然后访问/home，结果如图 9-21 所示。

图 9-21　应用自定义目录中的布局视图

9.7.6　_ViewImports 与_ViewStart 文件

_ViewImports 与_ViewStart 这两个文件的命名是固定的，因为它们有特殊用途。

_ViewImports 文件经常用于编写引入命名空间、添加标记帮助器（Tag Helper）等指令。_ViewImports 文件可以为多个视图文件共享导入命令。例如，以下 Razor 标记将引入两个命名空间。

```
@using MyWebApplication
@using MyWebApplication1.Models
```

此_ViewImports 文件将作用于与其处于同级目录以及子目录下的所有视图文件。这些视图文件不需要重复引入命名空间。

_ViewImports 文件支持以下 Razor 指令。

（1）@namespace：为多个视图定义命名空间，在生成代码时将自动套用该指令所提供的命名空间。

（2）@using：为多个视图引入命名空间。如果多个_ViewImports 文件作用于同一个视图文件，那么所有@using 指令将被合并执行，重复引入的命名空间会忽略。

（3）@addTagHelper 与@removeTagHelper：添加/移除标记帮助器。如果同一视图应用多个_ViewImports 文件，那么所有@addTagHelper 和@removeTagHelper 指令都会执行。

（4）@tagHelperPrefix：为标记帮助器指定前缀。如果多个_ViewImports 文件都有此指令，那么离当前视图最近的指令会替换其他_ViewImports 文件中的同名指令。

（5）@model：为多个视图关联模型类。如果多个_ViewImports 文件都使用了此指令，那么只有离视图最近的@model 指令将替换其他同名指令。

（6）@inherits：指定视图类型的基类。与当前视图文件最接近的_ViewImports 文件会替换其他文件中的同名指令。

（7）@inject：用于依赖注入，向服务容器请求指定的类型。与当前视图最近的_ViewImports 文件会替换其他文件中的@inject 指令。

_ViewStart 文件中的 Razor 代码将在视图呈现前执行，一般用于放置可重复使用的代码。例如，多个视图使用同一个布局文件，没必要在每个视图都设置一次 Layout 属性，而应该将代码写在_ViewStart 文件中。例如：

```
@{
    Layout = "_rootLayout";
}
```

所有位于_ViewStart 文件作用域内的视图都会使用_rootLayout 布局。

_ViewImports 和_ViewStart 文件的作用域都有层次性。与这两个文件同级或子目录中的视图都会执行这两个文件。若这两个文件位于项目的根目录下，那么当前项目中所有视图文件都在它们的作用域内。

9.7.7 示例：_ViewStart 文件的替换行为

假设有 Test1、Test2 控制器，视图文件分别位于/Views/Test1/和/Views/Test2 目录下，并且在/Views 目录和/Views/Test2 目录下各有一个_ViewStart 文件。本示例将验证位于最里层的_ViewStart 文件是否会替换外层的_ViewStart 文件。

（1）两个控制器的定义如下。

```
public class Test1Controller : Controller
{
    public IActionResult DoSomething()
    {
        return View("myView");
    }
}

public class Test2Controller : Controller
{
    public IActionResult DoSomething()
    {
        return View("myView");
    }
}
```

两个控制器的结构相同，都包含名为 DoSomething()的操作方法，返回的视图名称都是 myView。

（2）在项目中新建 Views 目录。

（3）在 Views 目录下新建两个子目录，分别命名为 Test1 和 Test2。

（4）在 Test1 和 Test2 目录下各新建一个视图文件，文件名都是 myView.cshtml。这两个 myView.cshtml 文件的内容一样。

```
<div>
    @Path
</div>
<hr/>
<div>
    Test Data = @ViewData["val"]
</div>
```

Path 属性表示当前视图文件相对于根目录的路径。ViewData 属性是一个弱类型的字典对象，上述代码将从中取出 Key 为 val 的值并呈现在 HTML 中。val 的值会在_ViewStart 文件中设置。

（5）在/Views 目录下新建_ViewStart 文件（完整文件名为_ViewStart.cshtml）。然后设置 ViewData 属性中 val 的值。

```
@{
    ViewData["val"] = "Layer A";
}
```

（6）在/Views/Test2 目录下新建_ViewStart 文件，内容如下。

```
@{
    ViewData["val"] = "Layer B";
}
```

此时，两个_ViewStart 文件的位置如图 9-22 所示。

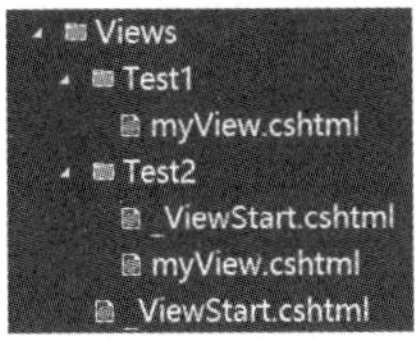

图 9-22　两个_ViewStart 文件的位置

运行示例程序，先访问/test1/dosomething，被执行的视图文件是/Views/Test1/myView.cshtml。对该视图有效的_ViewStart 文件位于/Views 目录下，因此 val 的值为 Layer A，如图 9-23 所示。

接着，访问/test2/dosomething，被执行的视图文件为/Views/Test2/myView.cshtml。由于 Test2 目录下存在_ViewStart 文件，它会替换 Views 目录下的_ViewStart 文件，因此 val 的值为 Layer B，如图 9-24 所示。

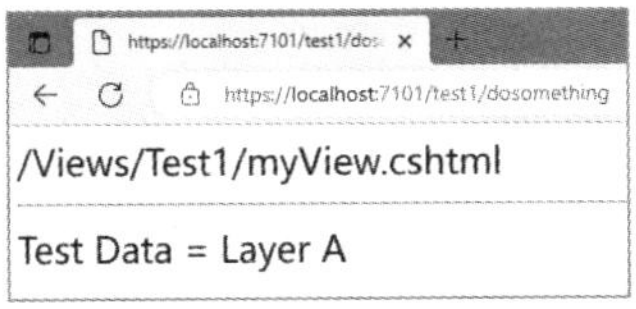

图 9-23　Views 目录下的_ViewStart 文件有效

图 9-24　Test2 目录下的_ViewStart 文件有效

_ViewImports 文件的替换行为与_ViewStart 文件相同，本书不再演示。

9.8　IViewLocationExpander 接口

前面已介绍过 ViewLocationFormats 的用法。该列表仅根据区域名称、控制器名称、视图名称这几个值生成视图的查找路径，但不能根据客户端的请求参数动态生成视图路径。例如，相同的用户界面，可以呈现为不同的语言。若查找路径为/Views/Docs/{0}.cshtml，当要呈现 Readme 视图时，实际查找路径为/Views/Docs/Readme.cshtml。为了呈现不同语言的文字，Readme 视图可能对应着多个 Razor 文档——呈现中文内容的文件名为 Readme_ZH.cshtml，呈现英文内容的文件名为 Readme_EN.cshtml。具体使用哪个文件，取决于客户端的请求。若用户希望显示中文内容，就执行 Readme_ZH.cshtml 中的代码。此场景单凭 ViewLocationFormats 列表是很难进行动态配置的。

于是，有必要在 ViewLocationFormats 的基础上进行增强，提供一种可以动态扩展视图查找路径的附加功能。这就要用到 IViewLocationExpander 接口了。

使用 IViewLocationExpander 接口需要实现以下两个方法。

```
public void PopulateValues (ViewLocationExpanderContext context);
public IEnumerable<string> ExpandViewLocations (ViewLocationExpanderContext
context, IEnumerable<string> viewLocations);
```

实现 PopulateValues()方法，可以向 ViewLocationExpanderContext.Values 属性填充所需要

的值（如果没有要填充的值，可以保留空白的方法体，但必须实现方法）。这些值随后可在ExpandViewLocations()方法中读取。

实现 ExpandViewLocations()方法，返回新生成的视图路径列表；或者将旧的路径列表与新生成的路径列表合并后返回。

实现 IViewLocationExpander 接口后，需要将这些类型的实例添加到选项类 RazorViewEngineOptions 的 ViewLocationExpanders 属性中。应用程序会缓存 ExpandViewLocations()方法所返回的路径列表。如果调用 PopulateValues()方法后，ViewLocationExpanderContext.Values 没有被更改，那么应用程序将使用缓存中的视图路径，不会调用 ExpandViewLocations()方法。如果调用 PopulateValues()方法后，ViewLocationExpanderContext.Values 属性中的数据有变动，那么应用程序就会调用 ExpandViewLocations()方法获取扩展后的视图路径列表，并建立新的缓存数据。

9.8.1 示例：多版本视图

本示例将实现根据请求的 URL 选择呈现不同版本的视图。例如，访问 http://abc.org/home/items/v1，就会执行/Views/Home/items/v1.cshtml 视图文件并返回 HTML 文档；如果访问的 URL 是 http://abc.org/home/Items/v2，就会执行/Views/Home/Items/v2.cshtml 文件并返回 HTML 文档。

具体实现步骤如下。

（1）定义 VersionViewLocationExpander 类，它实现了 IViewLocationExpander 接口。

```
public class VersionViewLocationExpander : IViewLocationExpander
{
    // 路由参数的键名
    private const string RouteValKey = "version";
    // 用于缓存视图路径的键名
    private const string PopValKey = "vs";

    public IEnumerable<string> ExpandViewLocations(ViewLocationExpanderContext
    context, IEnumerable<string> viewLocations)
    {
        // 取出 PopulateValues()方法中设置的值
        context.Values.TryGetValue(PopValKey, out string? version);
        // 生成新的路径
        foreach (var location in viewLocations)
        {
            if (!string.IsNullOrEmpty(version))
            {
                // 举例
                // 原格式：/Views/{1}/{0}.cshtml
                // 新格式：/Views/{1}/{0}/{version}.cshtml
                yield return location.Replace("{0}", "{0}/" + version);
            }
```

```
            else
            {
                // 返回旧的路径
                yield return location;
            }
        }
    }

    public void PopulateValues(ViewLocationExpanderContext context)
    {
        // 读取路由参数 - version
        if (context.ActionContext.RouteData.Values.TryGetValue
        ("version", out object? val))
        {
            // 将值转换为字符串
            string? verStr = val?.ToString();
            if (verStr != null)
            {
                // 添加到 Values 列表中
                context.Values[PopValKey] = verStr;
            }
        }
    }
}
```

在 PopulateValues()方法的实现代码中，先从路由参数列表中取出键名为 version 的值，然后将其添加到 ViewLocationExpanderContext.Values 属性中。

ExpandViewLocations()方法将从 ViewLocationExpanderContext.Values 属性读出已设置的值，然后对 viewLocations 中的路径列表进行修改，最后返回修改后的值。

新路径与旧路径的区别在于视图文件的名称上。如果默认的路径为

```
/Views/{1}/{0}.cshtml
```

那就修改为

```
/Views/{1}/{0}/{version}.cshtml
```

假设控制器名为 Home，视图名为 Items，version 的值为 v1，于是得到的新路径为 /Views/Home/Items/v1.cshtml。

（2）在初始化应用程序时，配置 RazorViewEngineOptions 选项类，将 VersionViewLocationExpander 实例添加到 ViewLocationExpanders 列表中。

```
...
builder.Services.AddControllersWithViews();
builder.Services.Configure<RazorViewEngineOptions>(opt =>
{
    opt.ViewLocationExpanders.Add(new VersionViewLocationExpander());
});
```

```
var app = builder.Build();
…
```

（3）添加 Home 控制器。

```
[Route("[controller]/[action]/{version?}")]
public class HomeController : Controller
{
    public IActionResult Items()
    {
        return View();
    }
}
```

路由参数 version 后面有“?”（英文问号）表示该参数是可选的。如果请求的 URL 中没有提供 version 参数，那么就按照默认的配置查找 Items.cshtml 视图文件；如果 version 参数的值为 v2，那么要查找的视图文件为 Items/v2.cshtml。

（4）在项目的根目录下新建 Views 目录。

（5）在 Views 目录下新建 Home 目录。

（6）在 Home 目录下新建 Items.cshtml 文件，为默认视图。

```
<h3>视图 - 默认</h3>

<div>欢迎使用</div>
```

（7）在 Home 目录下新建 Items 目录。版本化的视图文件将位于此目录下。

（8）在 Items 目录下新建 v1.cshtml 文件，为“版本 1”视图。

```
<h3>视图 - 版本 1</h3>

@for(int x = 1; x < 7; x++)
{
<p>项目 @x</p>
}
```

（9）在 Items 目录下新建 v2.cshtml 文件，为“版本 2”视图。

```
<h3>视图 - 版本 2</h3>

@for(int n = 1; n < 13; n++)
{
<p>任务 @n</p>
}
```

运行示例程序，当访问/home/items 时，version 参数被忽略，将呈现默认视图，如图 9-25 所示。

访问/home/items/v1，此时 version 参数的值为 v1，将呈现“版本 1”视图，如图 9-26 所示。

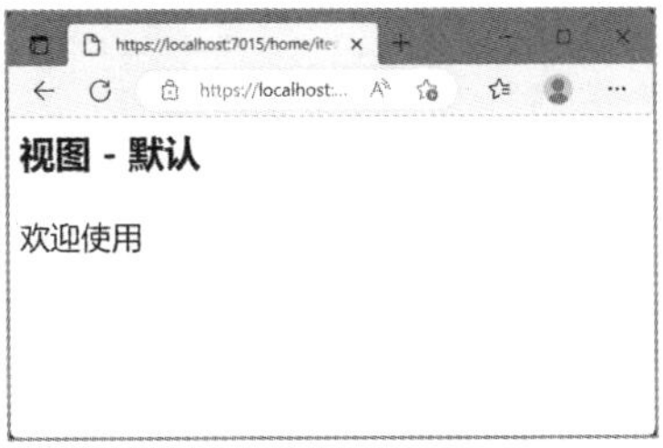

图 9-25　未版本化的视图

图 9-26　“版本 1”视图

访问/home/items/v2，将呈现“版本 2”视图，如图 9-27 所示。

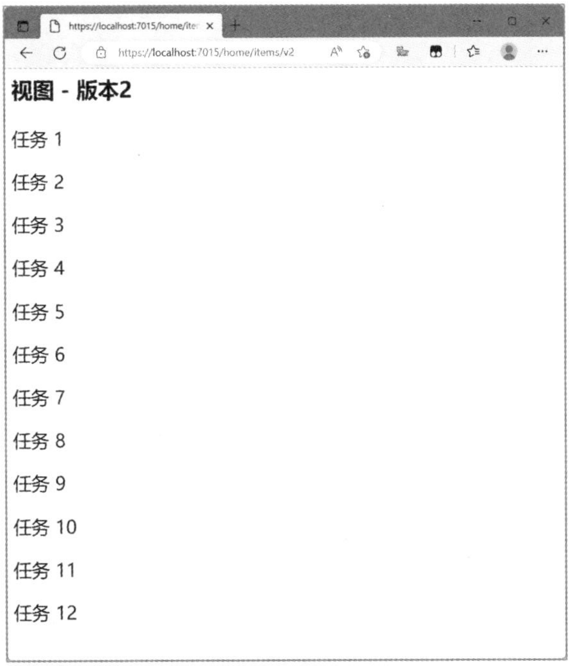

图 9-27　“版本 2”视图

9.8.2　示例：根据 URL 查询参数扩展视图路径

本示例将实现从 URL 的查询字符串中读取 style 参数，再根据该参数的值生成新的视图路径。示例使用的视图名称为 test，默认执行的 Razor 文件是 test.cshtml。如果请求的 URL 中带有 style 参数，如/gotoview?style=green，经过路径的扩展处理后，将查找的视图文件名改为 test_green.cshtml。

具体步骤如下。

（1）定义 StyleViewLocationExpander 类，它实现了 IViewLocationExpander 接口。

```
public class StyleViewLocationExpander : IViewLocationExpander
{
    public IEnumerable<string> ExpandViewLocations
    (ViewLocationExpanderContext context, IEnumerable<string> viewLocations)
    {
        context.Values.TryGetValue("my-style", out string? styleToUse);
        foreach(string oldLoc in viewLocations)
        {
            if(!string.IsNullOrWhiteSpace(styleToUse))
            {
                // 修改路径
                yield return oldLoc.Replace("{0}", "{0}_" + styleToUse);
            }
            else
            {
                // 不需要修改，返回旧路径
                yield return oldLoc;
            }
        }
    }

    public void PopulateValues(ViewLocationExpanderContext context)
    {
        // 从 URL 查询字符串中读取 style 字段的值
        string? style = context.ActionContext.HttpContext.Request.
        Query["style"];
        if(!string.IsNullOrEmpty(style))
        {
            // 添加到 Values 列表中
            context.Values.Add("my-style", style);
        }
    }
}
```

在 PopulateValues()方法中，从请求的 URL 中读出查询参数 style。如果顺利读出，就将它添加到 Values 列表中，键名为 my-style。

在 ExpandViewLocations()方法中，先读出 Values 列表中的 my-style 对应的值，然后通过 foreach 循环访问旧的视图路径，如果 style 参数的值有效就修改路径，否则返回旧的路径。

（2）在初始化应用程序的代码中，配置 RazorViewEngineOptions 选项类。将新的 StyleViewLocationExpander 实例插入 ViewLocationExpanders 列表的首位，使它的优先级最高，在其他 IViewLocationExpander 对象之前执行。

```
var builder = WebApplication.CreateBuilder(args);
builder.Services.AddControllersWithViews();
builder.Services.Configure<RazorViewEngineOptions>(o =>
```

```
{
    o.ViewLocationExpanders.Insert(0, new StyleViewLocationExpander());
});
var app = builder.Build();
…
```

（3）定义 Home 控制器。

```
public class HomeController : Controller
{
    public ActionResult Default()
    {
        return View();
    }

    [HttpGet("gotoview")]
    public ActionResult SetStyle()
    {
        return View("test");
    }

}
```

Default()方法返回主页，对应的视图文件为/Views/Home/Default.cshtml。

```
<h3>请选择一种风格</h3>
<hr/>
<style>
…
</style>

<a href="/gotoview?style=red">红色风格</a>
<a href="/gotoview?style=blue">蓝色风格</a>
<a href="/gotoview?style=yellow">黄色风格</a>
<a href="/gotoview?style=gray">灰色风格</a>
```

4 个<a>元素都链接到/gotoview（即调用 Home 控制器下的 SetStyle()操作方法），但传递的 style 参数不同。SetStyle()方法返回 test 视图。

（4）test 视图需要准备以下 4 个 Razor 文档。

```
// test_red.cshtml
<div style="color:red; font-size:28px">
    这是 red 风格
</div>

// test_blue.cshtml
<div style="color:blue; font-size:28px">
    这是 blue 风格
</div>

// test_yellow.cshtml
```

```
    <div style="color:yellow; font-size:28px">
        这是 yellow 风格
    </div>

    // test_gray.cshtml
    <div style="color:gray; font-size:28px">
        这是 gray 风格
    </div>
```

（5）另外，在/Views/Home 目录下还要创建一个 text.cshtml，用于作为默认视图（即没有提供 style 参数时呈现的内容）。

```
    <div style="font-size:28px">
        这是默认风格
    </div>
```

运行示例程序，将呈现主页面，其中有 4 个超链接，如图 9-28 所示。

单击页面上的超链接，即可根据 URL 查询参数呈现相应的视图，如图 9-29 所示。

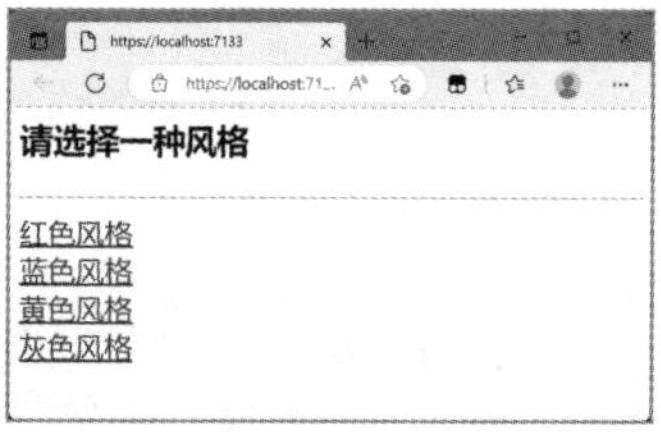

图 9-28　主页面

图 9-29　style 参数为 blue 时呈现的内容

9.8.3　LanguageViewLocationExpander 类

LanguageViewLocationExpander 是 ASP.NET Core 内置的视图路径扩展类（位于 Microsoft.AspNetCore.Mvc.Razor 命名空间），可以根据当前线程的语言/区域信息（CultureInfo）生成新的视图查找路径。

基于特定语言/区域的视图路径有两种格式，用 LanguageViewLocationExpanderFormat 枚举类型表示。

（1）Suffix：后缀，也就是将视图文件路径中的{0}替换为{0}.{culture}，即{0}.{culture}.cshtml，如 myview.zh-CN.cshtml（汉语，中国）、myview.en-US.cshtml（英语，美国）等。

（2）SubFolder：将语言/区域代码作为一个子目录，并将视图文件放在此目录下，如/Views/Demo/zh-HK/default.cshtml（汉语，中国香港）。

在使用 LanguageViewLocationExpander 类时，可以通过 AddViewLocalization()扩展方法来配置。

```
builder.Services.AddControllersWithViews()
.AddViewLocalization(format:LanguageViewLocationExpanderFormat.SubFolder);
```

也可以直接配置 RazorViewEngineOptions 选项类，把 LanguageViewLocationExpander 实例添加到列表中。

```
builder.Services.Configure<RazorViewEngineOptions>(opt =>
{
    opt.ViewLocationExpanders.Add(new LanguageViewLocationExpander
    (LanguageViewLocationExpanderFormat.SubFolder));
});
```

在组建 HTTP 管线时，要调用 UseRequestLocalization()方法，让 ASP.NET Core 根据 HttpContext 的信息自动设置当前线程的语言/区域。

```
app.UseRequestLocalization(…);
```

客户端在发起请求时，可用查询字符串、Cookies 等方法传递要使用的语言。例如/demo/someaction?culture=zh-TW&ui-culture=zh-TW，culture 字段用于设置当前线程或异步任务的语言特性，而 ui-culture 用于设置当前线程的用户界面所使用的语言。

下面的示例将使用子目录格式扩展视图路径。

```
var builder = WebApplication.CreateBuilder(args);
builder.Services.AddControllersWithViews();
builder.Services.Configure<RazorViewEngineOptions>(opt =>
{
    opt.ViewLocationExpanders.Add(new LanguageViewLocationExpander
    (LanguageViewLocationExpanderFormat.SubFolder));
});
var app = builder.Build();

RequestLocalizationOptions options = new RequestLocalizationOptions();
// 设置默认语言
options.SetDefaultCulture("zh-Hans");
// 设置支持的语言
options.AddSupportedCultures("zh-Hant");
options.AddSupportedUICultures("zh-Hant");
// 调用以下方法设置当前线程的语言
app.UseRequestLocalization(options);

app.MapControllerRoute("app", "{controller=Home}/{action=Index}");

app.Run();
```

在调用 UseRequestLocalization()扩展方法时，可以用 RequestLocalizationOptions 选项类设置应用程序所支持的语言/区域特性。上述代码配置应用程序支持繁体中文，并且默认语言是简体中文。

本示例已配置使用子目录的路径格式，因此在/Views/<控制器>目录下应先建立语言特性目录，再将视图文件放到此目录下。本示例的视图文件分布如图 9-30 所示，zh-Hans 表示简体中文，zh-Hant 表示繁体中文。

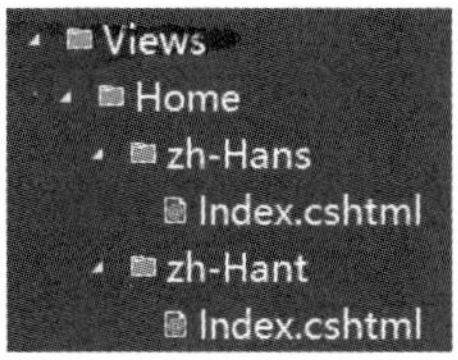

图 9-30　基于不同语言的视图文件

运行示例程序，访问以下 URL，将显示简体中文视图，如图 9-31 所示。

```
https://localhost/…/?culture=ch-hans
https://localhost/…/?ui-culture=ch-hans
https://localhost/…/?culture=zh-hans&ui-culture=ch-hans
```

访问以下 URL，将显示繁体中文视图，如图 9-32 所示。

```
https://localhost/…/?culture=zh-hant
https://localhost/…/?ui-culture=zh-hant
https://localhost/…/?ui-culture=zh-hant&culture=zh-hant
```

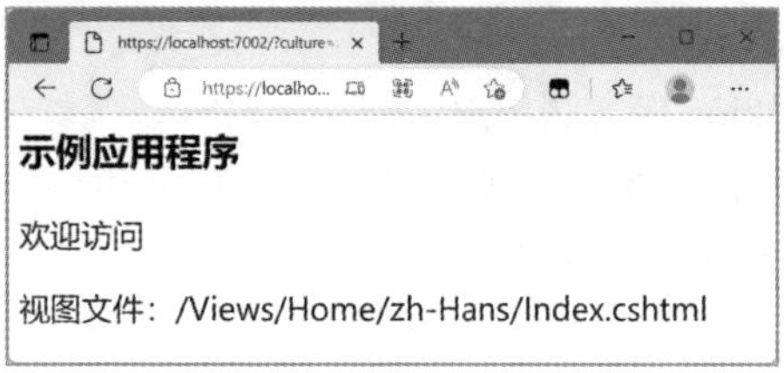

图 9-31　呈现简体中文视图

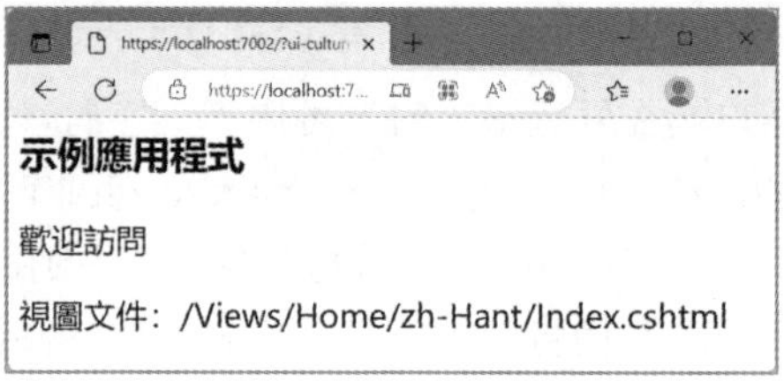

图 9-32　呈现繁体中文视图

9.9　局部视图

局部视图（Partial View，也译为“分部视图”，本书翻译为“局部视图”，语义上更易于理解）与普通视图相似，也是 Razor 文档。局部视图可以将一些需要重复使用的 HTML 以及 Razor 标记放到一个独立的文件中，组成一个小的部件。局部视图可以嵌套在普通视图或布局视图中。例如，可以将网站的导航菜单放到一个局部视图中，再将其作为布局视图的一部分。如果网站拥有多个布局视图，那么每个布局视图都可以引用这个导航菜单，免去了多次编写相同代码的麻烦。

局部视图的文件名一般会以“_”开头，如_test.cshtml、_colors.cshtml 等。这并不是一种强制性规则，但遵循这一约定，开发人员能够直观地区分普通视图和局部视图。

在控制器类中，局部视图可以通过操作方法返回。如果调用的是 View()方法，则与普通视图无异。View()方法既能执行普通视图，也能执行局部视图。如果调用的是 PartialView()方法，则局部视图不会执行_ViewStart 文件，但会执行_ViewImports 文件。

在 Razor 文档中呈现局部视图有以下几种方法。

（1）RenderPartialAsync()：此方法将局部视图的内容直接写入响应流中，性能较好。注意此方法不返回 HTML 内容，所以必须写在 Razor 代码块中。例如：

```
@{
    await Html.RenderPartialAsync("_myview", …);
}
```

Razor 代码块中使用的是标准的 C#代码，语句末尾必须有分号（英文）。

（2）PartialAsync()：此方法返回 HTML 内容，因此调用时在语句前加上@符号即可。语句末尾不用加分号。例如：

```
@await Html.PartialAsync("_myview", …)
```

（3）使用标记帮助器<partial>：类型为 PartialTagHelper（位于 Microsoft.AspNetCore.Mvc.TagHelpers 命名空间）。此方法使用起来如同普通 HTML 标记一样。例如：

```
<partial name="_myview"… />
```

name 属性指定视图名称或.cshtml 文件的路径。

另外，RenderPartialAsync()和 PartialAsync()方法有对应的同步方法 RenderPartial()和 Partial()。同步方法可能导致程序死锁，不建议使用。

9.9.1 示例：成绩单

本示例将使用局部视图呈现学生的考试成绩单。具体实现步骤如下。

（1）定义 Transcript 类，表示一名学生的成绩列表。

```
public class Transcript
{
    /// <summary>
    /// 学号
    /// </summary>
    public uint StudentID { get; set; }

    /// <summary>
    /// 学生姓名
    /// </summary>
    public string? StudentName { get; set; }

    /// <summary>
    /// 成绩表
    /// </summary>
    public IDictionary<string, float> Transcripts { get; set; } = new
    Dictionary<string, float>();
}
```

（2）定义 Demo 控制器，从 Reports()方法返回默认视图和成绩列表数据。

```
public class DemoController : Controller
{
    public IActionResult Reports()
    {
        // 准确示例数据
        Transcript[] list = new Transcript[]
        {
            new Transcript
            {
```

```
                StudentID = 2300151,
                StudentName = "张同学",
                Transcripts = new Dictionary<string, float>
                {
                    ["语文"] = 83.4f,
                    ["数学"] = 72.3f,
                    ["英语"] = 80.0f,
                    ["生物"] = 77.6f,
                    ["历史"] = 95f,
                    ["地理"] = 72f
                }
            },
            new Transcript
            {
                StudentID = 2300152,
                StudentName = "李同学",
                Transcripts = new Dictionary<string, float>
                {
                    ["语文"] = 81f,
                    ["数学"] = 69f,
                    ["英语"] = 78.5f,
                    ["生物"] = 85.2f,
                    ["历史"] = 88f,
                    ["地理"] = 71f
                }
            },
            new Transcript
            {
                StudentID = 2300153,
                StudentName = "吴同学",
                Transcripts = new Dictionary<string, float>
                {
                    ["语文"] = 97.4f,
                    ["数学"] = 82f,
                    ["英语"] = 83f,
                    ["生物"] = 70f,
                    ["历史"] = 89.5f,
                    ["地理"] = 84f
                }
            }
        };
        return View(list);
    }
}
```

View()方法在调用时没有指定视图名称，默认为 Reports。将 list 变量传递给 View()方法，以便在视图文件中可以访问 Transcript 对象数组。

（3）在项目的根目录下新建 Views 目录。

（4）在 Views 目录下新建 Demo 目录。

（5）在 Demo 目录下新建局部视图文件_transcript.cshtml，用于呈现一名学生的考试成绩。

```
@model Transcript

<style>
…
</style>

<table>
<caption style="text-align:left">@Model.StudentName   @Model.
StudentID</caption>
<tr>
        @foreach (var key in Model.Transcripts.Keys)
        {
            <td class="cellhd cell">@key</td>
        }
</tr>
<tr>
        @foreach (var val in Model.Transcripts.Values)
        {
            <td class="cell">@val</td>
        }
</tr>
</table>
```

@model 指令设置视图的模型对象为 Transcript 类型，随后 Model 属性引用的是表示某位同学考试成绩的 Transcript 实例。

（6）在 Demo 目录下添加 Reports.cshtml 视图文件，这个是普通视图，呈现完整的成绩单列表。

```
@model IEnumerable<Transcript>

<h3>三（1）班考试成绩单</h3>

@foreach (var trs in Model)
{
    @await Html.PartialAsync("_transcript", trs)
}
```

@model 指令设定模型为 IEnumerable<Transcript>类型，表示 Model 属性引用的是多个 Transcript 对象的列表(数组)。使用 foreach 循环遍历每个 Transcript 实例，并调用 PartialAsync() 方法呈现_transcript 局部视图，同时将当前 Transcript 实例传递进去。

（7）运行示例程序，结果如图 9-33 所示。

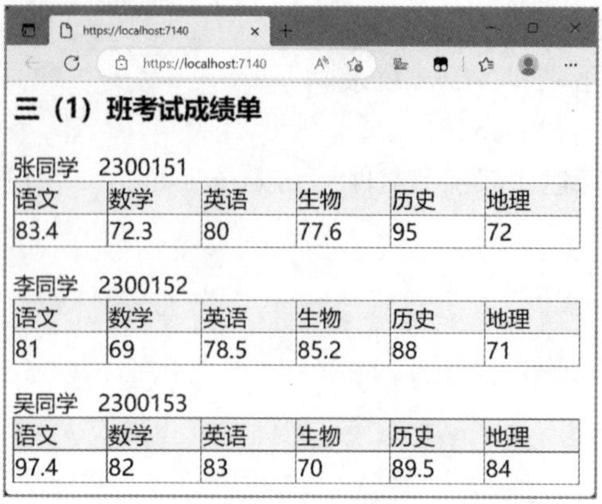

图 9-33 成绩单

9.9.2 示例：导航栏

本示例将使用局部视图制作导航栏_nav，然后在布局视图（_Layout）中通过<partial>标记帮助器引用_nav 视图。

具体实现步骤如下。

（1）定义 Home 控制器。

```
public class HomeController : Controller
{
    public IActionResult Index()
    {
        return View();
    }

    public IActionResult Contact()
    {
        return View();
    }

    public IActionResult Articles()
    {
        return View();
    }

    public IActionResult About()
    {
        return View();
    }
}
```

（2）在项目根目录下新建 Views 目录。

（3）在 Views 目录下新建 Home 目录。

（4）在 Home 目录下新建_nav.cshtml 文件。它是局部视图文件，用于制作导航栏。

```
<nav>
<ul class="nav">
<li><a href="/home/index">首页</a></li>
<li><a href="/home/contact">联系方式</a></li>
<li><a href="/home/articles">文章列表</a></li>
<li><a href="/home/about">关于本站</a></li>
</ul>
</nav>
```

导航栏有 4 个超链接，对应 Home 控制器中的 4 个操作方法。

（5）在 Home 目录下添加_Layout.cshtml 文件，作为布局视图。在此布局中，将引用局部视图_nav。

```
@addTagHelper Microsoft.AspNetCore.Mvc.TagHelpers.PartialTagHelper,
Microsoft.AspNetCore.Mvc.TagHelpers

<!DOCTYPE html>

<html>
<head>
<meta name="viewport" content="width=device-width" />
<title>@ViewBag.Title</title>
<style>
...
</style>
</head>
<body>
<div id="left">
<partial name="_nav" />
</div>
<div>
@RenderBody()
</div>
</body>
</html>
```

在文档的第 1 行要使用@addTagHelper 指令添加 PartialTagHelper 类的引用，即标记帮助器<partial>。页面上有两个<div>元素，第 1 个<div>元素位于页面左侧，用于呈现_nav 局部视图（导航栏），第 2 个<div>元素用于显示普通视图。

（6）在 Home 目录中添加_ViewStart.cshtml 文件，统一为各个视图引用布局文件。

```
@{
    Layout = "_Layout";
}
```

（7）在 Home 目录中依次添加 Index.cshtml、Contact.cshtml、Articles.cshtml、About.cshtml 这 4 个文件，分别对应 Home 控制器的操作方法。

```
// Index.cshtml
<h3>首页</h3>

// Contact.cshtml
<h3>联系方式</h3>
<p>地址：XX 省 YY 市 ZZ 区 KK 路 318 号</p>
<p>E-mail：dsbd@@126.net</p>

// Articles.cshtml
<h3>文章列表</h3>
<ul>
<li>文章 1</li>
<li>文章 2</li>
<li>文章 3</li>
<li>文章 4</li>
</ul>

// About.cshtml
<h3>关于本站</h3>
```

（8）运行示例程序，通过单击页面左侧的导航栏切换视图，如图 9-34 所示。

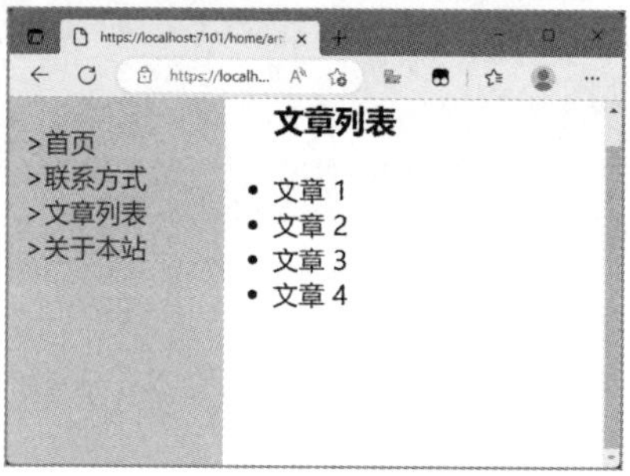

图 9-34　导航栏

9.10　视图组件

视图组件（View Component）使用起来要比局部视图灵活。它更像控制器，可以编写类代码处理程序逻辑，并返回指定视图文件。

符合以下条件的类，应用程序会识别为视图组件。

（1）公共非抽象类，不能包含泛型参数。

（2）类名以 ViewComponent 结尾，如 LoginViewComponent。

（3）类上应用了 ViewComponentAttribute 特性类。

（4）派生自 ViewComponent 类。

以下代码片段选自 ASP.NET Core 源代码，用于判断类型是否为视图组件。

```
public static bool IsComponent(TypeInfo typeInfo)
{
    if (typeInfo == null)
    {
        throw new ArgumentNullException(nameof(typeInfo));
    }

    if (!typeInfo.IsClass ||
        !typeInfo.IsPublic ||
        typeInfo.IsAbstract ||
```

```
        typeInfo.ContainsGenericParameters ||
        typeInfo.IsDefined(typeof(NonViewComponentAttribute)))
    {
        return false;
    }

    return
        typeInfo.Name.EndsWith(ViewComponentSuffix, StringComparison.
        OrdinalIgnoreCase) ||
        typeInfo.IsDefined(typeof(ViewComponentAttribute));
}
```

通常，编写视图组件只需要从 ViewComponent 类派生即可。作为视图组件的通用基类，ViewComponent 类已经应用了 ViewComponentAttribute 特性。如果不希望某个类被应用程序识别为视图组件，可以在类上应用 NonViewComponentAttribute 特性。

与中间件相似，视图组件必须包含名为 Invoke()或 InvokeAsync()的公共方法。

（1）Invoke()方法：同步调用，返回值为 IViewComponentResult 类型。

（2）InvokeAsync()方法：异步调用，返回值为 Task<IViewComponentResult>类型。

视图组件不直接处理 HTTP 请求，它可以在视图中调用，也可以在控制器中调用。

9.10.1　示例：一个简单的视图组件

本示例将完成一个 SayHello 视图组件，调用后会呈现文本“你好，MVC”。具体实现步骤如下。

（1）定义 SayHelloViewComponent 类，派生自 ViewComponent 类，表示该类是一个视图组件。

```
public class SayHelloViewComponent : ViewComponent
{
    public IViewComponentResult Invoke()
    {
        // 返回默认视图 Default
        return View();
    }
}
```

视图组件类中必须包含 Invoke()或 InvokeAsync()方法。本示例使用 Invoke()同步方法，返回类型为 IViewComponentResult。调用 View()方法可返回与该视图组件所使用的视图文件。上述代码中调用了无参数的 View()方法，表示返回的视图为默认命名 Default。

（2）定义 Home 控制器，其中包含 Test()方法，返回 Test 视图（此视图是普通视图，非视图组件的视图）。

```
public class HomeController : Controller
{
    public IActionResult Test()
    {
```

```
            // 返回 Test 视图
            return View();
        }
    }
```

（3）在项目根目录下新建 Views 目录。

（4）在 Views 目录下新建 Home 目录。

（5）在 Home 目录下新建 Test.cshtml 视图文件。

```
<div>
<h3>Test 视图</h3>
</div>
<p>以下为视图组件输出的内容：</p>
<div style="background-color:lightblue; padding:12px;">
    @await Component.InvokeAsync("SayHello")
</div>
```

调用 Component.InvokeAsync()方法可以执行一个视图组件，调用时要提供视图组件的名称（区分大小写）。

（6）在 Home 目录下新建 Components 目录。视图组件的视图文件将存放于此目录下。

（7）在 Components 目录下新建 SayHello 目录（视图组件的名称）。

（8）在 SayHello 目录下新建 Default.cshtml 文件（视图组件的默认视图名称）。

```
<span style="font-size:24px; color:purple">
你好，MVC</span>
```

运行示例程序，结果如图 9-35 所示。

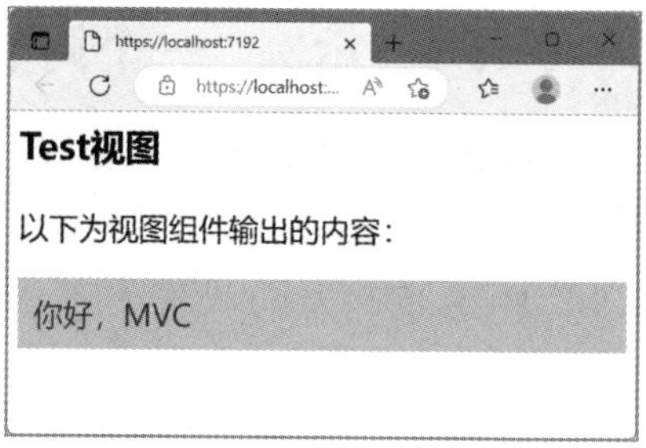

图 9-35　一个简单的视图组件

9.10.2　视图文件的查找路径

视图组件查找视图文件的路径是基于当前执行的普通视图路径生成的。例如，默认的视图查找路径为

```
/Views/{1}/{0}.cshtml
/Views/Shared/{0}.cshtml
```

假设当前执行的控制器为 Home，于是得到视图文件的查找路径为

```
/Views/Home/{0}.cshtml
```

视图组件的视图文件固定的相对路径为

```
Components/{0}/{1}
```

其中，{0}是视图组件的名称；{1}是视图名称。

然后，用这个相对路径替换普通视图路径中的{0}，得到视图组件查找视图的路径。

```
/Views/Home/Components/{0}/{1}.cshtml
```

假设视图组件的名称为 Foo，视图名称为 abc，那么视图文件将从以下路径查找。

```
/Views/Home/Components/Foo/abc.cshtml
```

又如，RazorViewEngineOptions.ViewLocationFormats 属性的配置如下。

```
/AppUI/{0}.cshtml
/AppUI/Common/{0}.cshtml
```

则视图组件将在以下路径查找视图文件。

```
/AppUI/Components/{0}/{1}.cshtml
/AppUI/Common/Components/{0}/{1}.cshtml
```

假设视图组件的名称为 PortList，视图名称为 Sorted，那么其查找路径为

```
/AppUI/Components/PortList/Sorted.cshtml
/AppUI/Common/Components/PortList/Sorted.cshtml
```

视图组件默认的视图名称为 Default，即在 Invoke()或 InvokeAsync()方法中调用 View()方法时，如果不明确指定视图名称或视图文件的路径，那么应用程序将认为视图名称为 Default。

9.10.3　示例：带参数的视图组件

如果视图组件需要参数，在定义 Invoke()或 InvokeAsync()方法时需要指定方法的参数列表。本示例将演示需要两个 int 类型参数的视图组件。组件执行后会计算两个参数的乘积，并把计算结果呈现到视图中。

具体实现步骤如下。

（1）定义 MultViewComponent 类，作为视图组件使用。组件名称为 Mult。

```
public class MultViewComponent : ViewComponent
{
    public IViewComponentResult Invoke(int num1, int num2)
    {
        // 计算结果
        int res = num1 * num2;
        return View("Result", res);
    }
}
```

Mult 视图组件在调用时需要 int 类型的参数 num1 和 num2，然后计算二者的乘积，最后作为视图模型传递给视图。

（2）定义 AppController 类，它是 MVC 控制器。

```
public class AppController : Controller
{
    public IActionResult Index() => View();
}
```

（3）在项目中新建 Views 目录，在 Views 目录下新建 App 目录。

（4）在 App 目录下新建 Index.cshtml 文件，它是 App 控制器中用到的视图。

```
@{
    var request = Context.Request;
    // 只有 POST 请求才需要处理
    if (request.Method == HttpMethods.Post)
    {
        // 读取 x、y 两个字段的值
        var x = request.Form["x"];
        var y = request.Form["y"];
        // 转换为 int 类型
        if (int.TryParse(x, out int n1) && int.TryParse(y, out int n2))
        {
            // 缓存这两个值，当重新呈现视图时用于还原输入框中的内容
            ViewData["num1"] = n1;
            ViewData["num2"] = n2;
            // 调用视图组件
            @await Component.InvokeAsync("Mult", new {
                num1 = n1,
                num2 = n2
            })
        }
    }
}

<div>
<p>输入参数</p>
<form method="post">
<fieldset>
<legend>第 1 个整数</legend>
<input type="number" id="x" name="x" value="@ViewData["num1"]"/>
</fieldset>
<fieldset>
<legend>第 2 个整数</legend>
<input type="number" id="y" name="y" value="@ViewData["num2"]"/>
</fieldset>

<button type="submit">计算</button>
</form>
</div>
```

视图代码先判断 HTTP 请求方式，如果为 HTTP-POST，就读出<form>表单提交的数值，最后调用 Mult 视图组件。

在调用视图组件时，将组件参数包装为一个匿名对象进行传递。

```
@await Component.InvokeAsync("Mult", new {
    num1 = n1,
    num2 = n2
})
```

匿名对象的属性名称要与视图组件的参数名称一一对应（即 num1 和 num2）。Component.InvokeAsync()方法调用语句前有@符号，表示此语句的执行结果会输出到视图内容中，因此行末不要加分号（;）。

（5）在 App 目录下新建 Components 目录。在 Components 目录下新建 Mult 目录（以视图组件的名称命名）。

（6）在 Mult 目录下新建 Result.cshtml 文件。这是 Mult 视图组件的视图文件。

```
@model int

<span>计算结果：@Model</span>
```

（7）运行示例程序，第 1 个整数输入 50，第 2 个整数输入 8，如图 9-36 所示。

（8）单击“计算”按钮，得到两个整数的乘积，如图 9-37 所示。

图 9-36　输入待计算数值

图 9-37　呈现计算结果

9.10.4　通过标记帮助器调用视图组件

在 Razor 文档中，除了调用视图文件的 Component.InvokeAsync()方法，还可以以标记帮助器（Tag Helper）的方式调用。HTML 标记格式为

```
<vc:[视图组件名称]></vc: [视图组件名称]>
```

例如，视图组件的名称为 Bar，那么标记格式为

```
<vc:Bar></vc:Bar>
```

也可以使用单闭合的标签，如

```
<vc:Bar />
```

如果视图组件有参数，可以使用 HTML 标签特性的格式。

```
<vc:Bar arg1=… arg2=… ></vc:Bar>
```

其中，arg1、arg2 就是视图组件的参数。

要想让 Razor 语言引擎将视图组件识别为标记帮助器，需要加上@addTagHelper 指令，如

```
@addTagHelper *, Abc
```

@addTagHelper 指令需要一个字符串类型的参数，表示要查找视图组件的程序集。*表示扫描并识别程序集 Abc 中所有视图组件类。

由于 Razor 语言服务会在编译时动态生成基于视图组件的标记帮助器，在编写代码时无法预知类型的名称，因此应当使用*使应用程序自动导入指定程序集中的所有视图组件。

9.10.5 示例：Greeting 视图组件

本示例将实现 Greeting 视图组件。该组件带有一个 name 参数，调用后会输出“Hello, {name}”。例如，name 参数的值为 Lucy，那么视图组件调用后将输出“Hello, Lucy”。

本示例主要演示用标记帮助器的方式调用 Greeting 视图组件。具体实现步骤如下。

（1）定义 GreetingViewComponent 类。

```
public class GreetingViewComponent : ViewComponent
{
    public IViewComponentResult Invoke(string name)
    {
        return Content($"Hello, {name}.");
    }
}
```

上述代码中，Invoke()方法中调用的是 Content()方法，它将返回指定的字符串，并输出到 HTML 文档中。类型为 ContentViewComponentResult，此返回结果不需要执行视图文件。

（2）定义 Demo 控制器。

```
public class DemoController : Controller
{
    public IActionResult Index()
    {
        return View();
    }
}
```

（3）以下为 Index.cshtml 视图文件的内容（位于/Views/Demo）。

```
@addTagHelper *, MyApp

<p>Index 视图</p>

<div style="padding:15px; background-
color:yellow">
<vc:greeting name="Bob"></vc:greeting>
</div>
```

第 1 行需要使用@addTagHelper 指令添加视图组件所在程序集引用（本例为 MyApp）。引用 Greeting 视图组件的标记为<vc:greeting>，此处传递给 name 参数的值为 Bob。

（4）运行示例程序，结果如图 9-38 所示。

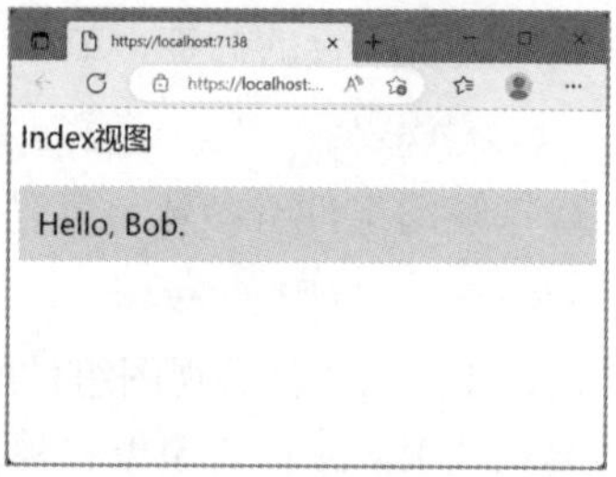

图 9-38 通过标记帮助器调用视图组件

9.10.6 示例：在 MVC 控制器中调用视图组件

本示例将演示如何在控制器中直接调用视图组件并返回结果。

首先，定义 RandViewComponent 类，视图组件名称为 Rand，调用后返回随机数值。

```
public class RandViewComponent : ViewComponent
{
    public IViewComponentResult Invoke()
    {
        var random = Random.Shared;
        // 生成随机数
        double v = random.NextDouble();
        // 将数值乘以 10000
        v *= 10000d;
        // 返回结果
        return Content($"随机数：{v:F3}");
    }
}
```

Random.Shared 属性是静态成员，可获取一个线程共享的 Random 实例。调用该实例的方法即可获取随机数值。上述代码中，NextDouble()方法生成 0 ~ 1 的双精度小数，然后乘以 10000。在返回结果时，使用了格式控制符 F3，表示保留 3 位小数的浮点数。

接着，定义 Test 控制器。在 GetNum()操作方法中调用 Rand 视图组件，并呈现结果。

```
public class TestController : Controller
{
    public IActionResult GetNum()
    {
        return ViewComponent("Rand");
    }
}
```

运行示例程序，访问/test/getnum，即可查看随机数值，如图 9-39 所示。

图 9-39 随机浮点数

9.10.7 两个特性类

在定义视图组件时，可以搭配使用以下两个特性类。

（1）ViewComponentAttribute：将该特性应用于某个类上，表明这个类为视图组件。同时，可通过 Name 属性自定义视图组件的名称。ViewComponent 类已经应用了此特性，所以从 ViewComponent 派生的类都会成为视图组件。例如：

```
[ViewComponent(Name = "Fly")]
public class SomeoneViewComponent : ViewComponent
{
    public IViewComponentResult Invoke() =>…
}
```

应用了 ViewComponentAttribute 后，SomeoneViewComponent 类将被标识为视图组件，并且组件名称为 Fly。

（2）NonViewComponentAttribute：应用了此特性的类不会被应用程序识别为视图组件。例如：

```
[NonViewComponent]
public class CleanUpViewComponent : ViewComponent
{
    …
}
```

调用 CleanUp 视图组件将失败，因为它不会被判定为视图组件。

9.11 识别其他程序集中的控制器

MVC 项目中存在“应用程序部件”的概念——用 ApplicationPart 类（位于 Microsoft.AspNetCore.Mvc.ApplicationParts 命名空间）表示。默认情况下，运行 ASP.NET Core 应用程序的主程序集已经是应用程序部件列表中的一员，即应用程序会在主程序集中查找控制器类。

若希望 ASP.NET Core 运行时能发现和识别其他程序集（此程序集可能被 ASP.NET Core 项目引用，也可能是已生成的类库文件），则需要将这些程序集添加到主应用程序的 ApplicationPart 列表中。

9.11.1 示例：使用 ApplicationPartAttribute 类

在主程序集中应用 ApplicationPartAttribute 特性类，并指定目标程序集的名称，就可以将这些程序集包含在 ApplicationPart 列表中。本示例即将演示 ApplicationPartAttribute 类的使用。

（1）新建一个类库项目，名称为 DemoLib。

（2）打开 DemoLib.csproj 项目文件，将 Project 节点的 Sdk 属性修改为 Microsoft.NET.Sdk.Web，输出类型设置为 Library（即类库项目）。然后保存并关闭文件。

```
<Project Sdk="Microsoft.NET.Sdk.Web">

<PropertyGroup>
…
<OutputType>Library</OutputType>
</PropertyGroup>

</Project>
```

（3）添加控制器类 DemoController。

```
public class DemoController : Controller
{
    public IActionResult Index()
    {
```

```
        return View();
    }
}
```

（4）创建 Views 目录。

（5）在 Views 目录下创建 Demo 子目录。

（6）在 Demo 目录下创建 Index.cshtml 视图文件。

```
<div>视图：@Path</div>

<p>来自程序集@(GetType().Assembly.GetName().Name)</p>
```

（7）创建 ASP.NET Core 应用程序项目，本示例的项目名称为 MyApp。

（8）打开 MyApp.csproj 项目文件，添加 ProjectReference 元素，引用 DemoLib 项目。

```
<Project Sdk="Microsoft.NET.Sdk.Web">

    …
    <ItemGroup>
        <ProjectReference Include="../DemoLib/DemoLib.csproj" />
    </ItemGroup>
</Project>
```

（9）打开 Program.cs 文件，在所有代码之前添加以下内容。

```
[assembly: ApplicationPart("DemoLib")]
```

ApplicationPartAttribute 是应用到程序集上的，因此必须写在所有程序代码语句之前（不包括 using 语句）。通过构造函数的参数提供程序集的名称。本示例中要定位控制器的程序集是 DemoLib。

（10）ASP.NET Core 应用程序的初始化代码如下。

```
var builder = WebApplication.CreateBuilder(args);
builder.Services.AddControllersWithViews();
var app = builder.Build();

app.MapControllerRoute("app",
"{controller=demo}/{action=index}");
app.Run();
```

（11）运行应用程序，结果如图 9-40 所示。

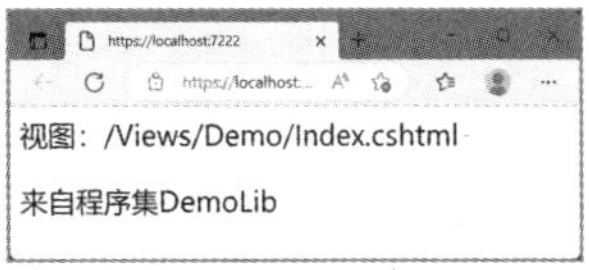

图 9-40　来自其他程序集的 Index 视图

9.11.2　示例：使用 AddApplicationPart()扩展方法

本示例将演示通过调用 AddApplicationPart()扩展方法添加 ApplicationPart，从而可以识别其他程序集中的 MVC 控制器。

（1）新建一个类库项目，命名为 CalcLib。

（2）打开 CalcLib.csproj 项目文件。将 Project 元素的 Sdk 属性修改为 Microsoft.NET.Sdk.Web，项目输出类型设置为 Library。

```
<Project Sdk="Microsoft.NET.Sdk.Web">

    <PropertyGroup>
        <TargetFramework>net7.0</TargetFramework>
        <Nullable>enable</Nullable>
        <ImplicitUsings>enable</ImplicitUsings>
        <OutputType>Library</OutputType>
    </PropertyGroup>

</Project>
```

（3）添加 Calculator 控制器。

```
[Route("cal/[action]")]
public class CalculatorController : Controller
{
    [HttpGet("{a:int?}/{b:int?}")]
    public IActionResult Work(int a = 0, int b = 0)
    {
        IList<string> lines = new List<string>();
        // 加
        lines.Add($"{a} + {b} = {a + b}");
        // 减
        lines.Add($"{a} - {b} = {a - b}");
        // 乘
        lines.Add($"{a} * {b} = {a * b}");
        // 除
        if(b != 0)     // 除数不能为 0
        {
            lines.Add($"{a} / {b} = {a / b}");
        }

        // 返回显示结果的视图
        return View("Result", lines);
    }
}
```

参数 a 和 b 均为 int 类型数值，Work()操作方法对两个参数作四则运算，最终返回 Result 视图。

（4）新建 Views 目录。在 Views 目录下新建 Calculator 子目录。

（5）在 Calculator 目录中新建 Result.cshtml 文件。

```
@model IEnumerable<string>

<p>计算完毕</p>

@foreach(string s in Model)
{
<div>@s</div>
}
```

（6）新建 ASP.NET Core 项目，本示例将项目命名为 MyApp。

（7）打开 MyApp.csproj 项目文件，添加 ProjectReference 元素，引用 CalcLib 程序集。

```
<Project Sdk="Microsoft.NET.Sdk.Web">

…

    <ItemGroup>
        <ProjectReference Include="../CalcLib/CalcLib.csproj"/>
    </ItemGroup>
</Project>
```

（8）在初始化代码中，调用 AddApplicationPart()扩展方法，将 CalcLib 程序集添加到 ApplicationPart 列表中。

```
builder.Services.AddControllersWithViews().
AddApplicationPart(Assembly.Load("CalcLib"));
```

（9）运行示例程序。按照 Calculator 控制器配置的路径规则，假设两个参数的值分别为 1200 和 25。访问的 URL 为/cal/work/1200/25，结果如图 9-41 所示。

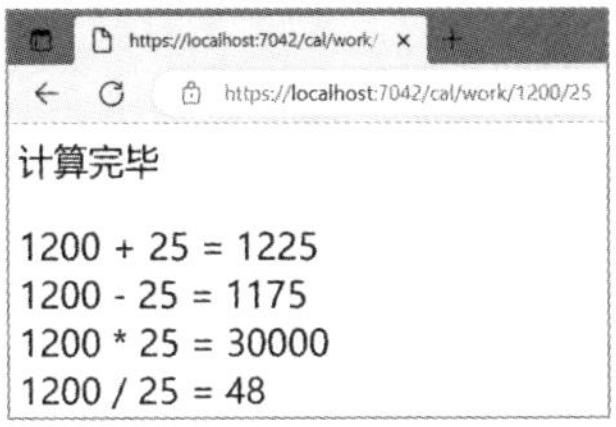

图 9-41　计算结果

9.11.3　示例：使用 ApplicationPartManager 类

ApplicationPartManager 类公开 ApplicationParts 属性，类型为 ApplicationPart 列表。而 AssemblyPart 派生自 ApplicationPart 类。因此，将 AssemblyPart 实例添加到 ApplicationParts 列表中，能让应用程序查找其他程序集中的 MVC 控制器。本示例演示 ApplicationParts 属性的使用。

（1）创建一个类库项目，命名为 TestLib。

（2）打开 TestLib.csproj 项目文件，将 Sdk 属性修改为 Microsoft.NET.Sdk.Web，输出类型设置为 Library，然后保存并关闭文件。

```
<Project Sdk="Microsoft.NET.Sdk.Web">

    <PropertyGroup>
        …
        <OutputType>Library</OutputType>
    </PropertyGroup>

</Project>
```

（3）添加 Demo 控制器。

```
public class DemoController : Controller
{
    public IActionResult Default()
    {
        return View("Index");
    }
}
```

（4）在 TestLib 项目的根目录下新建 Views 目录。

（5）在 Views 目录下新建 Demo 目录。

（6）在 Demo 目录下新建 Index.cshtml 视图文件。

```
<div>
    此视图来自@(GetType().Assembly.GetName().Name)程序集
</div>
```

（7）创建 ASP.NET Core 应用程序。打开项目文件，添加对 TestLib 项目的引用。

```
<Project Sdk="Microsoft.NET.Sdk.Web">

    …

    <ItemGroup>
        <ProjectReference Include="../TestLib/TestLib.csproj"/>
    </ItemGroup>
</Project>
```

（8）在应用程序的初始化代码中，调用 ConfigureApplicationPartManager()扩展方法，配置 ApplicationPartManager 对象。

```
var builder = WebApplication.CreateBuilder(args);
builder.Services.AddControllersWithViews().ConfigureApplicationPartManager
(appmgr =>
{
    appmgr.ApplicationParts.Add(new AssemblyPart(Assembly.Load("TestLib")));
});
…
```

（9）运行示例程序，结果如图 9-42 所示。

图 9-42　调用 TestLib 程序集中的控制器

第 10 章 模 型 绑 定

本章要点

- 使用内置的模型绑定器（自动绑定）
- 手动设置模型绑定的数据源
- 自定义 IValueProvider 和 IModelBinder 接口
- MVC 控制器与 Razor Pages 的属性绑定

10.1 概述

模型绑定（Model Binding）是一种自动化机制，从多种来源中检索数据，然后将检索结果转换为.NET 类型，最后赋值给 Razor Pages 模型或 MVC 控制器的方法参数和属性。

模型数据一般来自 HTTP 请求，如表单数据（form-data）、URL 查询字符串、HTTP 消息正文、URL 路由参数等。服务容器注册与 MVC 相关的服务时，默认会通过 MvcOptions 配置以下数据来源。

```
// 模型数据来自表单
options.ValueProviderFactories.Add(new FormValueProviderFactory());

// 模型数据来自路由参数
options.ValueProviderFactories.Add(new RouteValueProviderFactory());

// 模型数据来自 URL 查询字符串
options.ValueProviderFactories.Add(new QueryStringValueProviderFactory());

// 模型数据来自通过 JQuery 提交的表单数据（form-data）
options.ValueProviderFactories.Add(new JQueryFormValueProviderFactory());

// 模型数据来自表单的文件上传
options.ValueProviderFactories.Add(new FormFileValueProviderFactory());
```

上述*ValueProviderFactory 类型都实现了 IValueProviderFactory 接口。

```
public interface IValueProviderFactory
{
    Task CreateValueProviderAsync(ValueProviderFactoryContext context);
}
```

实现 CreateValueProviderAsync()方法为数据来源创建*ValueProvider 实例——数据提供者。而*ValueProvider 会实现 IValueProvider 接口。在模型绑定时，模型绑定器（ModelBinder）会从*ValueProvider 中获取数据，然后赋值给.NET 对象，完成绑定。

假设模型数据来自 URL 的查询字符串，那么对应的数据提供者是 QueryStringValueProvider 类，而负责创建 QueryStringValueProvider 实例的是 QueryStringValueProviderFactory 类。详细的类型对照可参考表 10-1。

表 10-1 各种数据来源的数据提供者类型

数据来源	*ValueProviderFactory	*ValueProvider
表单数据	FormValueProviderFactory	FormValueProvider
查询字符串	QueryStringValueProviderFactory	QueryStringValueProvider
JQuery提交的表单数据	JQueryFormValueProviderFactory	JQueryFormValueProvider
路由参数	RouteValueProviderFactory	RouteValueProvider
上传文件	FormFileValueProviderFactory	FormFileValueProvider

应用程序还可以使用 CompositeValueProvider 类将多个*ValueProvider 对象组合起来，可以从多个数据来源检索同一个数据项。

借助 ASP.NET Core 内部已实现的模型绑定机制，开发人员可免去编写代码读取 HTTP 请求再转换为.NET 类型的复杂过程，能更好地将精力投入程序逻辑和功能开发。

10.2 自动绑定

URL 查询字符串和表单数据是最常见的数据来源。这些数据会自动与 MVC 操作方法的参数进行绑定，开发人员无须进行额外的处理。

一般来说，HTTP-GET 请求不带 HTTP 消息正文，数据将通过查询字符串传递，如

```
https://test.org/track?id=5
```

上述 URL 在访问时向服务器传递了名为 id 的参数，其值是 5。如果有多个参数，使用&符号连接，如

```
https://abc.com/query?class=7&key=front
```

上述 URL 携带了两个查询参数：class=7，key=front。

如果客户端发送的是 HTTP-POST 请求，通常带有消息正文，并且正文内容是表单数据，其表示格式与 URL 查询字符串相似，如

```
a=1&b=2&c=3
```

当 HTTP-POST 方式发送的消息没有正文（no-body）时，也可以使用 URL 查询字符串传递数据。对于 HTTP-PUT、HTTP-DELETE 等请求方式，其数据传递方法也类似。

当 URL 路由规则中出现的字段（controller、action 等特殊字段除外）名称与 MVC 操作方法的参数名称相同时，也可以自动绑定，如

```
pictures/[action]/{title}
```

如果某个操作方法中存在名为 title 的参数，当访问以下 URL 时，会自动设置该参数的值为“春日晚景”。

```
http://mywds.net/pictures/load/春日晚景
```

10.2.1　示例：计算器

本示例通过 Calculator 控制器实现 3 个整数的四则运算。例如：

```
5+2+3
200-50-4
12×5×28
5000÷10÷50
```

3 个参数 x、y、z 对应参与运算的整数，4 个操作方法对应四则运算规则。

CalculatorController 类的实现代码如下。

```
public class CalculatorController : Controller
{
    [HttpGet("calc/add")]
    public IActionResult Addition(int x, int y, int z)
    {
        int result = default;
        result = x + y + z;
        return Content($"{x} + {y} + {z} = {result}");
    }

    [HttpGet("calc/sub")]
    public IActionResult Subtraction(int x, int y, int z)
    {
        int result = x - y - z;
        return Content($"{x} - {y} - {z} = {result}");
    }

    [HttpGet("calc/mul")]
    public IActionResult Multiplication(int x, int y, int z)
    {
        int result = x * y * z;
        return Content($"{x} * {y} * {z} = {result}");
    }

    [HttpGet("calc/div")]
```

```
        public IActionResult Division(int x, int y, int z)
        {
            if(y == 0 || z == 0)
            {
                return Content("0 不能作除数");
            }
            int result = x / y / z;
            return Content($"{x} / {y} / {z} = {result}");
        }
    }
```

HttpGet 特性指定操作方法仅接受 HTTP-GET 请求。

下面是应用程序初始化代码。

```
var builder = WebApplication.CreateBuilder(args);
builder.Services.AddControllers();
var app = builder.Build();

app.MapControllers();

// 设置应用程序的侦听 URL
app.Urls.Add("https://localhost:8275");

app.Run();
```

运行示例程序，然后打开命令行窗口，执行以下命令，用 http-repl 工具进行测试。

```
httprepl https://localhost:8275
```

若要进行加法运算，则访问的 URL 为/calc/add，参数 x、y、z 的值将通过查询字符串来传递。执行命令如下。

```
get /calc/add?x=100&y=50&z=30
```

Web 服务器将返回如图 10-1 所示的响应结果。

若要执行乘法运算，则应访问/calc/mul。命令如下。

```
get /calc/mul?x=5&y=10&z=7
```

计算结果如图 10-2 所示。

图 10-1　加法运算的响应结果

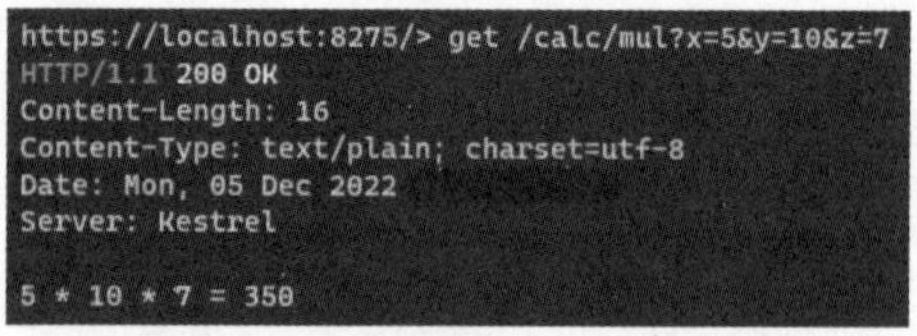

图 10-2　乘法运算的计算结果

10.2.2　示例：绑定数组类型的数据

本示例将演示客户端如何向服务器传递数组类型的数据。

Test 控制器的代码如下。

```
[Route("[controller]/[action]")]
public class TestController : Controller
{
    public IActionResult SetItems(string[] items)
    {
        string s = string.Join(", ", items);
        return Ok("你提交的内容：" + s);
    }
}
```

SetItems()方法的 items 参数是字符串数组类型，客户端调用时有以下两种比较简单且易于理解的传递格式。

（1）使用相同的字段名，且字段名与参数名称相同，如

```
items=a&items=b&items=c
```

（2）指定索引，如

```
items[0]=a&items[1]=b&items[2]=c
```

运行示例程序后，用 http-repl 工具测试。先用 HTTP-GET 方式调用一下 SetItems()操作方法，命令如下。

```
get /test/setitems?items[0]=香蕉&items[1]=桃子&items[2]=李子&items[3]=苹果
```

执行上述命令后，调用 SetItems()方法，并向 items 参数传递了 4 个元素，结果如图 10-3 所示。

图 10-3　以 HTTP-GET 方式调用 SetItems()方法

下面的命令将以 HTTP-POST 方式提交，内容格式为表单数据。

```
post /test/setitems -c "items=fox&items=cat&items=fish" -h "Content-Type=
application/x-www-form-urlencoded"
```

-h 参数指定 HTTP 消息头，此处要配置 Content-Type 消息头，提交表单字段默认的格式为 x-www-form-urlencoded，简称 form urlencoded。-c 参数指定要提交的内容，即 HTTP 消息的正文部分。指定 3 个字段，名称均为 items，就可以为 items 参数设置 3 个元素了，结果如图 10-4 所示。

在使用索引传递参数时，也可以省略参数名称。例如，以下命令以 HTTP-PUT 方式调用 SetItems()方法。

```
put /test/setitems?[0]=Tom&[1]=Jim&[2]=Bob --no-body
```

```
https://localhost:6025/> post /test/setitems -c "items=fox&items=cat&items=fish" -h "Content-Type=application/x-www-form-urlencoded"
HTTP/1.1 200 OK
Content-Type: text/plain; charset=utf-8
Date: Mon, [redacted] 2022 09:58:55 GMT
Server: Kestrel
Transfer-Encoding: chunked

你提交的内容：fox, cat, fish
```

图 10-4　以 HTTP-POST 方式调用 SetItems()方法

--no-body 参数表明此次发送的 HTTP 消息没有正文部分。服务器的响应结果如图 10-5 所示。

```
https://localhost:6025/> put /test/setitems?[0]=Tom&[1]=Jim&[2]=Bob --no-body
HTTP/1.1 200 OK
Content-Type: text/plain; charset=utf-8
Date: Mon, [redacted] 2022 10:13:13 GMT
Server: Kestrel
Transfer-Encoding: chunked

你提交的内容：Tom, Jim, Bob
```

图 10-5　以 HTTP-PUT 方式调用 SetItems()方法

10.2.3　示例：绑定复杂类

对于简单的基础类型，在传递数据时只要字段名称与操作方法的参数名称相同即可。而对于自定义的复杂类型，则需要考虑字段的前缀以及对象的成员路径。

本示例中的数据模型是两个自定义类：AddressInfo 和 Person。

```
public class AddressInfo
{
    public string? Province { get; set; }
    public string? City { get; set; }
}

public class Person
{
    public string? Name { get; set; }
    public string? Email { get; set; }
    public string? Phone { get; set; }
    public AddressInfo? Address { get; set; }
}
```

这两个类之间存在引用关系——Person 类的 Address 属性值是 AddressInfo 类型。

下面的代码定义了 Test 控制器，其中 NewData()方法的参数是 Person 类型。

```
[Route("[controller]/[action]")]
public class TestController : ControllerBase
{
    public IActionResult NewData(Person ps)
    {
        if (!ModelState.IsValid)
        {
```

```
            return Content("提交的数据无效");
        }
        string r = $"你提交的数据\n 姓名：{ps.Name ?? "未知"}\n";
        r += $"电邮：{ps.Email}\n";
        r += $"手机号：{ps.Phone}";
        if (ps.Address != null)
        {
            r += $"\n 来自：{ps.Address.Province ?? "未知"}省{ps.Address.
            City ?? "未知"}市";
        }
        return Content(r);
    }
}
```

在访问/test/newdata 时，字段名称的前缀为参数名称，即 ps。

```
POST /test/newdata HTTP/1.1
Content-Type: application/x-www-form-urlencoded

ps.name=小明&ps.email=coder@abc.org&ps.phone=1122334455&ps.address.
province=四川&ps.address.city=成都
```

假如要为 ps 参数的 Name 属性赋值，那么发送请求的字段名称为 ps.name（字段名称不区分大小写）。对于 Address 属性，由于它引用了 AddressInfo 类型的对象，所以要指定完整的属性路径。

```
ps.address.province
ps.address.city
```

服务器响应的结果如下。

```
HTTP/1.1 200 OK
Content-Type: text/plain; charset=utf-8
…

你提交的数据
姓名：小明
电邮：coder@abc.org
手机号：1122334455
来自：四川省成都市
```

由于 NewData()方法只有一个 ps 参数，因此在传递数据时可以省略 ps 前缀。

```
POST /test/newdata HTTP/1.1
Content-Type: application/x-www-form-urlencoded
…

name=小红&email=hong76@126.net&phone=15567898743&address.province=
湖北&address.city=武汉
```

这里要注意，Address 是属性成员，不是前缀名称，不能省略。请求发送后，服务器将响应以下内容。

```
HTTP/1.1 200 OK
Content-Length: 110
Content-Type: text/plain; charset=utf-8
…

你提交的数据
姓名：小红
电邮：hong76@126.net
手机号：15567898743
来自：湖北省武汉市
```

10.2.4　多个参数的模型绑定

当操作方法包含两个或两个以上参数时，可使用各个参数的名称作为字段前缀。假设某操作方法的签名如下。

```
public ActionResult Submit(A da, B db);
```

调用时数据传递方式为

```
da.prop1=xxx&da.prop2=yyy&db.prop3=aaa&db.prop4=bbb
```

字段的前缀也可以省略，即

```
prop1=xxx&prop2=yyy&prop3=aaa&prop4=bbb
```

不过，如果两个参数的类型中包含相同名称的成员（假设 A 类中存在 Len 属性，B 类中也存在 Len 属性），省略前缀会导致两个来自不同类型的同名成员被绑定相同的值。例如，向服务器提交以下数据。

```
len=105&chr=k
```

那么，不管操作方法有多少个参数，只要参数对象中有 Len 属性，都会被赋值为 105。因此，这种情况下不应该省略字段前缀。

```
a.len=100&b.len=200
```

这样就可以保证参数 a 的 Len 属性的值为 100，参数 b 的 Len 属性的值为 200。

10.2.5　示例：绑定 3 个参数

本示例演示 3 个参数的模型绑定。具体步骤如下。

（1）定义 Company 类，包含 Name、Contacts 属性。

```
public class Company
{
    public string? Name { get; set; }
    public string? Contacts { get; set; }
```

```
}
```

（2）定义 Order 类，包含 Qty、Cate、Price 属性。

```
public class Order
{
    public long Qty { get; set; }
    public string? Cate { get; set; }
    public decimal Price { get; set; }
}
```

（3）定义 Demo 控制器。

```
public class DemoController : ControllerBase
{
    [HttpPost("ods/addone")]
    public string PutData(Company c, Order o, string key)
    {
        StringBuilder bd = new StringBuilder();
        bd.AppendFormat("公司：{0}，联系人：{1}", c.Name, c.Contacts);
        bd.AppendFormat("\n订单分类：{0}，数量：{1}，单价：{2}", o.Cate, o.Qty,
        o.Price);
        bd.AppendFormat("\n密钥：{0}", key);
        return bd.ToString();
    }
}
```

Demo 控制器中包含 PutData()方法。该方法有 3 个参数：c 是 Company 类型；o 是 Order 类型；key 是字符串类型。调用后以普通文本格式返回给客户端。调用时应当传递类似以下格式的数据。

```
c.name=ccc&c.contacts=ddd&o.cate=eee&o.qty=nnn&o.price=xxx&key=fff
```

（4）初始化应用程序。

```
var builder = WebApplication.CreateBuilder(args);
builder.Services.AddControllers();
var app = builder.Build();

app.MapControllers();
// 设置应用程序 URL
app.Urls.Add("https://localhost:5810");

app.Run();
```

（5）新建一个控制台应用程序项目。

（6）使用 HttpClient 类向服务器提交测试数据。

```
// 请求 URL
Uri reqUrl = new("https://localhost:5810");

// 创建 HTTP 客户端对象
```

```
HttpClient client = new();
// 设置请求的基础地址
client.BaseAddress = reqUrl;

// 准备请求数据
IDictionary<string, string> formData = new Dictionary<string, string>();
formData["c.name"] = "Nimeral";
formData["c.contacts"] = "李先生";
formData["o.qty"] = "3000";
formData["o.cate"] = "电池组";
formData["o.price"] = "50";
formData["key"] = "e645cb17a92d2f3";
FormUrlEncodedContent form = new FormUrlEncodedContent(formData);

// 发送 POST 请求
var response = await client.PostAsync("/ods/addone", form);

// 显示结果
Console.WriteLine($"状态码: {response.StatusCode}");
Console.WriteLine($"服务器响应内容: \n{await response.Content.
ReadAsStringAsync()}");

Console.WriteLine("\n 按任意键退出……");
Console.ReadKey();
```

（7）先运行 ASP.NET Core 应用程序，再运行控制台应用程序。结果如图 10-6 所示。

图 10-6　绑定 3 个参数

10.2.6　字典类型的模型绑定

对于字典类型的参数，其模型绑定方法与数组相似，将索引值替换为字典的 Key 即可。例如：

```
arg[key1]=val1&arg[key2]=val2&arg[key3]=val3
```

用索引方式也是可行的，但要显式指定 Key 和 Value。

```
arg[0].key=key1&arg[0].value=val1&arg[1].key=key2&arg[1].value=val2
```

字段前缀也是可以省略的。

```
[key1]=val1&[key2]=val2
[0].key=key1&[0].value=val1&[1].key=key2&[1].value=val2
```

10.2.7　示例：绑定字典数据

本示例将演示参数类型为 Dictionary<int, string>的模型绑定。Demo 控制器的定义代码如下。

```
public class DemoController : Controller
{
    [HttpPost("demo/dic")]
    public IActionResult BindDic(Dictionary<int, string> data)
    {
        StringBuilder strbd = new();
        foreach(KeyValuePair<int, string> kp in data)
        {
            strbd.AppendLine($"{kp.Key}: {kp.Value}");
        }
        return Ok(strbd.ToString());
    }
}
```

BindDic()方法的访问 URL 为/demo/dic。

以下代码用于测试 BindDic()方法的调用。

```
// 请求 URL
Uri baseAddr = new("https://localhost:8026");

HttpClient client = new HttpClient();
// 设置请求的基础地址
client.BaseAddress = baseAddr;

// 准备数据
IDictionary<string, string> d = new Dictionary<string, string>()
{
    ["data[101]"] = "table",
    ["data[102]"] = "line",
    ["data[103]"] = "symbol"
};
FormUrlEncodedContent form = new(d);

// 发送请求
var response = await client.PostAsync("/demo/dic", form);

// 处理结果
Console.WriteLine($"状态码: {(int)response.StatusCode}");
Console.WriteLine($"服务器响应的内容: ");
Console.WriteLine(await response.Content.ReadAsStringAsync());

Console.WriteLine("\n 按任意键继续……");
Console.ReadKey();
```

由于字段的 data 前缀可以省略，所以上述代码中变量 d 也可以进行如下初始化。

```
IDictionary<string, string> d = new Dictionary<string, string>()
{
```

```
    ["[101]"] = "table",
    ["[102]"] = "line",
    ["[103]"] = "symbol"
};
```

运行示例程序，结果如图 10-7 所示。

状态码：200
服务器响应的内容：
101: table
102: line
103: symbol

图 10-7　绑定字典数据

10.2.8　示例：绑定 IFormCollection 类型

当操作方法的参数类型为 IFormCollection 时，可以直接读取表单数据。控制器代码如下。

```
public class HomeController : Controller
{
    public IActionResult Index()
    {
        return View();
    }

    [HttpPost]
    public IActionResult ProcPost(IFormCollection form)
    {
        // 存储从表单中读出的数据
        IList<string> readedItems = new List<string>();
        foreach(string key in form.Keys)
        {
            StringValues value = form[key];
            readedItems.Add($"{key} = {value}");
        }

        // 返回指定视图，并传递模型数据
        return View("Show", readedItems);
    }
}
```

ProcPost()方法的参数定义为 IFormCollection 类型，当接收到客户端提交的表单数据后，会自动填充该参数。随后程序代码就可以从 form 参数中读出各字段的值。

本示例还需要两个视图：第 1 个是 Index 视图，作为应用主页，用户可在主页上填写测试用的表单数据；第 2 个是 Show 视图，用于显示已提交的表单数据。

```
// 视图：/Views/Home/Index.cshtml
<form method="post" action="/home/procpost">
    @for (int n = 0; n < 5; n++)
    {
        <div>
        <label for="item_@n">字段 @(n + 1): </label>
        <input type="text" name="item_@n" id="item_@n" />
        </div>
    }
<button type="submit">提交</button>
```

```
</form>

// 视图：/Views/Home/Show.cshtml
@model IEnumerable<string>
@if(Model.Count() == 0)
{
    <p>无数据</p>
}
else
{
    foreach(string s in Model)
    {
        <div>@s</div>
    }
}
```

运行示例程序，依次在 5 个输入框中输入文本，如图 10-8 所示。然后单击“提交”按钮，便能看到刚输入的表单数据了，如图 10-9 所示。

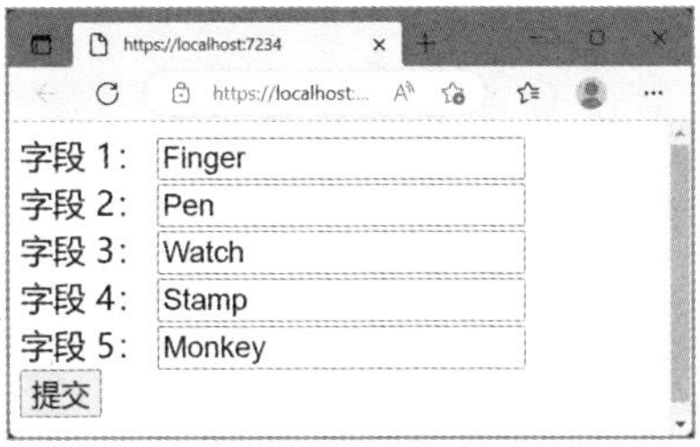

图 10-8　输入文本

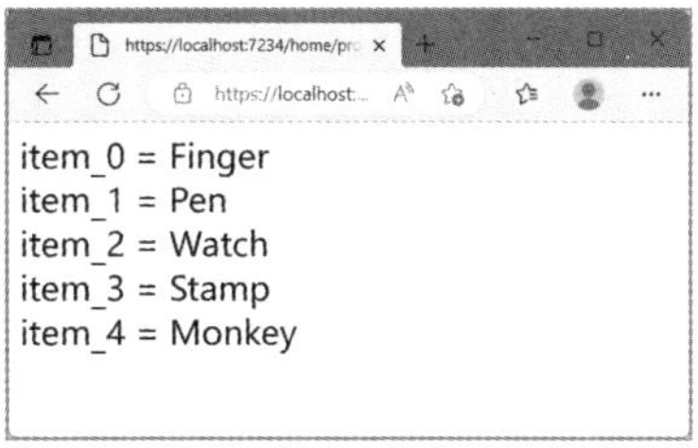

图 10-9　呈现表单中的数据

10.2.9　示例：MD5 计算器

字节数组（byte[]）在模型绑定中比较特殊。其他数组类型在绑定未成功时会提供一个默认值——空白数组。即数组实例不为 null，只是数组中没有元素。而 byte[]类型的参数则不同，若模型绑定不成功，参数的值就为 null。

本示例将实现一个简单的 MD5 哈希计算器，操作方法包含一个 byte[]类型的参数，用于存储客户端提交的数据。计算出来的哈希值将以十六进制字符串的形式返回。

Test 控制器代码如下。

```
public class TestController : ControllerBase
{
    [HttpPost("app/md5")]
    public string Md5(byte[] buffer)
    {
        // 如果是 null，直接返回
        if(buffer == null)
        {
```

```
            return "未提供数据";
        }
        // 计算 MD5
        byte[] result = MD5.HashData(buffer);
        // 将结果转换为字符串
        return $"计算结果：{Convert.ToHexString(result)}";
    }
}
```

当通过表单向 Md5()方法传递数据时，需要先将字节数组转换为 Base64 字符串。客户端测试代码如下。

```
// 服务器的根地址
Uri baseAddr = new("https://localhost:7425");
HttpClient client = new HttpClient();
// 设置基础地址
client.BaseAddress = baseAddr;

// 将字符串转换为字节数组
byte[] data = Encoding.UTF8.GetBytes("Hello All");
// 在发送前要将字节数组转换为 Base64 编码的字符串
Dictionary<string, string> dic = new Dictionary<string, string>
{
    ["buffer"] = Convert.ToBase64String(data)
};
FormUrlEncodedContent content = new FormUrlEncodedContent(dic);

// 发送请求
var response = await client.PostAsync("/app/md5", content);
// 处理结果
Console.WriteLine($"响应状态码：{(int)response.StatusCode}");
// 输出响应内容
Console.WriteLine(await response.Content.ReadAsStringAsync());
```

上述代码先将字符串“Hello All”转换为字节数组（使用 UTF-8 编码），再转换为 Base64 字符串，最后再提交给服务器。运行示例程序后将输出如下内容。

```
响应状态码：200
计算结果：A85AF5D07724BE331F1F3F89D5BB1897
```

10.2.10 绑定 IFormFile 和 IFormFileCollection 类型

上传文件时，客户端一般会设置 Content-Type 标头为 multipart/form-data。这种格式的表单数据可以划分为多部分，每部分可以包含独立的数据内容（文本、图像、音频等），可以单独设置 Content-Type 等标头。

由多部分组成的表单既可以上传一个文件，也可以上传多个文件。对于 MVC 的操作方法，如果只接收单个文件，参数类型可以声明为 IFormFile 类型；如果要接收多个文件，参数类型应当为 IFormFileCollection 类型。IFormFileCollection 是 IFormFile 的集合，一个 IFormFile 类型的对象表示一个文件实体。其中，Name 属性表示该文件在表单数据中的字段名称，FileName 属性表示文件名，Length 属性表示文件的长度（文件大小）。

要将上传文件的内容复制到其他流对象中，可以调用 CopyTo()方法（或 CopyToAsync()异步方法）；要读取文件的内容，可以调用 OpenReadStream()方法打开文件，并返回相关的流对象。

10.2.11　示例：上传一个文本文件

本示例将实现用户选择一个文本文件，然后将其上传到服务器，随后服务器会在响应消息中显示文件的名称、大小和内容。

FileUploader 控制器代码如下。

```
public class FileUploaderController : Controller
{
    [HttpGet("/")]
    public IActionResult Index()
    {
        return View();
    }

    [HttpPost("file/upload")]
    public async Task<IActionResult> UploadAsync(IFormFile txtFile)
    {
        if(txtFile == null)
        {
            return Content("未发现文件");
        }
        // 读取文件内容
        string filecontent = string.Empty;
        using(StreamReader reader = new StreamReader(txtFile.OpenReadStream()))
        {
            filecontent = await reader.ReadToEndAsync();
        }
        // 文件名
        string fn = txtFile.FileName;
        // 文件大小
        long size = txtFile.Length;
        return Content($"""
            文件名：{fn}
            大小：{size}字节
            文件内容：
            {filecontent}
            """);
    }
}
```

上述控制器需要一个 Index 视图。

```
// 文件：/Views/FileUploader/Index.cshtml
<form action="/file/upload" enctype="multipart/form-data" method="post">
<div>
<label for="txtFile">请选择一个文件：</label>
<input type="file" name="txtFile" id="txtFile" accept=".txt"/>
</div>
<button type="submit">上传</button>
</form>
```

注意，<form>元素的 enctype 属性的值要设置为 multipart/form-data。<input>元素的 name 和 id 属性的值均为 IFormFile 类型参数的名称，即 txtFile。

在任意目录下新建一个文本文件（假设命名为 abc.txt）。用文本编辑器打开文件，输入“示例文本文件”，然后保存并关闭文件。接着运行示例程序，选择刚创建的文本文件，如图 10-10 所示，单击“上传”按钮。服务器将返回文件的内容，如图 10-11 所示。

图 10-10　选择文本文件

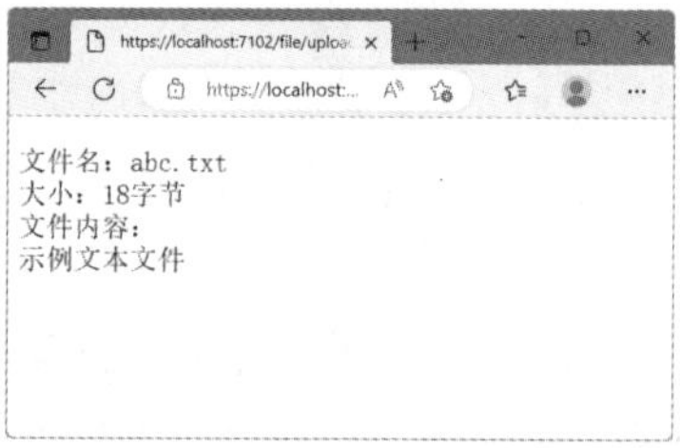

图 10-11　显示文件内容

10.2.12　示例：上传多个文件

本示例将演示向服务器批量上传 3 个文件。要接收多个文件，操作方法的参数应当声明为 IFormFileCollection 类型，其中单个元素都是 IFormFile 对象，代表一个文件。

MultFileUploader 控制器代码如下。

```
public class MultFileUploaderController : Controller
{
    [HttpPost("/file/upload")]
    public IActionResult Upload(IFormFileCollection fileList)
    {
        int count = fileList.Count;
        IEnumerable<string> files = from formFile in fileList select
        $"{formFile.FileName}，大小：
        {formFile.Length}字节，类型：{formFile.ContentType}";
        string msg = $"你上传了{count}个文件\n";
        msg += $"它们是：\n{string.Join('\n', files)}";
        return Content(msg);
    }
}
```

Upload()方法先读出上传文件的数量，接着通过 LINQ 语句轮询每个文件相关的 IFormFile 对象，并获取文件名、文件大小、内容格式等信息，最后把这些文件信息返回给客户端。

本示例的测试代码如下。

```
// 应用程序根地址
Uri baseaddr = new("http://localhost:9188");

HttpClient client = new() { BaseAddress = baseaddr };

// 第 1 个文件
byte[] buf1 = new byte[5];
Random.Shared.NextBytes(buf1);                    // 产生随机数据
MemoryStream stream1 = new MemoryStream(buf1);
StreamContent strcontent1 = new StreamContent(stream1);
// 设置 Content-Type 标头为 application/octet-stream
strcontent1.Headers.ContentType =
MediaTypeHeaderValue.Parse(MediaTypeNames.Application.Octet);

// 第 2 个文件
byte[] buf2= new byte[6];
Random.Shared.NextBytes(buf2);
MemoryStream stream2 = new MemoryStream(buf2);
StreamContent strcontent2 = new StreamContent(stream2);
strcontent2.Headers.ContentType =
MediaTypeHeaderValue.Parse(MediaTypeNames.Application.Octet);

// 第 3 个文件
byte[] buf3 = new byte[7];
Random.Shared.NextBytes(buf3);
MemoryStream stream3 = new MemoryStream(buf3);
StreamContent strcontent3 = new StreamContent(stream3);
strcontent3.Headers.ContentType =
MediaTypeHeaderValue.Parse(MediaTypeNames.Application.Octet);

// 表单
MultipartFormDataContent form =new MultipartFormDataContent();
// 注意：字段名称都是 fileList
form.Add(strcontent1, "fileList", "file_1");
form.Add(strcontent2, "fileList", "file_2");
form.Add(strcontent3, "fileList", "file_3");

// 发送请求
var response = await client.PostAsync("/file/upload", form);
// 处理响应结果
Console.WriteLine($"响应状态码：{(int)response.StatusCode}");
Console.WriteLine($"响应内容：\n{await response.Content.ReadAsStringAsync()}");
```

测试代码将向服务器上传 3 个文件。这 3 个文件的内容是通过 Random.Shared.NextBytes() 方法随机生成的。文件名依次为 file_1、file_2、file_3，内容类型（Content-Type）都是 application/octet-stream，代表二进制数据。

运行示例程序，结果如图 10-12 所示。

响应状态码：200
响应内容：
你上传了3个文件
它们是：
file_1，大小：5字节，类型：application/octet-stream
file_2，大小：6字节，类型：application/octet-stream
file_3，大小：7字节，类型：application/octet-stream

图 10-12　3 个文件的基本信息

10.3　设置模型绑定的来源

当模型绑定未从预期的来源获取数据时，就需要手动设定绑定来源。例如，开发者期待某个方法参数的值从表单数据（form-data）中获取。实际运行程序后发现，参数的值却是从 URL 查询字符串中读取的，或者参数没有获取到有效的值。这时开发者可以明确给该参数加上 FromForm 特性，使它正确地获取数据。

ASP.NET Core 已公开以下特性类，用于为模型绑定手动设定数据源。

（1）FromHeaderAttribute：数据来自 HTTP 消息头。

（2）FromQueryAttribute：数据来自 URL 的查询字符串。

（3）FromFormAttribute：数据来自表单数据。

（4）FromBodyAttribute：将 HTTP 消息的正文部分作为数据源。

（5）FromRouteAttribute：数据来自路由规则中某个字段的值。

（6）FromServicesAttribute：通过依赖注入从服务容器中获取数据。

10.3.1　示例：绑定 HTTP 消息头

本示例通过 FromHeader 特性，让模型绑定从 HTTP 消息头中获取数据。Test 控制器的定义代码如下。

```
public class TestController : Controller
{
    [HttpGet("test/getnums")]
    public ActionResult GetList([FromHeader(Name = "min-val")]int min,
    [FromHeader(Name = "max-val")]int max)
    {
        // 产生一组整数
        int n = min;
        HashSet<int> nums = new HashSet<int>();
        while(n <= max)
        {
```

```
            nums.Add(n);
            n++;
        }
        // JSON 格式
        return Json(nums);
    }
}
```

在应用 FromHeader 特性时，Name 属性可以设置要读取的 HTTP 消息头的名称。如果要读取的消息头的名称与参数名称相同，则不需要设置 Name 属性。在本示例中，min 参数的值将从名为 min-val 的消息头中获取，max 参数的值则从名为 max-val 的消息头中获取。

运行示例程序，然后用 http-repl 工具测试。

```
httprepl https://localhost:9453
```

发出 HTTP-GET 请求，注意要添加 min-val 和 max-val 消息头。

```
get /test/getnums -h min-val=5 -h max-val=9
```

服务器将响应一个包含 5 个元素的 JSON 数组，如图 10-13 所示。

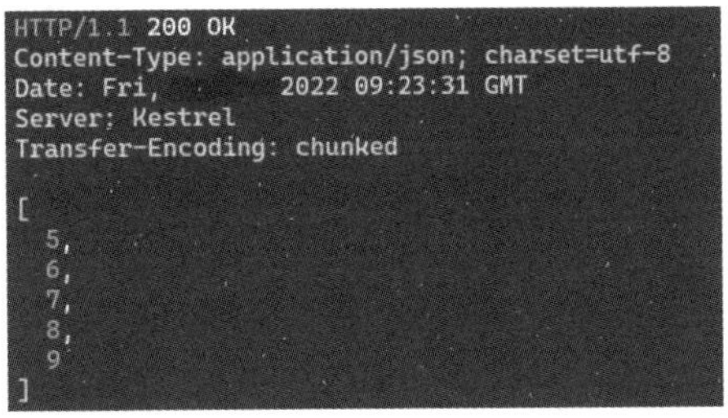

图 10-13 从服务器返回的 JSON 数组

10.3.2 示例：从 HTTP 消息正文提取数据

使用 BodyModelBinder 类将数据模型绑定到 HTTP 消息正文。该类需要通过 IInputFormatter 接口解析 HTTP 正文中的数据。MVC 应用程序默认使用 SystemTextJsonInputFormatter 类，只接收 JSON 格式的数据。如果需要解析非 JSON 格式的数据，可以自定义一个实现 IInputFormatter 接口的类，然后将此类的实例添加到 MvcOptions 选项类的 InputFormatters 集合中。

本示例只从 HTTP 消息正文中提取字符串类型的数据。客户端在发送数据时，必须使用 JSON 格式的数据。

JsonText 控制器的实现代码如下。

```
public class JsonTextController : Controller
{
    [HttpPost("json/senddata")]
    public IActionResult SetData([FromBody] string msg)
    {
        return Content($"来自客户端的内容：{msg}");
    }
}
```

msg 参数要应用 FromBody 特性后才能绑定到 HTTP 正文，否则应用程序默认会以表单数据的方式进行绑定。

下面的代码演示了一个简单测试客户端程序。

```
// 服务器根地址
Uri baseUrl = new("http://localhost:10050");

HttpClient client = new();
client.BaseAddress = baseUrl;

// 准备数据
JsonContent content = JsonContent.Create("千锤万凿出深山");
// 发送请求
var resp = await client.PostAsync("json/senddata", content);
// 处理响应消息
Console.WriteLine($"响应状态码: {(int)resp.StatusCode}");
Console.WriteLine(await resp.Content.ReadAsStringAsync());
```

JsonContent.Create()是静态方法，它会通过 JSON 序列化将传入对象转换为 JSON 格式的内容，同时设置 Content-Type 标头为 application/json，最后返回 JsonContent 类型的实例。

10.3.3 示例：与路由参数绑定

本示例将演示 FromRoute 特性的使用。Demo 控制器代码如下。

```
public class DemoController : Controller
{
    [HttpGet("api/test/{key}/{x}")]
    public IActionResult Work([FromRoute]string key, [FromRoute(Name =
    "x")]float scale)
    {
        string msg = $"key = {key}\nscale={scale}";
        return Content(msg);
    }
}
```

Work()方法的两个参数都应用了 FromRoute 特性，表示其值将来自路由 URL 中的参数字段。key 参数的值将从路由字段{key}中读取。路由 URL 中没有 scale 字段，因此，在应用 FromRoute 特性时需要通过 Name 属性指定数据将来自{x}字段。也就是说，如果操作方法的参数名称与路由参数的字段名称相同，FromRoute 特性不需要设置 Name 属性，否则就要显式设置 Name 属性。

运行示例程序后，向服务器发送以下 GET 请求。

```
/api/test/abc/5.22
```

根据 URL 路由规则，key 字段为 abc，x 字段为 5.22。模型绑定后，Work()方法的 key 参数为 abc，scale 参数的值为 5.22（已转换为浮点数）。

10.3.4 示例：FromServices 特性的使用

在操作方法的参数上应用 FromServices 特性后，当方法被调用时会通过依赖注入从服务容器中获取参数的值。

下面的代码创建一个处理字符串的服务。

```
// 服务接口
public interface ICustService
{
    string Process(string input);
}

// 服务实现类
public class MyCustService : ICustService
{
    public string Process(string input)
    {
        if(string.IsNullOrEmpty(input))
        {
            return string.Empty;
        }
        char[] chrs = new char[input.Length];
        for(int i = 0; i < input.Length; i++)
        {
            char c = input[i];
            // 如果是大写字母，就转换为小写字母
            // 如果是小写字母，就转换为大写字母
            if (char.IsUpper(c))
                chrs[i] = char.ToLower(c);
            else
                chrs[i] = char.ToUpper(c);
        }
        return new string(chrs);
    }
}
```

该服务的功能是反转字符串中的字母大小写——大写转换为小写，小写转换为大写。

在服务容器中注册服务类。

```
builder.Services.AddSingleton<ICustService, MyCustService>();
```

下面的代码定义 Home 控制器。

```
public class HomeController : ControllerBase
{
    [HttpGet("/test")]
    public IActionResult Test([FromServices]ICustService svc)
    {
        string src = "RtXydAMnq";
        // 调用服务
        string res = svc.Process(src);
```

```
        // 响应内容
        return Content($"{src} --> {res}");
    }
}
```

运行示例程序，访问/test，结果如图 10-14 所示。

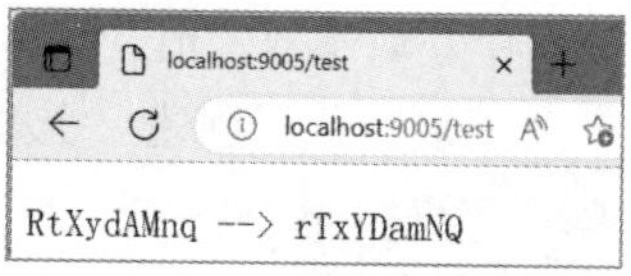

图 10-14　反转大小写

10.3.5　示例：混合使用 From*特性类

当一个操作方法有多个参数且参数的值有多个来源时，FromForm、FromHeader 等特性类可以混合使用。

本示例将混合使用 FromForm 和 FromQuery 特性进行模型绑定。Demo 控制器的类代码如下。

```
public class DemoController : Controller
{
    [HttpPost("/test/data")]
    public IActionResult SetData([FromForm]SomeData data,
    [FromQuery]int stars)
    {
        StringBuilder strbd = new();
        // SomeData 类的成员，来自 form-data
        strbd.AppendLine($"Flag = {data.Flag}");
        strbd.AppendLine($"Label = {data.Label}");
        // stars 参数的值，来自查询字符串
        strbd.AppendLine($"Stars = {stars}");
        return Content(strbd.ToString());
    }
}
```

SomeData 是自定义类，代码如下。

```
public class SomeData
{
    public long Flag { get; set; }
    public string? Label { get; set; }
}
```

SetData()方法有两个参数。其中，data 参数应用了 FromForm 特性，表明它的值来自表单提交的数据；stars 参数应用了 FromQuery 特性，表明它的值从 URL 查询字符串中获得。

新建一个控制台应用程序作为测试客户端，代码如下。

```
// 服务器根地址
Uri baseAddr = new("http://localhost:5475");

HttpClient client = new()
{
    BaseAddress = baseAddr
};
```

```
// 准备数据
IDictionary<string, string> data = new Dictionary<string, string>();
data["flag"] = "12345678";
data["label"] = "Sample Text";
FormUrlEncodedContent content = new(data);

// 发送请求
var response = await client.PostAsync("/test/data?stars=5", content);
// 处理响应结果
Console.WriteLine($"响应状态码: {(int)response.
StatusCode}");
Console.WriteLine($"响应内容: \n{await
response.Content.ReadAsStringAsync()}");
```

先启动 ASP.NET Core 应用程序，再启动测试客户端程序（也可以两者同时启动）。控制台输出的结果如图 10-15 所示。

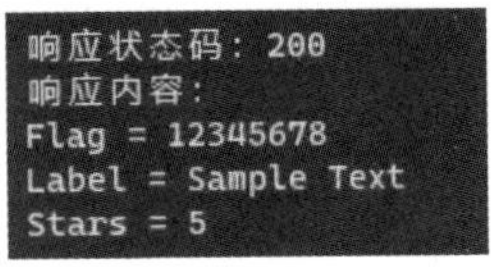

图 10-15　来自表单数据和查询字符串的数据

10.3.6　示例：将 From*特性类应用于属性成员

From*特性类不仅可以应用到方法的参数上，在定义模型类时，也可以应用到属性成员上。本示例定义了一个名为 Production 的模型类。

```
public class Production
{
    // 属性值来自路由参数
    [FromRoute]
    public int ID { get; set; }

    // 属性值来自表单数据
    [FromForm]
    public string? Cate { get; set; }

    // 属性值来自表单数据
    [FromForm]
    public string? Desc { get; set; }

    // 属性值来自表单数据
    [FromForm]
    public float Size { get; set; }
}
```

Production 类中只有 ID 属性的值是从路由参数中获取的，其他属性的值均取自表单数据。下面的代码将 Production 类用于 Home 控制器。

```
public class HomeController : ControllerBase
{
    [HttpPost("/home/test/{id}")]
    public IActionResult PostData(Production prod)
    {
```

```
            StringBuilder bd = new StringBuilder();
            bd.AppendLine($"来自路由参数的值: ID = {prod.ID}");
            bd.AppendLine("以下属性的值皆来自表单数据: ");
            bd.AppendLine($"Cate = {prod.Cate}, Desc = {prod.Desc}, Size =
            {prod.Size}");
            // 返回响应内容
            return Content(bd.ToString());
        }
    }
```

注意，给 PostData()方法指定的 URL 路由规则中要包含 id 字段，此路由参数对应 Production 类的 ID 属性。

运行示例程序，使用 http-repl 工具进行测试。

```
post /home/test/490043 -h Content-Type=application/x-www-form-urlencoded
-c cate=线材&desc=常规电子线&size=20
```

在请求地址/home/test/490043 中，490043 将传递给 ID 属性，其他属性将从表单数据中读取相应的值。请求发出后，服务器将响应以下内容。

```
HTTP/1.1 200 OK
Content-Length: 136
Content-Type: text/plain; charset=utf-8
…

来自路由参数的值: ID = 490043
以下属性的值皆来自表单数据:
Cate = 线材, Desc = 常规电子线, Size = 20
```

10.4 自定义 IValueProvider 接口

实现 IValueProvider 接口可以自定义对象值的提供方式。ASP.NET Core 内置一些默认的实现类，可从查询字符串（QueryStringValueProvider）、表单（FormValueProvider）等来源获取数据。

IValueProvider 接口包括两个方法：

```
bool ContainsPrefix(string prefix);
ValueProviderResult GetValue(string key);
```

ContainsPrefix()方法用来判断数据中的字段名称是否存在与 prefix 参数的值相同的前缀，若前缀存在就返回 true，否则返回 false。如果程序逻辑不需要考虑前缀，可以直接返回 false。GetValue()方法将根据 key 参数所指定的字段名称搜索对应的值。若能找到对应的值，就用 ValueProviderResult 对象实现封装搜索到的值；若没有相关的值，就直接返回 ValueProvider-Result.None。

应用程序代码实现 IValueProvider 接口只负责为模型绑定提供所需的值，关于数据如何赋值给模型对象，将交给 IModelBinder 接口去处理。

实现 IValueProvider 接口后，新类型需要添加到模型绑定的上下文中才能起作用。这就要用到另一个接口——IValueProviderFactory。该接口只定义了一个方法：

```
Task CreateValueProviderAsync(ValueProviderFactoryContext context);
```

此方法用于创建 IValueProvider 实例。context 参数有一个 ValueProviders 列表，自定义的 IValueProvider 对象需要添加到该列表中，才能在模型绑定中被应用程序调用。在模型绑定过程中，应用程序会循环访问所有 IValueProvider 对象，只要有一个能获取到有效的值，循环立即停止。

10.4.1　示例：由自定义字符串提供的值

本示例演示通过自定义的字符串产生用于绑定字符串数组（string[]）的值。内置的模型绑定机制会将以下格式的值识别为字符串数组。

```
[0]=abcd
[1]=open
[2]=close
[3]=AnyOne
```

但在本示例中，客户端提交的数据并不是标准的表单数据，而是由某个分隔符连接起来的文本，如（假设分隔符为#）

```
copy#cut#paste#over
```

这样的格式不适用于默认的模型绑定器，应用程序需要将其转换为

```
[0]=copy
[1]=cut
[2]=paste
[3]=over
```

随后模型绑定器会产生.NET 对象 string[4]{ "copy","cut","paste","over" }。

首先，定义实现 IValueProviderFactory 接口的 ArrayStringProviderFactory 类，用于创建 ArrayStringValueProvider 实例。

```
public class ArrayStringProviderFactory : IValueProviderFactory
{
    // 分隔符
    private readonly char _splitChar;

    // 构造函数
    public ArrayStringProviderFactory(char splitChar = ',')
    {
        _splitChar = splitChar;
    }

    public async Task CreateValueProviderAsync(ValueProviderFactoryContext
    context)
```

```
        {
            var request = context.ActionContext.HttpContext.Request;
            // 先判断一下 Content-Type 消息头
            if (request.ContentType != null && request.ContentType.Equals
            ("text/plain", StringComparison.InvariantCultureIgnoreCase))
            {
                IDictionary<string, string> resValues = new Dictionary<string,
                string>();
                // 读取正文
                string content;
                using(StreamReader reader = new(request.Body))
                {
                    content = await reader.ReadToEndAsync();
                }
                // 分隔字符串
                string[] parts = content.Split(_splitChar);
                // 生成字段键
                for(int x = 0; x < parts.Length; x++)
                {
                    resValues[$"[{x}]"] = parts[x];
                }
                // 创建 ArrayStringValueProvider 实例
                context.ValueProviders.Insert(0, new ArrayStringValueProvider
                (resValues));
            }

        }
    }
```

ArrayStringProviderFactory 类的构造函数带一个 splitChar 参数，用来指定文本的分隔符（默认为逗号）。在 CreateValueProviderAsync()方法中，程序代码将从客户端提交的消息正文中读出所有文本，然后用 splitChar 参数指定的分隔符拆分文本，最后将拆分后的文本重新组织为[0]=val1，[1]=val2，…的格式，并存储在一个字典对象中。在将 ArrayStringValueProvider 实例添加到 ValueProviders 列表时调用的是 Insert()方法，将此实例插入索引为 0 的地方，使其成为 ValueProviders 列表的第 1 个元素。这样，应用程序在模型绑定时具有最高的优先级。

下面的代码定义 ArrayStringValueProvider 类，它实现 IValueProvider 接口。

```
public class ArrayStringValueProvider : IValueProvider
{
    private readonly IDictionary<string, string> _values;

    // 构造函数
    public ArrayStringValueProvider(IDictionary<string, string> data)
    {
        _values = data;
    }

    public bool ContainsPrefix(string prefix)
```

```
        {
            // 此处不需要前缀，总是返回 false
            return false;
        }

        public ValueProviderResult GetValue(string key)
        {
            if (key.Length == 0)
            {
                return ValueProviderResult.None;
            }
            bool successed = _values.TryGetValue(key, out string? val);
            if (!successed)
            {
                return ValueProviderResult.None;
            }
            if (val == null || val.Length == 0)
            {
                return ValueProviderResult.None;
            }
            return new ValueProviderResult(val);
        }
    }
```

要让 ArrayStringProviderFactory 在 MVC 模型绑定中起作用，还需要配置 MvcOptions 选项类。

```
var builder = WebApplication.CreateBuilder(args);
builder.Services.AddControllers(opt =>
{
    opt.ValueProviderFactories.Insert(0, new ArrayStringProviderFactory
    ('|'));
});
var app = builder.Build();
...
```

MvcOptions.ValueProviderFactories 属性是一个 IValueProviderFactory 列表，自定义的 IValueProviderFactory 实现类必须将其实例添加到该列表中，才能在模型绑定过程中被调用。上述代码通过 ArrayStringProviderFactory 类的构造函数指定了文本分隔符为|。

下面的代码定义 Home 控制器，PostArrString()方法带有一个字符串数组类型的 arr 参数。该参数的值将由前面已实现的 ArrayStringValueProvider 类提供。调用后方法将数组实例直接返回。

```
public class HomeController : Controller
{
    [HttpPost("/data/post")]
    public string[] PostArrString(string[] arr)
    {
        return arr;
```

```
    }
}
```

运行示例程序，通过 http-repl 工具进行测试。

```
post /data/post -h "Content-Type:text/plain" -c "eye|ear|nose|mouth"
```

服务器接收到请求后，将回复以下 JSON 格式的数据。

```
HTTP/1.1 200 OK
Content-Type: application/json; charset=utf-8
…

[
"eye",
"ear",
"nose",
"mouth"
]
```

10.4.2 示例：CookieValueProvider

本示例将演示一个自定义的 IValueProvider 接口，由 Cookie 提供用于模型绑定的数据值。主要类型如下。

（1）CookieValueProvider：从 HTTP 请求消息中读取 Cookie。

（2）CookieVauleProviderFactory：负责创建 CookieValueProvider 实例。

（3）FromCookieAttribute：一个特性类，用于指明模型绑定的数据源为 Cookie。其作用与 FromForm、FromBody 等特性类相似。

具体实现步骤如下。

（1）定义 CookieValueProvider 类，基类为 BindingSourceValueProvider，同时实现 IEnumerableValueProvider 接口。

```
public class CookieValueProvider : BindingSourceValueProvider,
IEnumerableValueProvider
{
    private readonly PrefixContainer _prefixContainer;
    private readonly IRequestCookieCollection _cookies;

    // 构造函数
    public CookieValueProvider(BindingSource bindsource,
    IRequestCookieCollection cookies)
        :base(bindsource)
    {
        if(cookies == null)
        {
            throw new ArgumentNullException(nameof(cookies));
        }
        _cookies = cookies;
        _prefixContainer = new PrefixContainer(_cookies.Keys);
```

```
    }

    public override bool ContainsPrefix(string prefix)
    {
        return _prefixContainer.ContainsPrefix(prefix);
    }

    public IDictionary<string, string> GetKeysFromPrefix(string prefix)
    {
        return _prefixContainer.GetKeysFromPrefix(prefix);
    }

    public override ValueProviderResult GetValue(string key)
    {
        // 处理空白字段名
        if(string.IsNullOrEmpty(key))
        {
            return ValueProviderResult.None;
        }
        // 尝试取值
        _cookies.TryGetValue(key, out string? val);
        // 判断值是否有效
        if (string.IsNullOrEmpty(val))
        {
            return ValueProviderResult.None;
        }
        // 返回读到的值
        return new ValueProviderResult(val);
    }
}
```

BindingSourceValueProvider 是抽象类，已实现 IValueProvider 接口，同时它也实现了 IBindingSourceValueProvider 接口。调用 CookieValueProvider 的构造函数时，需要向基类（BindingSourceValueProvider）的构造函数传递一个 BindingSource 对象，以指定模型绑定的数据源。

CookieValueProvider 类的内部使用了 PrefixContainer 类。它是一个辅助类，用于处理数据字段的前缀。

（2）定义 CookieVauleProviderFactory 类，实现 IValueProviderFactory 接口。该类负责创建 CookieValueProvider 实例。

```
public class CookieVauleProviderFactory : IValueProviderFactory
{
    // 这个静态字段表示自定义的 BindingSource
    public static readonly BindingSource CookieBindingSource = new
    BindingSource(
                id: "Cookie",
                displayName: "Cookie Binding",
```

```
                isGreedy: false,
                isFromRequest: true
                );

    public Task CreateValueProviderAsync(ValueProviderFactoryContext
    context)
    {
        var request = context.ActionContext.HttpContext.Request;
        var cookies = request.Cookies;
        // 检查是否包含 Cookie
        if(cookies.Count > 0)
        {
            // 创建 CookieValueProvider 实例
            var cookieValueProvider = new CookieValueProvider
            (CookieBindingSource, cookies);
            context.ValueProviders.Add(cookieValueProvider);
        }

        return Task.CompletedTask;
    }
}
```

通过 HttpContext 对象的 Request.Cookies 能获取到随 HTTP 请求一起发送到服务器的 Cookie 数据。随后将 Cookies 集合传递给 CookieValueProvider 类的构造函数。

（3）定义 FromCookieAttribute 类。它是一个特性类，功能与 FromBody 等特性类相同。

```
[AttributeUsage(AttributeTargets.Parameter | AttributeTargets.Property,
AllowMultiple = false, Inherited = true)]
public class FromCookieAttribute : Attribute, IBindingSourceMetadata,
IModelNameProvider
{
    public BindingSource? BindingSource => CookieVauleProviderFactory.
    CookieBindingSource;
    // 指定模型的名称
    public string? Name { get; set; }
}
```

当应用程序不能正确地选择从 Cookie 中读取数据时，可以在操作方法的参数或模型类的属性上应用 FromCookie 特性。

FromCookieAttribute 类实现了两个接口。IBindingSourceMetadata 接口要求定义 BindingSource 属性，返回数据来源相关的 BindingSource 对象，本示例中可直接引用 CookieValueProviderFactory 类的静态字段 CookieBindingSource。实现 IModelNameProvider 接口需要公开一个 Name 属性，用于设置模型名称。当客户端提交的 Cookie 名称的前缀与操作方法的参数相同时，可以忽略 Name 属性；否则应设置 Name 属性为 Cookie 名称的前缀。

（4）创建一个 Demo 控制器，用于验证 CookieValueProvider 类是否正常使用。

```
public class DemoController : ControllerBase
{
```

```
    [HttpGet("/users/new")]
    public IActionResult Default(Register reg)
    {
        string s = $"你好，{reg.Name}（{reg.Phone}/{reg.Email}）已完成注册";
        return Content(s);
    }
}
```

Register 是一个自定义的模型类，代码如下。

```
public class Register
{
    public string? Name { get; set; }
    public string? Email { get; set; }
    public string? Phone { get; set; }
}
```

Default 方法中，reg 参数并没有应用 FromCookie 特性，因为客户端提交的 Cookie 名称已包含前缀 reg（如 reg.name、reg.email）。

（5）在初始化 ASP.NET Core 应用程序时，需要将 CookieVauleProviderFactory 实例添加到 MvcOptions 对象的 ValueProviderFactories 列表中。

```
var builder = WebApplication.CreateBuilder(args);
builder.Services.AddControllers(options =>
{
    options.ValueProviderFactories.Add(new CookieVauleProviderFactory());
});

var app = builder.Build();
…
```

（6）新建一个控制台应用程序项目，作为测试客户端。

```
// 服务器的根地址
Uri rootUrl = new("http://localhost:6268");

// 准备 Cookie
Cookie ckName = new Cookie("reg.name", "Mike", "/", rootUrl.Host);
Cookie ckEmail = new Cookie("reg.email", "Foobar@163.net", "/",
rootUrl.Host);
Cookie ckPhone = new Cookie("reg.phone", "15542380726", "/", rootUrl.Host);
CookieContainer container = new CookieContainer();
container.Add(ckName);
container.Add(ckEmail);
container.Add(ckPhone);

// HTTP 客户端
HttpClientHandler handler = new HttpClientHandler();
// 设置 Cookie
handler.CookieContainer = container;
HttpClient client = new HttpClient(handler);
```

```
client.BaseAddress = rootUrl;

// 发起请求
var response = await client.GetAsync("/users/new");
// 处理响应结果
Console.WriteLine($"响应状态码：{(int)response.StatusCode}");
Console.WriteLine($"响应内容：\n{await response.Content.ReadAsStringAsync()}");
```

（7）先运行服务器，后运行测试客户端。运行结果如图 10-16 所示。

响应状态码：200
响应内容：
你好，Mike (15542380726/Foobar@163.net) 已完成注册

图 10-16　从 Cookie 中获取的数据

（8）在 Demo 控制器中，尝试将操作方法的参数修改为 user，或者应用 FromCookie 特性将模型名称修改为 user。

```
[HttpGet("/users/new")]
public IActionResult Default([FromCookie(Name = "user")] Register reg)
{
    …
}
```

此时，客户端所使用的 Cookie 也应该将命名前缀改为 user。

```
…
// 准备 Cookie
Cookie ckName = new Cookie("user.name", "Mike", "/", rootUrl.Host);
Cookie ckEmail = new Cookie("user.email", "Foobar@163.net", "/",
rootUrl.Host);
Cookie ckPhone = new Cookie("user.phone", "15542380726", "/",
rootUrl.Host);
CookieContainer container = new CookieContainer();
container.Add(ckName);
container.Add(ckEmail);
container.Add(ckPhone);
…
```

重新运行服务器和测试客户端，也能得到正确的结果。

10.5　IModelBinder 接口

在模型绑定的前期阶段，应用程序通过各个 IValueProvider 对象收集数据。但这些数据都是文本内容，模型绑定的重要环节是将这些文本内容转换为各种.NET 类型。这个转换过程将由 IModelBinder 接口负责。

IModelBinder 接口只有一个待实现的方法：

```
public Task BindModelAsync (ModelBindingContext bindingContext);
```

在进入模型绑定过程前，应用程序会创建 ModelBindingContext 类（位于 Microsoft.AspNetCore.Mvc.ModelBinding 命名空间）的实例，携带所有与模型绑定相关的信息，传递给 bindingContext 参数。

ModelBindingContext 是抽象类，其默认的实现类为 DefaultModelBindingContext。其中，有几个属性最为常用，具体如下。

（1）ActionContext：获取与当前正在执行的操作方法相关的上下文对象。

（2）BinderModelName：如果方法参数或属性成员应用了 FromRoute、FromHeader、FromQuery 等特性，并且通过 Name 属性设置了自定义名称，那么 BinderModelName 属性将返回该名称。

（3）HttpContext：获取关联的 HttpContext 对象。

（4）Model：可以获取或设置当前正在处理的模型类的实例。该属性可以不设置，除非在上下文中传递状态数据。

（5）ModelName：获取或设置模型名称。该名称将在 IValueProvider 中查找数据值时使用。

（6）ModelType：获取模型类相关的 Type 对象。

（7）ValueProvider：获取或设置用来查找数据的 IValueProvider 对象。

（8）Result：不管绑定是否成功，尽可能设置该属性，以便其他程序代码能知道绑定状态。若绑定成功，可调用 Success()静态方法；若绑定失败，也可调用 Failed()静态方法。

实现 IModelBinder 接口可以自定义绑定器。如果绑定器要应用到全局范围（整个 MVC 应用程序中），还需要实现 IModelBinderProvider 接口，并将类实例添加到 MvcOptions.ModelBinderProviders 列表中。

如果自定义的 IModelBinder 接口只用于特定的模型类（局部应用），可以在方法参数、模型类或模型类的属性成员上应用 ModelBinder 特性（ModelBinderAttribute 类），并通过构造函数参数或 BinderType 属性设置自定义绑定器的 Type。

10.5.1 内置绑定器

ASP.NET Core 类库已针对各种.NET 类型（包括基础类型和复杂类型，以及自定义类型）实现默认的绑定器。以下源代码展示了 MVC 应用程序初始化时所添加的 IModelBinderProvider 列表（options 变量即为 MvcOptions 实例）。

```
options.ModelBinderProviders.Add(new BinderTypeModelBinderProvider());
options.ModelBinderProviders.Add(new ServicesModelBinderProvider());
options.ModelBinderProviders.Add(new BodyModelBinderProvider
(options.InputFormatters, _readerFactory, _loggerFactory, options));
options.ModelBinderProviders.Add(new HeaderModelBinderProvider());
options.ModelBinderProviders.Add(new FloatingPointTypeModelBinderProvider());
options.ModelBinderProviders.Add(new EnumTypeModelBinderProvider(options));
options.ModelBinderProviders.Add(new DateTimeModelBinderProvider());
options.ModelBinderProviders.Add(new TryParseModelBinderProvider());
```

```
options.ModelBinderProviders.Add(new SimpleTypeModelBinderProvider());
options.ModelBinderProviders.Add(new CancellationTokenModelBinderProvider());
options.ModelBinderProviders.Add(new ByteArrayModelBinderProvider());
options.ModelBinderProviders.Add(new FormFileModelBinderProvider());
options.ModelBinderProviders.Add(new FormCollectionModelBinderProvider());
options.ModelBinderProviders.Add(new KeyValuePairModelBinderProvider());
options.ModelBinderProviders.Add(new DictionaryModelBinderProvider());
options.ModelBinderProviders.Add(new ArrayModelBinderProvider());
options.ModelBinderProviders.Add(new CollectionModelBinderProvider());
options.ModelBinderProviders.Add(new ComplexObjectModelBinderProvider());
```

对应的绑定器类型有 BinderTypeModelBinder、ComplexObjectModelBinder、DateTimeModelBinder、ByteArrayModelBinder 等，这些类型都位于 Microsoft.AspNetCore.Mvc.ModelBinding.Binders 命名空间内，能够将文本数据转换为 string、int、float、double、DateTime、decimal 等.NET 基础类型，以及集合、数组、字典等数据类型。ComplexObjectModelBinder 能处理复杂类型的绑定，如自定义类，或者类的属性引用其他类型等情况都可以完成转换。

10.5.2 示例：AddressInfoModelBinder 类

ASP.NET Core 内置有许多模型绑定器，一般情况下开发者不需要自行实现 IModelBinder 接口。除非遇到一些特殊的绑定（转换），如本示例将文本信息转换为自定义的模型类。

本示例实现将文本表示的地址信息转换为 AddressInfo 类实例的功能。AddressInfo 类的定义如下。

```
public class AddressInfo
{
    /// <summary>
    /// 哪栋大楼
    /// </summary>
    public string Building { get; set; } = string.Empty;

    /// <summary>
    /// 哪一楼层
    /// </summary>
    public int Floor { get; set; } = default;

    /// <summary>
    /// 哪个科室/办公室
    /// </summary>
    public uint Room { get; set; } = default;
}
```

Building 属性表示地址位于哪栋大楼，Floor 属性表示楼层，Room 属性表示某一楼层中某一科室/办公室。

定义 AddressInfoModelBinder 类，实现 IModelBinder 接口。

```
public class AddressInfoModelBinder : IModelBinder
{
    public Task BindModelAsync(ModelBindingContext bindingContext)
    {
        if (bindingContext == null) throw new ArgumentNullException
        (nameof(bindingContext));

        // 尝试读取数据
        var value = bindingContext.ValueProvider.GetValue(bindingContext.
        ModelName);
        if (value == ValueProviderResult.None)
        {
            return Task.CompletedTask;
        }
        // 通过正则表达式找出需要的值
        string? textdata = value.FirstValue;
        if (string.IsNullOrEmpty(textdata))
        {
            return Task.CompletedTask;
        }
        var reg = Regex.Match(textdata, "([a-zA-Z0-9]{1,2})栋(\\d+)楼
        (\\d{1,3})室");
        if (reg.Success)
        {
            // 读取各分组的值
            AddressInfo addr = new AddressInfo();
            addr.Building = reg.Groups[1].Value;
            addr.Floor = int.Parse(reg.Groups[2].Value);
            addr.Room = uint.Parse(reg.Groups[3].Value);
            // 设置绑定结果
            bindingContext.Result = ModelBindingResult.Success(addr);
        }
        return Task.CompletedTask;
    }
}
```

IValueProvider 接口提供的值为单个字符串实例，格式形如“A 栋 3 楼 102 室”。映射到 AddressInfo 对象的属性后，即

```
Building = A
Floor = 3
Room = 102
```

数据转换时使用了正则表达式。匹配规则为

```
([a-zA-Z0-9]{1,2})栋(\\d+)楼(\\d{1,3})室
```

上述表达式将捕捉到 3 个分组。

（1）[a-zA-Z0-9]{1,2}：允许使用大小写字母和数字 0 ~ 9，如 A2、C、6E、4、5f 等。

（2）\d+：1 位以上数字，如 6、12、08 等。

（3）\d{1,3}：1～3 位数字，如 123、59 等。

绑定成功后要设置 bindingContext.Result 属性，否则应用程序会认为绑定失败而无法获取到正确的 AddressInfo 实例。

AddressInfoModelBinder 类是专为 AddressInfo 对象而编写的绑定器，因此不需要全局应用，只需要通过 ModelBinder 特性将其关联到 AddressInfo 类上即可。

```
[ModelBinder(typeof(AddressInfoModelBinder))]
public class AddressInfo
{
    …
}
```

下面的代码定义 Home 控制器。

```
public class HomeController : Controller
{
    [HttpGet("/home/test/{address}")]
    public IActionResult Test(AddressInfo address)
    {
        string str = $"""
            地址：
            建筑：{address.Building}栋
            楼层：{address.Floor}
            科室：{address.Room}
            """;
        return Content(str);
    }
}
```

在 Test()方法中，address 参数将从路由参数中获取数据，再由 AddressInfoModelBinder 类将其转换为 AddressInfo 实例。

运行示例程序，访问以下 URL 就能完成 AddressInfo 类的绑定，如图 10-17 所示。

```
https://<主机名>/home/test/
C栋7楼116室
```

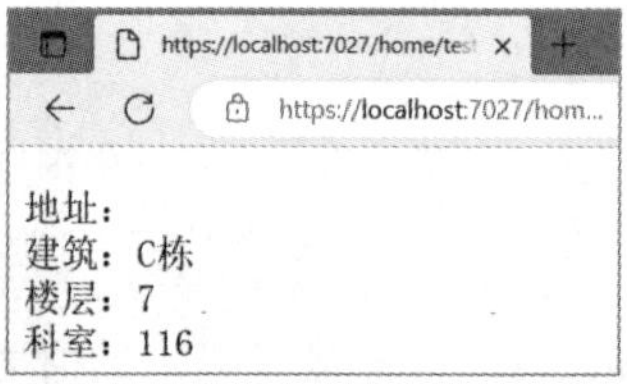

图 10-17　将文本参数转换为 AddressInfo 对象

10.6　BindRequiredAttribute 类与 BindNeverAttribute 类

BindRequiredAttribute 类与 BindNeverAttribute 类是两个特性类，可应用于操作方法的参数上，或者模型类及其属性成员上。

应用 BindRequired 特性表示在模型绑定过程中，IValueProvider 接口必须提供相应的值。如果在 IValueProvider 接口中找不到数据值，将生成错误信息，并且 ControllerBase 类（包括 Controller、PageModel 等类）的 ModelState.IsValid 属性将返回 false。

应用 BindNever 特性表示该对象不参与模型绑定，就算 IValueProvider 接口中存在相应的值也会被排除。

如果两个特性都不应用，那么目标成员将被视为“可选”——若 IValueProvider 中有需要的值就进行绑定，若无相关的值则保留目标成员的默认值。IsValid 属性不会返回 false。

下面的代码定义一个 Order 类。

```
public class Order
{
    /// <summary>
    /// 订单号
    /// </summary>
    [BindRequired]
    public long UCode { get; set; }

    /// <summary>
    /// 订货数量
    /// </summary>
    [BindRequired]
    public int Qty { get; set; }

    /// <summary>
    /// 货品单价
    /// </summary>
    [BindRequired]
    public decimal Price { get; set; }

    /// <summary>
    /// 下单时间
    /// </summary>
    public DateTime BuildTime { get; set; }

    /// <summary>
    /// 是否有效
    /// </summary>
    public bool Valid { get; set; } = true;

    /// <summary>
    /// 备注
    /// </summary>
    [BindNever]
    public string Remark { get; set; } = string.Empty;
}
```

其中，UCode、Qty、Price 属性在模型绑定中为必需项，若未提供相应的值，将绑定失败；Remark 属性被排除，不进行模型绑定。

假设有以下 Home 控制器，允许客户端提交 Order 的数据。

```
public class HomeController : Controller
```

```
{
    [HttpPost("/orders/add")]
    public IActionResult NewOrder(Order ord)
    {
        // 若绑定未成功
        if(ModelState.IsValid == false)
        {
            return StatusCode(StatusCodes.Status500InternalServerError,
            "模型绑定失败");
        }
        // 若绑定成功，以 JSON 格式返回 Order 对象
        return Json(ord);
    }
}
```

在发出 HTTP-POST 请求时，如果只提供了 UCode、Qty 属性的值，而未提供 Price 属性的值，模型绑定将失败。

```
POST /orders/add HTTP/1.1
Content-Length: 21
Content-Type: application/x-www-form-urlencoded
…

ucode=2566543&qty=300

// 服务器响应
HTTP/1.1 500 Internal Server Error
Content-Type: text/plain; charset=utf-8
…

模型绑定失败
```

Remark 属性不进行绑定，就算提供值也不会被使用。

```
POST /orders/add HTTP/1.1
Content-Length: 43
Content-Type: application/x-www-form-urlencoded
…

ucode=5590743&qty=320&price=100&remark=test

// 服务器响应
HTTP/1.1 200 OK
Content-Type: application/json; charset=utf-8
…

{
"uCode": 5590743,
"qty": 320,
"price": 100,
```

```
"buildTime": 0001/1/1 0:00:00,
"valid": true,
"remark": ""
}
```

10.7　绑定到属性成员

如果要进行模型绑定的对象不是操作方法的参数，而是控制器的属性，就要在属性上应用 BindProperty 特性（BindPropertyAttribute 类）。如果控制器类中定义的属性都需要模型绑定，可以在控制器类上应用 BindProperties 特性（BindPropertiesAttribute 类）。

BindProperty 特性与 ModelBinder 特性在使用方法上差不多，只不过 BindProperty 特性应用于属性成员，而 ModelBinder 特性多用在方法参数或数据模型类上。

10.7.1　示例：控制器的属性绑定

本示例将演示如何让控制器类的公共属性支持模型绑定。具体步骤如下。

（1）定义 Fan 类，表示一台电风扇的简单信息。

```
public class Fan
{
    // <summary>
    // 转速
    // </summary>
    public int Speed { get; set; }

    // <summary>
    // 颜色
    // </summary>
    public string? Color { get; set; }

    // <summary>
    // 工作电压
    // </summary>
    public float Voltage { get; set; }

    // <summary>
    // 高度
    // </summary>
    public float Height { get; set; }
}
```

（2）定义 Goods 控制器，它有一个 FanInfo 属性。属性应用了 BindProperty 特性，并设置其模型名称为 fan。

```
public class GoodsController : ControllerBase
{
    // 模型名称为 fan
```

```
    [BindProperty(Name = "fan")]
    public Fan? FanInfo { get; set; }
    …
}
```

（3）在 Goods 控制器中定义 SetData()操作方法。

```
[HttpPost("/goods/set")]
public IActionResult SetData()
{
    if(!ModelState.IsValid || FanInfo == null)
    {
        return StatusCode(StatusCodes.Status500InternalServerError,
        "模型数据无效");
    }
    // 返回文本
    return Content($"""
        风扇信息：
        转速：{FanInfo.Speed}RPM
        颜色：{FanInfo.Color}
        工作电压：{FanInfo.Voltage}V
        高度：{FanInfo.Height}CM
        """);
}
```

（4）运行示例程序，待服务器启动后，用 http-repl 工具进行 POST 请求，结果如图 10-18 所示。

```
post /goods/set -h Content-type=application/
x-www-form-urlencoded -c
"fan.speed=66&fan.color=Green&fan.height=
23.5&fan.voltage=5"
```

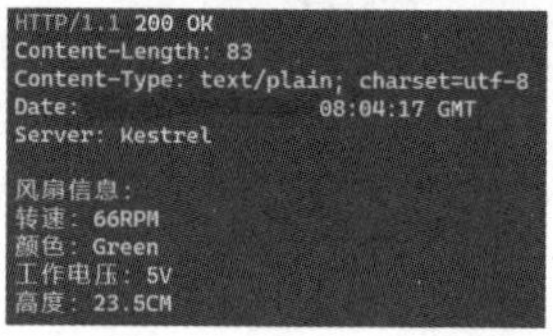

图 10-18　服务器返回的消息

10.7.2　示例：PageModel 中的属性绑定

在 Razor Pages 应用程序中，页面模型类中的以下成员也可以使用模型绑定。

（1）Handler（页面处理程序）的参数，其模型绑定原理与 MVC 控制器中的操作方法相同。

（2）公共属性，与 MVC 控制器中的属性绑定原理相同。

本示例将演示基于页面模型类（PageModel）的属性绑定。下面的代码定义模型类 User。

```
public class User
{
    [BindRequired, Display(Name = "用户名")]
    public string Name { get; set; } = string.Empty;

    [BindRequired, Display(Name = "密码"), DataType(DataType.Password)]
    public string Password { get; set; } = string.Empty;
```

```
    [Display(Name = "电子信箱")]
    public string Email { get; set; } = string.Empty;
}
```

BindRequired 指示在模型绑定时 IValueProvider 接口必须提供目标属性的值。Display 特性指定目标属性在用户界面上显示的文本（如 Password 属性在用户界面上会显示“密码”）。DataType(DataType.Password) 特性指定 Password 属性在与<input>元素绑定时自动设置 type="password"，如

```
<input type="password"… />
```

定义页面模型类，派生自 PageModel。

```
public class ReguserPageModel : PageModel
{
    [BindProperty]
    public User? UserData { get; set; }

    // 客户端发送 POST 请求时调用
    public void OnPost()
    {
        if(UserData != null && ModelState.IsValid)
        {
            string s = $"已注册的用户名：{UserData.Name}，关联的电子邮箱：
            {UserData.Email}";
            ViewData["added"] = s;
        }
    }
}
```

UserData 属性的值是 User 类型，并且应用 BindProperty 特性，使其支持模型绑定。如果模型绑定有效，OnPost()方法将生成一条字符串信息，并存储在 ViewData 属性中。随后应用程序重新加载 Razor 文件，就可以读取 ViewData 属性中的数据，然后呈现到 HTML 文档中。ViewData 属性是字典结构，可以用 Key-Value 的方式读写。

与页面模型对应的 Razor 文档如下，位于/Pages 目录下。

```
@page
@model ReguserPageModel

@if (ViewData.ContainsKey("added"))
{
    <div>
    <p>恭喜你，用户添加成功</p>
    <p>@ViewData["added"]</p>
    </div>
}
else
{
```

```
        <form method="post">
        <div>
        <label asp-for="UserData!.Name"></label>:
        <input asp-for="UserData!.Name"/>
        </div>
        <div>
        <label asp-for="UserData!.Password"></label>:
        <input asp-for="UserData!.Password"/>
        </div>
        <div>
        <label asp-for="UserData!.Email"></label>:
        <input asp-for="UserData!.Email"/>
        </div>
        <button type="submit">注册</button>
        </form>
    }
```

如果 ViewData 属性中能读出键名为 added 的数据，那么就在页面上显示它；否则呈现<form>元素，让用户输入信息。

运行示例程序，首先呈现输入表单，如图 10-19 所示。

输入用户信息后，单击“注册”按钮，Web 服务器将接收并处理数据，最后呈现新用户的摘要信息，如图 10-20 所示。

图 10-19　呈现表单元素

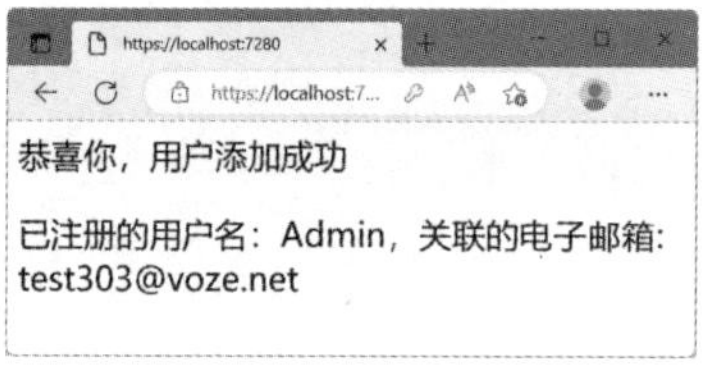

图 10-20　新用户摘要

10.7.3　示例：CancellationToken 类型的属性绑定

ASP.NET Core 服务器在处理客户端请求时，如果连接丢失（用户可能关闭了浏览器，或者客户端程序所在的网络断开），那么 HttpContext 类的 RequestAborted 属性就会被标记为“已取消”状态。为了便于访问，ASP.NET Core 类库实现了 CancellationTokenModelBinder 类。这表明 RequestAborted 属性将支持模型绑定。只要 MVC 或 Razor 页面模型存在 CancellationToken 类型的属性或方法参数，就可以自动获取 RequestAborted 属性的值。

本示例将通过在控制器的某个公共属性上应用 BindProperty 特性实现与 RequestAborted 绑定。

Home 控制器的定义代码如下。

```
public class HomeController : Controller
{
    // 此对象用于写入日志，通过依赖注入获取
```

```
    private readonly ILogger<HomeController> _logger;

    // 构造函数
    public HomeController(ILogger<HomeController> logger)
    {
        _logger = logger;
    }

    // 该属性将进行模型绑定
    [BindProperty(SupportsGet = true)]
    public CancellationToken CancelOpt { get; set; }

    [HttpGet("/")]
    public IActionResult Main()
    {
        return View();
    }

    [HttpGet("/home/test")]
    public async Task<IActionResult> Test()
    {
        // 模拟耗时操作，等待一段时间
        int x = 0;
        while (x < 60)
        {
            if (CancelOpt.IsCancellationRequested)
            {
                // 输出日志
                _logger.LogWarning("注意：客户端关闭了连接");
                break;                              // 跳出循环
            }
            await Task.Delay(100);
            x++;
        }
        return Content($"处理完毕 / {DateTime.Now.ToLongTimeString()}");
    }
}
```

CancelOpt 属性的类型为 CancellationToken 结构体，并且应用了 BindProperty 特性。此处要将 SupportsGet 属性设置为 true，因为 Test()方法中由 HttpGet 特性指定了它只接受 HTTP-GET 请求，而 BindProperty 特性默认不支持在处理 HTTP-GET 请求时进行绑定。

在 Test()方法中，结合 while 循环和 Task.Delay()方法模拟一个耗时操作（需要执行一段时间才能向客户端回应消息）。在 while 循环内部，随时检查 IsCancellationRequested 属性是否为 true。若为 true，表明客户端已断开连接，于是跳出 while 循环，并在日志中写入一条“警告”消息。

下面是 Main 视图的核心代码。

```
<html lang="zh-cn">
…
<body>
<p>…</p>
<button id="btn">开始测试</button>

<script type="text/javascript">
        // 获取<button>元素
        var button = document.getElementById("btn");
        // 为 click 事件添加侦听器
        button.addEventListener("click", async _ => {
            // 发送请求前禁用按钮
            button.setAttribute("disabled", true);
            // 以 GET 方式发出 HTTP 请求
            let response = await fetch("/home/test");
            // 请求结束后恢复按钮为可用状态
            button.removeAttribute("disabled");
            let text = await response.text();
            let div = document.getElementById("msg");
            if(div == null){
                div = document.createElement("div");
                div.setAttribute("id", "msg");
            }
            // 在动态插入的<div>元素中显示服务器返回的内容
            div.innerHTML = text;
            document.body.append(div);
        });
</script>
</body>
</html>
```

页面上定义了一个按钮（<button>元素），被单击后由 JavaScript 脚本访问 Home 控制器的 Test()方法并等待服务器处理。当服务器发回响应消息后，将响应的内容显示在一个<div>元素中。

运行示例程序，如图 10-21 所示。

单击“开始测试”按钮，此时按钮变成灰色（不可用状态）。在服务器返回消息前将 Web 浏览器关闭，此时服务器控制台会输出一条“警告”日志，如图 10-22 所示。

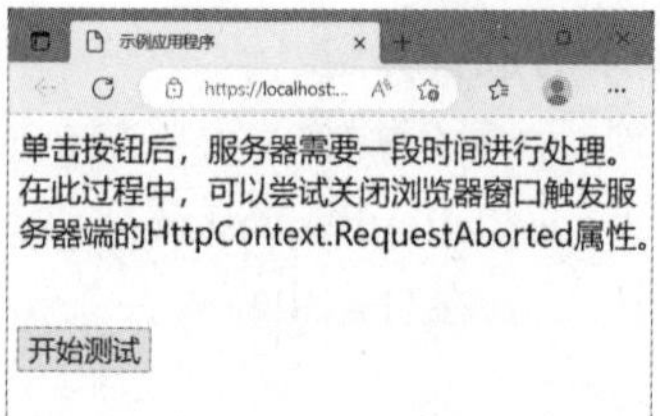
示例应用程序
https://localhost...
单击按钮后，服务器需要一段时间进行处理。在此过程中，可以尝试关闭浏览器窗口触发服务器端的HttpContext.RequestAborted属性。
开始测试

图 10-21　示例主页面

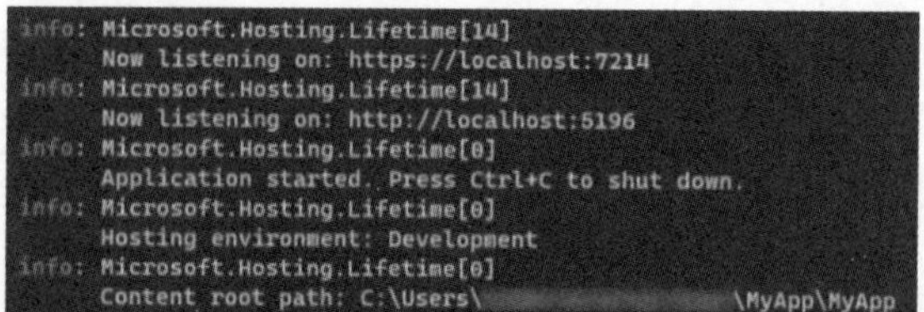
info: Microsoft.Hosting.Lifetime[14]
Now listening on: https://localhost:7214
info: Microsoft.Hosting.Lifetime[14]
Now listening on: http://localhost:5196
info: Microsoft.Hosting.Lifetime[0]
Application started. Press Ctrl+C to shut down.
info: Microsoft.Hosting.Lifetime[0]
Hosting environment: Development
info: Microsoft.Hosting.Lifetime[0]
Content root path: C:\Users\
\MyApp\MyApp
warn: MyApp.HomeController[0]
注意：客户端关闭了连接

图 10-22　输出“警告”日志

第 11 章

Web API

本章要点

- ❑ 定义 Web API
- ❑ API 的输入/输出格式
- ❑ 极小 API
- ❑ API 浏览功能
- ❑ Swagger 与文档生成

11.1 Web API 基础

Web API 即基于 Web 技术实现的应用程序接口。它是 MVC 应用程序的一种特殊表现形式。Web API 一般不直接与最终用户进行交互，也不呈现可见视图（不返回 HTML 文档），而是开放给其他开发人员调用。也就是说，Web API 仅负责数据的输入、处理和输出，并不考虑如何呈现数据。

例如，Jim 开发了一款可以查看全国各地天气预报的应用程序，Jim 的程序自身无法收集各地的气象数据，只能调用 K 公司提供的程序接口来读取数据，然后将梳理后的数据通过可视化界面（如窗口、网页等）呈现给最终用户。

11.1.1 ControllerBase 类与 Controller 类

在 MVC 项目中，从 ControllerBase 或 Controller 派生的类都可以被认为是控制器。两者的区别在于 Controller 可以调用 View()、PartialView()、ViewComponent()等方法成员返回视图，而 ControllerBase 类则没有相关的方法成员。由于 Web API 不返回视图，因此在定义控制器时选择 ControllerBase 作为基类会更合适。当然，如果某个控制器既提供 Web API 功能，又提供视图功能，则应该选择 Controller 作为基类。

11.1.2 ApiController 特性

如果一个控制器将完全作为 Web API 来使用，那么，将 ApiController 特性（ApiController-Attribute 类）应用到该类上会使开发体验变得更友好。

应用该特性后将对 MVC 控制器产生以下影响。

（1）Web API 不使用全局路由规则，而必须使用局部路由规则（Route 特性）。

（2）在模型绑定时自动推断数据来源。本书在讲述模型绑定时提到过可以使用 FromQuery、FromForm、FromBody 等特性显式标注方法参数/属性的数据来源。应用 ApiController 特性后，可以省略 FromBody 等特性，应用程序将自动推断数据来源。

（3）对于以 HTTP 消息正文提交的数据，应用程序将使用 BodyModelBinder 进行绑定，并由 IInputFormatter 接口负责读取正文内容。默认支持 JSON 格式和 XML 格式（需手动开启）的数据。

（4）如果模型绑定失败，将自动返回错误信息（由 ModelStateInvalidFilter 进行过滤）。

11.1.3 示例：一个简单的 Web API

本示例将演示一个简单的 Web API，调用后返回 Album 类的实例。Album 类的定义如下。

```
/// <summary>
/// 音乐专辑
/// </summary>
public class Album
{
    /// <summary>
    /// 标题
    /// </summary>
    public string Title { get; set; } = string.Empty;

    /// <summary>
    /// 发行年份
    /// </summary>
    public int Year { get; set; }

    /// <summary>
    /// 简介
    /// </summary>
    public string Description { get; set; } = string.Empty;

    /// <summary>
    /// 封面艺术家
    /// </summary>
    public string Artist { get; set; } = string.Empty;

    /// <summary>
```

```
    /// 曲目数量
    /// </summary>
    public int TrackNum { get; set; }
}
```

下面的代码定义 Test 控制器。

```
[Route("api/[controller]")]
[ApiController]
public class TestController : ControllerBase
{
    [HttpGet]
    public Album GetAlbum()
    {
        return new Album()
        {
            Title = "样品专辑",
            Year = 2012,
            Artist = "老胡",
            Description = "试听碟",
            TrackNum = 13
        };
    }
}
```

注意，Test 控制器已应用 ApiController 特性。Route 特性所指定的 URL 路由规则中只提供了 controller 字段，这使得访问 GetAlbum()操作方法时的 URL 为/api/test。示例程序将返回以下 JSON 格式的数据（默认格式）。

```
{
"title": "样品专辑",
"year": 2012,
"description": "试听碟",
"artist": "老胡",
"trackNum": 13
}
```

应用程序在访问上述URL时未指定action字段值的情况下能正确调用 GetAlbum()方法，是因为 HttpGet 特性也具备设置 URL 路由的功能，且默认的路由模板为空。如果 Test 控制器中同时存在以下操作方法，访问/api/test 将引发异常，如图 11-1 所示。此时，GetAlbum()和 GetNumber()方法同时匹配成功。

```
[HttpGet]
public int GetNumber() => 35600;
```

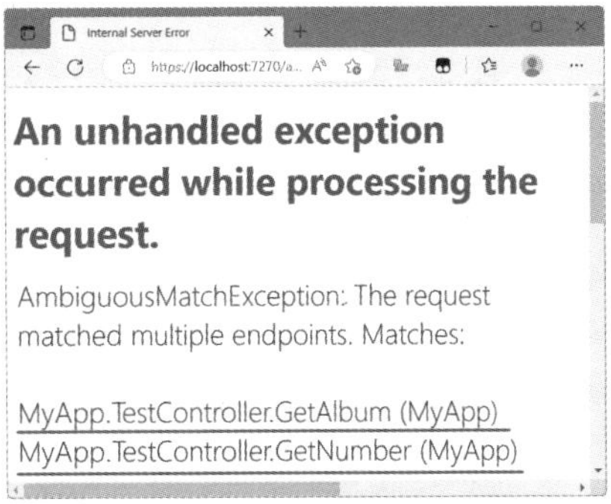

图 11-1　两个操作同时匹配成功

在初始化应用程序时，由于 Web API 项目不需要返回视图，因此，应该在服务容器上调用 AddControllers()

方法，并在 WebApplication 对象上调用 MapControllers()方法添加终结点映射，不需要指定全局 URL 路由。

```
var builder = WebApplication.CreateBuilder(args);
builder.Services.AddControllers();
var app = builder.Build();

app.MapControllers();

app.Run();
```

11.1.4 示例：以 POST 方式提交数据

Web API 默认使用 JSON 格式处理数据的输入与输出。若以 HTTP-POST 方式向服务器提交数据，应当用 JSON 格式的数据填充 HTTP 消息的正文（Body）部分，还要将 Content-Type 消息头设置为 application/json。

本示例将使用 Person 类作为数据模型。

```
public class Person
{
    /// <summary>
    /// 身份证号
    /// </summary>
    public long Id { get; set; }

    /// <summary>
    /// 姓名
    /// </summary>
    public string Name { get; set; } = string.Empty;

    /// <summary>
    /// 年龄
    /// </summary>
    public int Age { get; set; }
}
```

下面的代码定义 Person 控制器。

```
[ApiController]
[Route("api/[controller]")]
public class PersonController : ControllerBase
{
    [HttpPost]
    public object PostData(Person ps)
    {
        string msg = $"个人信息：{ps.Name}（{ps.Id}），{ps.Age}岁";
        // 响应内容
        var result = new
        {
```

```
            code = 0,
            message = msg
        };
        return result;
    }
}
```

PostData 接收 Person 对象作为输入参数（应用程序自动推断数据来源，从消息正文读取 JSON 格式的数据）。方法的返回值声明为 object，实际返回类型是匿名类型（result 变量）。此匿名类型包含两个属性：code 和 message。

下面的代码用于测试 Web API 的调用。

```
// 服务器根地址
Uri rootUrl = new("https://localhost:7025");

HttpClient client = new HttpClient();
client.BaseAddress = rootUrl;

// 准备数据
var dataObj = new
{
    Id = 544102885669,
    Name = "小明",
    Age = 30
};
JsonContent content = JsonContent.Create(dataObj, dataObj.GetType());

// 发送请求
var response = await client.PostAsync("/api/person", content);
// 输出服务器响应结果
Console.WriteLine($"响应状态码: {(int)response.StatusCode}");
Console.WriteLine("响应内容: ");
// 输出格式化后的 JSON
JsonDocument jdoc = JsonDocument.Parse(await response.Content.
ReadAsStringAsync());
// 将控制台的输出编码改为 UTF-8
Console.OutputEncoding = Encoding.UTF8;
Utf8JsonWriter writer = new(Console.OpenStandardOutput(), new
JsonWriterOptions
{
    // 包含缩进
    Indented = true,
    // 不转义字符
    Encoder = JavaScriptEncoder.UnsafeRelaxedJsonEscaping
});
// 写入 UTF-8 字符
jdoc.WriteTo(writer);
```

```
// 释放资源
writer.Dispose();
jdoc.Dispose();
```

在客户端代码中，不需要重新定义 Person 类，而是使用匿名类型（dataObj 变量）。只要匿名类型的属性名称（Id、Name、Age）与 Person 类相同即可。

为了让服务器返回的 JSON 格式的数据更直观，上述代码使用了 JsonDocument 和 Utf8JsonWriter 类将 JSON 格式的数据输出到控制台。注意下面这行代码，其作用是把控制台的输出编码修改为 UTF-8。因为有些操作系统（如现阶段的 Windows 系统）的默认编码不是 UTF-8，统一修改编码可防止汉字字符显示为乱码。

```
Console.OutputEncoding = Encoding.UTF8;
```

先运行 ASP.NET Core 服务器，再运行测试客户端。当客户端成功提交数据后，服务器返回的 JSON 格式的数据如图 11-2 所示。

图 11-2　服务器返回的 JSON 格式的数据

11.2　XML 格式

ASP.NET Core 的 Web API 项目默认支持传输 JSON 格式的数据（即使用 JSON 序列化）。以下代码片段摘录自 ASP.NET Core 源代码。

```
// Set up default input formatters.
options.InputFormatters.Add(new SystemTextJsonInputFormatter(_jsonOptions.
Value, _loggerFactory.CreateLogger<SystemTextJsonInputFormatter>()));

// Media type formatter mappings for JSON
options.FormatterMappings.SetMediaTypeMappingForFormat("json",
MediaTypeHeaderValues.ApplicationJson);

// Set up default output formatters.
options.OutputFormatters.Add(new HttpNoContentOutputFormatter());
options.OutputFormatters.Add(new StringOutputFormatter());
options.OutputFormatters.Add(new StreamOutputFormatter());

var jsonOutputFormatter = SystemTextJsonOutputFormatter.CreateFormatter
(_jsonOptions.Value);
options.OutputFormatters.Add(jsonOutputFormatter);
```

通过对源代码的分析可知，当 MVC 操作方法被调用时，应用程序会使用 SystemTextJsonInputFormatter 类读客户端发送的内容（输入数据）；当操作方法返回时，应用程序使用 SystemTextJsonOutputFormatter、StringOutputFormatter 或 StreamOutputFormatter 等类型写入响应内容（输出数据）。

ASP.NET Core 支持两种 XML 序列化方案，如表 11-1 所示。

表 11-1 Web API的XML序列化方案

序列化方案	输入（InputFormatter）	输出（OutputFormatter）
XmlSerializer	XmlSerializerInputFormatter	XmlSerializerOutputFormatter
DataContractSerializer	XmlDataContractSerializerInputFormatter	XmlDataContractSerializerOutputFormatter

ASP.NET Core 应用程序默认未开启 XML 序列化的功能，需要手动设置。

（1）要使用 XmlSerializer 序列化方案，应在服务容器上调用 AddXmlSerializerFormatters() 扩展方法。

（2）要使用 DataContractSerializer 序列化方案，应当在服务容器上调用 AddXmlDataContractSerializerFormatters()扩展方法。

11.2.1 示例：常规的 XML 序列化方案

本示例将演示使用 XmlSerializer 序列化方案的 Web API。具体步骤如下。

（1）定义 TestModel 模型类。

```
[XmlRoot(ElementName = "app")]
public class TestModel
{
    [XmlElement(ElementName = "val_a")]
    public double ValueA { get; set; }

    [XmlElement(ElementName = "val_b")]
    public long ValueB { get; set; }

    [XmlElement(ElementName = "val_c")]
    public Guid ValueC { get; set; }
}
```

上述代码在定义类时使用了用于个性化定制 XML 文档的特性。XmlRoot 特性指定序列化后 XML 文档根元素的名称。如果不应用此特性，那么 XML 文档的根元素将与类名相同。同理，XmlElement 特性用于设置 TestModel 类的属性成员映射到 XML 文档中的元素名称。如果不使用 XmlElement 特性，则 XML 元素的名称与属性相同。

经过 XmlRoot、XmlElement 特性的修饰后，TestModel 类在序列化后，生成的 XML 文档的根元素名为 app，ValueA 属性对应的 XML 元素名为 val_a。ValueB 和 ValueC 属性也与上述类似。

（2）定义 Demo 控制器。

```
[Route("api/[controller]/[action]")]
[ApiController]
public class DemoController : ControllerBase
```

```
{
    [HttpPost]
    public void CheckOne(TestModel obj, ILogger<DemoController> logger)
    {
        string msg = $"""
            客户端提交的内容：
            Value A: {obj.ValueA}
            Value B: {obj.ValueB}
            Value C: {obj.ValueC}
        """;
        // 记录日志
        logger.LogInformation(msg);
    }
}
```

CheckOne()方法有两个参数，第 1 个参数的类型是 TestModel，应用程序自动推断出它将从请求消息的正文处读取数据；而 logger 参数则从服务窗口中通过依赖注入获取相关实例的引用。logger 参数用于输出日志记录。

（3）在 ASP.NET Core 应用程序的初始化代码中，需要手动添加对 XML 序列化的支持（调用 AddXmlSerializerFormatters()扩展方法）。

```
var builder = WebApplication.CreateBuilder(args);
builder.Services.AddControllers().AddXmlSerializerFormatters();
var app = builder.Build();
…
```

接下来将编写一个测试客户端程序。

（4）新建一个控制台应用程序。

（5）重新定义模型类，类名与属性名称不必与服务器项目中的 TestModel 类一致，只要应用的 XmlRoot、XmlElement 等特性的设置参数一致即可。

```
[XmlRoot(ElementName = "app")]
public class DataObj
{
    [XmlElement(ElementName = "val_a")]
    public double PropA { get; set; }

    [XmlElement(ElementName = "val_b")]
    public long PropB { get; set; }

    [XmlElement(ElementName = "val_c")]
    public Guid PropC { get; set; }
}
```

有参数相同的 XmlRoot 和 XmlElement 特性，就算 TestModel 和 DataObj 的类名与属性名称不同，也能进行序列化和反序列化。

为了展示 XmlRoot 等特性类的功能，本示例采用服务器与客户端单独定义模型类的方式。

在实际开发中，也可以把模型类的定义放在一个公共的类库项目中，然后服务器和客户端项目都引用这个类库项目。这样一来，模型类只定义一次即可。

（6）下面的代码将测试调用 Web API。

```
// 准备数据
DataObj data = new()
{
    PropA = 152.002511d,
    PropB = 52365589,
    PropC = Guid.NewGuid()
};
// 序列化
XmlSerializer xs = new XmlSerializer(typeof(DataObj));
Console.WriteLine("序列化后的 XML 文档：");
MemoryStream mmStream = new();
xs.Serialize(mmStream, data);
// 将 XML 也输出到控制台
using (Stream csout = Console.OpenStandardOutput())
{
    // 重置流的当前位置
    mmStream.Seek(0, SeekOrigin.Begin);
    mmStream.CopyTo(csout);
}
// HTTP 正文
mmStream.Seek(0, SeekOrigin.Begin);
StreamContent content = new(mmStream);
// 内容类型：application/xml
content.Headers.ContentType = MediaTypeHeaderValue.Parse(MediaTypeNames.
Application.Xml);
// 发起请求
HttpClient client = new HttpClient();
client.BaseAddress = rootUrl;
var response = await client.PostAsync("/api/demo/checkone", content);
// 输出响应结果
Console.WriteLine($"\n\n 响应状态：{response.StatusCode}");
```

发起请求前，先进行 XML 序列化并将结果写入内存流（mmStream 变量）中。再基于内存流创建 StreamContent 实例。服务器开启对 XML 格式的支持，客户端发起请求时，要将 Content-Type 消息头设置为 application/xml。

序列化生成的 XML 文档如下。

```
<?xml version="1.0" encoding="utf-8"?>
<app xmlns:xsi="http://www.w3.org/2001/XMLSchema-instance" xmlns:xsd=
"http://www.w3.org/2001/XMLSchema">
<val_a>152.002511</val_a>
<val_b>52365589</val_b>
<val_c>3222493b-fd74-4944-adc9-add53cc78b85</val_c>
</app>
```

服务器在接收到客户端请求后，在日志中写入一条记录，如图 11-3 所示。

```
info: MyApp.DemoController[0]
      客户端提交的内容：
      Value A: 152.002511
      Value B: 52365589
      Value C: 3222493b-fd74-4944-adc9-add53cc78b85
```

图 11-3　服务器接收到的内容

11.2.2　示例：使用 XmlDataContractSerializer 方案

本示例将演示基于数据协定（Data Contract）的 XML 序列化方案，即使用 DataContract-Serializer 类进行 XML 序列化和反序列化。

定义 Home 控制器。

```
[Route("api/[controller]")]
[ApiController]
public class HomeController : ControllerBase
{
    [HttpPost]
    public void PostDic(IDictionary<int, string> dic, ILogger<HomeController>
    logger)
    {
        string msg = "客户端提交的数据：\n";
        foreach(var pair in dic)
        {
            msg += $"\t{pair.Key} = {pair.Value}\n";
        }
        logger.LogInformation(msg);
    }
}
```

PostDic()方法的第 1 个参数是字典类型，属于.NET 现有的 CLR 类型，不需要自定义模型类；第 2 个参数将通过依赖注入从服务容器中获取，用于写入日志。

在 ASP.NET Core 应用程序的初始化代码中，调用 AddXmlDataContractSerializer-Formatters()扩展方法添加对数据协定序列化的支持。

```
var builder = WebApplication.CreateBuilder(args);
builder.Services.AddControllers().AddXmlDataContractSerializerFormatters();
var app = builder.Build();
...
```

以下是用于调用 PostDic()方法的测试代码。

```
// 服务器根地址
Uri baseUrl = new Uri("http://localhost:5219");
// 准备数据
Dictionary<int, string> data = new();
data[1] = "A";
data[2] = "B";
```

```
data[3] = "C";
using MemoryStream mmStream = new MemoryStream();
// 将数据序列化
DataContractSerializer sz = new DataContractSerializer(data.GetType());
sz.WriteObject(mmStream, data);
// 将 XML 文档打印到控制台
using (var stream = Console.OpenStandardOutput())
{
    mmStream.Seek(0, SeekOrigin.Begin);
    mmStream.CopyTo(stream);
}
Console.WriteLine("\n\n");
// 发送请求
HttpClient client = new HttpClient();
client.BaseAddress = baseUrl;
mmStream.Seek(0, SeekOrigin.Begin);
StreamContent content = new(mmStream);
// 设置 Content-Type 标头
content.Headers.ContentType = MediaTypeHeaderValue.Parse("application/
xml");
var response = await client.PostAsync("/api/home", content);
if(response.IsSuccessStatusCode)
{
    Console.WriteLine("数据提交成功");
}
else
{
    Console.WriteLine("数据提交失败,状态码:{0}", (int)response.StatusCode);
}
```

数据序列化后保存到内存流中，再通过 StreamContent 类封装为 HTTP 消息正文。为了让服务器能够正确接收到消息，需要将 Content-Type 消息头的值设置为 application/xml。

序列化时使用了 DataContractSerializer 类，会生成以下 XML 文档。

```
<ArrayOfKeyValueOfintstring xmlns="http://schemas.microsoft.com/2003/10/
Serialization/Arrays" xmlns:i="http://www.w3.org/2001/XMLSchema-instance">
    <KeyValueOfintstring>
        <Key>1</Key>
        <Value>A</Value>
    </KeyValueOfintstring>
    <KeyValueOfintstring>
        <Key>2</Key>
        <Value>B</Value>
    </KeyValueOfintstring>
    <KeyValueOfintstring>
        <Key>3</Key>
        <Value>C</Value>
    </KeyValueOfintstring>
</ArrayOfKeyValueOfintstring>
```

服务器接收到 HTTP 消息后，将输出如图 11-4 所示的日志信息。

```
info: MyApp.HomeController[0]
      客户端提交的数据：
        1 = A
        2 = B
        3 = C
```

图 11-4　服务器接收到的数据

11.3　选择响应格式

Web API 默认使用 JSON 格式返回数据。如果 API 支持多种格式（如 XML 或自定义格式），那么调用者可以选择需要的格式。

11.3.1　示例：通过 Accept 消息头选择响应格式

本示例所演示的 API 将返回 Pet 类的实例，其数据格式将通过设置 Accept 消息头进行选择。Pet 类定义如下。

```
public class Pet
{
    public string Nick { get; set; } = string.Empty;
    public string Owner { get; set; } = string.Empty;
}
```

以下是 Pets 控制器的实现代码。

```
[Route("api/[controller]")]
[ApiController]
public class PetsController : ControllerBase
{
    [HttpGet]
    public Pet GetPet()
    {
        return new Pet
        {
            Nick = "Tommy",
            Owner = "Jim"
        };
    }
}
```

在应用程序配置代码中，添加对 XML 格式的支持。

```
var builder = WebApplication.CreateBuilder(args);
builder.Services.AddControllers().AddXmlDataContractSerializerFormatters();
var app = builder.Build();
...
```

本示例使用的是基于数据协定的 XML 格式。

运行示例程序后，在以 HTTP-GET 方式访问/api/pets 时添加以下消息头：

```
Accept: application/xml
```

Pet 实例将以 XML 格式返回。

```
<Pet xmlns:i="http://www.w3.org/2001/XMLSchema-instance" xmlns="http://
schemas.datacontract.org/2004/07/MyApp">
<Nick>Tommy</Nick>
<Owner>Jim</Owner>
</Pet>
```

在发出请求时若加上以下消息头：

```
Accept: application/json
```

Pet 实例将以 JSON 格式返回。

```
{
"nick": "Tommy",
"owner": "Jim"
}
```

11.3.2　示例：使用格式过滤器

格式过滤器（Formatter Filter，或称为格式筛选器）也可以用于选择 Web API 的响应格式。方法是在 URL 上使用 format 字段指定格式。

format 字段可以作为查询字符串来提供，如

```
http://someurl/api/test?format=json
```

也可以作为路由参数提供，如

```
http://someurl/api/test/json
```

本示例将以路由参数的方式使用格式过滤器。

定义 record（记录）类型 Point。

```
public record Point
{
    public Point() { }
    public Point(int x, int y)
    {
        X = x;
        Y = y;
    }
    public int X { get; set; }
    public int Y { get; set; }
}
```

以下代码实现 Point 控制器。

```
[ApiController]
public class PointsController : ControllerBase
{
```

```
    [HttpGet("api/points/{format?}"), FormatFilter]
    public Point[] GetPoints()
    {
        return new Point[]
        {
            new Point(10, 50),
            new Point(30, 5),
            new Point(80, 65),
            new Point(25, 40)
        };
    }
}
```

要启用格式过滤器，需要在操作方法上应用 FormatFilter 特性。

在应用程序的初始化代码中，添加对 XML 格式的支持。

```
var builder = WebApplication.CreateBuilder(args);
builder.Services.AddControllers().AddXmlSerializerFormatters();
var app = builder.Build();
…
```

运行示例程序，当访问/api/points/json 时，调用结果以 JSON 格式返回。

```
[
    {
    "x": 10,
    "y": 50
    },
    {
    "x": 30,
    "y": 5
    },
    {
    "x": 80,
    "y": 65
    },
    {
    "x": 25,
    "y": 40
    }
]
```

当访问/api/points/xml 时，调用结果将以 XML 格式返回。

```
<ArrayOfPoint xmlns:xsi="http://www.w3.org/2001/XMLSchema-instance"
xmlns:xsd="http://www.w3.org/2001/XMLSchema">
    <Point>
        <X>10</X>
        <Y>50</Y>
    </Point>
    <Point>
        <X>30</X>
```

```
        <Y>5</Y>
    </Point>
    <Point>
        <X>80</X>
        <Y>65</Y>
    </Point>
    <Point>
        <X>25</X>
        <Y>40</Y>
    </Point>
</ArrayOfPoint>
```

11.4　自定义格式

当内置的格式（JSON 和 XML）不能满足特定需求时，开发人员可以考虑为应用程序增加自定义的格式。

在 Microsoft.AspNetCore.Mvc.Formatters 下包含两个与格式化数据有关的通用接口。

（1）IInputFormatter 接口：表示输入数据的格式。即从请求正文中读取客户端提交的数据，并转换为应用程序所需要的.NET 类型。

（2）IOutputFormatter 接口：表示输出数据的格式。在将结果返回给客户端前，将.NET 类型转换为特定的格式，再写入响应流中。

一般情况下，开发者不需要直接实现上述接口，而是寻找一个合适的抽象类，然后实现它。这样能减少开发工作量，提高效率。如果数据是基于二进制的（非文本），可以选择实现 InputFormatter 和 OutputFormatter 类；如果数据是文本内容，应当实现 TextInputFormatter 和 TextOutputFormatter 类。例如内置的 JSON 格式，其内容属于文本型，因此 SystemTextJson-InputFormatter 类实现了抽象类 TextInputFormatter，SystemTextJsonOutputFormatter 类实现了 TextOutputFormatter 抽象类。InputFormatter、OutputFormatter 类都公开了 SupportedMediaTypes 属性。该属性是集合类型，用于添加正在实现的格式化程序所支持的媒体类型（用 MIME 类型表示，如 image/png）。不管是输入还是输出格式，控制器将通过媒体类型来选择数据处理格式。

在初始化应用程序时，需要配置 MvcOptions 选项类，InputFormatters 属性对应的是 InputFormatter 的集合，OutputFormatters 属性对应的是 OutputFormatter 集合。将自定义的 *Formatter 实例添加到相应的集合中即可。

11.4.1　示例：CustDataInputFormatter 类

本示例演示一个自定义的输入格式化程序：CustDataInputFormatter。该程序将从以下格式的文本中解析出数据类型所需要的属性名称和属性值（不区分大小写）。

```
[属性 1]值 1,[属性 2]值 2,[属性 3]值 3
```

举个例子，假如客户端提交了以下内容：

```
[name]小明,[city]北京,[qq]1234567
```

就会解析出以下属性/值列表。

```
Name = 小明
City = 北京
QQ = 1234567
```

中括号内为属性名称，紧跟在中括号后面的是属性的值。属性之间用英文逗号分隔。

具体实现步骤如下。

（1）定义 CustDataInputFormatter 类，实现 TextInputFormatter 抽象类，用于处理输入格式。

```
public class CustDataInputFormatter : TextInputFormatter
{
    public CustDataInputFormatter()
    {
        // 设置此格式支持的文本编码
        SupportedEncodings.Add(UTF8EncodingWithoutBOM);
        // 设置此格式支持的媒体类型
        SupportedMediaTypes.Add("application/cust");
    }

    protected override bool CanReadType(Type type)
    {
        // 目标类型匹配时才可读
        return type == typeof(MyData);
    }

    public override async Task<InputFormatterResult> ReadRequestBodyAsync
    (InputFormatterContext context, Encoding encoding)
    {
        if (context == null)
        {
            throw new ArgumentNullException(nameof(context));
        }
        if (encoding == null)
        {
            throw new ArgumentNullException(nameof(encoding));
        }
        // 读取正文
        var reader = context.ReaderFactory(context.HttpContext.Request.
        Body, encoding);
        string? text = await reader.ReadToEndAsync();
        // 分割字符串
        string[] parts = text.Split(',');
        // 创建对象实例
        Type dataType = context.ModelType;
        object dataObject = Activator.CreateInstance(dataType)!;
```

```
// 获取属性列表
var propList = dataType.GetProperties(BindingFlags.Instance |
BindingFlags.Public);
foreach (string p in parts)
{
    var (ok, prop, val) = ParseValues(p.Trim());
    if (ok)
    {
        // 查找属性
        var theProperty = propList.FirstOrDefault(x => x.Name.
        Equals(prop,StringComparison.OrdinalIgnoreCase));
        if(theProperty != null)
        {
            try
            {
                // 设置属性的值
                if (theProperty.PropertyType.IsAssignableFrom
                (val.GetType()))
                {
                    // 类型兼容，可直接赋值
                    theProperty.SetValue(dataObject, val);
                }
                else
                {
                    // 需要类型转换
                    object? tval = Convert.ChangeType(val,
                    theProperty.PropertyType);
                    if (tval != null)
                    {
                        theProperty.SetValue(dataObject, tval);
                    }
                }
            }
            catch
            {
                continue;
            }
        }
    }
}
return InputFormatterResult.Success(dataObject);

//---------------------------------------------------
static (bool ok, string? prop, string? val) ParseValues(string
input)
{
    // 找出“[”和“]”的位置
    int s1 = input.IndexOf('[');
    int s2 = input.IndexOf("]");
```

```
            if (s1 != 0 || s2 < 2)
            {
                return (false, null, null);
            }
            // 读取属性名
            string propName = input.Substring(s1 + 1, s2 - s1 - 1);
            // 读取属性值
            string propValue = input.Substring(s2 + 1);
            return (true, propName, propValue);
        }
    }
}
```

在类的构造函数中，向 SupportedEncodings 列表添加对 UTF-8 文本编码的支持；向 SupportedMediaTypes 列表添加支持的媒体类型，本示例支持 application/cust。即客户端提交数据时需要将 Content-Type 消息头设置为 application/cust 才会被 CustDataInputFormatter 处理。

然后重写 CanReadType()方法。

```
protected override bool CanReadType(Type type)
{
    // 目标类型匹配时才可读
    return type == typeof(MyData);
}
```

上述代码表示当前 InputFormatter 只为 MyData 类型读取数据。如果要为任意类型读取数据，可以直接返回 true。

最核心的代码是实现 ReadRequestBodyAsync()方法。其功能就是从请求正文中读取数据，进行分析处理，最终产生.NET 类型实例并为其属性赋值，完成模型绑定。数据处理完毕后必须返回 InputFormatterResult 对象，同时传递模型类的实例。

```
return InputFormatterResult.Success(dataObject);
```

（2）实现 Cust 控制器。

```
[Route("api/[controller]")]
[ApiController]
public class CustController : ControllerBase
{
    [HttpPost]
    public string PostCustdata(MyData data)
    {
        return $"Size = {data.Size}, Flag = {data.Flag}, Label =
        {data.Label}";
    }
}
```

（3）在应用程序初始化代码中配置自定义的 CustDataInputFormatter 类。

```
var builder = WebApplication.CreateBuilder(args);
builder.Services.AddControllers().AddMvcOptions(opt=>
```

```
{
    opt.InputFormatters.Add(new CustDataInputFormatter());
});
…
```

运行示例程序，以 HTTP-POST 方式向服务器提交以下数据（Content-Type 标头一定要设置为 application/cust）。

```
POST /api/cust HTTP/1.1
Content-Length: 36
Content-Type: application/cust
…

[flag]25,[label]测试,[size]122.075
```

服务器响应如下。

```
HTTP/1.1 200 OK
Content-Type: text/plain; charset=utf-8
…

Size = 122.075, Flag = 25, Label = 测试
```

11.4.2 示例：BytesToHexOutputFormatter 类

本示例实现的是自定义输出格式。字节数组经格式化处理后，将变成类似以下形式的文本。

```
0xaa,0xbb,0xcc,0x12
```

BytesToHexOutputFormatter 类的完整代码如下。

```
public class BytesToHexOutputFormatter : TextOutputFormatter
{
    public BytesToHexOutputFormatter()
    {
        // 设置支持的文本编码
        SupportedEncodings.Add(Encoding.UTF8);
        // 设置支持的媒体类型
        SupportedMediaTypes.Add("text/hex");
    }

    protected override bool CanWriteType(Type? type)
    {
        // 只作用于 byte[]类型
        return type == typeof(byte[]);
    }

    public override async Task WriteResponseBodyAsync
    (OutputFormatterWriteContext context, Encoding selectedEncoding)
    {
```

```
        if(context == null || selectedEncoding == null)
        {
            throw new ArgumentNullException();
        }

        // 将字节数组转换为十六进制字符串数组
        byte[]? arr = context.Object as byte[];
        if(arr == null)
        {
            return;
        }
        // 格式：0xab
        var hexs = arr.Select(b => $"0x{b:x2}");
        // 用逗号拼接
        string res = string.Join(',', hexs);
        // 写入响应流
        var writer = context.WriterFactory(context.HttpContext.Response.
        Body, selectedEncoding);
        await writer.WriteAsync(res);
        await writer.FlushAsync();
    }
}
```

此输出格式只处理 byte[]类型（字节数组）的数据，因此，需要重写 CanWriteType()方法，只有当 type 参数表示的类型是 byte[]时才返回 true。此输出格式会设置响应消息头 Content-Type 为 text/hex。

在初始化应用程序的代码中，把 BytesToHexOutputFormatter 实例添加到 OutputFormatters 集合中。

```
var builder = WebApplication.CreateBuilder(args);
builder.Services.AddControllers()
    .AddXmlSerializerFormatters()
    .AddMvcOptions(options =>
        {
            options.OutputFormatters.Add(new BytesToHexOutputFormatter());
            options.FormatterMappings.SetMediaTypeMappingForFormat
            ("hex", "text/hex");
        });
```

FormatterMappings 是一个媒体类型的映射表，在使用 FormatterFilter（格式过滤器）时，应用程序将从 FormatterMappings 中查找格式名称对应的媒体类型。应用程序默认会添加 JSON、XML 格式的映射。

```
json 对应 application/json
xml 对应 application/xml
```

本示例添加名为 hex 的格式映射，对应的媒体类型为 text/hex。

下面的代码实现 API 控制器。

```
[ApiController]
public class TestController : ControllerBase
{
    [HttpGet("api/bytes/{format?}"), FormatFilter]
    public byte[] GetHex()
    {
        byte[] data = new byte[]
        {
            1, 2, 3, 4, 5, 6
        };
        return data;
    }
}
```

GetHex()方法返回的类型为 byte[]，应用程序会使用 BytesToHexOutputFormatter 格式化处理输出数据。同时，GetHex()方法使用了格式过滤器，可在 URL 中设置 format 字段的值来选择输出格式。

运行示例程序，访问/api/bytes/json，选择以 JSON 格式输出，服务器返回以下结果。

```
"AQIDBAUG"
```

访问/api/bytes/xml，将选择 XML 格式输出，结果如下。

```
<base64Binary>AQIDBAUG</base64Binary>
```

访问/api/bytes/hex，将使用本示例实现的自定义格式输出数据，结果如下。

```
0x01,0x02,0x03,0x04,0x05,0x06
```

11.5 极小 API

极小 API（即 Minimal API）是通过在终结点上添加路由映射而产生的 Web API，无须定义控制器，使用起来便捷高效。读者已熟知的如 MapGet()、MapPost()等扩展方法，正是用来建立极小 API 的。

11.5.1 示例：一些简单的极小 API 例子

本示例将演示一组简单但具有代表性的极小 API 范例。

（1）以 HTTP-GET 方式访问，路径为根 URL，返回字符串“Hello World!”。

```
app.MapGet("/", () =>"Hello World!");
```

调用时直接以 GET 方式访问，服务器将作出以下响应。

```
HTTP/1.1 200 OK
Content-Type: text/plain; charset=utf-8
…

Hello World!
```

（2）以 HTTP-GET 方式访问，路径为/number，返回 int 值。

```
app.MapGet("/number", () => 55600);
```

服务器将响应以下内容。

```
HTTP/1.1 200 OK
Content-Type: application/json; charset=utf-8
…

55600
```

（3）以 HTTP-POST 方式访问，用表单（form-data）形式提交数据。

```
app.MapPost("/newitem", (HttpRequest request) =>
{
    // 读取表单中的内容
    foreach(var f in request.Form)
    {
        Console.WriteLine("Name={0}, Value={1}", f.Key, f.Value);
    }
    return Results.Ok();
});
```

调用时提交 3 组数据：a=5，b=12，c=25。

```
POST /newitem HTTP/1.1
Content-Length: 13
Content-Type: application/x-www-form-urlencoded
…

a=5&b=12&c=25
```

服务器向控制台输出以下内容。

```
Name=a, Value=5
Name=b, Value=12
Name=c, Value=25
```

（4）以 HTTP-DELETE 方式访问，由 URL 查询字符串提供输入参数。

```
app. MapDelete ("/artin", (string token) => Results.Ok(token));
```

以 DELETE 方式调用，token 参数的值为 abcdefghijk。

```
DELETE /artin?token=abcdefghijk HTTP/1.1
```

服务器响应如下。

```
HTTP/1.1 200 OK
Content-Type: application/json; charset=utf-8
…

"abcdefghijk"
```

11.5.2　示例：在极小 API 上使用数据源特性

极小 API 也具有自动推断数据来源的功能。同时，FromHeader、FromQuery、FromBody 等特性也适用，可以手动指定模型绑定的来源。

（1）下面的代码定义以 HTTP-GET 方式访问的 API。其中，tag 参数的值将从 HTTP 消息头获得，key1 和 key2 的值均来自 URL 查询字符串。

```
app.MapGet("/test1",  ([FromHeader(Name = "x-tag")]string tag,
                        [FromQuery]string key1,
                        [FromQuery]string key2) =>
{
    Console.WriteLine($"tag:    {tag}");
    Console.WriteLine($"key1:    {key1}");
    Console.WriteLine($"key2:    {key2}");
    return Results.Ok();
});
```

接下来将使用 Python 脚本调用测试。

```
import requests

# 消息头
headers = {
    'x-tag': "data from header"
}

# URL 参数
q = {
    'key1': 'part one',
    'key2': 'part two'
}

# 发送请求
r = requests.get('http://localhost:4428/test1', params=q, headers=headers)
print(r.status_code)
```

脚本执行后，服务器控制台将输出以下内容。

```
tag:     data from header
key1:    part one
key2:    part two
```

（2）下面的代码将定义以 HTTP-POST 方式访问的 API，其参数为自定义类 User。参数的模型绑定将从 HTTP 消息的正文部分读取数据，默认使用 JSON 格式。

```
app.MapPost("/test2", ([FromBody]User u) =>
{
    Console.WriteLine($"用户名: {u.UserName}");
    Console.WriteLine($"密码: {u.Password}");
```

```
    // 返回的状态码
    return Results.StatusCode(205);
});

// User 类的定义
public class User
{
    public string? UserName { get; set; }
    public string? Password { get; set; }
}
```

以下 Python 脚本用于调用测试。

```
import requests

# JSON 数据
body = {
    'username': 'admin',
    'password': '11223344'
}

r = requests.post("http://localhost:4428/test2", json = body)
print(r.status_code)
```

脚本执行后，服务器控制台输出如下。

```
用户名：admin
密码：11223344
```

11.5.3 上传文件

在极小 API 上读取上传的文件，可采用以下 3 种方案。

（1）定义 IFormFile 类型的参数，客户端以表单数据的形式上传文件。此方案只能读取一个文件。

（2）定义 IFormFileCollection 类型的参数，客户端以表单数据的形式上传文件。此方案可以读取多个文件。

（3）定义 Stream 类型的参数，直接用流读取文件内容。

11.5.4 示例：直接读取文件流

本示例将定义 Stream 类型的参数，直接读取被上传文件的内容。API 定义如下。

```
app.MapPost("/upload", async (Stream inStream, string fileName) =>
{
    // 获取当前目录
    string currDir = app.Environment.ContentRootPath;
    // 组成新路径
    string filePath = Path.Combine(currDir, fileName);
```

```
    // 如果已存在同名文件
    if(File.Exists(filePath))
    {
        return Results.Content($"{fileName}已存在");
    }
    // 保存文件
    using(var fsOut = File.OpenWrite(filePath))
    {
        await inStream.CopyToAsync(fsOut);
    }
    return Results.Ok();
});
```

inStream 参数表示 HTTP 消息正文，用流读取。在本示例中，调用 CopyToAsync()方法把已上传文件的内容直接复制到输出文件流。fileName 参数用于指定新文件名，此参数将自动推断出由 URL 查询字符串提供。

下面的代码将使用 HttpClient 对象模拟上传文本文件。

```
// 请求的根 URL
Uri rootUrl = new("http://localhost:1005");
// 实例化 HTTP 客户端
HttpClient client = new HttpClient()
{
    BaseAddress = rootUrl
};

// 生成文本数据流
string text = """
    江林多秀发，
    云日复相鲜。
    征路那逢此，
    春心益渺然。
    """;

byte[] bindata = Encoding.UTF8.GetBytes(text);
using MemoryStream stream = new MemoryStream(bindata);
StreamContent content = new StreamContent(stream);
// 设置消息头
content.Headers.ContentType = MediaTypeHeaderValue.Parse(MediaTypeNames.
Application.Octet);

// 用 GUID 产生文件名
string fileName = $"{Guid.NewGuid()}.txt";
// 发起请求
var response = await client.PostAsync("/upload?filename=" + fileName,
content);
```

```
if(response.IsSuccessStatusCode)
{
    Console.WriteLine("上传成功");
}
else
{
    Console.WriteLine("上传失败，响应：{0}", await
    response.Content.ReadAsStringAsync());
}
```

已上传文本文件的内容如图 11-5 所示。

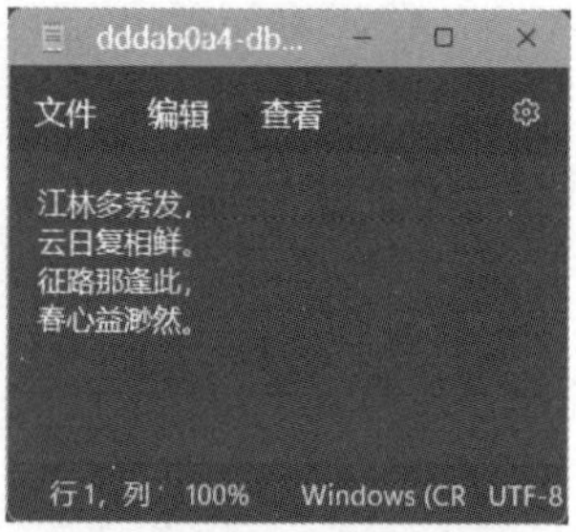

图 11-5 已上传的文件

11.5.5 示例：上传多个文件

在极小 API 中，参数类型声明为 IFormFile 或 IFormFileCollection 都可以自动识别文件上传操作。IFormFile 类型的参数仅能读取单个文件，而 IFormFileCollection 类型的参数则可以读取多个文件。

本示例将通过极小 API 实现多个文件的上传。API 实现代码如下。

```
app.MapPost("/upload", async (IFormFileCollection formfiles) =>
{
    int fileCount = formfiles.Count;
    if(fileCount == 0)
    {
        return Results.Content("未选择任何文件");
    }
    // 创建存放文件的目录
    string saveDir = Path.Combine(app.Environment.ContentRootPath,
    "Uploads");
    if(!Directory.Exists(saveDir))
    {
        Directory.CreateDirectory(saveDir);
    }
    StringBuilder strbuilder = new();
    strbuilder.AppendLine($"文件数量：{fileCount}\n");
    // 开始读取文件
    for(int i = 0; i < fileCount; i++)
    {
        IFormFile file = formfiles[i];
        strbuilder.AppendLine("---------------");
        strbuilder.AppendLine($"文件：{file.FileName}");
        strbuilder.AppendLine($"大小：{file.Length}");
        // 保存文件时生成新文件名，不建议使用客户端提供的文件名
        // 获取原文件的扩展名
        string fileExt = Path.GetExtension(file.FileName) ?? ".tmp";
        // 用 GUID 产生新文件名
```

```
        string newFileName = $"{Guid.NewGuid()}{fileExt}";
        // 组成新的路径
        string newFilePath = Path.Combine(saveDir, newFileName);
        using(var outStream = File.OpenWrite(newFilePath))
        {
            // 直接复制文件内容
            await file.CopyToAsync(outStream);
        }
        strbuilder.Append("状态：已保存\n\n");
    }
    return Results.Content(strbuilder.ToString());
});
```

首先通过 Count 属性检查上传文件的数量，若为 0，表示文件列表为空，可以直接返回。随后在 for 循环中，使用索引读出文件信息，并把文件内容复制到服务器上（即将文件保存）。一般来说，服务器在保存文件时应重新生成文件名（本示例将通过 GUID 产生文件名），而不是使用客户端提供的文件名。这是出于安全考虑，上传者可能通过文件名植入恶意代码。

单个 IFormFile 对象表示一个文件的信息。其中，FileName 属性（Name 属性代表的是表单中的字段名称，非文件名）代表文件名；Length 属性可获取文件的大小（字节）。若要以流的形式读取，请调用 OpenReadStream()方法；若只希望将客户端上传的文件保存到服务器上，可以调用 CopyTo()或 CopyToAsync()方法，将文件内容直接复制到另一个流对象中。

本示例将使用一个 Razor 页面触发文件上传操作。

```
@page

<!DOCTYPE html>
<html lang="zh-cn">
<head>
<meta charset="utf-8" />
<title>上传多个文件示例</title>
</head>
<body>
<form action="/upload" method="post" enctype="multipart/form-data">
<input type="file" name="formfiles" id="formfiles" multiple />
<button type="submit">上传</button>
</form>
</body>
</html>
```

注意，<form>元素要将编码类型设置为 multipart/form-data；<input>元素的 type 属性设置为 file，并加上 multiple 属性，这样才能支持同时上传多个文件。

运行示例程序，单击页面上的“选择文件”按钮，然后同时选择多个文件。选择文件后单击“上传”按钮。服务器在保存文件后会返回文件列表信息，如图 11-6 所示。

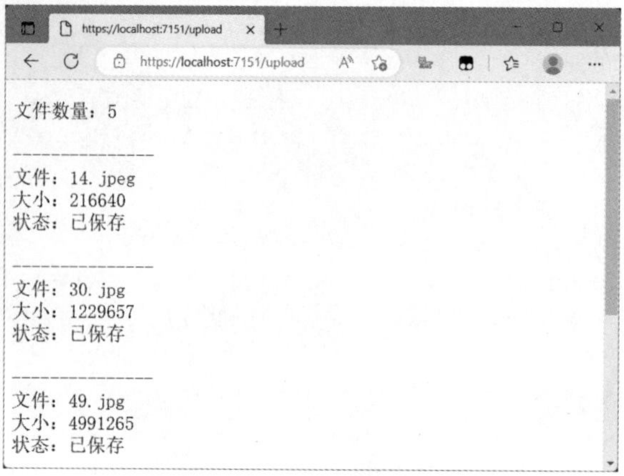

图 11-6　已上传的文件列表

11.5.6　IResult 接口

IResult 接口公开以下方法。

```
public Task ExecuteAsync (HttpContext httpContext);
```

实现此方法可以向响应流写入 HTTP 终结点的执行结果。对于极小 API，可以直接返回.NET 类型（JSON 序列化后写入响应流），也可以通过 IResult 接口返回特殊的结果（如返回某个状态码，或 HTML 文本）。

开发者一般不需要实现 IResult 接口，Results 类已经公开了最常用的静态方法，直接调用即可。

（1）Ok()方法：返回 200 状态码。

（2）Accepted()或 AcceptedAtRoute()方法：返回 202 状态码。

（3）Content()或 Text()方法：返回 HTML 文本。

（4）NoContent()方法：返回 204 状态码，无 HTML 文本内容。

（5）StatusCode()方法：返回一个自定义的状态码。

（6）NotFound()方法：返回 404 状态码，表示未找到资源。

（7）Redirect()或 RedirectToRoute()方法：跳转（重定向）到指定 URL。

（8）Created()或 CreatedAtRoute()方法：返回 201 状态码。

（9）Bytes()方法：返回字节数组。

（10）BadRequest()方法：返回 400 状态码。

（11）Stream()方法：返回流数据，直接写入响应消息的正文部分。

（12）File()方法：和 Bytes()方法类似，返回文件内容（可实现文件下载）。

（13）Json()方法：返回 JSON 格式的数据。

（14）Forbid()方法：默认返回 403 状态码。某些身份验证方案可能会返回其他状态码并重定向到指定的路径（如跳转到登录页面）。

11.5.7　示例：Results 类的使用

本示例将演示在极小 API 中调用 Results 类的静态方法返回特定的结果。

（1）返回普通文本。

```
app.MapGet("/", () =>
{
    return Results.Content("你好，世界");
});
```

这里也可以调用 Results.Text()方法。

（2）返回 404 状态码。

```
app.MapGet("/about", ()=> Results.NotFound());
```

（3）返回 JSON 格式的数据。

```
app.MapMethods("/stores", new string[] {"GET"}, () =>
{
    var obj = new
    {
        Id = 50050,
        Key = "X-V-S",
        Qty = 20,
        Unit = "CM"
    };
    return Results.Json(obj);
});
```

访问/stores 后得到的 JSON 对象如下。

```
{
"id": 50050,
"key": "X-V-S",
"qty": 20,
"unit": "CM"
}
```

（4）返回 200 状态码，并附加文本 Done。

```
app.MapDelete("/flip", () =>
{
    return Results.Ok("Done");
});
```

11.6　API 浏览功能

Web API 浏览（API Explorer）可发现应用程序内各个 MVC 所公开的 API 列表，并可以获取单个 API 的详细信息，包括调用 API 的相对路径、关联的 MVC 操作方法、参数表列、响应格式等。

11.6.1 IApiDescriptionGroupCollectionProvider 接口

API 浏览功能的核心接口是 IApiDescriptionGroupCollectionProvider，默认实现类是 ApiDescriptionGroupCollectionProvider。该接口可返回只读的 ApiDescriptionGroup 列表，其中包含已分组的 API 信息。

Web API 默认情况下仅存在一个未命名的分组（组名为空字符串）。若需要对 API 进行分组，可以在定义控制器时在类或操作方法上应用 ApiExplorerSettings 特性，并通过 GroupName 属性设置分组名称。

每个 API 分组下均包含一个 ApiDescription 列表。ApiDescription 类用于描述单个 API 的基本信息，包括如下内容。

（1）GroupName：此 API 所在的分组名称。

（2）RelativePath：调用此 API 的路径（相对于根路径）。

（3）HttpMethod：支持的 HTTP 请求方式，如 GET、POST 等。

（4）ParameterDescriptions：返回一个 ApiParameterDescription 对象列表。一个 ApiParameterDescription 对象用于描述一个参数的信息，包括参数名称、参数类型、参数默认值等。

（5）SupportedRequestFormats：受支持的请求格式列表，包括格式名称（如 JSON、XML 等）和关联的 IInputFormatter。此列表可能为空。

（6）SupportedResponseTypes：受支持的响应格式列表，包括响应状态码、返回值的.NET 类型等。其中，ApiResponseFormats 属性返回格式名称与 IOutputFormatter 列表。

在服务容器上调用 AddEndpointsApiExplorer()扩展方法后，将向容器注册 IApiDescriptionGroupCollectionProvider 类型的服务。在代码中可通过依赖注入或主动获取服务引用的方式进行访问。

11.6.2 示例：列出已定义的 Web API

本示例运行后，将获取应用程序内定义的 Web API 信息，然后呈现在 Razor 页面上。

（1）定义 Demo 控制器，其中包含 3 个操作方法。

```
[ApiController]
public class DemoController : ControllerBase
{
    [HttpGet("/api/dm/test1")]
    public string Hello() =>"Hello!";

    [HttpPost("/api/dm/test2")]
    public int Compute(int n) => n * n * 2;

    [HttpGet("/api/dm/test3")]
    public string GetID() => Guid.NewGuid().ToString();
}
```

（2）定义 Books 控制器，其中包含两个操作方法。

```
[ApiController]
public class BooksController : ControllerBase
{
    [HttpGet("/api/books/list")]
    public IEnumerable<Book> GetList()
    {
        Book[] books = new Book[3];
        books[0] = new()
        {
            Title = "图书-A",
            Author = "小王",
            ISBN = "511097210"
        };
        books[1] = new()
        {
            Title = "图书-B",
            Author = "小高",
            ISBN = "79010053"
        };
        books[2] = new()
        {
            Title = "图书-C",
            Author = "小谢",
            ISBN = "625315430"
        };
        return books;
    }

    [HttpPost("/api/books/add")]
    public void AddNew(Book book)
    {

    }
}
```

上述代码用到了 Book 类，其声明如下。

```
public class Book
{
    public string? Title { get; set; }
    public string? Author { get; set; }
    public string? ISBN { get; set; }
}
```

（3）在项目目录下创建 Pages 目录。

（4）在 Pages 目录下新建 Razor 页面文件，命名为 Index.cshtml。

```
@page "/"
```

```
@using Microsoft.AspNetCore.Mvc.ApiExplorer
@inject IApiDescriptionGroupCollectionProvider apiGroupsProvider

@if (apiGroupsProvider.ApiDescriptionGroups.Items.Count == 0)
{
<div>无 API</div>
}
else
{
    @foreach (ApiDescriptionGroup g in apiGroupsProvider.
ApiDescriptionGroups.Items)
    {
        @foreach (ApiDescription apidesc in g.Items)
        {
            <div style="margin-top:15px;margin-bottom:10px; border-color:
            blue; border-style:dashed">
            <div>
                    @*获取 HTTP 请求方式*@
            <strong>@apidesc.HttpMethod</strong>: 
                    @*获取相对 URL*@
                    @apidesc.RelativePath
            </div>
                @if (apidesc.SupportedResponseTypes.Count > 0)
                {
            <div>
                        @*列出返回值信息*@
                        @foreach (ApiResponseType resptype in apidesc.
                        SupportedResponseTypes)
                        {
                            <div>
                                状态码：@resptype.StatusCode
                                @if(resptype.ApiResponseFormats.Count > 0)
                                {
                                    <ul>
                                        @*列出输出格式*@
                                        @foreach(ApiResponseFormat format in
                                        resptype.ApiResponseFormats)
                                        {
                                            <li>@format.MediaType</li>
                                        }
                                    </ul>
                                }
                            </div>
                        }
                    </div>
                }
            </div>
        }
    }
}
```

上述代码中，使用@inject 指令通过依赖注入获取 IApiDescriptionGroupCollectionProvider 类型的对象引用。随后找出描述 API 信息的 ApiDescription 对象，并呈现每个 API 的基本信息：HTTP 请求方式、相对路径、返回格式。

（5）在应用程序的初始化代码中，添加 MVC 控制器和 API 浏览功能。

```
builder.Services.AddControllers();
builder.Services.AddEndpointsApiExplorer();
```

运行示例程序，结果如图 11-7 所示。

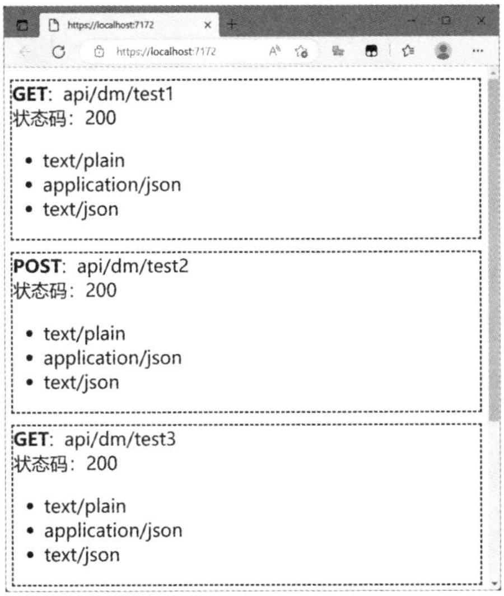

图 11-7　获取 API 信息

11.6.3　API 约定

API 约定是一个静态类，其中定义一组静态方法成员。这些静态方法可以应用 ProducesResponseType 特性指定 API 可能会返回的状态码或数据类型。此静态类就成了 API 约定，如

```
public static class CustApiConventions
{
    [ApiConventionNameMatch(ApiConventionNameMatchBehavior.Prefix)]
    [ProducesResponseType(StatusCodes.Status200OK)]
    [ProducesResponseType(StatusCodes.Status204NoContent)]
    [ProducesResponseType(StatusCodes.Status201Created)]
    [ProducesResponseType(StatusCodes.Status400BadRequest)]
    [ProducesDefaultResponseType]
    public static void New(
        [ApiConventionNameMatch(ApiConventionNameMatchBehavior.Any)]
        [ApiConventionTypeMatch(ApiConventionTypeMatchBehavior.Any)]
        object obj)
```

```
        { }

        [ApiConventionNameMatch(ApiConventionNameMatchBehavior.Prefix)]
        [ProducesResponseType(StatusCodes.Status200OK)]
        [ProducesResponseType(StatusCodes.Status204NoContent)]
        [ProducesResponseType(StatusCodes.Status401Unauthorized)]
        [ProducesResponseType(StatusCodes.Status403Forbidden)]
        [ProducesResponseType(StatusCodes.Status404NotFound)]
        [ProducesDefaultResponseType]
        public static void Remove(
            [ApiConventionNameMatch(ApiConventionNameMatchBehavior.Suffix)]
            [ApiConventionTypeMatch(ApiConventionTypeMatchBehavior.Any)]
            object id)
        { }
    }
```

当 CustApiConventions 类被作为 API 约定使用时，它的两个方法成员将代表两条匹配规则。

（1）New()方法：ApiConventionNameMatch 特性设置了 ApiConventionNameMatchBehavior.Prefix 匹配方式。即 New 将作为前缀来匹配，MVC 控制器中以 New 开头的操作方法均可参与匹配。参数 obj 的匹配方式为 Any，表示参数为任意类型、任意名称皆能匹配。一旦 MVC 控制器中有操作方法匹配成功，API 浏览功能会把 ProducesDefaultResponseType、ProducesResponseType 特性所指定的状态码（或返回类型）应用到 MVC 控制器的操作方法上。即 API 浏览功能会认为目标操作方法支持返回 200、201、204、400 等状态码。

（2）Remove()方法，名称匹配规则为 Prefix，也属于前缀匹配。参数的类型为任意类型，名称规则为 Suffix，表示后缀匹配。即参与匹配的参数命名要以 id 结尾。注意，此后缀匹配是按照单词边界匹配的（叙述性词汇），当参数的名称中含有其他单词时，id 的首字母要改为大写，如 studentId、imageId 等。

假设有以下 MVC 控制器。

```
[Route("api/[controller]")]
[ApiController]
[ApiConventionType(typeof(CustApiConventions))]
public class TestController : ControllerBase
{
    // 匹配 API 约定
    [HttpPost("add")]
    public void NewOrder(Order ord)
    {

    }

    // 不匹配
    [HttpGet("getone")]
    public Order GetSingle() => new Order();
```

```
        // 匹配 API 约定
        [HttpGet("del/{orderId}")]
        public void RemoveOrder(int orderId)
        {

        }
    }
```

ApiConventionType 特性用于指定该控制器将与上文中定义的 API 约定（CustApi-Conventions）关联。API 浏览功能在分析控制器时会进行约定匹配检测。上述控制器中，NewOrder()方法以 New 开头，RemoveOrder()方法以 Remove 开头且参数 orderId 在命名上也与 id 后缀匹配。如图 11-8 所示，NewOrder()和 RemoveOrder()方法与约定匹配，将自动附加 API 约定所指定的响应格式设置。GetSingle()方法与约定不匹配，不会自动添加响应格式。

图 11-8　与约定匹配的 API 将自动添加响应格式

11.6.4　Swagger 框架

Swagger 是一个基于 Open API 规范的开源框架，用于自动生成人机可读，且与编程语言无关的 API 文档。当后端 API 发生变更后，也会自动生成最新版本的文档。开发人员不需要手动撰写文档。

Swashbuckle.AspNetCore 类库以 API 浏览功能为基础来获取 API 信息，再依据 Open API 规范自动生成 API 文档。此类库包含三大组件，具体如下。

（1）Swashbuckle.AspNetCore.Swagger：包含公开 Swagger 文档的中间件。

（2）Swashbuckle.AspNetCore.SwaggerGen：包含文档模型、架构生成器、服务容器扩展方法等功能。

（3）Swashbuckle.AspNetCore.SwaggerUI：包含可视化的 API 浏览、测试功能。

11.6.5 示例：使用 Swagger 生成 API 文档

本示例将演示 Swagger 的基本使用方法。

（1）在项目所在的目录下执行以下命令，安装 Swashbuckle.AspNetCore 类库（Nuget 包）。

```
dotnet add package Swashbuckle.AspNetCore
```

也可以直接打开项目文件(*.csproj)，手动添加以下内容引用 Swashbuckle.AspNetCore 库。

```
<ItemGroup>
<PackageReference Include="Swashbuckle.AspNetCore" Version="6.4.0" />
</ItemGroup>
```

（2）在应用程序初始化代码中，需要启用 API 浏览功能。

```
builder.Services.AddEndpointsApiExplorer();
```

（3）还要启用 Swagger 文档生成功能。

```
builder.Services.AddSwaggerGen();
```

（4）调用 AddControllers()扩展方法，启用 MVC 控制器功能，以便编写 Web API。

```
builder.Services.AddControllers();
```

（5）应用程序构建后，在 app 对象上调用 UseSwagger()、UseSwaggerUI()扩展方法，向 HTTP 管线加入 Swagger 相关的中间件，使应用程序能够返回生成的 API 文档以及可视化的测试页面。

```
…
var app = builder.Build();
app.UseSwagger();
app.UseSwaggerUI();
```

通常，浏览和测试 API 仅在项目开发阶段使用，待项目正式上线后一般不再使用此功能。因此，可以将上述代码改为仅在应用程序处于开发阶段时才添加 Swagger 中间件。

```
if(app.Environment.IsDevelopment())
{
    app.UseSwagger();
    app.UseSwaggerUI();
};
```

（6）添加 MVC 终结点映射。

```
app.MapControllers();
```

（7）定义 Demo 控制器。

```
[ApiController]
public class DemoController : ControllerBase
```

```
    {
        [HttpPost("api/test/compute/{a:int}/{b:int}")]
        public string DoCompute(int a, int b)
        {
            // 四则运算
            int r1 = a + b;
            int r2 = a - b;
            int r3 = a * b;
            int r4 = a / b;
            // 拼接返回字符串
            StringBuilder strbuilder = new StringBuilder();
            strbuilder.AppendLine($"{a} + {b} = {r1}")
                      .AppendLine($"{a} - {b} = {r2}")
                      .AppendLine($"{a} × {b} = {r3}")
                      .AppendLine($"{a} ÷ {b} = {r4}");
            return strbuilder.ToString();
        }

        [HttpGet("api/test/num")]
        public double GetNumber()
        {
            return Random.Shared.NextDouble() * 1000.0d;
        }
    }
```

（8）运行示例程序，访问/swagger 路径，即可看到可视化的 API 浏览/测试页面，如图 11-9 所示。

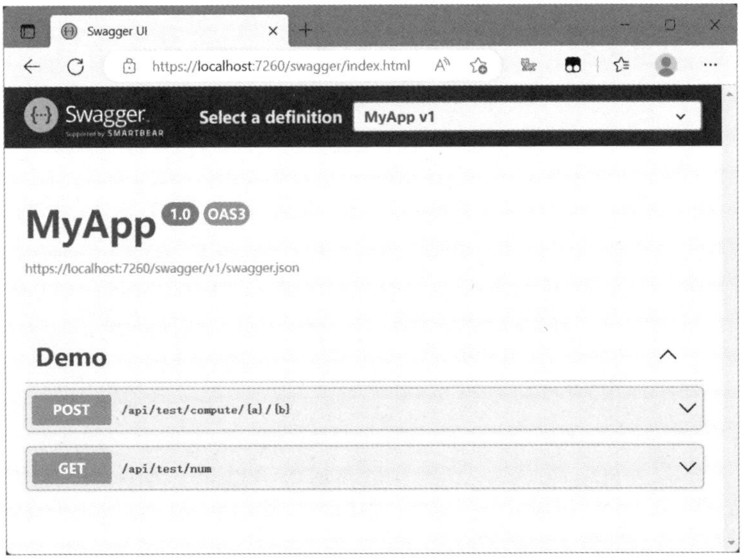

图 11-9　可视化 API 浏览/测试页面

（9）展开 API 信息，单击 Try it out 按钮，输入参数 a 和 b 的值，如图 11-10 所示。单击

Execute 按钮，测试 API 调用并返回结果，如图 11-11 所示。

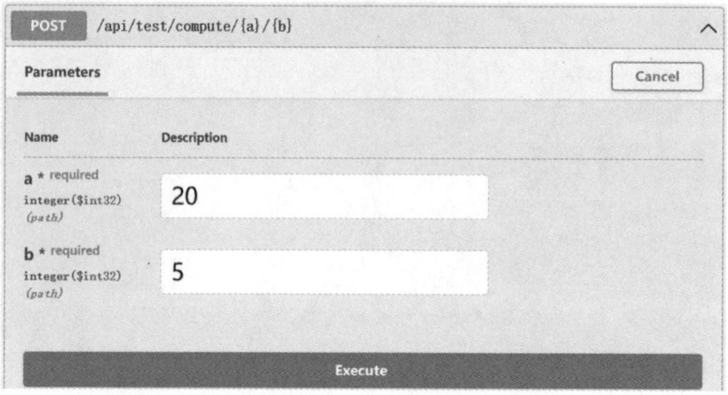

图 11-10　输入 API 参数

图 11-11　API 调用结果

（10）如果希望从根 URL（/）打开可视化 API 浏览/测试页面，在调用 UseSwagger()、UseSwaggerUI()方法时需要做一些选项配置。

```
app.UseSwagger(opt =>
{
    opt.RouteTemplate = "{documentName}/swagger.{json|yaml}";
});
app.UseSwaggerUI(opt =>
{
    opt.RoutePrefix = string.Empty;
});
```

在调用 UseSwaggerUI()方法时，将 RoutePrefix 选项设置为空字符串，即去掉了 swagger 前缀，使访问 URL 由/swagger 变为/。

由于 URL 去掉了 swagger 前缀，会导致指向描述 API 的 JSON 文档的路径错误——由 swagger/v1/swagger.json 变为 v1/swagger.json。所以，在调用 UseSwagger()方法时，要重新设置 RouteTemplate 选项，将默认的 swagger/{documentName}/swagger.{json|yaml}改为{documentName}/swagger.{json|yaml}。

（11）再次运行示例程序，就可以在根路径下打开 API 浏览/测试页面了。

第 12 章　过　滤　器

本章要点

- 过滤器的执行过程
- IFilterFactory 接口
- 常用的过滤器特性类
- 同步与异步过滤器

12.1　过滤器的执行过程

过滤器（Filter，也可以译为筛选器）用于 Razor Pages 和 MVC 控制器，在执行某个目标（如操作方法）之前或之后都可以运行过滤器。过滤器可以修改输入参数以及返回的结果。多个过滤器会构成一个调用管道，类似于中间件，当一个过滤器执行完后会调用下一个过滤器。

所有过滤器类型都实现一个公共接口——IFilterMetadata。此接口未包含任何要实现的成员，仅作为一个标记，表示它是一个过滤器。根据过滤器的功能，IFilterMetadata 接口派生出一组新的接口，并且这些接口在过滤器管道中会按照特定的顺序执行。

（1）IAuthorizationFilter 接口：表示授权过滤器。授权过滤器首先运行，以确定用户是否拥有访问特定资源的权限。

（2）IResourceFilter 接口：异步版本为 IAsyncResourceFilter 接口。在授权过滤器之后、模型绑定之前执行。在此过滤器中可以修改用于模型绑定的 IValueProviderFactory 列表，如添加新的 IValueProvider 类型，或移除默认的 IValueProvider 类型。

（3）IActionFilter 接口：异步版本为 IAsyncActionFilter 接口。在数据绑定后，MVC 操作方法执行之前、之后执行。

（4）IExceptionFilter 接口：异步版本为 IAsyncExceptionFilter 接口。只有在操作方法内发生异常时才会运行该过滤器。若未发生异常，则该过滤器不会执行。

（5）IResultFilter 接口：异步版本为 IAsyncResultFilter 接口。只有当操作方法顺利执行（未发生异常）后并产生操作结果时才运行该过滤器。可以在需要时修改操作方法的结果。

有些过滤器存在同步调用和异步调用版本，开发者仅实现其中任一个版本即可。如果同步、异步版本的接口都实现了，那么应用程序会选择异步版本的过滤器。

12.1.1 示例：观察过滤器的运行顺序

前面已介绍了过滤器的运行过程，本节将通过一个示例更直观地了解各类过滤器的运行顺序。本示例将实现这些接口：IAuthorizationFilter、IResourceFilter、IActionFilter、IExceptionFilter、IResultFilter。这些过滤器都继承了 Attribute 类，成为特性类，可以直接应用到 MVC 控制器的操作方法上。

（1）实现授权过滤器 CustAuthorizationFilterAttribute。

```
[AttributeUsage(AttributeTargets.Method, AllowMultiple = false, Inherited
= true)]
public class CustAuthorizationFilterAttribute : Attribute,
IAuthorizationFilter
{
    public void OnAuthorization(AuthorizationFilterContext context)
    {
        var logger = context.HttpContext.RequestServices.GetRequiredService
        <ILogger<CustAuthorizationFilterAttribute>>();
        logger.LogInformation("Authorization Filter");
    }
}
```

该类需要实现 IAuthorizationFilter 接口的 OnAuthorization()方法。由于本示例的功能是观察过滤器的运行顺序，因此，仅通过 ILogger 对象输出一条日志消息，未进行其他处理。以下各个过滤器的实现也类似。

（2）实现 IResourceFilter 接口，完成一个资源过滤器。

```
[AttributeUsage(AttributeTargets.Method, AllowMultiple = false, Inherited
= true)]
public class CustResourceFilterAttribute : Attribute, IResourceFilter
{
    public void OnResourceExecuted(ResourceExecutedContext context)
    {
        var logger = context.HttpContext.RequestServices.GetRequiredService
        <ILogger<CustResourceFilterAttribute>>();
        logger.LogInformation("Resource Filter - executed");
    }

    public void OnResourceExecuting(ResourceExecutingContext context)
    {
        var logger = context.HttpContext.RequestServices.GetRequiredService
        <ILogger<CustResourceFilterAttribute>>();
        logger.LogInformation("Resource Filter - executing");
    }
}
```

OnResourceExecuting()方法在剩余的过滤器之前运行，而 OnResourceExecuted()方法则是在剩余的过滤器之后运行。

（3）实现操作方法过滤器 CustActionFilterAttribute。

```
[AttributeUsage(AttributeTargets.Method, AllowMultiple = false, Inherited
= true)]
public class CustActionFilterAttribute : Attribute, IActionFilter
{
    public void OnActionExecuted(ActionExecutedContext context)
    {
        var logger = context.HttpContext.RequestServices.GetRequiredService
        <ILogger<CustActionFilterAttribute>>();
        logger.LogInformation("Action Filter - executed");
    }

    public void OnActionExecuting(ActionExecutingContext context)
    {
        var logger = context.HttpContext.RequestServices.GetRequiredService
        <ILogger<CustActionFilterAttribute>>();
        logger.LogInformation("Action Filter - executing");
    }
}
```

OnActionExecuting()方法在操作方法调用之前运行，OnActionExecuted()方法在操作方法调用后立即运行。

（4）实现 IExceptionFilter 接口，处理在操作方法中抛出的异常。

```
[AttributeUsage(AttributeTargets.Method, AllowMultiple = false, Inherited
= true)]
public class CustExceptionFilterAttribute : Attribute, IExceptionFilter
{
    public void OnException(ExceptionContext context)
    {
        // 标注此异常已被处理
        context.ExceptionHandled = true;
        // 设置自定义的响应结果
        context.Result = new StatusCodeResult(400);
        var logger = context.HttpContext.RequestServices.GetRequiredService
        <ILogger<CustExceptionFilterAttribute>>();
        logger.LogInformation("Exception Filter");
    }
}
```

将 ExceptionHandled 属性设置为 true，表明此异常已经处理过了，应用程序不需要再次抛出此异常。

（5）实现 CustResultFilterAttribute 类，在操作方法成功返回结果时运行。

```
[AttributeUsage(AttributeTargets.Method, AllowMultiple = false, Inherited
= true)]
public class CustResultFilterAttribute : Attribute, IResultFilter
{
    public void OnResultExecuted(ResultExecutedContext context)
    {
        var logger = context.HttpContext.RequestServices.GetRequiredService
        <ILogger<CustResultFilterAttribute>>();
        logger.LogInformation("Result Filter - executed");
    }

    public void OnResultExecuting(ResultExecutingContext context)
    {
        var logger = context.HttpContext.RequestServices.GetRequiredService
        <ILogger<CustResultFilterAttribute>>();
        logger.LogInformation("Result Filter - executing");
    }
}
```

OnResultExecuting()方法在执行操作结果前运行，OnResultExecuted()方法在操作结果执行后立即运行。

（6）定义 Demo 控制器，然后将上面实现的 5 个过滤器以特性类的方式应用到 Index()方法上。

```
public class DemoController : Controller
{
    [CustActionFilter]
    [CustExceptionFilter]
    [CustResultFilter]
    [CustResourceFilter]
    [CustAuthorizationFilter]
    public IActionResult Index()
    {
        // 随机引发异常
        int n = Random.Shared.Next(1, 100);
        if(n % 3 == 0)
        {
            throw new Exception("test");
        }
        return Content("Index");
    }
}
```

异常过滤器只有在操作方法抛出异常时才会运行。上述代码通过生成一个随机整数，然后以该整数能否被 3 整除作为条件决定是否抛出异常。

如果发生异常，CustExceptionFilter 过滤器会运行，CustResultFilter 过滤器不会运行；如果不发生异常，CustExceptionFilter 过滤器不会运行，CustResultFilter 过滤器会运行。

（7）打开应用配置文件 appsettings.json。将 Microsoft.AspNetCore 与 Microsoft.Hosting.Lifetime 两个类别的日志输出模式设置为 None，即禁止输出日志。这样做能减少干扰信息，方便查看过滤器输出的日志。

```
{
"Logging": {
"LogLevel": {
"Default": "Information",
"Microsoft.AspNetCore": "None",
"Microsoft.Hosting.Lifetime": "None"
    }
  },
…
}
```

运行示例程序，若 Index()操作方法顺利运行，则应用程序会输出以下内容。

```
info: MyApp.CustAuthorizationFilterAttribute[0]
      Authorization Filter
info: MyApp.CustResourceFilterAttribute[0]
      Resource Filter - executing
info: MyApp.CustActionFilterAttribute[0]
      Action Filter - executing
info: MyApp.CustActionFilterAttribute[0]
      Action Filter - executed
info: MyApp.CustResultFilterAttribute[0]
      Result Filter - executing
info: MyApp.CustResultFilterAttribute[0]
      Result Filter - executed
info: MyApp.CustResourceFilterAttribute[0]
      Resource Filter - executed
```

图 12-1 将展示上述代码的执行过程。

如果执行 Index()方法的过程中发生异常，那么应用程序就会输出以下内容。

```
info: MyApp.CustAuthorizationFilterAttribute[0]
      Authorization Filter
info: MyApp.CustResourceFilterAttribute[0]
      Resource Filter - executing
info: MyApp.CustActionFilterAttribute[0]
      Action Filter - executing
info: MyApp.CustActionFilterAttribute[0]
      Action Filter - executed
info: MyApp.CustExceptionFilterAttribute[0]
      Exception Filter
info: MyApp.CustResourceFilterAttribute[0]
      Resource Filter - executed
```

图 12-2 将展示发生异常时过滤器的执行过程。

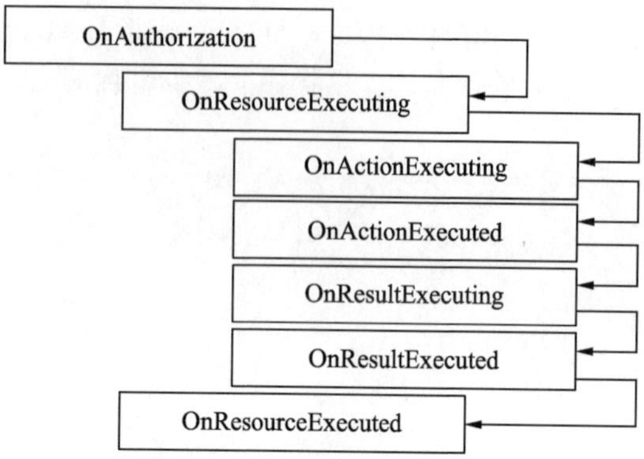

图 12-1 Index()方法顺利执行时过滤器的运行顺序

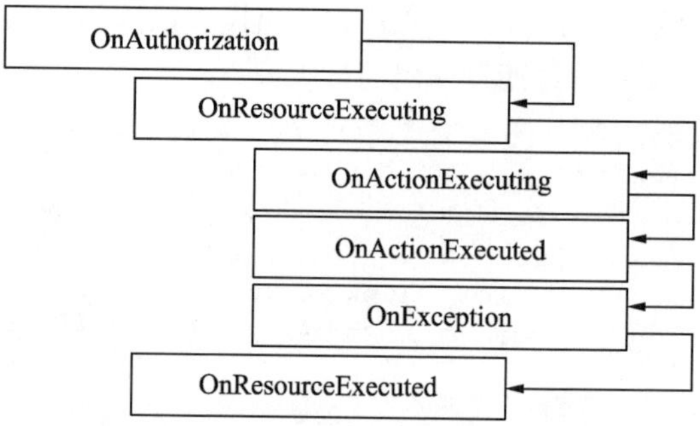

图 12-2 Index()方法抛出异常后过滤器的运行顺序

12.1.2 示例：同时实现多个接口

在编写过滤器类型时，可以一次性实现多个接口。例如，同时实现 IActionFilter 和 IResultFilter 接口，让过滤器类既可以在操作方法调用前后运行，又能在执行返回结果前后运行。

下面的代码定义 MyFilterAttribute 过滤器，它实现 3 个接口：IResourceFilter、IActionFilter、IResultFilter。

```
[AttributeUsage(AttributeTargets.Method, AllowMultiple = false, Inherited
= true)]
public class MyFilterAttribute : Attribute, IResourceFilter,
IActionFilter, IResultFilter
{
    public void OnActionExecuted(ActionExecutedContext context)
    {
        Console.WriteLine(nameof(OnActionExecuted));
    }

    public void OnActionExecuting(ActionExecutingContext context)
    {
        Console.WriteLine(nameof(OnActionExecuting));
    }

    public void OnResourceExecuted(ResourceExecutedContext context)
    {
        Console.WriteLine(nameof(OnResourceExecuted));
    }

    public void OnResourceExecuting(ResourceExecutingContext context)
    {
        Console.WriteLine(nameof(OnResourceExecuting));
```

```
    }

    public void OnResultExecuted(ResultExecutedContext context)
    {
        Console.WriteLine(nameof(OnResultExecuted));
    }

    public void OnResultExecuting(ResultExecutingContext context)
    {
        Console.WriteLine(nameof(OnResultExecuting));
    }
}
```

MyFilterAttribute 同时扮演资源过滤器、操作方法过滤器和结果过滤器 3 种角色。因此，在 MVC 控制器的方法上应用 MyFilterAttribute 时，只需要使用一次就可以了。

```
public class HomeController : Controller
{
    [MyFilter]
    public IActionResult Index()
    {
        return Content("Hello");
    }
}
```

12.2　过滤器的作用域

在应用程序初始化时通过 MvcOptions 选项类配置的过滤器，将作用于整个应用程序内的 MVC 控制器或 Razor 页面。而作为特性类使用的过滤器（如 12.1 节中的示例），仅作用于其应用目标上（如控制器类、控制器的操作方法）。

12.2.1　示例：全局过滤器

在应用程序初始化阶段通过添加到 MvcOptions.Filters 集合注册的过滤器，会作用于整个应用程序。下面的代码创建一个用于演示的过滤器类。

```
public class DilliServerHeaderFilter : IResultFilter
{
    public void OnResultExecuted(ResultExecutedContext context)
    {
        // 修改 HTTP 头
        context.HttpContext.Response.Headers.Server = "Dilli-Server";
        context.HttpContext.Response.Headers.Date = new DateOnly(2015, 2,
        1).ToString("yyyy-M-d");
    }

    public void OnResultExecuting(ResultExecutingContext context)
```

```
    {

    }
}
```

上述过滤器实现了 IResultFilter 接口，在操作方法的返回结果执行后修改响应消息的 HTTP 头。OnResultExecuting()方法在本示例中不需要任何处理，因此保留空白（空方法体）即可。

在初始化应用程序时，将 DilliServerHeaderFilter 过滤器添加到 Filters 集合中。

```
var builder = WebApplication.CreateBuilder(args);
builder.Services.AddControllers(options =>
{
    options.Filters.Add<DilliServerHeaderFilter>();
});
...
```

下面的代码定义 Test 控制器。

```
public class TestController : ControllerBase
{
    [HttpGet("/")]
    public IActionResult Home() => Content("你好，世界");
}
```

Test 控制器用于验证 DilliServerHeaderFilter 过滤器是否有效。运行示例程序，访问根 URL 即可调用 Test 控制器的 Home()方法。在返回客户端的 HTTP 消息中，将出现被过滤器修改后的 Server、Date 消息头，如图 12-3 所示。

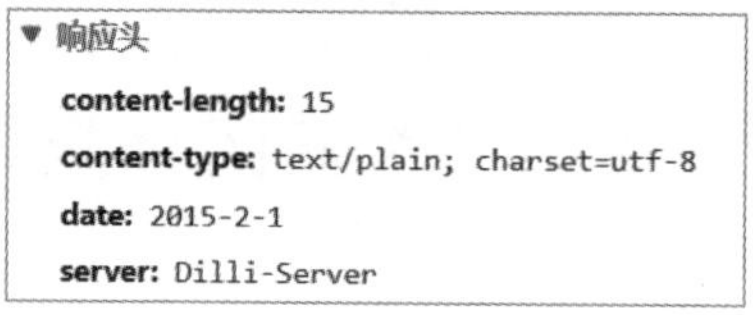

图 12-3 被过滤器修改的消息头

12.2.2 示例：特性化的过滤器

被定义为特性类的过滤器，可以应用于 MVC 控制器、控制器操作方法或 Razor 页面类，成为局部过滤器。

本示例所演示的过滤器实现 IActionFilter 接口，可在 MVC 操作方法执行之前、之后立即运行；同时，它派生自 Attribute 类，成为特性类。

```
[AttributeUsage(AttributeTargets.Class | AttributeTargets.Method,
AllowMultiple = false, Inherited = false)]
public class DemoFilterAttribute : Attribute, IActionFilter
{
    public void OnActionExecuted(ActionExecutedContext context)
    {
        // 生成 12 个随机字节
        byte[] buffer = new byte[12];
        Random.Shared.NextBytes(buffer);
        // 转换为 Base64 字符串
```

```
        string bs64 = Convert.ToBase64String(buffer);

        var response = context.HttpContext.Response;
        // 删除请求时提供的 Cookie
        response.Headers.Remove(HeaderNames.Cookie);
        // 设置新的 Cookie
        response.Cookies.Append("_token", bs64);
        // 修改 Server 头
        response.Headers.Server = "Cool-Server";
        // 修改 Date 头
        response.Headers.Date = new DateTime(2017, 5, 28, 17, 20, 15).
        ToString("yyyy-M-d HH:mm:ss");
    }

    public void OnActionExecuting(ActionExecutingContext context)
    {

    }
}
```

由于在操作方法执行之前无需任何处理，因此，OnActionExecuting()方法保留空的方法体即可。在 OnActionExecuted()方法中，代码修改响应消息的 Server 和 Date 头，并设置名为_token 的 Cookie。

接下来将 DemoFilterAttribute 应用到 MVC 控制器上。

```
[DemoFilter]
public class MainController : ControllerBase
{
    [HttpGet("/op1")]
    public IActionResult Operat1()
    {
        return Ok("Done");
    }

    [HttpGet("/op2")]
    public IActionResult Operat2()
    {
        return NoContent();
    }

    [HttpGet("/op3")]
    public IActionResult Operat3()
    {
        return Content("Finished");
    }
}
```

DemoFilter 特性应用到控制器类上，这使得 3 个操作方法都在过滤器的作用域内。所以，运行示例程序后，访问以下任意一个地址均可以触发 DemoFilterAttribute 过滤器。

```
/op1
/op2
/op3
```

访问/op3 后得到的响应消息头如图 12-4 所示。

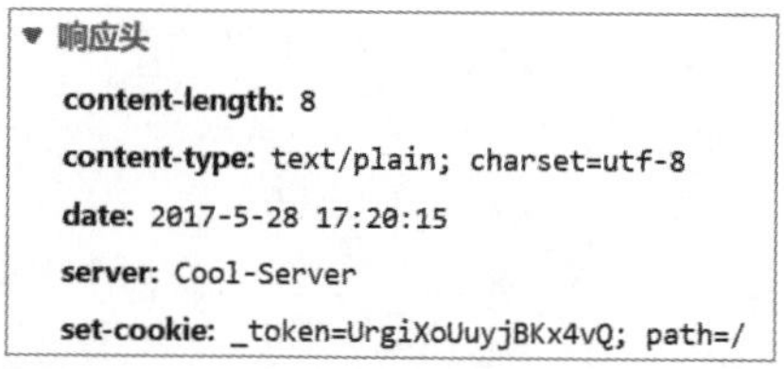

图 12-4　访问/op3 后服务器返回的消息头

12.3　在 Razor Pages 中使用过滤器

对于 Razor Pages（Razor 页面）应用程序，可以在 Razor 标记文件或页面模型类（Page Model）上使用过滤器，但不能把过滤器直接应用到页面处理程序方法（Handler）上。

用于 Razor Pages 的过滤器不需要实现 IActionFilter 或 IAsyncActionFilter 接口，因为它们仅适用于 MVC 控制器的操作方法，在 Razor Pages 中不会运行。

对于全局过滤器，和 MVC 应用一样，也是通过 MvcOptions.Filters 属性配置；而局部过滤器也是以特性类的方式应用到 Razor 标记页或页面模型类上。

12.3.1　示例：在 Razor 标记页和页面模型类上应用过滤器

本示例将定义两个特性化的过滤器类型，分别命名为 DemoFilter1 和 DemoFilter2。DemoFilter1 过滤器应用到 Razor 标记页上，DemoFilter2 过滤器则应用到页面模型类上。

两个过滤器的定义如下。

```
[AttributeUsage(AttributeTargets.Class, AllowMultiple = false, Inherited
= true)]
public class DemoFilter1Attribute : Attribute, IResourceFilter
{
    public void OnResourceExecuted(ResourceExecutedContext context)
    {
        Console.WriteLine("DemoFilter1 - OnResourceExecuted");
    }

    public void OnResourceExecuting(ResourceExecutingContext context)
    {
        Console.WriteLine("DemoFilter1 - OnResourceExecuting");
    }
}

[AttributeUsage(AttributeTargets.Class, AllowMultiple = false, Inherited
= true)]
public class DemoFilter2Attribute : Attribute, IResultFilter
{
    public void OnResultExecuted(ResultExecutedContext context)
    {
        Console.WriteLine("DemoFilter2 - OnResultExecuted");
```

```
    }

    public void OnResultExecuting(ResultExecutingContext context)
    {
        Console.WriteLine("DemoFilter2 - OnResultExecuting");
    }
}
```

新建 Razor 页面，命名为 Index.cshtml。

```
@page
@model MyApp.Pages.IndexModel
@attribute [MyApp.DemoFilter1]

<h1>欢迎光临</h1>
```

在 Razor 标记中，可以使用@attribute 指令应用特性类。上述代码应用了 DemoFilter1 过滤器。

随后，在 IndexModel 模型类上应用另一个滤过器 DemoFilter2。

```
[DemoFilter2]
public class IndexModel : PageModel
{
    …
}
```

运行示例程序，控制台将输出以下信息。

```
DemoFilter1 - OnResourceExecuting
DemoFilter2 - OnResultExecuting
DemoFilter2 - OnResultExecuted
DemoFilter1 - OnResourceExecuted
```

通过该示例可知，将过滤器定义为特性类，在 Razor 标记页和页面模型类上均可以使用，并且可以同时使用。

12.3.2　示例：在 Razor Pages 中应用全局过滤器

本示例将使用以下过滤器。

```
public class CustFilter : IResourceFilter
{
    public void OnResourceExecuted(ResourceExecutedContext context)
    {
        Console.WriteLine($"{nameof(OnResourceExecuted)}方法被调用");
    }

    public void OnResourceExecuting(ResourceExecutingContext context)
    {
        Console.WriteLine($"{nameof(OnResourceExecuting)}方法被调用");
    }
}
```

在初始化应用程序时，调用 AddRazorPages()扩展方法后立即调用 AddMvcOptions()方法，通过 MvcOptions 选项类将 CustFilter 注册为全局过滤器。该过滤器会在应用程序内任意 Razor 页面被访问时运行。

```
var builder = WebApplication.CreateBuilder(args);
builder.Services.AddRazorPages().AddMvcOptions(options =>
{
    options.Filters.Add<CustFilter>();
});
var app = builder.Build();
...
```

12.3.3 页面处理程序的过滤器

在页面处理程序执行前后会运行 IPageFilter 过滤器（异步版本为 IAsyncPageFilter 接口）。IPageFilter 接口包含以下方法成员。

（1）OnPageHandlerSelected()：在选择处理程序方法后、模型绑定前运行。

（2）OnPageHandlerExecuting()：在模型绑定后、页面处理程序执行前运行。

（3）OnPageHandlerExecuted()：在页面处理程序执行后运行。

实现 IPageFilter 接口（或 IAsyncPageFilter 接口）的过滤器类型可以注册为全局过滤器（通过 MvcOptions.Filters 属性），也可以声明为特性类，应用到包含处理程序方法的类型上（通常是页面模型类），不能应用到处理程序方法上。

12.3.4 示例：实现 IPageFilter 接口

本示例将实现一个简单的页面处理程序过滤器——CustPageHandlerFilterAttribute。应用程序选择页面处理程序后输出被选中的方法成员的名称，在执行处理程序方法前后也输出相关信息。

CustPageHandlerFilterAttribute 类的实现代码如下。

```
[AttributeUsage(AttributeTargets.Class, AllowMultiple = false)]
public class CustPageHandlerFilterAttribute : Attribute, IPageFilter
{
    public void OnPageHandlerExecuted(PageHandlerExecutedContext context)
    {
        string? method = context.HandlerMethod?.MethodInfo.Name;
        if (method != null)
        {
            Console.WriteLine($"{method}方法执行完毕");
        }
    }

    public void OnPageHandlerExecuting(PageHandlerExecutingContext context)
    {
        string? method = context.HandlerMethod?.MethodInfo.Name;
        if(method is not null)
        {
```

```
            Console.WriteLine($"即将调用{method}方法");
        }
    }

    public void OnPageHandlerSelected(PageHandlerSelectedContext context)
    {
        string? methodName = context.HandlerMethod?.MethodInfo.Name;
        if(methodName is not null)
        {
            Console.WriteLine($"被选中的页面处理方法: {methodName}");
        }
    }
}
```

表示过滤器运行上下文的 3 个类（PageHandlerSelectedContext、PageHandlerExecutingContext 以及 PageHandlerExecutedContext）都有一个 HandlerMethod 属性。此属性包含要调用的处理程序方法的相关信息。本示例只通过 HandlerMethod.MethodInfo.Name 获取被调用方法成员的名称。

在示例项目中添加新页面，命名为 Index.cshtml，关联的代码文件为 Index.cshtml.cs，即 IndexModel 类。

```
[CustPageHandlerFilter]
public class IndexModel : PageModel
{
    public void OnGet()
    {
        Console.WriteLine("正在执行{0}方法", nameof(OnGet));
    }
}
```

在 IndexModel 类中定义 OnGet()方法，此方法将作为页面的默认处理程序，以 HTTP-GET 方式访问时会自动调用。最后把 CustPageHandlerFilterAttribute 过滤器也应用到 IndexModel 类上。

运行示例程序，只要 Index 页面被执行，控制台就会输出以下信息。

```
被选中的页面处理方法: OnGet
即将调用 OnGet 方法
正在执行 OnGet 方法
OnGet 方法执行完毕
```

综上所述，CustPageHandlerFilterAttribute 类中方法成员被调用的顺序为：OnPageHandlerSelected()，OnPageHandlerExecuting()，OnPageHandlerExecuted()。

12.4 异步过滤器接口

实现异步过滤器接口，在调用时支持异步等待。异步接口与同步接口有着对应关系。例如，IExceptionFilter 接口对应的异步接口为 IAsyncExceptionFilter，IAuthorizationFilter 接口对应的异步接口为 IAsyncAuthorizationFilter。

12.4.1 示例：实现异步授权过滤器

实现异步授权过滤器需要实现 IAsyncAuthorizationFilter 接口，定义如下。

```
public interface IAsyncAuthorizationFilter : IFilterMetadata
{
    Task OnAuthorizationAsync(AuthorizationFilterContext context);
}
```

OnAuthorizationAsync()是异步方法，需要返回表示异步任务的 Task 对象。

本示例定了名为 DemoAsyncAuthorizationFilter 的过滤器，代码如下。

```
public class DemoAsyncAuthorizationFilter : IAsyncAuthorizationFilter
{
    public Task OnAuthorizationAsync(AuthorizationFilterContext context)
    {
        Console.WriteLine($"{nameof(OnAuthorizationAsync)}方法被调用");
        return Task.CompletedTask;
    }
}
```

在上述代码中，由于 OnAuthorizationAsync()方法内未执行任何异步等待的代码，所以方法最后要用静态属性 Task.CompletedTask 返回一个 Task 对象。

下面的代码将 DemoAsyncAuthorizationFilter 注册为全局过滤器。

```
var builder = WebApplication.CreateBuilder(args);
builder.Services.AddControllers(mvcOptions =>
{
    mvcOptions.Filters.Add<DemoAsyncAuthorizationFilter>();
});
var app = builder.Build();
…
```

定义一个 Web API 控制器，当它被访问时，就会运行 DemoAsyncAuthorizationFilter 过滤器。

```
[ApiController]
public class TestController : ControllerBase
{
    [HttpGet("/")]
    public string SayHello() =>"Hello All";
}
```

运行示例程序，控制台会输出以下内容。

```
OnAuthorizationAsync 方法被调用
```

12.4.2 示例：实现异步资源过滤器

异步资源过滤器需要实现 IAsyncResourceFilter 接口，定义如下。

```
public interface IAsyncResourceFilter : IFilterMetadata
{
    Task OnResourceExecutionAsync(ResourceExecutingContext context,
    ResourceExecutionDelegate next);
}
```

在 IResourceFilter 同步接口中，包含两个同步调用的方法：

```
void OnResourceExecuting(ResourceExecutingContext context);
void OnResourceExecuted(ResourceExecutedContext context);
```

在过滤器管道的剩余过滤器运行之前运行 OnResourceExecuting()方法，在剩余过滤器运行之后运行 OnResourceExecuted()方法。但是，在异步接口中，只有一个 OnResourceExecutionAsync()方法。也就是说，要在一个异步方法中同时完成 OnResourceExecuting()和 OnResourceExecuted()方法的功能。其流程如下。

（1）在运行剩余的过滤器之前，若需要处理，可以通过 context 参数来实现，如修改 ValueProviderFactories 列表。这相当于调用 OnResourceExecuting()方法。

（2）next 参数是委托类型，表示剩余过滤器管道中的下一个过滤器实例。执行此委托实例就相当于运行剩余的过滤器，然后该委托实例返回一个 ResourceExecutedContext 对象。

（3）这个被返回的 ResourceExecutedContext 对象就相当于调用 OnResourceExecuted()方法。此时，剩余的过滤器已执行完毕，若需要处理，可以通过此 ResourceExecutedContext 对象完成。

下面的代码实现 SetHeaderAsyncFilter 过滤器，其作用是在资源对象执行完后向 HTTP 响应消息添加自定义头。

```
public class SetHeaderAsyncFilter : IAsyncResourceFilter
{
    public async Task OnResourceExecutionAsync(ResourceExecutingContext
    context, ResourceExecutionDelegate next)
    {
        //---- 此时相当于调用 OnResourceExecuting()方法 ----
        // 例如：
        // context.ValueProviderFactories.Clear();
        //-------------------------------------------------
        // 现在调用剩余的过滤器
        ResourceExecutedContext executedContext = await next();
        //---- 此时相当于调用 OnResourceExecuted()方法 ----
        // 添加 HTTP 消息头
        executedContext.HttpContext.Response.Headers.Add("app-env",
        "asp-dotnet-core");
    }
}
```

在 next 委托执行完成后（即在资源执行完成后），通过其返回的 ResourceExecutedContext 对象设置 HTTP 消息头。

随后，将 SetHeaderAsyncFilter 注册为全局过滤器。

```
var builder = WebApplication.CreateBuilder(args);
builder.Services.AddControllers(mvcOpt =>
{
    mvcOpt.Filters.Add<SetHeaderAsyncFilter>();
});
…
```

运行示例程序，当客户端完成一次请求后，响应消息中将带有自定义的消息头，如图 12-5 所示。

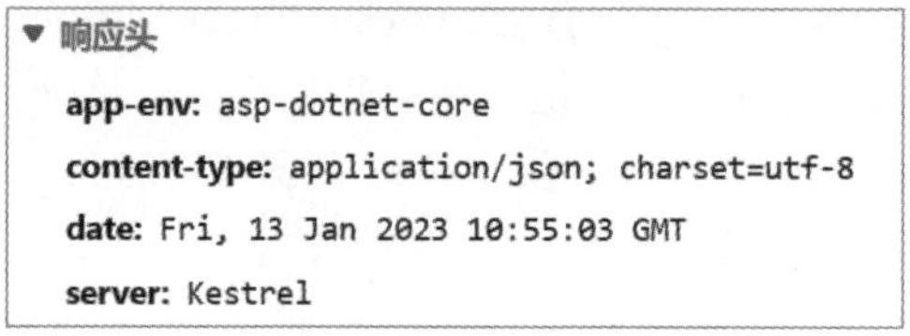

图 12-5 过滤器添加的 app-env 消息头

12.5 IAlwaysRunResultFilter 接口

结果（实现 IResultFilter 或 IAsyncResultFilter 接口的类型）过滤器遇到以下情况将不会执行。

（1）代码抛出异常，此时会运行异常过滤器（IExceptionFilter 或 IAsyncExceptionFilter 接口的实现类型），但不会运行结果过滤器。

（2）运行授权过滤器，但授权失败，此时也不会运行结果过滤器。

如果应用程序需要一个在任何时候都能运行的结果过滤器，那么就要实现 IAlwaysRun-ResultFilter 接口（异步版本为 IAsyncAlwaysRunResultFilter 接口）。该接口派生自 IResultFilter 接口，因此包含 OnResultExecuting()和 OnResultExecuted()方法。

下面的代码实现了 3 个过滤器。

```
[AttributeUsage(AttributeTargets.Method, AllowMultiple = false, Inherited
= true)]
public class DemoAuthorizationFilterAttribute : Attribute,
IAuthorizationFilter
{
    public void OnAuthorization(AuthorizationFilterContext context)
    {
        Console.WriteLine("执行{0}", nameof(OnAuthorization));
        // 设置 Result 会导致“短路”，结果过滤器不会执行
        context.Result = new UnauthorizedResult();
    }
}

[AttributeUsage(AttributeTargets.Method, AllowMultiple = false, Inherited
= true)]
```

```
public class DemoResultFilterAttribute : Attribute, IResultFilter
{
    public void OnResultExecuted(ResultExecutedContext context)
    {
        Console.WriteLine("执行{0}.{1}", nameof(DemoResultFilterAttribute),
        nameof(OnResultExecuted));
    }

    public void OnResultExecuting(ResultExecutingContext context)
    {
        Console.WriteLine("执行{0}.{1}", nameof(DemoResultFilterAttribute),
        nameof(OnResultExecuting));
    }
}

[AttributeUsage(AttributeTargets.Method, AllowMultiple = false, Inherited
= true)]
public class DemoAlwaysRunResultFilterAttribute : Attribute,
IAlwaysRunResultFilter
{
    public void OnResultExecuted(ResultExecutedContext context)
    {
        Console.WriteLine($"执行{nameof(DemoAlwaysRunResultFilterAttribute)}.
        {nameof(OnResultExecuted)}");
    }

    public void OnResultExecuting(ResultExecutingContext context)
    {
        Console.WriteLine($"执行{nameof(DemoAlwaysRunResultFilterAttribute)}.
        {nameof(OnResultExecuting)}");
    }
}
```

在 OnAuthorization()方法中，直接设置了 context.Result 属性，这会使 HTTP 请求被“短路”，即直接返回 UnauthorizedResult，此时 DemoResultFilterAttribute 不会运行，但 DemoAlwaysRunResultFilterAttribute 仍然会运行。

下面的代码将上述 3 个过滤器应用到 MVC 操作方法上。

```
public class HomeController : Controller
{
    [DemoAuthorizationFilter]
    [DemoResultFilter]
    [DemoAlwaysRunResultFilter]
    [HttpGet("/")]
    public IActionResult Index()
    {
        return Content("Hello");
    }
}
```

运行应用程序，控制台会输出以下信息。

```
执行 OnAuthorization
执行 DemoAlwaysRunResultFilterAttribute.OnResultExecuting
执行 DemoAlwaysRunResultFilterAttribute.OnResultExecuted
```

从输出结果可以看到，DemoResultFilter 过滤器没有运行，但 DemoAlwaysRunResultFilter 过滤器运行了。这证明了实现 IAlwaysRunResultFilter 接口的过滤器在任何情况下都会运行。

12.6 IFilterFactory 接口

过滤器若实现 IFilterFactory 接口，可用来创建另一个过滤器的实例。这种过滤器适用于以下场景。

（1）构造函数的参数需要通过依赖注入赋值的过滤器。

（2）获取已注册到服务容器中的过滤器实例。

12.6.1 示例：访问服务容器中的过滤器

本示例定义了 SampleFilter 过滤器，该过滤器将在服务容器中注册为单实例服务。还需要定义一个实现 IFilterFactory 接口的特性化过滤器，用于从服务容器中获取 SampleFilter 的实例。

（1）定义 SampleFilter 过滤器，它实现 IResultFilter 接口。

```
public class SampleFilter : IResultFilter
{
    public void OnResultExecuted(ResultExecutedContext context)
    {
        Console.WriteLine($"{nameof(SampleFilter)} - {nameof
        (OnResultExecuted)}调用");
    }

    public void OnResultExecuting(ResultExecutingContext context)
    {
        Console.WriteLine($"{nameof(SampleFilter)} - {nameof
        (OnResultExecuting)}调用");
    }
}
```

（2）再定义一个 SampleFilterFactAttribute 过滤器类，实现 IFilterFactory 接口。此过滤器用于获取 SampleFilter 实例。

```
[AttributeUsage(AttributeTargets.Class | AttributeTargets.Method,
AllowMultiple = true, Inherited = true)]
public class SampleFilterFactAttribute : Attribute, IFilterFactory
{
    public bool IsReusable => true;
```

```
    public IFilterMetadata CreateInstance(IServiceProvider serviceProvider)
    {
        return serviceProvider.GetRequiredService<SampleFilter>();
    }
}
```

IsReusable 属性指示 SampleFilter 实例是否可以在多个 HTTP 请求中重复使用。由于 SampleFilter 类会在服务容器中注册为单实例服务，可以在多个请求之间共享实例，所以 IsReusable 属性应返回 true。

CreateInstance()方法的输入参数是 IServiceProvider 类型，调用 GetRequiredService()方法可以轻松地获得 SampleFilter 实例的引用。

（3）把 SampleFilterFactAttribute 应用到控制器类上。

```
[SampleFilterFact]
public class HomeController : Controller
{
    …
}
```

（4）在应用程序的初始化代码中，向服务容器注册 SampleFilter 类。

```
var builder = WebApplication.CreateBuilder(args);
builder.Services.AddControllers();
builder.Services.AddSingleton<SampleFilter>();
…
```

12.6.2　示例：使用 TypeFilterAttribute 类创建过滤器实例

TypeFilterAttribute 是 ASP.NET Core 内置的特性类（位于 Microsoft.AspNetCore.Mvc 命名空间），且实现 IFilterFactory 接口。该特性类可为需要依赖注入的过滤器创建类型实例。

如果被创建的目标过滤器的构造函数无参数，或者参数的值均来自依赖注入，则可以忽略 Arguments 属性；如果目标过滤器的构造函数需要非依赖注入的参数值，就需要通过 Arguments 属性来设置。例如：

```
// 过滤器类型
public class MyFilter : IActionFilter
{
    // tag 参数的值并非来自依赖注入
    public MyFilter(string tag)
    {
        …
    }

    public void OnActionExecuted(ActionExecutedContext context)
    {
        …
    }
```

```
    public void OnActionExecuting(ActionExecutingContext context)
    {
        …
    }
}

// 在应用 TypeFilterAttribute 时，Arguments 属性将参数值传递给 tag 参数
[TypeFilter(typeof(MyFilter), Arguments = new[] { "test" })]
```

本示例所演示的 SetHostEnvFilter 过滤器需要通过依赖注入获取 IHostEnvironment 类型的服务，再通过该服务获得当前应用程序的运行环境名称。最后把此运行环境名称设置到 HTTP 响应消息头中。

下面的代码定义 SetHostEnvFilter 过滤器。

```
public class SetHostEnvFilter : IResultFilter
{
    private readonly IHostEnvironment _environment;

    public SetHostEnvFilter(IHostEnvironment hostenv)
    {
        // 通过依赖注入获取服务引用
        _environment = hostenv;
    }

    public void OnResultExecuted(ResultExecutedContext context)
    {
        // 设置自定义消息头
        var response = context.HttpContext.Response;
        if (!response.HasStarted)
        {
            response.Headers.Add("host-environment", _environment.
            EnvironmentName);
        }
    }

    public void OnResultExecuting(ResultExecutingContext context)
    {
        // 此处不需要处理代码
    }
}
```

将 TypeFilterAttribute 应用到 MVC 操作方法上，用来创建 SetHostEnvFilter 实例。

```
public class HomeController : Controller
{
    [HttpGet("/")]
    [TypeFilter(typeof(SetHostEnvFilter))]
    public IActionResult Index()
    {
```

```
        …
    }
}
```

运行示例程序，访问根 URL 后，服务器会返回包含以下头部的 HTTP 消息。

```
host-environment: Development
```

12.6.3　示例：使用 ServiceFilterAttribute 类访问服务容器中的过滤器

ServiceFilterAttribute 也是 ASP.NET Core 内置的特性类（位于 Microsoft.AspNetCore.Mvc 命名空间），它用于访问服务容器中的过滤器。

下面的代码定义 ActionLoggerFilter 过滤器。

```
public class ActionLoggerFilter : IActionFilter
{
    private readonly ILogger<ActionLoggerFilter> _logger;

    public ActionLoggerFilter(ILogger<ActionLoggerFilter> logger)
    {
        _logger = logger;
    }

    public void OnActionExecuted(ActionExecutedContext context)
    {
        // 获取当前操作方法的标识
        string actName = context.ActionDescriptor.DisplayName ?? "None";
        _logger.LogInformation("过滤器在{0}操作之后运行", actName);
    }

    public void OnActionExecuting(ActionExecutingContext context)
    {
        string actionName = context.ActionDescriptor.DisplayName ?? "None";
        _logger.LogInformation("过滤器在{0}操作之前运行", actionName);
    }
}
```

应用程序的初始化代码中，将 ActionLoggerFilter 过滤器注册到服务容器中。

```
var builder = WebApplication.CreateBuilder(args);
builder.Services.AddControllers();
builder.Services.AddScoped<ActionLoggerFilter>();
…
```

在控制器类上应用 ServiceFilterAttribute，并指定目标类型为 ActionLoggerFilter。

```
[ServiceFilter(typeof(ActionLoggerFilter))]
public class HomeController : Controller
{
    …
}
```

运行示例程序，访问 Home 控制器中的任意操作方法，ActionLoggerFilter 过滤器都会运行。控制台的输出如图 12-6 所示。

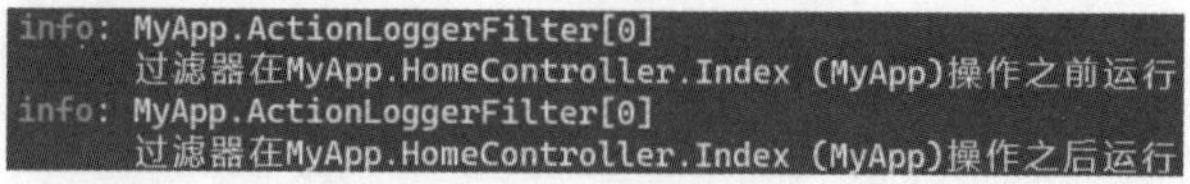

info: MyApp.ActionLoggerFilter[0]
过滤器在MyApp.HomeController.Index (MyApp)操作之前运行
info: MyApp.ActionLoggerFilter[0]
过滤器在MyApp.HomeController.Index (MyApp)操作之后运行

图 12-6　运行 ActionLoggerFilter 过滤器后输出的日志

12.7　过滤器的运行顺序

IOrderedFilter 接口提供了 Order 属性，它是一个 int 类型的整数值，用于修改过滤器运行的相对顺序。实现了该接口的过滤器，将以 Order 属性进行升序排列，即 Order 属性值小的过滤器先运行。

这里所说的顺序是同一分类下过滤器之间的相对顺序。例如，A、B 都是资源过滤器（Resource Filter），它们可以实现 IOrderedFilter 接口，用 Order 属性确定运行顺序；而授权过滤器与资源过滤器之间是无法通过 Order 属性确定运行顺序的，毕竟授权过滤器必须在其他过滤器之前运行。

12.7.1　示例：过滤器的作用域与运行顺序

本示例将定义 3 个过滤器，它们都实现 IAsyncActionFilter 接口，即它们都是同一类过滤器（MVC 操作方法过滤器）。示例将检验它们的运行顺序。

（1）定义 GlobalFilter 类，该类将用于全局过滤器。

```
public class GlobalFilter : IAsyncActionFilter
{
    public async Task OnActionExecutionAsync(ActionExecutingContext
    context, ActionExecutionDelegate next)
    {
        var filterName = nameof(GlobalFilter);
        Console.WriteLine($"{filterName} Executing");
        _ = await next();
        Console.WriteLine($"{filterName} Executed");
    }
}
```

（2）SomeFilterAttribute 声明为特性类，可作为局部过滤器应用于操作方法上。

```
[AttributeUsage(AttributeTargets.Method, AllowMultiple = true, Inherited
= true)]
public class SomeFilterAttribute : Attribute, IAsyncActionFilter
{
    public async Task OnActionExecutionAsync(ActionExecutingContext
    context, ActionExecutionDelegate next)
    {
        Console.WriteLine($"{nameof(SomeFilterAttribute)} Executing");
```

```
        _ = await next();
        Console.WriteLine($"{nameof(SomeFilterAttribute)} Executed");
    }
}
```

（3）Home 控制器类同时实现了 IActionFilter 和 IAsyncActionFilter 接口，所以，控制器自身也可以作为操作方法过滤器使用。

```
public class HomeController : Controller
{
    #region 实现过滤器功能
    public override async Task OnActionExecutionAsync
    (ActionExecutingContext context, ActionExecutionDelegate next)
    {
        Console.WriteLine($"{nameof(HomeController)} Filter Executing");
        _ = await next();
        Console.WriteLine($"{nameof(HomeController)} Filter Executed");
    }
    #endregion

    [SomeFilter, HttpGet("/")]
    public IActionResult Index()
    {
        return Ok("Index");
    }
}
```

上述代码只重写了 OnActionExecutionAsync()方法，实现异步版本的操作过滤器。若需要实现同步版本，可重写 OnActionExecuting()和 OnActionExecuted()方法。代码还在 Index()方法上应用了 SomeFilterAttribute 局部过滤器。

（4）在应用程序初始化代码中，调用 AddControllers()扩展方法注册 MVC 功能，并将 GlobalFilter 过滤器添加到 MvcOptions.Filters 列表中，成为全局过滤器。

```
var builder = WebApplication.CreateBuilder(args);
builder.Services.AddControllers(opt =>
{
    opt.Filters.Add<GlobalFilter>();
});
…
```

运行应用程序，控制台将输出以下内容。

```
HomeController Filter Executing
GlobalFilter Executing
SomeFilterAttribute Executing
SomeFilterAttribute Executed
GlobalFilter Executed
HomeController Filter Executed
```

由输出结果可以发现，Controller 或其派生类（控制器自身实现的过滤器）先于全局过滤器运行。这是因为一个名为 ControllerActionFilter 的内部类实现了 IOrderedFilter 接口，并且将 Order 属性设置为 int.MinValue，即 int 类型的最小值，这使得控制器类自身实现的过滤器先于其他同类过滤器运行。下面是 ControllerActionFilter 类的部分源代码。

```
internal sealed class ControllerActionFilter : IAsyncActionFilter,
IOrderedFilter
{
    public int Order { get; set; } = int.MinValue;

    public Task OnActionExecutionAsync(
        ActionExecutingContext context,
        ActionExecutionDelegate next)
    {
        …

        var controller = context.Controller;
        if (controller == null)
        {
            throw new InvalidOperationException(Resources.
            FormatPropertyOfTypeCannotBeNull(
                    nameof(context.Controller),
                    nameof(ActionExecutingContext)));
        }

        if (controller is IAsyncActionFilter asyncActionFilter)
        {
            return asyncActionFilter.OnActionExecutionAsync(context, next);
        }
        else if (controller is IActionFilter actionFilter)
        {
            return ExecuteActionFilter(context, next, actionFilter);
        }
        else
        {
            return next();
        }
    }

    private static async Task ExecuteActionFilter(
        ActionExecutingContext context,
        ActionExecutionDelegate next,
        IActionFilter actionFilter)
    {
        actionFilter.OnActionExecuting(context);
        if (context.Result == null)
        {
            actionFilter.OnActionExecuted(await next());
        }
    }
}
```

首先，通过 context.Controller 属性获得关联控制器的实例。接着，验证控制器类所实现的接口。若实现了 IAsyncActionFilter 接口，就调用控制器的 OnActionExecutionAsync()方法；若实现了 IActionFilter 接口，就依次调用控制器的 OnActionExecuting()、OnActionExecuted()方法。

同理，如果控制器类实现了 IResultFilter、IAsyncResultFilter 接口，也可以将自身作为过滤器使用。内部类型 ControllerResultFilter 用于将控制器自身作为结果过滤器使用。

```
internal sealed class ControllerResultFilter : IAsyncResultFilter,
IOrderedFilter
{
    public int Order { get; set; } = int.MinValue;

    public Task OnResultExecutionAsync(
        ResultExecutingContext context,
        ResultExecutionDelegate next)
    {
        …

        var controller = context.Controller;
        if (controller == null)
        {
            throw new InvalidOperationException(Resources.
            FormatPropertyOfTypeCannotBeNull(
                        nameof(context.Controller),
                        nameof(ResultExecutingContext)));
        }

        if (controller is IAsyncResultFilter asyncResultFilter)
        {
            return asyncResultFilter.OnResultExecutionAsync(context, next);
        }
        else if (controller is IResultFilter resultFilter)
        {
            return ExecuteResultFilter(context, next, resultFilter);
        }
        else
        {
            return next();
        }
    }

    private static async Task ExecuteResultFilter(
        ResultExecutingContext context,
        ResultExecutionDelegate next,
        IResultFilter resultFilter)
    {
        resultFilter.OnResultExecuting(context);
        if (!context.Cancel)
        {
            resultFilter.OnResultExecuted(await next());
```

```
        }
    }
}
```

注意，其 Order 属性也是设置为 int.MinValue，使其先于其他结果过滤器运行。

12.7.2 示例：自定义过滤器的运行顺序

本示例将通过 Order 属性修改资源过滤器的运行顺序，使局部资源过滤器在全局资源过滤器之前运行。

以下是 GlobalResourceFilter 类的定义，它将注册为全局资源过滤器。

```
public class GlobalResourceFilter : IResourceFilter
{
    public void OnResourceExecuted(ResourceExecutedContext context)
    {
        Console.WriteLine($"{nameof(GlobalResourceFilter)} - {nameof
        (OnResourceExecuted)}调用");
    }

    public void OnResourceExecuting(ResourceExecutingContext context)
    {
        Console.WriteLine($"{nameof(GlobalResourceFilter)} - {nameof
        (OnResourceExecuting)}调用");
    }
}
```

全局过滤器不用实现 IOrderedFilter 接口，因为 MvcOptions.Filters 属性（FilterCollection 类型）内部会使用 TypeFilterAttribute 过滤器来封装。其运行顺序将通过 Add()方法传递。例如，下面的代码将 GlobalResourceFilter 注册为全局过滤器，并设置运行顺序为 2。

```
var builder = WebApplication.CreateBuilder(args);
builder.Services.AddControllers(opt =>
{
    opt.Filters.Add<GlobalResourceFilter>(2);
});
...
```

如果调用 Add()方法时不指定顺序，则默认为 0。

下面的代码定义 LocalResourceFilterAttribute 局部过滤器。

```
[AttributeUsage(AttributeTargets.Class | AttributeTargets.Method,
AllowMultiple = true, Inherited = true)]
public class LocalResourceFilterAttribute : Attribute, IResourceFilter,
IOrderedFilter
{
    // 相对顺序
    public int Order => 1;

    public void OnResourceExecuted(ResourceExecutedContext context)
```

```
    {
        Console.WriteLine($"{nameof(LocalResourceFilterAttribute)} -
        {nameof(OnResourceExecuted)}调用");
    }

    public void OnResourceExecuting(ResourceExecutingContext context)
    {
        Console.WriteLine($"{nameof(LocalResourceFilterAttribute)} -
        {nameof(OnResourceExecuting)}调用");
    }
}
```

LocalResourceFilterAttribute 过滤器实现了 IOrderedFilter 接口，并让 Order 属性返回 1，即其运行顺序为 1。

将 LocalResourceFilterAttribute 过滤器应用到控制器上。

```
[LocalResourceFilter]
public class HomeController : Controller
{
    …
}
```

此时，GlobalResourceFilter 全局过滤器的运行顺序为 2，LocalResourceFilterAttribute 局部过滤器的运行顺序为 1。应用程序先运行 LocalResourceFilterAttribute，后运行 GlobalResource-Filter。因此，控制台将输出以下信息。

```
LocalResourceFilterAttribute - OnResourceExecuting 调用
GlobalResourceFilter - OnResourceExecuting 调用
GlobalResourceFilter - OnResourceExecuted 调用
LocalResourceFilterAttribute - OnResourceExecuted 调用
```

12.8　抽象的过滤器特性类

除了实现相关的过滤器接口外，实现抽象类也能完成自定义过滤器。ASP.NET Core 内置了一组抽象类。这些抽象类都继承了 Attribute 类，成为特性类，同时实现了过滤器接口。开发人员可以直接从这些抽象类派生，再加入自定义的代码逻辑，即可创建自定义过滤器。

主要的过滤器抽象类如下。

（1）ActionFilterAttribute：同时实现了 5 个接口，即 IActionFilter、IAsyncActionFilter、IResultFilter、IAsyncResultFilter、IOrderedFilter。也就是说，该抽象类同时包含同步和异步版本的操作过滤器方法。因此，在重写抽象方法时要注意，同步版本和异步版本的方法不应该同时重写。例如，若重写了 OnActionExecuting()和 OnActionExecuted()方法，就不要重写 OnAction-ExecutionAsync()方法；若重写了 OnActionExecutionAsync()方法，就不要重写 OnActionExecuting()和 OnActionExecuted()方法。

（2）ResultFilterAttribute：该抽象类实现了 3 个接口，即 IResultFilter、IAsyncResultFilter、IOrderedFilter。所以，它包含同步和异步版本的结果过滤器方法。与 ActionFilterAttribute 类一样，在重写 ResultFilterAttribute 类的方法时，同步版本与异步版本不应同时重写。

（3）ExceptionFilterAttribute：它实现了 3 个接口，即 IExceptionFilter、IAsyncExceptionFilter、IOrderedFilter。因此，该抽象类包含同步和异步的异常过滤器方法。派生类在重写该类的方法时，也要注意不应该同时重写同步方法和异步方法。

12.8.1 示例：重写 ActionFilterAttribute 类

本示例将实现 DemoActionFilterAttribute 过滤器，该类只实现异步版本的操作过滤方法和结果过滤方法。具体代码如下。

```
public class DemoActionFilterAttribute : ActionFilterAttribute
{
    public override async Task OnActionExecutionAsync(ActionExecutingContext
    context, ActionExecutionDelegate next)
    {
        Console.WriteLine($"Demo：操作过滤器在{context.ActionDescriptor.
        DisplayName ?? "null"}之前运行");
        _ = await next();
        Console.WriteLine($"Demo：操作过滤器在{context.ActionDescriptor.
        DisplayName ?? "null"}之后运行");
    }

    public override async Task OnResultExecutionAsync(ResultExecutingContext
    context, ResultExecutionDelegate next)
    {
        Console.WriteLine($"Demo：结果过滤器在{context.Result.GetType().
        Name}之前运行");
        _ = await next();
        Console.WriteLine($"Demo：结果过滤器在{context.Result.GetType().
        Name}之后运行");
    }
}
```

将 DemoActionFilterAttribute 应用到控制器类上即可生效。

```
[DemoActionFilter]
public class HomeController : Controller
{
    public IActionResult Index()
    {
        return Ok();
    }
}
```

运行示例程序，控制台将输出以下内容。

```
Demo：操作过滤器在 MyApp.HomeController.Index (MyApp)之前运行
Demo：操作过滤器在 MyApp.HomeController.Index (MyApp)之后运行
Demo：结果过滤器在 OkResult 之前运行
Demo：结果过滤器在 OkResult 之后运行
```

12.8.2 示例：重写 ExceptionFilterAttribute 类

本示例将实现自定义的 CustExceptionFilterAttribute 异常过滤器，代码如下。

```
public class CustExceptionFilterAttribute : ExceptionFilterAttribute
{
    public override void OnException(ExceptionContext context)
    {
        if(context.ExceptionHandled == false)
        {
            var exception = context.Exception;
            Console.WriteLine($"异常：{exception.Message}");
            context.ExceptionHandled = true;
        }
    }
}
```

如果 ExceptionHandled 属性的值为 true，表示异常已被处理过（可能其他异常过滤器已处理）。因此，只有当 ExceptionHandled 属性为 false 时才进行异常处理。

将 CustExceptionFilterAttribute 应用于控制器类或操作方法上都可以。

```
[CustExceptionFilter]
public class HomeController : Controller
{
    public IActionResult Index()
    {
        throw new Exception("错误信息");
    }
}
```

异常过滤器只有在发生异常的情况下才会运行。为了让 CustExceptionFilterAttribute 过滤器能够运行，Index()方法直接抛出异常。

第 13 章　标记帮助器

本章要点

- ❑ 标记帮助器的基本用法
- ❑ ASP.NET Core 内置的标记帮助器
- ❑ 标记帮助器组件

13.1　标记帮助器简介

标记帮助器（Tag Helpers）是实现了 ITagHelper 接口的.NET 类，用于 Razor 文档。标记帮助器能够根据.NET 代码生成 HTML 标记以及标记的属性和内容，既可作用于现有的 HTML 标记，也可以是自定义的 HTML 标记。有了标记帮助器，开发人员在编写 Razor 文档时会更方便。

ITagHelper 接口派生自 ITagHelperComponent 接口，它没有定义新的成员，只继承了 ITagHelperComponent 接口的成员，具体如下。

（1）Init()方法：标记帮助器初始化时调用。若没有相关的属性或数据需要初始化，Init()方法可留空（方法体为一对空大括号）。

（2）ProcessAsync()方法：此方法实现标记帮助器的核心功能——生成 HTML 标记或 HTML 内容。

（3）Order 属性：一个整数值，表示当前标记帮助器优先级。属性值越小，优先级越高。

ITagHelper 接口仅作为一个标志，其作用是应用程序可通过此接口识别某个类型是否属于标记帮助器。在编写标记帮助器时，开发人员通常不需要直接实现 ITagHelper 接口，而是实现 TagHelper 抽象类。ASP.NET Core 自带了许多现成的标记帮助器，如面向<form>元素的 FormTagHelper、面向<label>元素的 LabelTagHelper、面向<input>元素的 InputTagHelper 等。

标记帮助器的名称一般以 TagHelper 结尾（即后缀），如 InputTagHelper。这只是一种约定，便于人们快速辨别，并不是强制语法规则。即自定义标记帮助器时，类名可以不使用 TagHelper 后缀。

在 Razor 文档中，需要用@addTagHelper 指令引用标记帮助器，然后才能在 HTML 中使用帮助器。该指令的语法如下。

```
@addTagHelper <标记帮助器>, <程序集>
```

指令后面的字符串分为两部分，用英文的逗号隔开。第 1 部分是标记帮助器的名称（即帮助器类的名称，包括命名空间的名称）；第 2 部分是标记帮助器所在程序集的名称。例如，要引入 FormTagHelper 帮助器，则@addTagHelper 指令如下。

```
@addTagHelper Microsoft.AspNetCore.Mvc.TagHelpers.FormTagHelper,
Microsoft.AspNetCore.Mvc.TagHelpers
```

Microsoft.AspNetCore.Mvc.TagHelpers.FormTagHelper 是完整的类名，Microsoft.AspNetCore.Mvc.TagHelpers 是程序集名称。

@addTagHelper 指令在选择帮助器名称时支持通配符（*）。例如，要引入 Demo 程序集内的所有标记帮助器，则指令可以写为

```
@addTagHelper *, Demo
```

如果要引用 Microsoft.AspNetCore.Mvc.TagHelpers 命名空间下的所有标记帮助器，则指令可以写为

```
@addTagHelper Microsoft.AspNetCore.Mvc.TagHelpers.*,
Microsoft.AspNetCore.Mvc.TagHelpers
```

下面的用法可以引用 Demo 命令空间下所有以 Div 开头的标记帮助器。

```
@addTagHelper Demo.Div*, Demo
```

@addTagHelper 指令也可以放到_ViewImports.cshtml 文件中，使其作用于多个文件。

另外，若希望从已引入的标记帮助器中删除指定的帮助器，可以使用@removeTagHelper 指令。它的参数与@addTagHelper 指令相同——指定帮助器名称和程序集名称。

13.1.1　示例：为<span>标记添加“加粗”功能

本示例通过自定义的标记帮助器为<span>标记添加 is-bold 属性，若属性值为 true，则<span></span>标签内的内容呈现为加粗文本，否则呈现为常规文本。

SpanBoldTagHelper 标记帮助器的实现代码如下。

```
[HtmlTargetElement("span")]
public class SpanBoldTagHelper : TagHelper
{
    [HtmlAttributeName("is-bold")]
    public bool Bold { get; set; } = false;

    public override void Process(TagHelperContext context, TagHelperOutput
    output)
    {
        if(Bold)
        {
```

```
                // 在<span>标记的内容前面插入 HTML 标记
                output.PreContent.SetHtmlContent("<strong>");
                // 在<span>标记的内容后面插入 HTML 标记
                output.PostContent.SetHtmlContent("</strong>");
            }
        }
    }
```

在 SpanBoldTagHelper 类上应用 HtmlTargetElement 特性，用来指定此帮助器是针对哪个 HTML 标记的（本示例是<span>标记）。Bold 属性所绑定的 HTML 属性为 is-bold。

上述代码重写了 Process()方法，在<span>标记的内容之前插入<strong>元素的开始标记，在内容之后插入 strong 元素的结束标记。即当 Bold 属性为 true 时，将<span>标记原有的内容变为

```
<span><strong>内容</strong></span>
```

在 Razor 文档中，使用 addTagHelper 指令添加标记引用。

```
@addTagHelper DemoApp.SpanBoldTagHelper, DemoApp
```

为了能直观对比，下面的 HTML 代码中用到两个段落标记（<p>）。第 1 个段落中包含加粗文本，第 2 个段落中均为常规文本。

```
<p>
    该段落包含<span is-bold="true">加粗的文本
</span>
</p>

<p>
    该段落包含<span is-bold="false">常规的文
本</span>
</p>
```

示例的运行效果如图 13-1 所示。

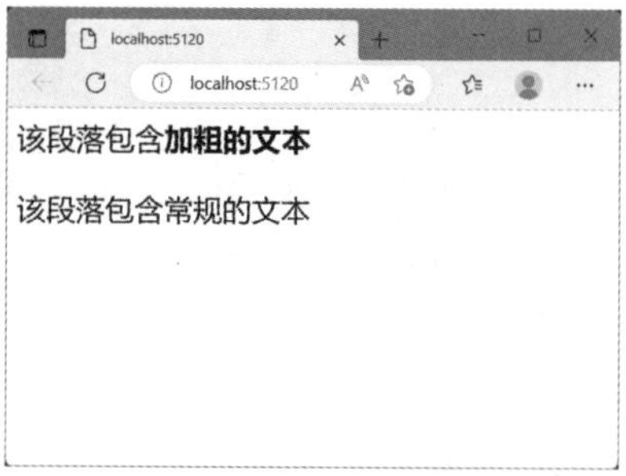

图 13-1　加粗文本与常规文本

13.1.2　示例：<url>标记帮助器

本示例将实现一个<url>标记（并非 HTML 现有的标记）的帮助器，生成<a>标记，并把<url>与</url>之间的内容设置为 href 属性的值。

例如，在 Razor 文档中使用<url>标记。

```
<url>http://abc.org</url>
```

运行后会输出为<a>标记。

```
<a href="http://abc.org">http://abc.org</url>
```

UrlTagHelper 类的定义如下。

```
[HtmlTargetElement(TAG_NAME)]
public class UrlTagHelper : TagHelper
```

```
    {
        // 标记名称
        const string TAG_NAME = "url";

        public override async Task ProcessAsync(TagHelperContext context,
        TagHelperOutput output)
        {
            // 将<url>替换为<a>
            output.TagName = "a";
            // 先执行当前标记的子级（即<url>与</url>之间的内容）并获取其内容
            string content = (await output.GetChildContentAsync()).GetContent();
            // 将标记内容设置为 href 属性的值
            output.Attributes.SetAttribute("href", content);
        }
    }
```

处理标记生成时，上述代码使用了异步等待，因此，UrlTagHelper 类重写的是 ProcessAsync()方法，即异步版本。

output.GetChildContentAsync()方法的调用是为了先让<url>标记的子级元素先执行，然后才能获取到有效的文本。此处<url>标记的子级就是<url>和</url>之间的文本。获取到子级内容后，把它设置为 href 属性的值。

在 Razor 文档中，使用@addTagHelper 指令添加<url>标记帮助器（MyApp 是当前示例的程序集名称）。

```
@addTagHelper MyApp.UrlTagHelper, MyApp
```

随后就可以使用<url>标记了。

```
<url>https://www.baidu.com</url>
```

运行示例程序，实际生成的 HTML 如下。

```
<a href="https://www.baidu.com">https://www.baidu.com</a>
```

13.1.3　示例：使用标记帮助器设置 HTML 元素的文本样式

本示例演示了 TextStyleTagHelper 标记帮助器，它可以使用便捷的属性设置 HTML 元素中文本内容的样式，包括字体名称、字体大小、文本颜色、是否加粗等。

TextStyleTagHelper 类的代码如下。

```
[HtmlTargetElement("*")]
public class TextStyleTagHelper : TagHelper
{
    // 设置字体名称
    [HtmlAttributeName("font-family")]
    public string? FontFamily{ get; set; }

    // 设置字体大小
    [HtmlAttributeName("font-size")]
```

```
    public float FontSize{ get; set; }

    // 是否加粗
    [HtmlAttributeName("bold")]
    public bool Bold{ get; set; }

    // 是否为斜体
    [HtmlAttributeName("italic")]
    public bool Italic{ get; set; }

    // 文本颜色
    [HtmlAttributeName("text-color")]
    public string? TextColor{ get; set; }

    public override void Process(TagHelperContext context, TagHelperOutput
    output)
    {
        StringBuilder strbuilder = new();
        // 拼接 CSS 样式
        if(!string.IsNullOrWhiteSpace(FontFamily))
        {
            strbuilder.Append($"font-family: {FontFamily};");
        }
        if(FontSize > 0f)
        {
            strbuilder.Append($"font-size: {FontSize}px;");
        }
        if(Bold)
        {
            strbuilder.Append("font-weight: bold;");
        }
        if(Italic)
        {
            strbuilder.Append("font-style: italic;");
        }
        if(!string.IsNullOrWhiteSpace(TextColor))
        {
            strbuilder.Append($"color: {TextColor};");
        }
        // 设置 style 属性
        output.Attributes.SetAttribute("style", strbuilder.ToString());
    }
}
```

HtmlTargetElement 特性指定了*为目标标记，表示该帮助器对任何 HTML 标记都有效。HtmlAttributeName 特性指定了.NET 属性与 HTML 属性之间的映射，如 FontFamily 对应的 HTML 属性为 font-family。示例代码使用 StringBuilder 类拼接 CSS 样式，最后设置为 style 属性的值。

在 Razor 视图中，使用@addTagHelper 指令引入帮助器（MyApp 是示例的程序集名称）。

```
@addTagHelper MyApp.TextStyleTagHelper, MyApp
```

下面几个 HTML 元素将用标记帮助器设置文本样式。

```
<p font-family="楷体" font-size="45" bold text-color="orange">
    测试文本
</p>
<p text-color="#F722A8" font-family="仿宋" font-size="36">
    测试文本
</p>
<span font-family="隶书" font-size="42" text-color="green" bold>
    测试文本
</span>
<div font-family="黑体" font-size="38" italic text-color="pink">
    测试文本
</div>
```

各文本样式如图 13-2 所示。

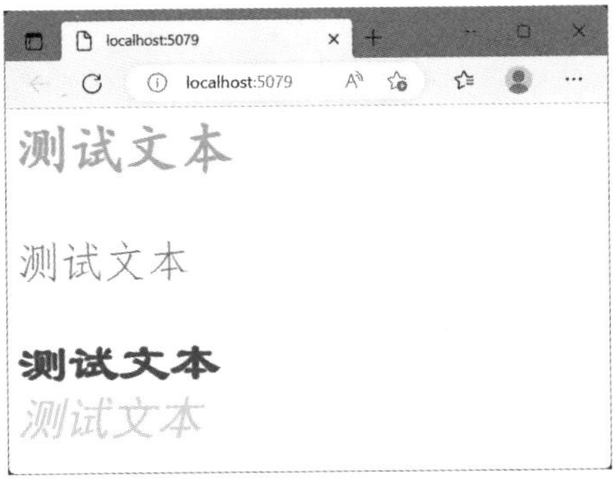

图 13-2　标记帮助器设置的文本样式

13.2　将标记帮助器注册到服务容器

标记帮助器每次使用时都会创建新的实例，对于不需要记录状态数据的帮助器，反复实例化会带来额外的性能开销，尤其是初始化代码比较多时。于是，可以考虑将标记帮助器注册为单个实例的服务。如此一来，在整应用程序的生命周期内，标记帮助器只实例化一次，在反复调用时不需要创建新实例。

可以用一个示例进行实验。

定义 DivTagHelper 标记帮助器，该标记帮助器将作用于<div>标记。

```
[HtmlTargetElement("div")]
public class DivTagHelper : TagHelper
{
```

```
    public DivTagHelper()
    {
        System.Console.WriteLine("Div 帮助器实例化 - {0}", Guid.NewGuid());
    }

    // 水平坐标
    public float X{ get; set; }

    // 垂直坐标
    public float Y{ get; set; }

    // 宽度
    public float Width{ get; set; }

    // 高度
    public float Height{ get; set;}

    // 背景颜色
    public string? BackColor{ get; set; }

    public override void Process(TagHelperContext context, TagHelperOutput
    output)
    {
        // 拼接 CSS 样式
        StringBuilder sbd = new();
        sbd.Append("position: absolute;");
        sbd.Append("overflow: hidden;");
        if(X > 0.0f)
        {
            sbd.AppendFormat("left: {0}px;", X);
        }
        if(Y > 0f)
        {
            sbd.AppendFormat("top: {0}px;", Y);
        }
        if(Width > 0f)
        {
            sbd.AppendFormat("width: {0}px;", Width);
        }
        if(Height > 0f)
        {
            sbd.AppendFormat("height: {0}px;", Height);
        }
        if(BackColor is not (null or {Length: 0}))
        {
            sbd.AppendFormat("background-color: {0};", BackColor);
        }
        // 设置 HTML 属性
        output.Attributes.SetAttribute("style", sbd.ToString());
```

```
    }
}
```

该标记帮助器通过 X、Y、Width、Height、BackColor 属性设置<div>标记在 HTML 页面中的绝对坐标，以及宽度、高度和背景颜色。在 Razor 文档中，可以这样使用：

```
<div x="150" y="60" back-color="yellow" width="135" height="130">
    桃子
</div>

<div x="15" y="209" height="65" width="300" back-color="#20F785">
    葡萄
</div>

<div x="340" y="140" width="87" height="36" back-color="#94A5DE">
    板栗
</div>
```

在 DivTagHelper 类的构造函数中调用了 Console.WriteLine()方法输出 GUID 值，只要有新的 DivTagHelper 实例被创建，控制台就会输出全新的 GUID。上述示例中用到了 3 个<div>标记，当页面呈现时会创建 3 个 DivTagHelper 实例，因此生成了 3 个 GUID，如图 13-3 所示。

图 13-3　生成 3 个 GUID

如果刷新页面，便会创建更多的 DivTagHelper 实例，控制台将输出更多的 GUID，如图 13-4 所示。

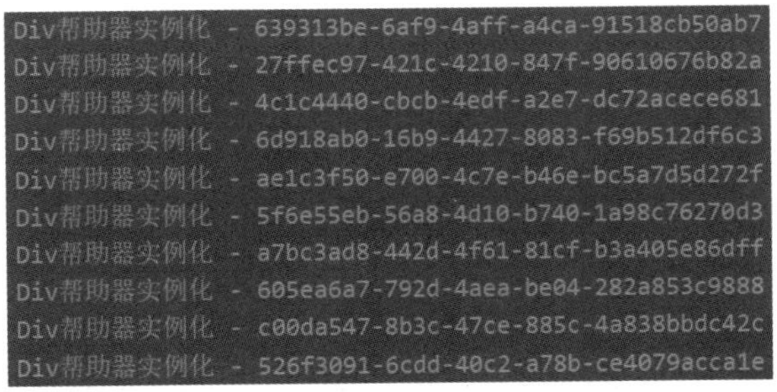

图 13-4　输出更多的 GUID

在应用程序的初始化代码中，将 DivTagHelper 注册为单实例服务。

```
builder.Services.AddSingleton<DivTagHelper>();
```

调用 AddTagHelpersAsServices()扩展方法，让应用程序从服务容器中获取标记帮助器实例。

```
builder.Services.AddControllersWithViews().AddTagHelpersAsServices();
```

再次运行示例程序。此时，DivTagHelper 标记帮助器只创建一个实例，无论被访问多少次，它始终引用用的是同一个实例。

13.3 内置的标记帮助器

ASP.NET Core 自身内置了许多有用的标记帮助器，其中常用的如下。

（1）FormTagHelper：面向<form>元素。可以使用 asp-action、asp-controller、asp-route 等属性指定表单提交时要执行的操作方法（或 Razor 页面的 Handler 方法）。

（2）FormActionTagHelper：面向<button>元素，以及 type=image 或 type=submit 的<input>元素。使用 asp-action、asp-page-handler、asp-controller 等属性触发指定的控制器、操作方法、Razor 页面 Handler 等。

（3）InputTagHelper：面向<input>元素。可使用 asp-for 属性设置模型绑定表达式。常用于将<input>元素与模型类的某个属性绑定。

（4）ComponentTagHelper：面向<component>元素，用于呈现 Razor 组件。常用于 Blazor 项目中。

（5）CacheTagHelper：其标记为<cache>。<cache>和</cache>之间的内容会被缓存，直到过期。enabled 属性可以控制启用或禁用缓存功能。

（6）EnvironmentTagHelper：标记名称为<environment>，可以根据当前应用程序所运行的环境决定是否呈现 HTML 内容。

（7）PartialTagHelper：标记名称为<partial>，用于呈现局部视图，可通过 name 属性指定视图的名称或路径。

（8）AnchorTagHelper：面向<a>元素。通过 asp-page、asp-controller、asp-action 等属性生成 URL。

（9）ScriptTagHelper：面向<script>元素。扩展<script>标记的功能，当 src 属性所指定的脚本文件无效时，可以使用 asp-fallback-src 属性指定的备用路径。当指定的脚本文件位于内容分发网络（Content Delivery Network，CDN）服务器上时，有可能出现无法访问的问题，此时 asp-fallback-src 属性指定的备用路径就派上用场了。

（10）TextAreaTagHelper：面向<textarea>元素。可通过 asp-for 属性设置模型绑定。

（11）ValidationSummaryTagHelper 和 ValidationMessageTagHelper：在模型验证时用于呈现错误信息。

13.3.1 示例：缓存当前时间

本示例使用<cache>标记缓存当前时间，并设置在 12s 后过期。

```
@page
@addTagHelper *, Microsoft.AspNetCore.Mvc.TagHelpers
```

```
<cache expires-after="@TimeSpan.FromSeconds(12)">
    当前时间：@DateTime.Now.ToLongTimeString()
</cache>
```

运行示例程序，呈现当前时间，如图 13-5 所示。

只要缓存尚未过期，不管刷新页面多少次，所显示的时间都是不变的，因为时间是从缓存中读取的。等待一段时间后（等待时间应大于 12s），再次刷新页面。由于上一次缓存的内容已过期，页面会缓存新的内容，此时页面会呈现最新的时间，如图 13-6 所示。

图 13-5　被缓存的时间

图 13-6　缓存的内容已更新

13.3.2　示例：用<button>元素提交表单

本示例演示使用 Button 标记帮助器提交表单。

Home 控制器的实现代码如下。

```
public class HomeController:Controller
{
    public IActionResult Index() => View();

    public IActionResult Test() => Content("Test 操作已执行");
}
```

在 Index 视图中，将<button>元素放在<form>元素内，并使用 asp-controller 属性指定提交后要访问的控制器名称，asp-action 属性指定要执行的操作方法名称。

```
@addTagHelper *, Microsoft.AspNetCore.Mvc.TagHelpers

<div>
    请单击下面的按钮：
</div>

<form>
<button asp-action="Test" asp-controller="Home">测试按钮</button>
</form>
```

在服务器上运行后，会生成以下 HTML。

```
<form>
<button formaction="/Home/Test">测试按钮</button>
</form>
```

此时单击“测试按钮”按钮，就会执行 Test()方法。

13.3.3 示例：asp-for 属性的使用

<lable>和<input>等表单元素均可使用 asp-for 属性指定要进行模型绑定的目标。一般来说，指定模型类的某个属性名称即可。

对于<label>元素，进行模型绑定后，可以从模型属性上所应用的 Display 特性获取要呈现的字段名称。例如：

```
[Display(Name = "姓名")]
public string? Name { get; set; }
```

<label>元素若与 Name 属性绑定，那么应用程序会自动将“姓名”填充到<label>与</label>之间的内容中，变成

```
<label for="Name">姓名</label>
```

对于<input>元素，如果与 Name 属性绑定，那么文本框所输入的内容在提交到服务器时传递给模型对象的 Name 属性。服务器生成的 HTML 为

```
<input type="text" id="Name" name="Name" value="">
```

假设输入 Tom，在表单提交给服务器后，模型对象的 Name 属性会自动被赋值为 Tom。

本示例将演示使用 asp-for 属性让单表元素与 Student 类进行模型绑定。

（1）定义 Student 类。

```
// 学员信息
public class Student
{
    [Display(Name = "学号")]
    public int ID { get; set; }

    [Display(Name = "姓名")]
    public string? Name { get; set; }

    [Display(Name = "手机号")]
    public string? Phone { get; set; }

    [Display(Name = "课程")]
    public string? Course { get; set; }
}
```

属性成员应用了 Display 特性，作用是提供在用户界面上显示的文本（如“手机号”“学号”等）。

（2）定义 Student 控制器。它包含两个操作方法：Home()方法在应用程序运行后自动执行，返回 AddNew 视图；NewStudent()方法用于接收并处理客户端提交的数据。

```
public class StudentController : Controller
{
    public IActionResult NewStudent(Student s)
```

```
    {
        string msg = $"学号: {s.ID}\n" +
                     $"姓名: {s.Name}\n" +
                     $"手机号: {s.Phone}\n" +
                     $"课程: {s.Course}";
        return Ok("添加成功\n\n" + msg);
    }

    public IActionResult Home()
    {
        return View("AddNew");
    }
}
```

（3）AddNew 视图的 Razor 代码如下。

```
@using MyApp
@addTagHelper *, Microsoft.AspNetCore.Mvc.TagHelpers
@model Student
@{
    // 设置布局页
    Layout = "_Layout";
}

<form method="post" asp-controller="Student" asp-action="NewStudent">
<div class="table">
<div class="row">
<div class="cell left">
<label asp-for="ID"></label>:
</div>
<div class="cell right">
<input asp-for="ID" />
</div>
</div>
<div class="row">
<div class="cell">
<label asp-for="Name"></label>:
</div>
<div class="cell">
<input asp-for="Name" />
</div>
</div>
<div class="row">
<div class="cell">
<label asp-for="Phone"></label>:
</div>
<div class="cell">
<input asp-for="Phone" />
</div>
```

```
</div>
<div class="row">
<div class="cell">
<label asp-for="Course"></label>:
</div>
<div class="cell">
<input asp-for="Course" />
</div>
</div>
</div>
<button type="submit">提交</button>
</form>
```

@model 指令设置与此视图关联的模型类，本示例中是 Student 类。asp-for 属性在使用时直接指定要绑定的属性名称即可（Student 类的属性成员）。<label>元素的内容不需要设置，运行阶段将自动填充。

（4）运行示例程序，在表单中填写相关的信息，如图 13-7 所示。

（5）单击“提交”按钮，数据将被发送至服务器并进行处理，如图 13-8 所示。

图 13-7　输入数据

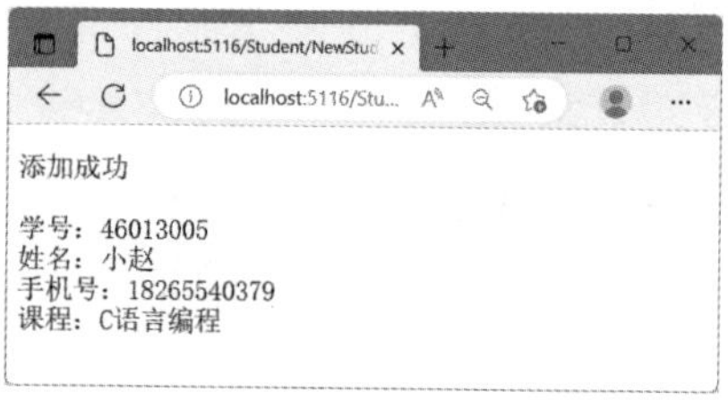

图 13-8　服务器接收到的数据

13.3.4　示例：呈现验证信息

本示例将演示使用 ValidationMessageTagHelper 类在 HTML 中呈现错误信息（模型验证失败后会呈现）。

ValidationMessageTagHelper 帮助器应用的目标是<span>标记。For 属性（对应的 HTML 属性是 asp-validation-for）用于指定针对要绑定模型所应用的表达式，通常指定目标模型类的属性名即可。

本示例用到的模型类如下。

```
public class News
{
    public string? Nid { get; set; }

    [Display(Name = "标题")]
    [StringLength(15)]
    [Required(AllowEmptyStrings = false, ErrorMessage = "标题不能为空")]
    public string? Title { get; set; }
```

```
    [Display(Name = "正文")]
    [Required(AllowEmptyStrings = false, ErrorMessage = "正文不能为空")]
    [MinLength(10, ErrorMessage = "{0}至少要{1}个字符")]
    public string? Body { get; set; }

    [Display(Name = "发布时间")]
    public DateTime? PubTime { get; set; }
}
```

假设 News 类表示一条新闻的数据；Nid 表示该数据的唯一标识；Title 属性表示新闻标题；Body 属性表示新闻内容；PubTime 属性表示新闻的发布时间。

Title 属性应用了 StringLength 特性，限制了新闻标题的长度不能超过 15 个字符。Body 属性则应用了 MinLength 特性，限制新闻内容的长度不小于 10 个字符。

下面的代码定义 Home 控制器。

```
public class HomeController : Controller
{
    [AcceptVerbs("GET")]
    public IActionResult Index()
    {
        return View();
    }

    [AcceptVerbs("POST")]
    public IActionResult Index(News data)
    {
        // 验证未通过
        if(!ModelState.IsValid)
        {
            return View();
        }
        // 设置 ID 值
        data.Nid = Guid.NewGuid().ToString();
        string s = "新闻发布成功\n" +
                   $"标题：{data.Title}";
        // 返回处理结果
        return Ok(s);
    }
}
```

Index 视图的布局如下。

```
…
@model News
@addTagHelper *, Microsoft.AspNetCore.Mvc.TagHelpers

<form method="post">
```

```
<div class="table">
<div class="tr">
<div class="td left">
<label asp-for="Title"></label>
</div>
<div class="td right">
<input asp-for="Title" />
<span asp-validation-for="Title" class="err"></span>
</div>
</div>
<div class="tr">
<div class="td left">
<label asp-for="Body"></label>
</div>
<div class="td right">
<input asp-for="Body" />
<span asp-validation-for="Body" class="err"></span>
</div>
</div>
<div class="tr">
<div class="td left">
<label asp-for="PubTime"></label>
</div>
<div class="td right">
<input asp-for="PubTime" />
<span asp-validation-for="PubTime" class="err"></span>
</div>
</div>
</div>
<button type="submit">确定</button>
</form>
```

由于 ValidationMessageTagHelper 类作用于<span>标记，因此在 Razor 文档中只能在<span>元素中使用 asp-validation-for 属性。当模型验证失败后，ASP.NET Core 会将错误信息呈现在<span>元素内，如图 13-9 所示。

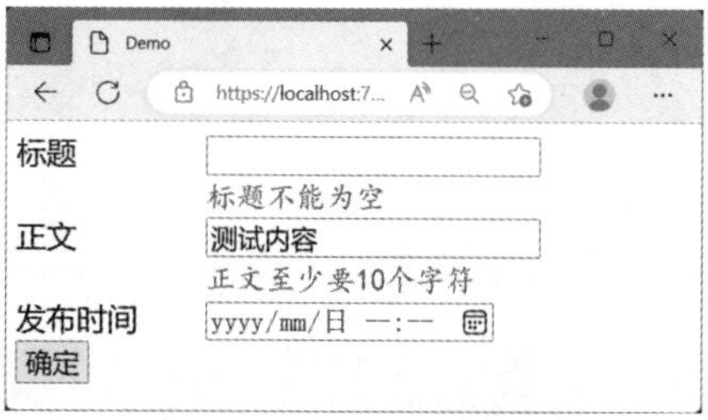

图 13-9　呈现错误信息

13.4　标记帮助器组件

Tag Helper Component 可以译为“组件化的帮助器”或“标记帮助器组件”，其抽象基类为 TagHelperComponent，该类实现了 ITagHelperComponent 接口。ITagHelper 也实现 ITagHelper-Component 接口，所以，常规的标记帮助器也是一种特殊的标记帮助器组件。

若需要自定义的标记帮助器组件，一般做法是实现 TagHelperComponent 抽象类。标记帮助器组件不使用@addTagHelper 指令引入，而是注册到服务容器中或添加到 ITagHelper-

ComponentManager 对象的 Components 列表中。

按照使用约定，标记帮助器组件常用于需要大面积修改的 HTML 标记上（主要针对<head>和<body>标记），如在<body>标记内插入脚本代码。

13.4.1　示例：在<body>元素内插入 CSS 样式

本示例将定义这样的标记帮助器组件：作用于<body>标记，在标记的所有子级内容之前插入 CSS 样式。

（1）AddStyleToBodyTaghelperComponent 类的定义如下，它派生自 TagHelperComponent 抽象类。

```
public class AddStyleToBodyTaghelperComponent : TagHelperComponent
{
    // 优先级
    public override int Order => 5;

    // 要插入 HTML 中的样式
    const string style = """
    <style type="text/css">
        div {
            position: absolute;
        }
        div.t1 {
            left: 25px;
            top: 5px;
            width: 125px;
            height: 58px;
            background-color: green;
        }
        div.t2 {
            left: 200px;
            top: 35px;
            height: 88px;
            width: 80px;
            background-color: blue;
        }
        div.t3 {
            left: 60px;
            top: 177px;
            width: 65px;
            height: 45px;
            background-color: peru;
        }
    </style>
    <style type="text/css">
        div {
            position: absolute;
        }
```

```
        div.t1 {
            left: 25px;
            top: 5px;
            width: 125px;
            height: 58px;
            background-color: green;
        }
        div.t2 {
            left: 200px;
            top: 35px;
            height: 88px;
            width: 80px;
            background-color: blue;
        }
        div.t3 {
            left: 60px;
            top: 177px;
            width: 65px;
            height: 45px;
            background-color: peru;
        }
    </style>
    """;

    public override void Process(TagHelperContext context, TagHelperOutput
    output)
    {
        // 只考虑<body>标记
        if(output.TagName.Equals("body",StringComparison.
        OrdinalIgnoreCase))
        {
            // 在<body>标记的所有内容之前插入样式
            output.PostContent.AppendHtml(style);
        }
    }
}
```

（2）将 AddStyleToBodyTaghelperComponent 类注册到服务容器中。

```
var builder = WebApplication.CreateBuilder(args);
builder.Services.AddControllersWithViews();
builder.Services.AddTransient<ITagHelperComponent, AddStyleToBody
TaghelperComponent>();
…
```

（3）添加一个 Home 控制器。

```
public class HomeController : Controller
{
    public IActionResult Index()
    {
        return View();
```

```
    }
}
```

（4）新建视图目录 Views。在 Views 目录下创建 Home 目录。

（5）在 Home 目录下新建_Layout.cshtml 布局文件。

```
<!DOCTYPE html>

<html>
<head>
<meta name="viewport" content="width=device-width" />
<title>@ViewBag.Title</title>
</head>
<body>
    @*此处会插入样式*@
<div>
        @RenderBody()
</div>
</body>
</html>
```

（6）在 Home 目录下新建 Index 视图。

```
@{
    Layout = "_Layout";
}

<div class="t1"></div>

<div class="t2"></div>

<div class="t3"></div>
```

上述 HTML 中，3 个<div>元素分别引用由 AddStyleToBodyTaghelperComponent 插入的 CSS 样式类（t1、t2、t3）。

（7）运行示例程序，结果如图 13-10 所示。

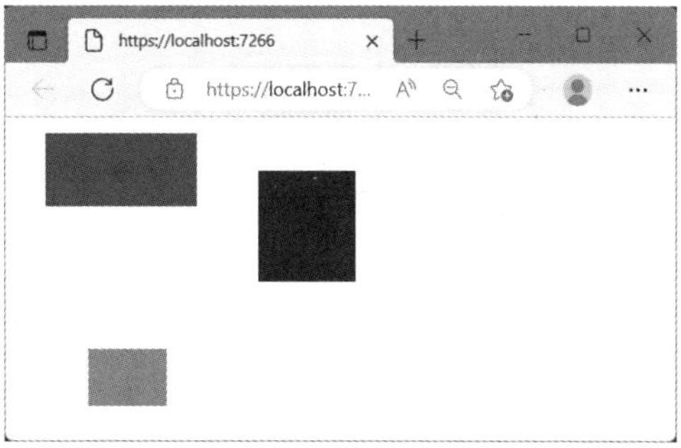

图 13-10　由标记帮助器组件插入的样式

13.4.2　示例：使用 ITagHelperComponentManager 对象注册标记帮助器组件

本示例将演示如何通过 ITagHelperComponentManager 对象在 Razor 文档中注册自定义的标记帮助器组件。

（1）定义 SetListStyleTagHelperComponent 类，它继承 TagHelperComponent 类。在<head>标记内插入自定义的 CSS 样式。

```
public class SetListStyleTagHelperComponent : TagHelperComponent
{
```

```
    // 自定义样式
    const string CUST_STYLE = """
    <style>
                ul[set-style] {
                    list-style-type: '❋ ';
                    color: rgb(245, 4, 205);
                }
    </style>
    """;

    public override void Process(TagHelperContext context, TagHelperOutput
    output)
    {
        if(output.TagName.Equals("head", StringComparison.OrdinalIgnoreCase))
        {
            output.PostContent.AppendHtml(CUST_STYLE);
        }
    }
}
```

自定义的样式作用于带有 set-style 属性的<ul>标记（无序列表）。

（2）在项目中新建 Pages 目录。

（3）在 Pages 目录下新建 Index.cshtml 文件。

```
@page
@using Microsoft.AspNetCore.Mvc.Razor.TagHelpers
@using MyApp
@inject ITagHelperComponentManager helperManager

@{
    // 实例化自定义标记帮助器组件
    var helper = new SetListStyleTagHelperComponent();
    // 添加到 Components 列表中
    helperManager.Components.Add(helper);
}

<!DOCTYPE html>
<html>
<head>
<title>Demo App</title>
<meta charset="utf-8" />
</head>
<body>
<p>默认列表：</p>
<ul>
<li>红茶</li>
<li>白茶</li>
<li>绿茶</li>
</ul>
```

```
<hr />
<p>应用了样式的列表：</p>
<ul set-style>
<li>菊花茶</li>
<li>桂花茶</li>
<li>玫瑰花茶</li>
</ul>
</body>
</html>
```

上述 Razor 代码在首行使用了@page 指令，表明 Index.cshtml 文件为 Razor 页面。@inject 指令将通过依赖注入获取 ITagHelperComponentManager 类型的服务。随后实例化 SetListStyleTagHelperComponent 类，并添加到 Components 列表中，完成注册。

（4）在应用程序的初始化代码中，添加 Razor Pages 相关功能和终结点映射。

```
var builder = WebApplication.CreateBuilder(args);
builder.Services.AddRazorPages();
var app = builder.Build();

app.MapRazorPages();

app.Run();
```

（5）运行示例程序，结果如图 13-11 所示。

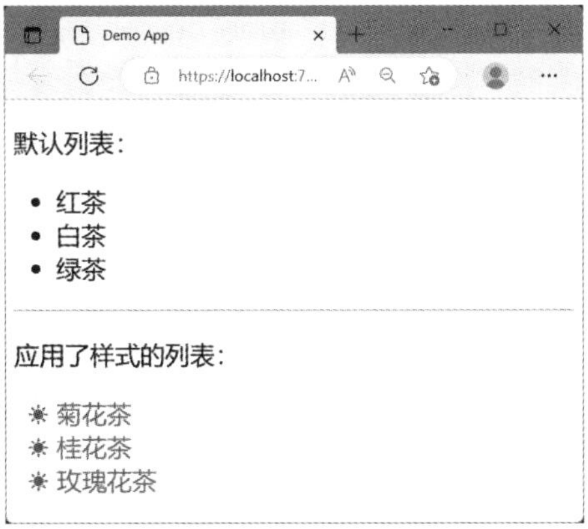

图 13-11　两组列表对比

第 14 章　静 态 文 件

本章要点

- ❑ 静态文件
- ❑ 目录浏览
- ❑ 文件服务

14.1　静态文件简介

在 ASP.NET Core 应用程序中，像 Razor 文档那样，需要先在服务器上执行 C#代码，再将生成的 HTML 提供给客户端，可以称之为“动态文件”。而静态文件则是直接提供给客户端的，不需要执行服务器代码。

常见的静态文件如下。

（1）图像文件，如.png、.jpg、.gif 等。

（2）音频/视频文件，如.mp3、.mp4、.aac 等。

（3）CSS 样式表。

（4）脚本文件，文件扩展名一般为.js、.ts。

（5）JSON/XML 文件或文本文件。

（6）数据库文件，如.db、.mdb、.accdb 等。

（7）网页（静态 HTML 文档），扩展名为.htm 或.html。

（8）其他不需要执行服务器代码的文件。

Web 服务器会公开一个 URL 作为基础地址（如/、/files/等），被访问的静态文件的位置将相对于此基础路径而产生。例如，基础地址为 https://test/static/，静态文件目录中存在名为 cool.jpg 的图像文件，那么该图像文件的访问地址为 https://test/static/cool.jpg。又如，静态文件目录下存在 styles 子目录，styles 目录下有 abc.css 文件，于是该文件的访问地址就是 https://test/static/styles/abc.css。

在 HTML 文档或 Razor 文档中也可以引用静态文件，如<img>元素要呈现某个图像，就可

以写为

```
<img src="/abc/opq/rst.png" />
```

上述路径为相对路径，也可以使用绝对路径（引用其他站点的文件时，需要完整地址）。

```
<img src="http://picstore.net/x3.png" />
```

还有一种访问方式——目录浏览。这种访问方式类似于本地文件浏览器，URL 会指定某个服务器上的目录，Web 浏览器可以查看此目录下的文件与子目录列表，包括文件名、文件大小、修改时间等信息。单击目录链接可以进入下一层目录；也可以单击导航链接返回上一级目录。图 14-1 所示为 Linux/Debian 发行版的国内镜像服务器截图，这是目录浏览功能的典型案例。

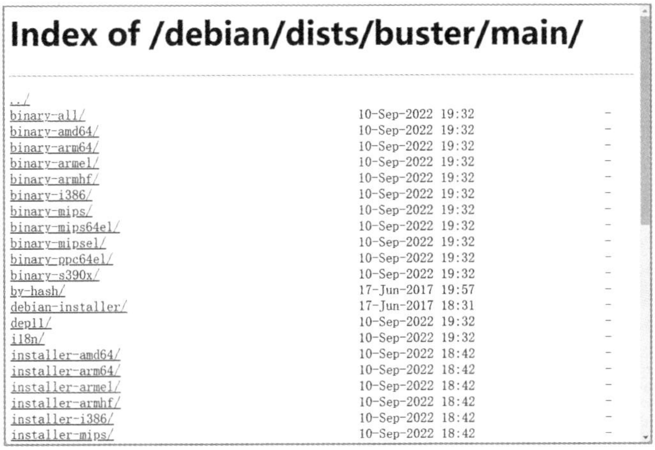

图 14-1　目录浏览功能

14.2　使用静态文件

在 ASP.NET Core 应用程序中开启静态文件功能不需要向服务容器注册额外的类型，只需要调用 UseStaticFiles()扩展方法在 HTTP 管道上添加 StaticFileMiddleware 中间件即可。

默认配置下，静态文件的存放位置为当前项目下的 wwwroot 目录。可以通过 WEBROOT 配置项更改路径。如果静态文件不在 WEBROOT 配置所指定的目录中，可以在调用 UseStaticFiles()方法时，通过 StaticFileOptions 选项类指定自定义的路径。StaticFileOptions 类也可以为静态文件目录分配一个自定义的 URL（默认为空白字符串，即应用根地址/）。

14.2.1　示例：访问图像文件

本示例将在页面上呈现两幅图像——通过<img>元素引用 wwwroot 目录内的图像文件。具体步骤如下。

（1）在项目目录内新建文件夹，命名为 wwwroot。

（2）在 wwwroot 目录下新建 images 目录。

（3）将两个图像文件复制到 images 目录下，分别重命名为 001.jpg、002.jpg（如果是.png 文件，可重命名为 001.png、002.png）。图像文件可以随意，此处仅用于演示。

（4）在项目中新建 Pages 目录，用于存放 Razor 页面文件。

（5）在 Pages 目录中新建 Index.cshtml 文件。

```
@page

<!DOCTYPE html>

<html>
<head>
<meta name="viewport" content="width=device-width" />
<title>Demo App</title>
</head>
<body>
<style type="text/css">
        img {
            display:inline;
            margin:12px;
            border-style:outset;
            border-width:10px;
            border-color:goldenrod;
            border-radius:8px;
            width:165px;
            height:165px;
        }
</style>
<div>
<img src="~/images/001.jpg" />
<img src="~/images/002.jpg" />
</div>
</body>
</html>
```

<img>元素的 src 属性指定图像文件的位置。此处用到了~符号，它表示一个虚拟路径，即静态文件的存放目录——wwwroot。

（6）在应用程序的初始代码中启用 Razor Pages 功能，然后调用 UseStaticFiles()和 MapRazorPages()方法。

```
var builder = WebApplication.CreateBuilder(args);
builder.Services.AddRazorPages();
var app = builder.Build();

// 处理静态文件的请求
app.UseStaticFiles();
// 映射 Razor Pages 终结点
app.MapRazorPages();

app.Run();
```

运行示例程序，最终效果如图 14-2 所示。

图 14-2　呈现两个图像文件

14.2.2　示例：修改 WEBROOT 路径

本示例通过修改 WEBROOT 配置项的值用 statics 目录替换默认的 wwwroot 目录。

（1）在项目目录下新建 statics 目录。

（2）在 statics 目录下新建 appStyle.css 文件（CSS 样式表），并定义以下 3 个样式类。

```
.test1 {
    font-size:24px;
    color:lightcoral;
    text-decoration-line:underline;
    text-decoration-color:green;
}

.test2 {
    font-size:32px;
    color:deeppink;
}

.test3 {
    font-size: 46px;
    color: chocolate;
    text-shadow: yellow 1px 1px 10px;
}
```

（3）在项目中新建 Views 目录用于存放 MVC 视图。

（4）在 Views 目录下新建 Home 子目录。

（5）在 Home 目录下新建 Index.cshtml 文件，其内容如下。

```
<html>
<head>
<title>Demo App</title>
<link href="~/appStyle.css" rel="stylesheet" />
</head>

<body>
<p class="test1">第 1 段文本</p>
```

```
<p class="test2">第 2 段文本</p>
<p class="test3">第 3 段文本</p>
</body>
</html>
```

在 HTML 文档中，使用<link>元素引用 appStyle.css 样式表。

（6）定义 Home 控制器。

```
public class HomeController : Controller
{
    public IActionResult Index()
    {
        return View();
    }
}
```

（7）在应用程序的初始化代码中启用带视图的 MVC 功能，并在 HTTP 管道上添加静态文件中间件，以及映射 MVC 终结点。

```
WebApplicationOptions appopts = new()
{
    WebRootPath = Path.Combine(Directory.GetCurrentDirectory(), "statics"),
    Args = args
};

var builder = WebApplication.CreateBuilder(appopts);
builder.Services.AddControllersWithViews();
var app = builder.Build();

app.UseStaticFiles();
app.MapDefaultControllerRoute();
app.Run();
```

在实例化应用程序对象前，先创建 WebApplicationOptions 实例，并且通过 WebRootPath 属性修改 WEBROOT 配置项的值为 statics 目录。接着将 WebApplicationOptions 实例传递给 WebApplication.CreateBuilder()方法。在应用程序初始化阶段，会应用新的 WEBROOT 值。

运行示例程序，效果如图 14-3 所示。

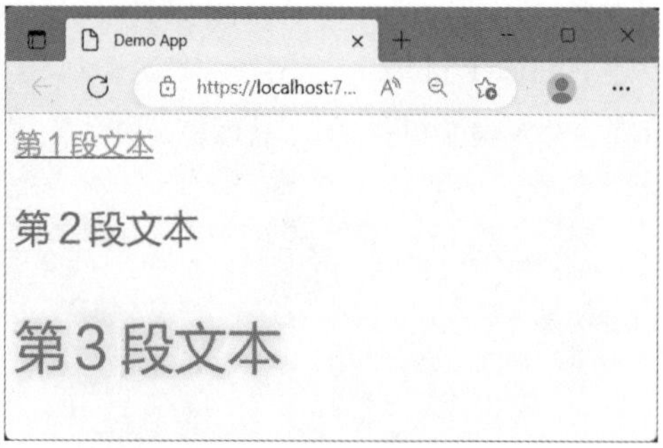

图 14-3　3 个段落都应用了 appStyle 文件中声明的样式

14.2.3 示例：统计输入的字符数量

本示例将实现在 Web 页面上实时显示文本框输入的字符个数。在本示例中，主页面会引用静态的脚本文件。而且本示例也会演示如何为静态文件设置自定义的请求 URL。

（1）在项目中新建 wwwroot 目录，这是静态文件的默认存放位置。

（2）在 wwwroot 目录中新建 scripts 子目录。

（3）在 scripts 目录下新建 JavaScript 脚本文件，并命名为 demo.js。代码如下。

```
var txt = document.getElementById('data');
var counter = document.getElementById('counter');

txt.addEventListener('input', ev => {
    let str = ev.target.value;
    counter.textContent = `已输入${str.length}个字符`;
});
```

txt 变量指的是 HTML 页中的<textarea>元素，counter 是<p>元素。侦听 txt 对象的 input 事件，当用户在 txt 对象中输入字符时会触发此事件。在响应 input 事件的代码中先获取 txt 对象中输入的文本，然后再通过 length 属性获取字符串的长度，将结果赋值给 counter 对象的 textContent 属性，使其呈现在 HTML 页面上。

（4）在项目中新建 Pages 目录，这是 Razor Pages 存放页面文件的默认位置。

（5）在 Pages 目录下新建 Index.cshtml 文件，即 Index 页面。

```
@page

<!DOCTYPE html>
<html>
<head>
<title>Demo App</title>
<meta charset="utf-8" />
</head>
<body>
<textarea id="data" name="data" rows="5" cols="20" placeholder="请输入文本"></textarea>
<p id="counter" name="counter"></p>

    @*引用脚本文件*@
<script src="~/myfiles/scripts/demo.js"></script>
</body>
</html>
```

<script>元素的 src 属性在引用脚本文件时，URL 带有/myfiles 前缀，这是通过 StaticFileOptions.RequestPath 属性配置的 URL 前缀（请看下一步骤）。

（6）在初始化应用程序时，启用 Razor Pages 功能，并配置 HTTP 管道处理静态文件访问。

```
var builder = WebApplication.CreateBuilder(args);
builder.Services.AddRazorPages();
```

```
var app = builder.Build();

StaticFileOptions opts = new()
{
    // 自定义前缀，即静态文件目录的基础地址
    // 为/myfiles
    RequestPath = "/myfiles"
};
app.UseStaticFiles(opts);
app.MapRazorPages();
app.Run();
```

运行示例程序，在文本区域中输入一些测试内容。文本区域下方会实时统计输入的字符数量，如图 14-4 所示。

图 14-4　实时统计输入的字符数量

14.2.4　示例：合并多个目录

本示例中包含两个存放静态文件的目录：images1 目录包含 01.jpg 和 02.jpg 文件；images2 目录包含 03.jpg 和 04.jpg 文件。在使用静态文件时将两个目录合并，直接通过/01.jpg、/02.jpg、/04.jpg 等 URL 就可以访问这些文件。

（1）在项目中新建两个目录，依次命名为 images1、images2。

（2）在 images1 目录下放置两个文件：01.jpg、02.jpg。

（3）在 images2 目录下放置两个文件：03.jpg、04.jpg。

（4）在项目中新建 Pages 目录，用于存放 Razor Pages 文件。

（5）在 Pages 目录下新建 Index.cshtml 文件。

```
@page

<!DOCTYPE html>
<html>

<head>
<title>Demo App</title>
<meta charset="utf-8" />
<link href="~/styles/app.css" rel="stylesheet" />
</head>

<body>
<div class="table">
<div class="tr">
<div class="td left">
            来自 images1 目录的文件
</div>
<div class="td right">
<div class="tr">
<div class="td">
```

```
<img src="~/01.jpg" />
</div>
</div>
<div class="tr">
<div class="td">
<img src="~/02.jpg" />
</div>
</div>
</div>
</div>
<div class="tr">
<div class="td left">
            来自 images2 目录的文件
</div>
<div class="td right">
<div class="tr">
<div class="td">
<img src="~/03.jpg" />
</div>
</div>
<div class="tr">
<div class="td">
<img src="~/04.jpg" />
</div>
</div>
</div>
</div>
</div>
</body>

</html>
```

（6）在应用程序的初始化代码中启用 Razor Pages 功能。

```
var builder = WebApplication.CreateBuilder(args);
builder.Services.AddRazorPages();
var app = builder.Build();
```

（7）在使用静态文件时，通过 StaticFileOptions.FileProvider 属性分别设置 images1 和 images2 目录。

```
// 拼接两个静态文件目录
string path1 = Path.Combine(app.Environment.ContentRootPath, "images1");
string path2 = Path.Combine(app.Environment.ContentRootPath, "images2");
// 为两个路径分别创建 FileProvider 实例
IFileProvider path1Provider = new PhysicalFileProvider(path1);
IFileProvider path2Provider = new PhysicalFileProvider(path2);
// 配置静态文件选项
StaticFileOptions opt1 = new()
{
    FileProvider = path1Provider
```

```
    };
    StaticFileOptions opt2 = new()
    {
        FileProvider = path2Provider
    };
    // 添加静态文件中间件
    app.UseStaticFiles();                        // 默认
    app.UseStaticFiles(opt1);
    app.UseStaticFiles(opt2);
    app.MapRazorPages();
    app.Run();
```

UseStaticFiles()方法可多次调用，以便传递不同的 StaticFileOptions 实例。分别添加 opt1 和 opt2，将合并两个选项类中的配置。

运行示例程序，结果如图 14-5 所示。

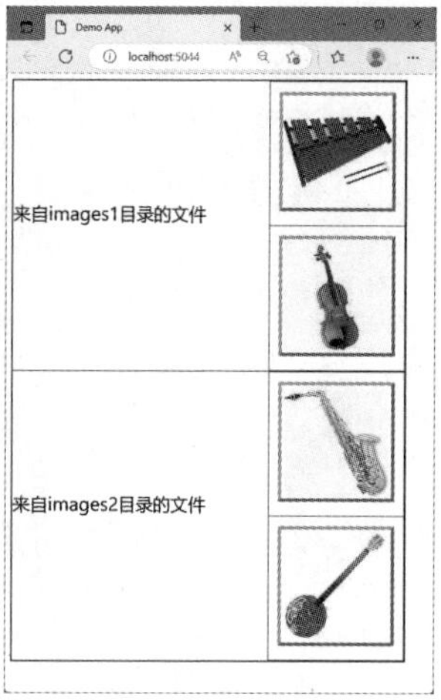

图 14-5　来自两个目录的静态文件

14.3　目录浏览

要使用目录浏览功能，一般建议先在服务容器上调用 AddDirectoryBrowser()扩展方法。其实该方法内部仅仅调用了 AddWebEncoders()方法。下面是 AddDirectoryBrowser()方法的源代码。

```
public static IServiceCollection AddDirectoryBrowser(this
IServiceCollection services)
{
    if (services == null)
    {
        throw new ArgumentNullException(nameof(services));
    }
    services.AddWebEncoders();
    return services;
}
```

AddWebEncoders()方法的作用是注册各种文本编码器（包含编码、解码功能），如 UrlEncoder、HtmlEncoder 等。AddWebEncoders()方法在其他扩展方法中也会调用，如 AddRazorPages()、AddControllers()。因此，如果应用程序中调用了 AddRazorPages()等方法，也可以不调用 AddDirectoryBrowser()方法。

在 HTTP 管道上，需要调用 UseDirectoryBrowser()方法插入 DirectoryBrowserMiddleware 中间件。若需要进行额外的配置，应先实例化 DirectoryBrowserOptions 选项类，设置好相关属性后，将 DirectoryBrowserOptions 实例传递给 UseDirectoryBrowser()方法。DirectoryBrowserOptions 类派生自 SharedOptionsBase 类，因此，它和 StaticFileOptions 类一样，也公开了 RequestPath 和 FileProvider 属性，以支持自定义 URL 前缀，以及浏览路径。

14.3.1　示例：浏览外部目录

本示例将允许 Web 浏览器查看应用程序以外的目录和文件。目录浏览功能仅允许查看目录，而不能访问文件内容。在实际开发中，都应该同时允许客户端查看目录和文件。因此，在使用目录浏览功能的同时也要使用静态文件功能，即要依次调用 UseStaticFiles()和 UseDirectoryBrowser()方法。

（1）在服务容器上，调用 AddDirectoryBrowser()方法注册相关功能。

```
var builder = WebApplication.CreateBuilder(args);
builder.Services.AddDirectoryBrowser();
…
```

（2）使用 PhysicalFileProvider 类（该类实现了 IFileProvider 接口）指定一个当前计算机上的物理路径（本示例指定的路径为 D:\Codes\Wol，读者可根据实际情况修改路径）。

```
IFileProvider fileProv = new PhysicalFileProvider(@"D:\Codes\Wol");
// 此变量指定请求的基础 URL
string requestPath = "/test";
```

（3）配置并使用静态文件。

```
StaticFileOptions staticOpt = new
StaticFileOptions{
    RequestPath = requestPath,
    FileProvider = fileProv
};
app.UseStaticFiles(staticOpt);
```

（4）配置与使用目录浏览。

```
DirectoryBrowserOptions dirbrowserOpt
=new DirectoryBrowserOptions{
    RequestPath = requestPath,
    FileProvider = fileProv
};
app.UseDirectoryBrowser(dirbrowserOpt);
```

运行示例程序，结果如图 14-6 所示。

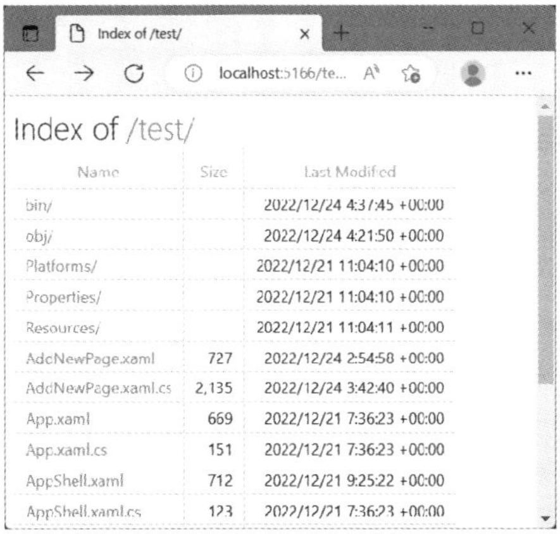

图 14-6　浏览外部目录

14.3.2　示例：自定义文件类型映射

在默认情况下，并不是所有静态文件都能访问。对于一些不常用的文件扩展名或自定义的文件扩展名，如果不存在 Content-Type 映射，访问时服务器会返回 404 错误。

若希望指定的文件类型能被访问，需要配置文件扩展名与 MIME 类型的映射。其中用到的类型为 FileExtensionContentTypeProvider（位于 Microsoft.AspNetCore.StaticFiles 命名空间），它有一个 Mappings 属性，以字典的形式存放文件类型映射。

在调用 FileExtensionContentTypeProvider 类的无参数构造函数时，它会创建一个默认的文件类型映射表，包含常见文件的 MIME 类型。源代码如下。

```
public FileExtensionContentTypeProvider()
    : this(new Dictionary<string, string>(StringComparer.OrdinalIgnoreCase)
    {
            { ".323", "text/h323" },
            { ".3g2", "video/3gpp2" },
            { ".3gp2", "video/3gpp2" },
            { ".3gp", "video/3gpp" },
            { ".3gpp", "video/3gpp" },
            { ".aac", "audio/aac" },
            { ".aaf", "application/octet-stream" },
            { ".aca", "application/octet-stream" },
            { ".accdb", "application/msaccess" },
            { ".accde", "application/msaccess" },
            { ".accdt", "application/msaccess" },
            { ".acx", "application/internet-property-stream" },
            { ".adt", "audio/vnd.dlna.adts" },
            { ".adts", "audio/vnd.dlna.adts" },
            { ".afm", "application/octet-stream" },
            { ".ai", "application/postscript" },
            { ".aif", "audio/x-aiff" },
            { ".aifc", "audio/aiff" },
            { ".aiff", "audio/aiff" },
            { ".appcache", "text/cache-manifest" },
            { ".application", "application/x-ms-application" },
            { ".art", "image/x-jg" },
            { ".asd", "application/octet-stream" },
            { ".asf", "video/x-ms-asf" },
            { ".asi", "application/octet-stream" },
            { ".asm", "text/plain" },
            { ".asr", "video/x-ms-asf" },
            { ".asx", "video/x-ms-asf" },
            { ".atom", "application/atom+xml" },
            { ".au", "audio/basic" },
            { ".avi", "video/x-msvideo" },
            { ".axs", "application/olescript" },
            { ".bas", "text/plain" },
            { ".bcpio", "application/x-bcpio" },
            { ".bin", "application/octet-stream" },
            { ".bmp", "image/bmp" },
            { ".c", "text/plain" },
            { ".cab", "application/vnd.ms-cab-compressed" },
            { ".calx", "application/vnd.ms-office.calx" },
            { ".cat", "application/vnd.ms-pki.seccat" },
            { ".cdf", "application/x-cdf" },
            { ".chm", "application/octet-stream" },
            { ".class", "application/x-java-applet" },
            { ".clp", "application/x-msclip" },
            { ".cmx", "image/x-cmx" },
            { ".cnf", "text/plain" },
```

```
{ ".cod", "image/cis-cod" },
{ ".cpio", "application/x-cpio" },
{ ".cpp", "text/plain" },
{ ".crd", "application/x-mscardfile" },
{ ".crl", "application/pkix-crl" },
{ ".crt", "application/x-x509-ca-cert" },
{ ".csh", "application/x-csh" },
{ ".css", "text/css" },
{ ".csv", "text/csv" },
{ ".cur", "application/octet-stream" },
{ ".dcr", "application/x-director" },
{ ".deploy", "application/octet-stream" },
{ ".der", "application/x-x509-ca-cert" },
{ ".dib", "image/bmp" },
{ ".dir", "application/x-director" },
{ ".disco", "text/xml" },
{ ".dlm", "text/dlm" },
{ ".doc", "application/msword" },
{ ".docm", "application/vnd.ms-word.document.macroEnabled.12" },
{ ".docx", "application/vnd.openxmlformats-officedocument.
wordprocessing document" },
{ ".dot", "application/msword" },
{ ".dotm", "application/vnd.ms-word.template.macroEnabled.12" },
{ ".dotx", "application/vnd.openxmlformats-officedocument.
wordprocessing template" },
{ ".dsp", "application/octet-stream" },
{ ".dtd", "text/xml" },
{ ".dvi", "application/x-dvi" },
{ ".dvr-ms", "video/x-ms-dvr" },
{ ".dwf", "drawing/x-dwf" },
{ ".dwp", "application/octet-stream" },
{ ".dxr", "application/x-director" },
{ ".eml", "message/rfc822" },
{ ".emz", "application/octet-stream" },
{ ".eot", "application/vnd.ms-fontobject" },
{ ".eps", "application/postscript" },
{ ".etx", "text/x-setext" },
{ ".evy", "application/envoy" },
{ ".exe", "application/vnd.microsoft.portable-executable" },
{ ".fdf", "application/vnd.fdf" },
{ ".fif", "application/fractals" },
{ ".fla", "application/octet-stream" },
{ ".flr", "x-world/x-vrml" },
{ ".flv", "video/x-flv" },
{ ".gif", "image/gif" },
{ ".gtar", "application/x-gtar" },
{ ".gz", "application/x-gzip" },
{ ".h", "text/plain" },
{ ".hdf", "application/x-hdf" },
{ ".hdml", "text/x-hdml" },
{ ".hhc", "application/x-oleobject" },
```

```
            { ".hhk", "application/octet-stream" },
            { ".hhp", "application/octet-stream" },
            { ".hlp", "application/winhlp" },
            { ".hqx", "application/mac-binhex40" },
            { ".hta", "application/hta" },
            { ".htc", "text/x-component" },
            { ".htm", "text/html" },
            { ".html", "text/html" },
            { ".htt", "text/webviewhtml" },
            { ".hxt", "text/html" },
            { ".ical", "text/calendar" },
            { ".icalendar", "text/calendar" },
            { ".ico", "image/x-icon" },
            { ".ics", "text/calendar" },
            { ".ief", "image/ief" },
            { ".ifb", "text/calendar" },
            { ".iii", "application/x-iphone" },
            { ".inf", "application/octet-stream" },
            { ".ins", "application/x-internet-signup" },
            { ".isp", "application/x-internet-signup" },
            { ".IVF", "video/x-ivf" },
            { ".jar", "application/java-archive" },
            { ".java", "application/octet-stream" },
            { ".jck", "application/liquidmotion" },
            { ".jcz", "application/liquidmotion" },
            { ".jfif", "image/pjpeg" },
            { ".jpb", "application/octet-stream" },
            { ".jpe", "image/jpeg" },
            { ".jpeg", "image/jpeg" },
            { ".jpg", "image/jpeg" },
            { ".js", "text/javascript" },
            { ".json", "application/json" },
            { ".jsx", "text/jscript" },
            { ".latex", "application/x-latex" },
            { ".lit", "application/x-ms-reader" },
            { ".lpk", "application/octet-stream" },
            { ".lsf", "video/x-la-asf" },
            { ".lsx", "video/x-la-asf" },
            { ".lzh", "application/octet-stream" },
            { ".m13", "application/x-msmediaview" },
            { ".m14", "application/x-msmediaview" },
            { ".m1v", "video/mpeg" },
            { ".m2ts", "video/vnd.dlna.mpeg-tts" },
            { ".m3u", "audio/x-mpegurl" },
            { ".m4a", "audio/mp4" },
            { ".m4v", "video/mp4" },
            { ".man", "application/x-troff-man" },
            { ".manifest", "application/x-ms-manifest" },
            { ".map", "text/plain" },
            { ".markdown", "text/markdown" },
            { ".md", "text/markdown" },
```

```
{ ".mdb", "application/x-msaccess" },
{ ".mdp", "application/octet-stream" },
{ ".me", "application/x-troff-me" },
{ ".mht", "message/rfc822" },
{ ".mhtml", "message/rfc822" },
{ ".mid", "audio/mid" },
{ ".midi", "audio/mid" },
{ ".mix", "application/octet-stream" },
{ ".mjs", "text/javascript" },
{ ".mmf", "application/x-smaf" },
{ ".mno", "text/xml" },
{ ".mny", "application/x-msmoney" },
{ ".mov", "video/quicktime" },
{ ".movie", "video/x-sgi-movie" },
{ ".mp2", "video/mpeg" },
{ ".mp3", "audio/mpeg" },
{ ".mp4", "video/mp4" },
{ ".mp4v", "video/mp4" },
{ ".mpa", "video/mpeg" },
{ ".mpe", "video/mpeg" },
{ ".mpeg", "video/mpeg" },
{ ".mpg", "video/mpeg" },
{ ".mpp", "application/vnd.ms-project" },
{ ".mpv2", "video/mpeg" },
{ ".ms", "application/x-troff-ms" },
{ ".msi", "application/octet-stream" },
{ ".mso", "application/octet-stream" },
{ ".mvb", "application/x-msmediaview" },
{ ".mvc", "application/x-miva-compiled" },
{ ".nc", "application/x-netcdf" },
{ ".nsc", "video/x-ms-asf" },
{ ".nws", "message/rfc822" },
{ ".ocx", "application/octet-stream" },
{ ".oda", "application/oda" },
{ ".odc", "text/x-ms-odc" },
{ ".ods", "application/oleobject" },
{ ".oga", "audio/ogg" },
{ ".ogg", "video/ogg" },
{ ".ogv", "video/ogg" },
{ ".ogx", "application/ogg" },
{ ".one", "application/onenote" },
{ ".onea", "application/onenote" },
{ ".onetoc", "application/onenote" },
{ ".onetoc2", "application/onenote" },
{ ".onetmp", "application/onenote" },
{ ".onepkg", "application/onenote" },
{ ".osdx", "application/opensearchdescription+xml" },
{ ".otf", "font/otf" },
{ ".p10", "application/pkcs10" },
{ ".p12", "application/x-pkcs12" },
{ ".p7b", "application/x-pkcs7-certificates" },
```

```
{ ".p7c", "application/pkcs7-mime" },
{ ".p7m", "application/pkcs7-mime" },
{ ".p7r", "application/x-pkcs7-certreqresp" },
{ ".p7s", "application/pkcs7-signature" },
{ ".pbm", "image/x-portable-bitmap" },
{ ".pcx", "application/octet-stream" },
{ ".pcz", "application/octet-stream" },
{ ".pdf", "application/pdf" },
{ ".pfb", "application/octet-stream" },
{ ".pfm", "application/octet-stream" },
{ ".pfx", "application/x-pkcs12" },
{ ".pgm", "image/x-portable-graymap" },
{ ".pko", "application/vnd.ms-pki.pko" },
{ ".pma", "application/x-perfmon" },
{ ".pmc", "application/x-perfmon" },
{ ".pml", "application/x-perfmon" },
{ ".pmr", "application/x-perfmon" },
{ ".pmw", "application/x-perfmon" },
{ ".png", "image/png" },
{ ".pnm", "image/x-portable-anymap" },
{ ".pnz", "image/png" },
{ ".pot", "application/vnd.ms-powerpoint" },
{ ".potm", "application/vnd.ms-powerpoint.template.
 macroEnabled.12" },
{ ".potx", "application/vnd.openxmlformats-officedocument.
 presentationm template" },
{ ".ppam", "application/vnd.ms-powerpoint.addin.macroEnabled.12" },
{ ".ppm", "image/x-portable-pixmap" },
{ ".pps", "application/vnd.ms-powerpoint" },
{ ".ppsm", "application/vnd.ms-powerpoint.slideshow.
 macroEnabled.12" },
{ ".ppsx", "application/vnd.openxmlformats-officedocument.
 presentationm slideshow" },
{ ".ppt", "application/vnd.ms-powerpoint" },
{ ".pptm", "application/vnd.ms-powerpoint.presentation.
 macroEnabled.12"
{ ".pptx", "application/vnd.openxmlformats-officedocument.
 presentationm presentation" },
{ ".prf", "application/pics-rules" },
{ ".prm", "application/octet-stream" },
{ ".prx", "application/octet-stream" },
{ ".ps", "application/postscript" },
{ ".psd", "application/octet-stream" },
{ ".psm", "application/octet-stream" },
{ ".psp", "application/octet-stream" },
{ ".pub", "application/x-mspublisher" },
{ ".qt", "video/quicktime" },
{ ".qtl", "application/x-quicktimeplayer" },
{ ".qxd", "application/octet-stream" },
{ ".ra", "audio/x-pn-realaudio" },
{ ".ram", "audio/x-pn-realaudio" },
```

```
{ ".rar", "application/octet-stream" },
{ ".ras", "image/x-cmu-raster" },
{ ".rf", "image/vnd.rn-realflash" },
{ ".rgb", "image/x-rgb" },
{ ".rm", "application/vnd.rn-realmedia" },
{ ".rmi", "audio/mid" },
{ ".roff", "application/x-troff" },
{ ".rpm", "audio/x-pn-realaudio-plugin" },
{ ".rtf", "application/rtf" },
{ ".rtx", "text/richtext" },
{ ".scd", "application/x-msschedule" },
{ ".sct", "text/scriptlet" },
{ ".sea", "application/octet-stream" },
{ ".setpay", "application/set-payment-initiation" },
{ ".setreg", "application/set-registration-initiation" },
{ ".sgml", "text/sgml" },
{ ".sh", "application/x-sh" },
{ ".shar", "application/x-shar" },
{ ".sit", "application/x-stuffit" },
{ ".sldm", "application/vnd.ms-powerpoint.slide.
 macroEnabled.12" },
{ ".sldx", "application/vnd.openxmlformats-officedocument.
 presentationm slide" },
{ ".smd", "audio/x-smd" },
{ ".smi", "application/octet-stream" },
{ ".smx", "audio/x-smd" },
{ ".smz", "audio/x-smd" },
{ ".snd", "audio/basic" },
{ ".snp", "application/octet-stream" },
{ ".spc", "application/x-pkcs7-certificates" },
{ ".spl", "application/futuresplash" },
{ ".spx", "audio/ogg" },
{ ".src", "application/x-wais-source" },
{ ".ssm", "application/streamingmedia" },
{ ".sst", "application/vnd.ms-pki.certstore" },
{ ".stl", "application/vnd.ms-pki.stl" },
{ ".sv4cpio", "application/x-sv4cpio" },
{ ".sv4crc", "application/x-sv4crc" },
{ ".svg", "image/svg+xml" },
{ ".svgz", "image/svg+xml" },
{ ".swf", "application/x-shockwave-flash" },
{ ".t", "application/x-troff" },
{ ".tar", "application/x-tar" },
{ ".tcl", "application/x-tcl" },
{ ".tex", "application/x-tex" },
{ ".texi", "application/x-texinfo" },
{ ".texinfo", "application/x-texinfo" },
{ ".tgz", "application/x-compressed" },
{ ".thmx", "application/vnd.ms-officetheme" },
{ ".thn", "application/octet-stream" },
{ ".tif", "image/tiff" },
```

```
                { ".tiff", "image/tiff" },
                { ".toc", "application/octet-stream" },
                { ".tr", "application/x-troff" },
                { ".trm", "application/x-msterminal" },
                { ".ts", "video/vnd.dlna.mpeg-tts" },
                { ".tsv", "text/tab-separated-values" },
                { ".ttc", "application/x-font-ttf" },
                { ".ttf", "application/x-font-ttf" },
                { ".tts", "video/vnd.dlna.mpeg-tts" },
                { ".txt", "text/plain" },
                { ".u32", "application/octet-stream" },
                { ".uls", "text/iuls" },
                { ".ustar", "application/x-ustar" },
                { ".vbs", "text/vbscript" },
                { ".vcf", "text/x-vcard" },
                { ".vcs", "text/plain" },
                { ".vdx", "application/vnd.ms-visio.viewer" },
                { ".vml", "text/xml" },
                { ".vsd", "application/vnd.visio" },
                { ".vss", "application/vnd.visio" },
                { ".vst", "application/vnd.visio" },
                { ".vsto", "application/x-ms-vsto" },
                { ".vsw", "application/vnd.visio" },
                { ".vsx", "application/vnd.visio" },
                { ".vtx", "application/vnd.visio" },
                { ".wasm", "application/wasm" },
                { ".wav", "audio/wav" },
                { ".wax", "audio/x-ms-wax" },
                { ".wbmp", "image/vnd.wap.wbmp" },
                { ".wcm", "application/vnd.ms-works" },
                { ".wdb", "application/vnd.ms-works" },
                { ".webm", "video/webm" },
                { ".webmanifest", "application/manifest+json" },
                { ".webp", "image/webp" },
                { ".wks", "application/vnd.ms-works" },
                { ".wm", "video/x-ms-wm" },
                { ".wma", "audio/x-ms-wma" },
                { ".wmd", "application/x-ms-wmd" },
                { ".wmf", "application/x-msmetafile" },
                { ".wml", "text/vnd.wap.wml" },
                { ".wmlc", "application/vnd.wap.wmlc" },
                { ".wmls", "text/vnd.wap.wmlscript" },
                { ".wmlsc", "application/vnd.wap.wmlscriptc" },
                { ".wmp", "video/x-ms-wmp" },
                { ".wmv", "video/x-ms-wmv" },
                { ".wmx", "video/x-ms-wmx" },
                { ".wmz", "application/x-ms-wmz" },
                { ".woff", "application/font-woff" },
                { ".woff2", "font/woff2" },
                { ".wps", "application/vnd.ms-works" },
                { ".wri", "application/x-mswrite" },
```

```
            { ".wrl", "x-world/x-vrml" },
            { ".wrz", "x-world/x-vrml" },
            { ".wsdl", "text/xml" },
            { ".wtv", "video/x-ms-wtv" },
            { ".wvx", "video/x-ms-wvx" },
            { ".x", "application/directx" },
            { ".xaf", "x-world/x-vrml" },
            { ".xaml", "application/xaml+xml" },
            { ".xap", "application/x-silverlight-app" },
            { ".xbap", "application/x-ms-xbap" },
            { ".xbm", "image/x-xbitmap" },
            { ".xdr", "text/plain" },
            { ".xht", "application/xhtml+xml" },
            { ".xhtml", "application/xhtml+xml" },
            { ".xla", "application/vnd.ms-excel" },
            { ".xlam", "application/vnd.ms-excel.addin.macroEnabled.12" },
            { ".xlc", "application/vnd.ms-excel" },
            { ".xlm", "application/vnd.ms-excel" },
            { ".xls", "application/vnd.ms-excel" },
            { ".xlsb", "application/vnd.ms-excel.sheet.binary.
             macroEnabled.12" },
            { ".xlsm", "application/vnd.ms-excel.sheet.macroEnabled.12" },
            { ".xlsx", "application/vnd.openxmlformats-officedocument.
             spreadsheetml sheet" },
            { ".xlt", "application/vnd.ms-excel" },
            { ".xltm", "application/vnd.ms-excel.template.macroEnabled.12" },
            { ".xltx", "application/vnd.openxmlformats-officedocument.
             spreadsheetml template" },
            { ".xlw", "application/vnd.ms-excel" },
            { ".xml", "text/xml" },
            { ".xof", "x-world/x-vrml" },
            { ".xpm", "image/x-xpixmap" },
            { ".xps", "application/vnd.ms-xpsdocument" },
            { ".xsd", "text/xml" },
            { ".xsf", "text/xml" },
            { ".xsl", "text/xml" },
            { ".xslt", "text/xml" },
            { ".xsn", "application/octet-stream" },
            { ".xtp", "application/octet-stream" },
            { ".xwd", "image/x-xwindowdump" },
            { ".z", "application/x-compress" },
            { ".zip", "application/x-zip-compressed" },
    })
}
```

本示例将添加.cs、.py、.vb、.php 这 4 个扩展的映射，MIME 类型都是 text/plain（普通文本）。

```
// 请求的 URL 前缀
```

```
string requestPrefix = "/cust";
// 要浏览的目录
IFileProvider dir = new PhysicalFileProvider("D:\\Samples");
// 定义文件类型映射
var filetypes = new FileExtensionContentTypeProvider();
filetypes.Mappings[".cs"] = "text/plain";
filetypes.Mappings[".py"] = "text/plain";
filetypes.Mappings[".vb"] = "text/plain";
filetypes.Mappings[".php"] = "text/plain";
// 配置静态文件
app.UseStaticFiles(new StaticFileOptions{
    RequestPath = requestPrefix,
    FileProvider = dir,
    ContentTypeProvider = filetypes
});
// 使用目录浏览
app.UseDirectoryBrowser(new DirectoryBrowserOptions{
    RequestPath = requestPrefix,
    FileProvider = dir
});
```

通过 FileExtensionContentTypeProvider 对象配置好文件扩展名映射后，一定要赋值给 StaticFileOptions.ContentTypeProvider 属性后才能起作用。

示例代码中设置的浏览目录为 D:\Samples，请读者按实际情况修改路径。

运行示例程序，结果如图 14-7 所示。

此时扩展名为.cs 的文件就能正常访问了。

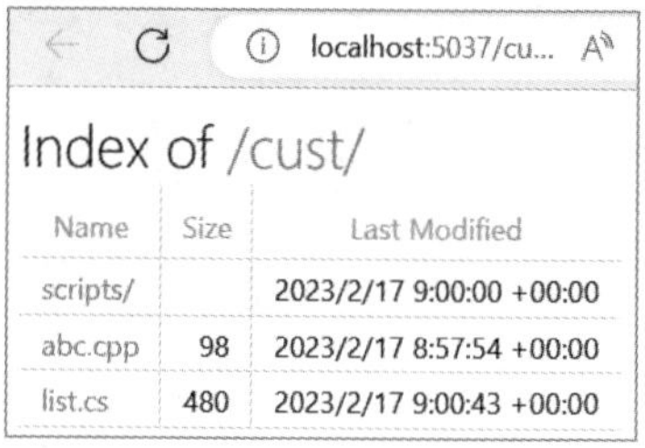

图 14-7 使用自定义的文件扩展名

14.4 文件服务

文件服务即 File Server，或者叫文件服务器，它实际上是以下 3 个功能的集合。

（1）静态文件（UseStaticFiles）。

（2）目录浏览（UseDirectoryBrowser）。

（3）默认文件名（UseDefaultFiles）。默认文件名是指请求 URL 所指向的目录中若包含 index.html、default.html 等文件（可自定义其他名称），就直接返回这些文件。例如，访问 http://test.net/maps/，如果存在 index.html 文件，那么就自动重定向为 http://test.net/maps/index.html，并返回 index.html 文件。

使用文件服务需要调用 UseFileServer()扩展方法，与其搭配的选项类是 FileServerOptions。FileServerOptions 是 StaticFileOptions、DirectoryBrowserOptions 和 DefaultFilesOptions 这 3 个类的组合。

FileServerOptions 类有两个属性值得注意。EnableDefaultFiles 属性指示是否启用默认文件名处理，默认是开启的。EnableDirectoryBrowsing 属性指示是否启用目录浏览功能，该属性默认为 false，即禁用目录浏览功能。因此，调用无参数的 UseFileServer()方法是无法进行目录浏览的，只能调用以下重载。

```
app.UseFileServer(enableDirectoryBrowsing: true)
```

enableDirectoryBrowsing 参数为布尔类型，调用方法时要传递 true 值。也可以调用以下重载，通过 FileServerOptions.EnableDirectoryBrowsing 属性进行配置。

```
app.UseFileServer(new FileServerOptions
{
    EnableDirectoryBrowsing = true
});
```

下面给出一个完整的示例。

```
var builder = WebApplication.CreateBuilder(args);
// 添加 WebEncoder 相关服务
builder.Services.AddDirectoryBrowser();
var app = builder.Build();

// 要浏览的本地目录，需要根据实际情况修改
IFileProvider localDir = new PhysicalFileProvider("E:\\");
// 使用文件服务
app.UseFileServer(new FileServerOptions
{
    FileProvider = localDir,
    // 开启目录浏览功能
    EnableDirectoryBrowsing = true
});

app.Run();
```

第 15 章

路 由 约 束

本章要点

- ❑ 内置的路由约束
- ❑ 自定义的路由约束

15.1 路由约束的作用

URL 路由在匹配时，路由参数均作为字符串处理，在某些应用场景中需要更高精度的匹配方案。例如，一个表示产品 ID 的路由参数，它的值要求必须为整数值。而客户端传递的路由参数值可能包含不合要求的字符，如 1234s、152Q5 这些值是不能转换为整数值的。如果这些不符合要求的值匹配成功，会产生两个问题。

（1）应用程序中若存在两个规则相似的 URL 路由，如/abc/{count}和/abc/{name}，尽管它们的结构相同，但含义是不同的，count 参数期待的值应为整数值，而 name 参数所期待的值是字符串。如果路由参数不进一步约束，那么这两条路由会发生冲突，应用程序无法精确地作出决定。这是因为路由参数默认是按字符串处理的，假设客户端发送的请求是/abc/112233，112233 只被看作字符串，此时/abc/{count}能成功匹配，/abc/{name}也可以匹配，所以会发生错误。若是限制 count 参数必须是整数值，其匹配精度就提高了，那么这个冲突问题就能解决。

（2）假设路由/store/{amount}中，amount 参数要求必须是双精度数值，但路由参数默认是作为字符串处理的，如果客户端发出的请求是/store/xyxy，xyxy 无法转换为双精度数值，可能导致应用程序发生错误。应用程序每次接收到 amount 参数的值后都要做一番验证工作，开发人员会增加许多不必要的工作量。

路由约束可以给参数附加限制条件，使其能够更精准地匹配。

15.2 IRouteConstraint 接口

路由约束要求实现 IRouteConstraint 接口，该接口只声明了 Match()方法。

```
public interface IRouteConstraint : IParameterPolicy
{
    bool Match(
        HttpContext? httpContext,
        IRouter? route,
        string routeKey,
        RouteValueDictionary values,
        RouteDirection routeDirection);
}
```

如果被约束的参数值符合要求，Match()方法就返回 true，否则返回 false。其中，routeKey 表示路由参数的名称，values 表示关联的路由字典。在实现 Match()方法时，可以先从 values 中查找一下是否包含 routeKey 的值。如果不存在 routeKey 对应的值，说明客户端在发出请求时并没有提供路由参数的值，一般可直接返回 false。如果 routeKey 的值存在，再做进一步处理，如是否能转换为期待的类型、是否满足特定格式要求等。

15.3　内置的路由约束

路由约束是通过 RouteOptions 选项类（位于 Microsoft.AspNetCore.Routing 命名空间）来配置的。该选项类公开名为 ConstraintMap 的属性，类型为 IDictionary<string, Type>。它是一个字典集合，Key 为字符串类型，Value 为 Type 类型，即路由约束类型的 Type。

ASP.NET Core 应用程序默认会添加常用的内置路由约束，详情可参考以下源代码。

```
private static IDictionary<string, Type> GetDefaultConstraintMap()
{
    var defaults = new Dictionary<string, Type>(StringComparer.
    OrdinalIgnoreCase);
    // Type-specific constraints
    AddConstraint<IntRouteConstraint>(defaults, "int");
    AddConstraint<BoolRouteConstraint>(defaults, "bool");
    AddConstraint<DateTimeRouteConstraint>(defaults, "datetime");
    AddConstraint<DecimalRouteConstraint>(defaults, "decimal");
    AddConstraint<DoubleRouteConstraint>(defaults, "double");
    AddConstraint<FloatRouteConstraint>(defaults, "float");
    AddConstraint<GuidRouteConstraint>(defaults, "guid");
    AddConstraint<LongRouteConstraint>(defaults, "long");
    // Length constraints
    AddConstraint<MinLengthRouteConstraint>(defaults, "minlength");
    AddConstraint<MaxLengthRouteConstraint>(defaults, "maxlength");
    AddConstraint<LengthRouteConstraint>(defaults, "length");
    // Min/Max value constraints
    AddConstraint<MinRouteConstraint>(defaults, "min");
    AddConstraint<MaxRouteConstraint>(defaults, "max");
    AddConstraint<RangeRouteConstraint>(defaults, "range");
    // Regex-based constraints
    AddConstraint<AlphaRouteConstraint>(defaults, "alpha");
    AddConstraint<RegexInlineRouteConstraint>(defaults, "regex");
```

```
        AddConstraint<RequiredRouteConstraint>(defaults, "required");
        // Files
        AddConstraint<FileNameRouteConstraint>(defaults, "file");
        AddConstraint<NonFileNameRouteConstraint>(defaults, "nonfile");
        return defaults;
    }
```

整理内置的路由约束，如表 15-1 所示。

表 15-1　内置的路由约束

名　称	类　型	说　明
int	IntRouteConstraint	要求路由参数的值为整数值
bool	BoolRouteConstraint	要求布尔类型的值，即true、false
datetime	DateTimeRouteConstraint	日期类型，如2018-10-22
decimal	DecimalRouteConstraint	十进制定点数，如1.25
double	DoubleRouteConstraint	双精度浮点数值，如7.0355
float	FloatRouteConstraint	单精度浮点数值
guid	GuidRouteConstraint	路由参数值为GUID的字符串表示形式，如936DA01F-9ABD-4d9d-80C7-02AF85C822A8
long	LongRouteConstraint	长整型数值，即long类型
minlength	MinLengthRouteConstraint	指定字符串的最小长度
maxlength	MaxLengthRouteConstraint	指定字符串的最大长度
length	LengthRouteConstraint	限制字符串的长度范围。例如，length(3,8)表示字符串的长度至少为3个字符，且不能超过8个字符
min	MinRouteConstraint	要求是整数值（int或long类型），并且指定最小值，如min(7)，即提供的值不能小于7
max	MaxRouteConstraint	整数值的最大值，如max(10)表示提供的值不能超过10
range	RangeRouteConstraint	整数值的范围，如range(2,5)表示整数值不能小于2，且不能超过5
alpha	AlphaRouteConstraint	提供的字符串必须由字母组成，并且区分大小写。Tom与TOM被认为是两个不同的值
regex	RegexInlineRouteConstraint	通过正则表达式匹配路由参数
required	RequiredRouteConstraint	表示路由参数是必需的，在请求的URL中必须包含其值
file	FileNameRouteConstraint	提供的字符串是文件路径
nonfile	NonFileNameRouteConstraint	字符串不是文件路径

另外，当应用程序启用 MVC 相关功能后，ConstraintMap 中还会添加一个 KnownRouteValueConstraint 约束（位于 Microsoft.AspNetCore.Mvc.Routing 命名空间），键名为 exists。该约束一般用于路由规则中 controller、action 等参数上，如{controller:exists}/{action:exists}，表示 URL 中必须提供 controller 和 action 参数的值。

在路由规则模板中使用约束时，只需要引用 ConstraintMap 中的键名（Key）即可，并且以英文的冒号（:）为前缀。例如，要约束路由参数 id 为 int 类型，在路由规则中的格式如下。

```
{controller}/{action}/{id:int}
```

要使用多个约束，只要每个约束都以冒号开头即可。

```
news/{action}/{title:string:maxlength(15)}
```

上述规则要求 title 参数为字符串类型，并且最大长度为 15 个字符。

15.3.1　示例：双精度数值约束

本示例将演示双精度数值（double）约束的使用。应用了此约束的路由参数只能接收双精度数值。

```
app.MapGet("/test/{number:double}", (double Number) =>
{
    return $"{nameof(Number)} = {Number}";
});
```

运行示例程序，请求 URL 为 http://localhost:5102/test/105.00721，Number 参数的值为 105.00721，如图 15-1 所示。

也可以传递负值，如图 15-2 所示。

当传递字符串 0.25ks 时，服务器返回 404 错误，如图 15-3 所示。这是因为 0.25ks 无法转换为 double 类型的值，路由规则匹配失败。

图 15-1　双精度正值

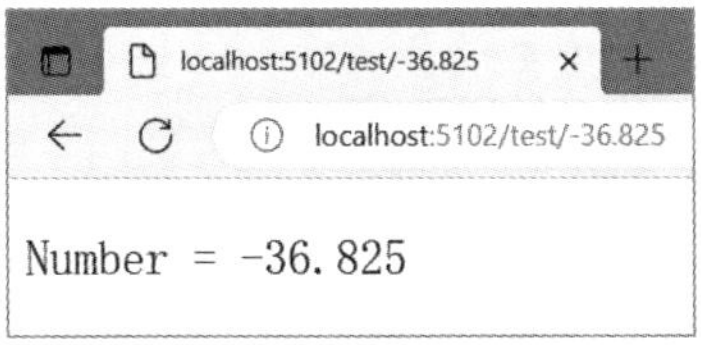

图 15-2　双精度负值

图 15-3　路由参数无效

15.3.2　示例：限制字符串长度

本示例是一个 Razor Pages 项目，在页面上通过@page 指令定义了一条路由规则。其中包含名为 title 的参数，并使用 length 约束限制字符串的长度至少 3 个字符，最多不超过 7 个字符。

```
@page "/mypage/{title:length(3,7)}"
```

```
<html>
<head>
<title>@RouteData.Values["title"]</title>
</head>
<body>
<h5>@RouteData.Values["title"]</h5>
</body>
</html>
```

如果匹配成功就会执行该页面，并且 title 参数的内容会出现在 HTML 的文档标题和<h5>元素中。

访问 http://localhost:5230/mypage/一片一片又一片，title 参数对应的字符串的长度正好是 7，路由匹配成功，结果如图 15-4 所示。

图 15-4　title 参数正好包含 7 个字符

当访问 http://localhost:5230/mypage/goooooood 时，服务器返回 404 错误。这是因为字符串 goooooood 的长度超过了 7，路由匹配失败。

15.3.3　示例：特定格式的订单号

本示例将演示一种有格式要求的订单号：

（1）只能以字母 F、M、Q 开头；

（2）首字母后紧接一个-字符；

（3）-字符后紧接着 5 ~ 8 位数字，如 123456、77887752。

综上所述，以下订单号的格式是可以匹配路由规则的。

```
Q-20585
Q-1008004
F-6661112
```

以下订单号是不能匹配路由规则的。

```
X-4912735                      // 字母 X 不满足要求
Q-8812508193                   // -字符后的数字字符已超过 8 个
```

下面的代码将用到正则表达式约束 oid 参数的内容。

```
app.MapGet("/orders/{oid:regex(^(F|M|Q)-\\d{{5,8}}$)}", (string oid) =>
{
    return $"你访问的订单号为：{oid}";
});
```

其中，^(F|M|Q)是组合选择，即以 F、M、Q 任意一个字母开头均可（不区分大小写）；\\d 即\d（\\是需要字符转义），表示一个数字字符（0 ~ 9）；后面的{{5,8}}也是经过转义的字符串，即{5,8}，表示前面的字符（指数字字符）可出现 5 ~ 8 次。

在路由规则模板中书写正则表达式时，遇到“{”“}”“[”“]”“\”等字符都需要转义，方法都是将被转义字符重复两次，如“{{”代表“{”，“]]”代表“]”。

运行示例程序，访问 http://localhost:5250/orders/M-4386290，结果如图 15-5 所示。

15.3.4　示例：限制整数值的范围

请看下面的代码。

```
app.MapGet("/person/{name:alpha}/{age:range(15,80)}", (string name, int age) =>
{
    return string.Format("{0}, {1}岁", name, age);
});
```

上述路由模板包含两个参数。name 参数使用了 alpha 约束，使其只能接受由字母组成的字符串（区分大小写）；age 参数使用了 range 约束，将其值限定为 15 ~ 80 的整数值。

当请求地址为 http://localhost:6008/person/Jim/39 时，name 参数填充的值为 Jim，age 参数的值为 39，结果如图 15-6 所示。

图 15-5　有特殊格式的订单号

图 15-6　name 和 age 参数的值均有效

如果请求地址为 http://localhost:6008/person/Jim/94，服务器将返回 404 错误，整数 94 已超过 range 约束所限制的值。

15.4　自定义路由约束

除了使用内置的路由约束，开发人员也可以编写自定义的约束，只须实现 IRouteConstraint 接口，并将其添加到 RouteOptions 选项类的 ConstraintMap 属性中即可。

下面给出一个示例。DemoRouteConstraint 类是自定义的路由约束，它会在路由参数的值里面查找 password、passwd、pwd 关键词。如果包含这些关键词就会匹配失败，否则匹配成功。

```
public class DemoRouteConstraint : IRouteConstraint
{
    // 关键词
    readonly string[] KeyWords = { "password", "pwd", "passwd"};

    public bool Match(HttpContext? httpContext, IRouter? route, string routeKey, RouteValueDictionary values, RouteDirection routeDirection)
    {
        if(values.TryGetValue(routeKey, out var val) && val != null)
        {
            // 先把值转换为字符串
            string str = Convert.ToString(val, CultureInfo.InvariantCulture)!;
```

```
            // 若不包含关键词，则匹配成功
            if(!KeyWords.Any(x => str.Contains(x)))
            {
                return true;
            }
        }

        return false;
    }
}
```

然后，将此自定义约束添加到 ConstraintMap 属性中。

```
var builder = WebApplication.CreateBuilder(args);
builder.Services.Configure<RouteOptions>(opt =>
{
    opt.ConstraintMap.Add("demo", typeof(DemoRouteConstraint));
});
```

DemoRouteConstraint 约束对应的键名为 demo，路由模板中将使用此名称。

下面的代码验证 DemoRouteConstraint 约束是否起作用。

```
app.MapGet("/test/{p:demo}", (string p) => $"p = {p}");
```

当请求的 URL 为/test/mypasswdabc 时，服务器将返回 404 状态码，因为 mypasswdabc 中包含了关键词 passwd。同理，请求/test/xxeepwd 也会返回 404 状态码。而请求/test/car 则能顺利匹配，因为 car 中不包含特定关键词。

第 16 章 SignalR

本章要点

- ❑ WebSocket
- ❑ 连接与调用 SignalR 中心
- ❑ 强类型 Hub

16.1 WebSocket

WebSocket 是构建于传输控制协议（Transmission Control Protocol，TCP）之上的全双工协议，让服务器与客户端之间能够方便地交换数据（通信）。Web 浏览器与服务器之间只需要通过一次握手建立持久连接，即可进行收发数据。

由于 WebSocket 要先建立连接再进行通信，就不用像 HTTP 那样每次通信都要携带状态数据（可能包含很长的 HTTP 头部），节省网络带宽资源。

16.1.1 示例：用 JavaScript 实现客户端

本示例将使用 JavaScript 编写 WebSocket 的客户端代码。WebSocket 的默认协议为 ws，即连接服务器时需要使用这样的 URL：ws://localhost/something。

ASP.NET Core 应用程序中使用 WebSocket 的方法是在 HTTP 管道上调用 UseWebSockets()方法。

```
app.UseWebSockets();
```

接着使用一个自定义的中间件处理 WebSocket 客户端的请求，并完成消息的接收/发送。

```
app.Use(async (context, next) =>
{
    if (context.Request.Path == "/test"&& context.WebSockets.
    IsWebSocketRequest)
    {
        // 接受客户端连接请求
```

```
            var socket = await context.WebSockets.AcceptWebSocketAsync();
            // 发送消息
            byte[] data = Encoding.UTF8.GetBytes("你好，我是服务器");
            await socket.SendAsync(
                // 要发送的数据
                buffer: data,
                // 消息的类型，如文本数据、二进制数据
                messageType: WebSocketMessageType.Text,
                // true 表示 buffer 中的数据是最后发送的部分
                // 由于数据较小，只需一次就能发送完成，因此 endOfMessage 参数为 true
                endOfMessage: true,
                CancellationToken.None
            );
            // 关闭 Socket
            await socket.CloseAsync(WebSocketCloseStatus.NormalClosure, "通信
            结束", CancellationToken.None);
        }
        else
        {
            await next();
        }
    });
```

客户端请求的任意 URL 都能激活该中间件，但本示例要求 URL 为/test。调用 AcceptWebSocketAsync()方法接受客户端的连接请求并返回一个 WebSocket 实例。该实例将用于与客户端进行通信。调用 ReceiveAsync()方法可接收来自客户端的消息；调用 SendAsync()方法可向客户端发送消息。为了使代码更易于理解，本示例在接受客户端连接后，只向客户端发送一条文本消息，并未接收任何来自客户端的消息。

发送完消息后，如果不需要继续通信，可调用 CloseAsync()方法关闭连接。该方法的声明如下。

```
Task WebSocket.CloseAsync(WebSocketCloseStatus closeStatus, string?
statusDescription, CancellationToken cancellationToken);
```

其中，closeStatus 参数指定关闭连接的状态，通常应选用正常关闭，即 NormalClosure，代码为 1000；statusDescription 参数为 closeStatus 参数的值提供一个简短的文字描述（关闭连接的理由，本示例指定为“通信结束”），可方便客户端排查错误；cancellationToken 参数关联一个 CancellationToken 结构体实例，可用于取消异步操作，若不需要，直接赋值 CancellationToken.None 即可。

JavaScript 客户端代码可以使用 WebSocket 对象与服务器通信。在 HTML 文档加载时会自动执行。

```
@page

<!DOCTYPE html>
```

```
<html>
<head>
<title>Demo</title>
<meta charset="utf-8"/>
</head>
<body>
<div>
<span id="display"></span>
<ul id="states" style="color:red"></ul>
</div>
<script>
          var txt = document.getElementById('display');
          var states = document.getElementById('states');
          // 建立连接
          var socket = new WebSocket('ws://localhost:5084/test');
          // 侦听 open 事件
          socket.addEventListener('open', ev => {
             var item = document.createElement('li');
             item.textContent = '连接成功';
             states.appendChild(item);
          });
          // 侦听 close 事件
          socket.addEventListener('close', ev => {
             var item = document.createElement('li');
             var msg = `连接关闭（代码：${ev.code}，理由：${ev.reason}）`;
             item.textContent = msg;
             states.appendChild(item);
          });
          // 侦听 message 事件，收到消息时触发
          socket.addEventListener('message', ev => {
             txt.textContent = `来自服务器的消息：${ev.data}`;
          });
</script>
</body>
</html>
```

在上述 HTML 文档中，<span>元素用于显示来自服务器的消息内容；在<ul>元素中可以添加子项（<li>元素），以列表形式显示状态信息。

WebSocket 对象在调用构造函数时可直接传递服务器的 URL，随后会自动尝试与服务器建立连接。指定 URL 时要注意，地址是以 ws://开头的，而不是 http://。

WebSocket 对象在使用时可能需要处理以下事件。

（1）open（对应 onopen 属性）：表示与服务器的连接已建立，可以收发消息了。

（2）close（对应 onclose 属性）：当与服务器的连接关闭后发生。

（3）error（对应 onerror 属性）：出现错误时发生。

（4）message（对应 onmessage 属性）：接收到新消息时发生。可以从事件参数的 data 属性

获取消息内容。

本示例只用到了 3 个事件。

（1）open：连接建立后向<ul>元素添加一条状态信息。

（2）close：连接关闭后向<ul>元素添加一条状态信息。

（3）message：接收到消息时在<span>元素内显示消息。

运行示例程序，结果如图 16-1 所示。

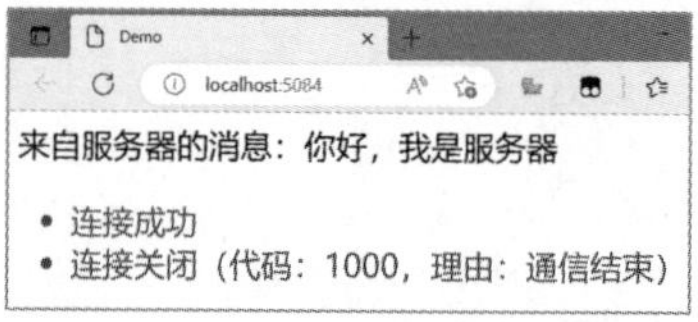

图 16-1　在 JavaScript 中使用 WebSocket

16.1.2　示例：用.NET 控制台实现 WebSocket 客户端

本示例将演示通过.NET 代码连接 WebSocket 服务器并向服务器发送消息。

服务器部分是 ASP.NET Core 应用，通过在 HTTP 管道中插入自定义中间件完成与客户端通信的逻辑。来自客户端的消息将写入日志中。

```
var builder = WebApplication.CreateBuilder(args);
var app = builder.Build();

app.UseWebSockets();

app.Run(async context =>
{
    ILoggerFactory logfac = context.RequestServices.GetRequiredService
    <ILoggerFactory>();
    ILogger logger = logfac.CreateLogger("Web Socket Demo");
    // 如果是 WebSocket 请求
    if(context.WebSockets.IsWebSocketRequest)
    {
        // 接受客户端连接请求
        WebSocket socket = await context.WebSockets.AcceptWebSocketAsync();
        // 循环接收客户端消息
        WebSocketReceiveResult result;
        byte[] buffer = new byte[4 * 1024];
        StringBuilder strbuilder = new();
        result = await socket.ReceiveAsync(buffer, CancellationToken.None);
        while(result.CloseStatus.HasValue == false)
        {
            // 读取数据
            string _t = Encoding.UTF8.GetString(buffer, 0, result.Count);
            strbuilder.Append(_t);
            // 如果一条完整的消息已接收完毕，就输出日志
            if(result.EndOfMessage)
            {
                logger.LogInformation("客户端：" + strbuilder.ToString());
                strbuilder.Clear();
```

```
            }
            // 接收下一条消息
            result = await socket.ReceiveAsync(buffer, CancellationToken.
            None);
        }
        // 关闭连接
        await socket.CloseAsync(WebSocketCloseStatus.NormalClosure, "通信
        结束", CancellationToken.None);
    }
    else
    {
        context.Response.StatusCode = StatusCodes.Status400BadRequest;
    }
});

app.Run("http://localhost:4900");
```

当客户端应用程序忽然退出或连接丢失，服务器会引发异常，同时日志会记录下异常。在本示例中，为了避免输出的日志给查看 WebSocket 的通信记录带来不便，需要禁用 Microsoft.AspNetCore.Server.Kestrel 和 Microsoft.AspNetCore.Diagnostics.DeveloperExceptionPageMiddleware 条目的日志输出。方法是打开 appsettings.json 文件，将上述两个条目的日志类型改为 None。

```
{
"Logging": {
"LogLevel": {
"Default": "Information",
"Microsoft.AspNetCore": "Warning",
"Microsoft.AspNetCore.Server.Kestrel": "None",
"Microsoft.AspNetCore.Diagnostics.DeveloperExceptionPageMiddleware": "None"
    }
  },
…
}
```

客户端部分是.NET 控制台应用。建立 WebSocket 连接要用到 ClientWebSocket 类。实例化之后调用 ConnectAsync()方法发出连接请求，等待连接建立后就可以正常收发消息了。要关闭连接，请调用 CloseAsync()方法。

```
// 服务器 URL
Uri server = new("ws://localhost:4900");
// 实例化 WebSocket 客户端
var socket = new ClientWebSocket();
// 建立连接
await socket.ConnectAsync(server, CancellationToken.None);
// 循环发送消息
Console.Write(">");
string? input = Console.ReadLine();
```

```
    while(!string.IsNullOrEmpty(input))
    {
        byte[] data = Encoding.UTF8.GetBytes(input);
        await socket.SendAsync(data, WebSocketMessageType.Text, true,
        CancellationToken.None);
        // 继续读取键盘输入
        Console.Write(">");
        input = Console.ReadLine();
    }

    // 关闭 socket
    await socket.CloseAsync(WebSocketCloseStatus.NormalClosure, "退出任务",
    CancellationToken.None);
```

上述程序是循环发送消息的。消息内容来自键盘输入（Console.ReadLine()方法负责读取），每轮发送完毕后都会等待输入，如果输入的是空白字符串（未输入字符，直接按 Enter 键），就退出循环并关闭 WebSocket 对象。

在测试本示例时，先进入 Server 目录，执行 dotnet run 命令运行服务器应用程序。然后再进入 Client 目录，同样执行 dotnet run 命令运行客户端应用程序。在客户端控制台窗口中输入测试消息，如图 16-2 所示，按 Enter 键，消息发送。在服务器端的控制台窗口中能看到来自客户端的消息，如图 16-3 所示。

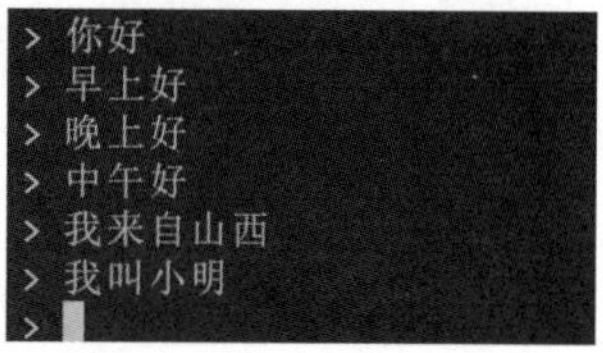

图 16-2　在客户端输入要发送的消息

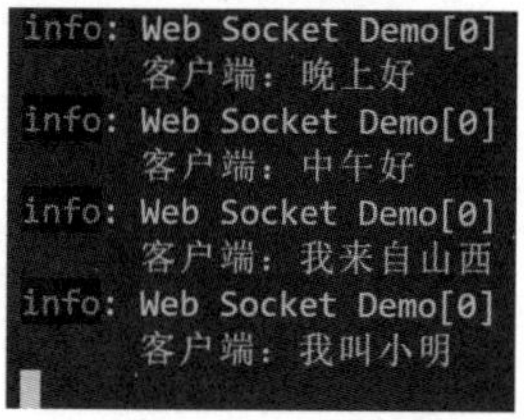

图 16-3　服务器记录了来自客户端的消息

16.1.3　子协议

WebSocket 允许使用子协议——字符串类型，可自定义名称，默认为空字符串。服务器在调用 AcceptWebSocketAsync()方法接受客户端连接时可以指定子协议名称；客户端在请求连接时（握手）也要附带相同的子协议，否则无法建立连接。子协议可以为服务器实现多种交互方式，如 A 协议实现文本聊天，B 协议实现图像传输。也可以模仿 HTTP 消息中的 Content-Type 头部，不同的协议传输不同格式的数据。

服务器在调用 AcceptWebSocketAsync()方法时可以指定需要的子协议，客户端在初始化时需要附带相同的子协议。

下面的示例定义了两个子协议：文本通信（text）和语音通信（audio）。在 HTML 文档中通过单选按钮选择子协议类型，单击“连接”按钮后使用被选中的子协议向服务器发起连接。连接成功后接收来自服务器的消息。

```
<!DOCTYPE html>

<html lang="zh-cn">

<head>
<meta charset="utf-8" />
<title>Demo</title>
<link href="/styles/main.css" rel="stylesheet" />
</head>
<body>
<div>
        请选择子协议：
</div>
<div>
<form id="myform">
<div>
<input name="subprot" id="op1" value="text" type="radio" checked />
<label for="op1">文本</label>
</div>
<div>
<input name="subprot" id="op2" value="audio" type="radio" />
<label for="op2">语音</label>
</div>
<div>
<input type="submit" value="连接"/>
</div>
</form>
</div>
<div>
<ul id="logs"></ul>
</div>

<script>
        // 获取<form>元素
        var form = document.getElementById("myform");
        // 获取<ul>元素
        var logs = document.getElementById("logs");
        // 添加事件处理
        form.addEventListener('submit', (e) => {
            var data = new FormData(form);
            // 获取单选项的值
            var selval = data.get("subprot");
            // 建立连接
            var socket = new WebSocket("ws:// ${window.location.host}/
            skapp", selval);
            // 连接成功后触发
            socket.addEventListener('open', oe => {
                var li = document.createElement('li');
```

```
                li.textContent = "已建立连接";
                li.style.color = "green";
                logs.appendChild(li);
            });
            // 连接关闭后触发
            socket.addEventListener('close', ce => {
                var li = document.createElement('li');
                li.style.color = "red";
                li.textContent = "已关闭连接" + `(${ce.code} - ${ce.reason})`;
                logs.appendChild(li);
            });
            // 接收到消息时触发
            socket.addEventListener('message', me => {
                var msg = me.data;
                var li = document.createElement('li');
                li.style.color = "blue";
                li.textContent = `来自服务器的消息：${msg}`;
                logs.appendChild(li);
            });
            e.preventDefault();
        });
</script>
</body>
</html>
```

<input type="radio">表示一个单选按钮，name 属性相同的单选按钮视为同一分组。同一分组内的单选按钮之间是互斥关系，只能选择其中一个。

在 ASP.NET Core 应用中，通过 Map()方法映射一个自定义终结点，然后在此终结点中接受客户端连接。

```
app.UseWebSockets();

app.Map("/skapp", async context =>
{
    if(context.WebSockets.IsWebSocketRequest)
    {
        var prots = context.WebSockets.WebSocketRequestedProtocols;

        // 分别接受不同子协议的连接
        foreach(string p in prots)
        {
            WebSocket socket =
            await context.WebSockets.AcceptWebSocketAsync(p);
            // 向客户端发送一条消息
            byte[] msg = Encoding.UTF8.GetBytes($"已接受连接，子协议：{p}");
            await socket.SendAsync(msg, WebSocketMessageType.Text, true,
            CancellationToken.None);
            // 关闭连接
```

```
            await socket.CloseAsync(WebSocketCloseStatus.NormalClosure,
            "通信结束", CancellationToken.None);
        }
    }
    else
    {
        context.Response.StatusCode = StatusCodes.Status400BadRequest;
    }
});
```

context.WebSockets.WebSocketRequestedProtocols 属性返回一个字符串列表。该列表中包含当前尝试连接的子协议列表。程序代码可以根据这个子协议列表分析哪些协议允许连接，哪些协议不允许连接。本示例将允许所有子协议进行连接，因此代码中使用了 foreach 循环，枚举出所有正尝试连接的协议，随后调用 AcceptWebSocketAsync()方法接受连接（通过方法参数指定需要的子协议名称）。

运行示例程序，在网页上选择一种子协议，然后单击“连接”按钮，页面上就会显示连接信息，以及来自服务器的消息，如图 16-4 所示。

图 16-4　以特定的子协议连接服务器

16.2　SignalR 基础

SignalR 是一个开源库，基于现有技术进行封装，主要用途是解决 Web 技术中实时通信的数据更新问题，尤其是实现服务器频繁向客户端推送内容的功能，要比客户端主动轮询更节省资源和数据流量。

SignalR 适合于实时聊天、游戏、数据监测等应用场景。

ASP.NET Core 提供的 SignalR 具备以下功能。

（1）自动管理连接，包括连接丢失后重新连接。

（2）可以向所有连接的某个客户端、所有客户端或已分组的客户端发送消息。

（3）通过 SignalR 中心（Hub）协议，客户端可以调用服务器上的方法，服务器也可以调用客户端的方法。

SignalR 支持以下传输方式。

（1）WebSocket；

（2）Server-Sent Events（服务器发送事件）；

（3）Long Polling（长轮询）。

SignalR 会自动选择最优传输方案。若不支持 WebSocket，将使用 Server-Sent Events；若不支持 Server-Sent Events，就退而选用 Long Polling。这些选择都是自动完成的，开发人员可以不关注传输方面的事情，而专注于完成接收/发送消息的代码逻辑。

16.2.1 SignalR 中心

中心（Hub）是 SignalR 的核心部分。客户端与服务器之间可以相互调用——服务器可以调用客户端定义的方法，反之亦然。服务器与多个客户端之间的连接将由 Hub 自动调度，开发人员不需要自行编写代码进行维护。

在服务器应用中，从 Hub 类（位于 Microsoft.AspNetCore.SignalR 命名空间）派生，即可实现自定义的 SignalR 中心。随后可以添加新的方法成员，这些方法可以在 SignalR 客户端调用。

完成自定义的 Hub 后，需要在服务容器中注册与 SignalR 相关的服务，一般调用 AddSignalR()扩展方法即可。在 HTTP 管道上还要调用 MapHub()方法为自定义的 Hub 添加终结点映射。

16.2.2 示例：简易计算器

本示例将使用 SignalR 实现一个简单的乘法计算器。客户端用 HTML + JavaScript 的方式实现。示例需要用到 signalr.js 脚本。为了使操作更简单，不需要安装其他程序包管理工具和客户端库，直接从 https://cdn.bootcdn.net/ajax/libs/microsoft-signalr/7.0.3/signalr.js 下载 signalr.js 文件，然后放到项目中即可。

（1）服务器代码先要实现一个自定义的 SignalR 中心，此处命名为 CalculateHub。

```
public class CalculateHub : Hub
{
    // 计算两个数值的乘积
    public ValueTask<double> Calculate(double num1, double num2)
    {
        double res = num1 * num2;
        return ValueTask.FromResult(res);
    }
}
```

类中包含一个 Calculate()方法，接受两个双精度数值，最后返回它们的乘积。

（2）在服务器容中启用 SignalR 功能。

```
var builder = WebApplication.CreateBuilder(args);
builder.Services.AddSignalR();
var app = builder.Build();
```

（3）在 HTTP 管道上为 CalculateHub 映射终结点。

```
app.MapHub<CalculateHub>("/testhub");
```

映射终结点时指定了一个 URL，客户端连接时需要指定此 URL。

（4）客户端是一个 HTML 文档。

```
<!DOCTYPE html>
<html>
```

```
<head>
<meta charset="utf-8" />
<title>SignalR Demo</title>
</head>

<body>
<table>
<tr>
<td><label for="num1">数值 1: </label></td>
<td><input id="num1" type="number" /></td>
</tr>
<tr>
<td><label for="num2">数值 2: </label></td>
<td><input id="num2" type="number" /></td>
</tr>
</table>
<button id="btn">计算</button>
<p id="result"></p>

<script src="scripts/signalr.js"></script>
<script>
      var btn = document.getElementById("btn");
      var result = document.getElementById("result");
      var txtnum1 = document.getElementById("num1");
      var txtnum2 = document.getElementById("num2");

      const conn = new signalR.HubConnectionBuilder()
         .withUrl("/testhub")
         .withAutomaticReconnect()
         .build();

      // 尝试连接
      try {
         conn.start();
      }
      catch (err) {
         console.log(err);
      }

      btn.addEventListener("click", async event => {
         if( conn.state !== signalR.HubConnectionState.Connected)
         {
            return;
         }
         // 转换为浮点数值
         let f1 = parseFloat(txtnum1.value);
         let f2 = parseFloat(txtnum2.value);
```

```
            // 向服务器发送消息
            var data = await conn.invoke("Calculate", f1, f2);
            // 显示结果
            result.textContent = `计算结果：${data}`;
        });
    </script>
    </body>

    </html>
```

调用 JavaScript 版本的 API 时，要加上 signalR 前缀。例如，要访问 HubConnectionBuilder 类，其完整名称为 signalR.HubConnectionBuilder。

HubConnectionBuilder 类用于创建 HubConnection 实例。withUrl()方法用于指定服务器上 SignalR 中心的地址，withAutomaticReconnect()方法设置当连接断开后将自动重新连接。最后 build()方法产生 HubConnection 实例。所有通信任务都在 HubConnection 对象上完成。

调用 start()方法发起连接。invoke()方法用来调用服务器上自定义 Hub 的方法。在本示例中，待调用的服务器方法是 Calculate()。invoke()方法的第 1 个参数用字符串指定服务器上被调用方法的名称，从第 2 个参数起是数量可变的参数，即传递给服务器方法的参数。在本示例中，Calculate()方法需要两个双精度浮点数值。因此，上述代码中，将 f1、f2 依次传递给 Calculate()方法的 num1、num2 参数。

注意<input>元素的值返回的是字符串类型，为了避免调用 Calculate()方法时出错，需要调用 parseFloat()方法转换为 number 类型。

（5）在项目中新建 wwwroot 目录。

（6）在 wwwroot 目录下新建 scripts 子目录，将 signalr.js 文件放到 scripts 目录下。

（7）访问 HTML 文件和 signalr.js 文件需要静态文件功能，因此要在 HTTP 管道上调用 UseFileServer()方法。

```
app.UseFileServer();
```

（8）运行示例程序，在页面上输入两个数值，然后单击“计算”按钮，就会呈现相应的结果，如图 16-5 所示。

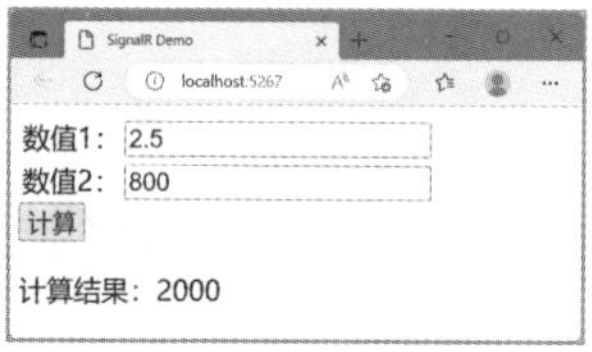

图 16-5 显示计算结果

16.2.3 示例：使用面向.NET 的 SignalR 库

在.NET 应用中（如控制台、Windows Forms、WPF 或 MAUI 应用程序），有专用的 SignalR 封装库——Microsoft.AspNetCore.SignalR.Client。ASP.NET Core 应用自身也可以使用该库，如访问其他服务器上的 SignalR 功能，或者在 Blazor 应用中使用。

本示例将使用控制台应用程序调用服务器上的 SignalR 方法。调用时传递一个字符串实例，随后服务器把该字符串进行重复拼接后返回。例如，传递的字符串为 Any，得到的结果就是 AnyAnyAnyAnyAny。

首先实现 ASP.NET Core 应用（服务器）。

```
var builder = WebApplication.CreateBuilder(args);
builder.Services.AddSignalR();
var app = builder.Build();

app.MapHub<TestHub>("/test");

app.Run();
```

下面是自定义 Hub 的代码。

```
public class TestHub : Hub
{
    public Task<string> RepeatString(string input)
    {
        string newstr = string.Empty;
        for(int n = 0; n < 5; n++)
        {
            newstr += input;
        }
        return Task.FromResult(newstr);
    }
}
```

然后实现控制台应用（客户端）。

客户端需要引用 Microsoft.AspNetCore.SignalR.Client 库。可以执行 dotnet add package 命令来添加。

```
dotnet add package Microsoft.AspNetCore.SignalR.Client
```

或者直接修改项目文件（*.csproj），添加<PackageReference>元素。

```
<Project Sdk="Microsoft.NET.Sdk">
...

<ItemGroup>
<PackageReference Include="Microsoft.AspNetCore.SignalR.Client" Version=
"7.0.3" />
</ItemGroup>

</Project>
```

客户端代码如下。

```
// 服务器 URL
Uri serverAddr = new("http://localhost:5112/test");

// 建立连接
HubConnection conn = new HubConnectionBuilder()
                    .WithUrl(serverAddr)
                    .Build();
```

```
    try
    {
        await conn.StartAsync();
    }
    catch(Exception err)
    {
        Console.WriteLine($"错误：{err.Message}");
    }

    // 如果连接未建立，则退出
    if(conn.State != HubConnectionState.Connected)
    {
        return;
    }

    // 调用服务器方法
    Console.Write("请输入任意字符串：");
    string? line = Console.ReadLine();
    if (!string.IsNullOrEmpty(line))
    {
        var result = await conn.InvokeAsync<string>("RepeatString", line);
        Console.WriteLine($"服务器返回：{result}");
    }

    Console.ReadKey();                          // 按任意键继续
    // 关闭连接
    await conn.StopAsync();
    await conn.DisposeAsync();
```

使用 HubConnectionBuilder 实例构建 HubConnection 对象。随后调用 HubConnection 对象的 StartAsync()方法建立连接。

连接建立后，调用 Console.ReadLine()方法读取键盘输入的字符串，并调用 InvokeAsync()方法执行服务器上的 RepeatString()方法。本示例使用 InvokeAsync()方法的以下重载。

```
Task<TResult> InvokeAsync<TResult>(this HubConnection hubConnection,
string methodName, object? arg1, CancellationToken cancellationToken =
default(CancellationToken));
```

这是 HubConnection 类的扩展方法。methodName 参数用字符串指定要调用的服务器方法名称（此处为 RepeatString）；arg1 参数用于传递参数（本示例中为字符串实例）。RepeatString()方法的返回值发回到客户端后，由 InvokeAsync()方法返回。TResult 类型参数正是用来设置返回值的类型（本示例中是 string）。

运行示例程序，结果如图 16-6 所示。

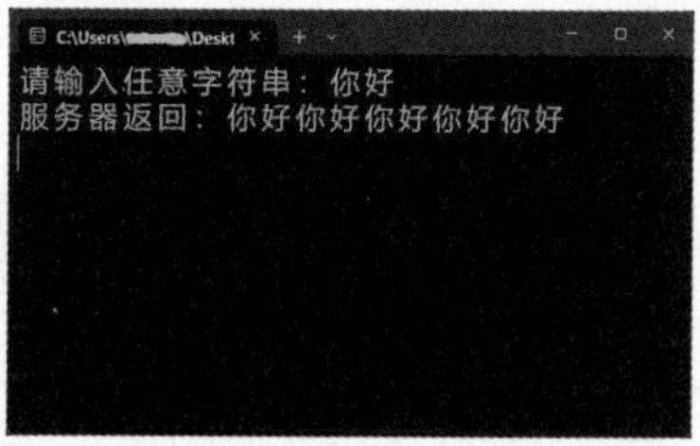

图 16-6　使用 SignalR 的.NET 客户端

16.3　调用客户端

一般使用 SendAsync()方法在服务器上调用客户端。例如：

```
public class MyHub : Hub
{
    public async Task SayHello(string name)
    {
        await Clients.Caller.SendAsync("callBack", $"Hello, {name}");
    }
}
```

或

```
public class MyHub : Hub
{
    public async Task SayHello(string name)
    {
        await Clients.All.SendAsync("callBack", $"Hello, {name}");
    }
}
```

Hub.Clients 属性提供了 3 种调用客户端的方案。

（1）All：调用所有已连接的客户端。

（2）Caller：调用当前客户端，即正在调用服务器的客户端。

（3）Others：除正在调用服务器以外的所有客户端。可以认为是在 All 的客户端列表中排除 Caller。

此外，Hub.Clients 还包含一些增强性的方法成员，如下。

（1）AllExcept()：该方法需要指定一个连接 ID，调用除此 ID 以外的所有客户端。连接 ID 可以唯一标识一个客户端。

（2）Client()：调用指定的客户端（单个）。

（3）Clients()：调用指定的客户端（多个）。

16.3.1　示例：聊天室

本示例将实现一个简单的多人聊天室。每位用户打开主页面后，需要填写自己的昵称，然后输入要发送的消息，单击“发送”按钮提交消息。所有在线的用户都会收到聊天信息。

（1）在服务器端实现自定义的 Hub。

```
public class ChatHub : Hub
{
    public async Task Speak(string nickname, string message)
    {
        // 当前时间
        var curTime = DateTime.Now;
        // 把收到的消息转发给所有客户端
```

```
        await Clients.All.SendAsync("recv", nickname, message, curTime.
        ToShortTimeString());
    }
}
```

Speak()方法由客户端调用，将聊天信息传输到服务器。聊天记录应该让所有参与的用户都能看到，所以此处在调用客户端时应使用 Clients.All 属性。recv 是由客户端定义的方法名称。

（2）Web 页面的 HTML 文档如下。

```
<table>
<tr>
<td><label for="nick">我的昵称：</label></td>
<td><input type="text" id="nick"/></td>
</tr>
<tr>
<td><label for="msg">消息：</label></td>
<td><textarea id="msg" rows="6" cols="23"></textarea></td>
</tr>
<tr>
<td colspan="2">
<button id="btn">发送</button>
</td>
</tr>
</table>
<div style="overflow-y: scroll; max-height: 200px">
<ul id="msglist" style="color:green"></ul>
</div>
```

聊天记录会呈现在<ul>元素内，通过 JavaScript 代码动态添加<li>元素。

（3）客户端的 JavaScript 代码如下。

```
<script src="~/scripts/signalr.min.js"></script>
<script>
    // 获取需要的 HTML 元素对象
    var btn = document.getElementById("btn");
    var txtNick = document.getElementById("nick");
    var txtMsg = document.getElementById("msg");
    var msglist = document.getElementById("msglist");
    // 连接服务器
    var conn = new signalR.HubConnectionBuilder()
            .withUrl("/chat")
            .build();
    // 注册客户端方法，由服务器调用
    conn.on("recv", (nick, msg, time) => {
        // 显示收到的消息
        let item = document.createElement("li");
        item.textContent = `[${time}]${nick}: ${msg}`;
        msglist.appendChild(item);
    });
```

```
        // 开始连接
        try{
            conn.start();
        }
        catch(err)
        {
            console.log(err);
        }
        // 发送消息
        btn.addEventListener("click", async evt => {
            // 如果未建立连接，就取消发送
            if(conn.state !== signalR.HubConnectionState.Connected)
            {
                return;
            }
            let _name = txtNick.value;
            let _msg = txtMsg.value;
            if(_name.length > 0 && _msg.length > 0)
            {
                // 调用服务器方法
                await conn.invoke("speak", _name, _msg);
            }
            // 禁用事件的默认行为
            evt.preventDefault();
        });
    </script>
```

（4）ASP.NET Core 应用程序初始化时开启 SignalR 功能。

```
builder.Services.AddSignalR();
```

（5）在 HTTP 管道上映射 ChatHub 的地址。

```
app.MapHub<ChatHub>("/chat");
```

运行示例程序，可以打开多个浏览器窗口（或标签）进行测试，每个窗口代表不同的用户。每个页面都要输入各自的昵称，输入消息后即可发送，如图 16-7 所示。

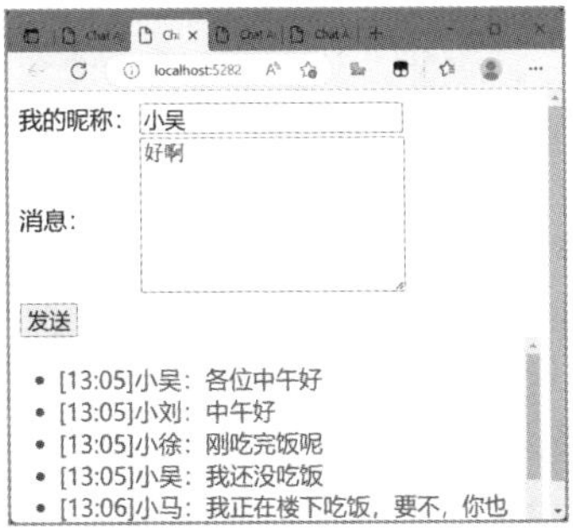

图 16-7　聊天页面

16.3.2　将客户端定义为接口

通过 SendAsync()方法调用客户端时都是用字符串常量指定客户端方法名称的。在书写代码时容易出现拼写错误（如将 FetchData 误写为 FetchData）。如果将客户端方法都声明在一个接口类型中，在编写代码时，Visual Studio 和 Visual Studio Code 编辑器会有代码提示，既提高效率，又可以避免方法名称拼写错误，而且在生成阶段编译器也会检查出不正确的成员名称。

把客户端的方法定义到接口类型中仅仅是用于提取方法名称，因此该接口不需要实现代码。

在编写客户端的 JavaScript 脚本代码时，要调用的服务器方法以及要注册的客户端方法也是使用字符串指定名称的。这种情况可以考虑把脚本代码写在 Razor 文档中，这样可以通过 C#语言的 nameof 运算符直接获取到服务器方法和客户端方法的名称，也能避免名称拼写错误。例如：

```
conn.invoke("@(nameof(DemoHub.Fly))", …);
conn.on("@(nameof(IClient.Callback))", …);
```

在程序运行阶段，会生成这样的代码：

```
conn.invoke("Fly", …);
conn.on("Callback", …);
```

当把客户端方法声明到接口类型后，在自定义 Hub 时需要继承 Hub<T>类（称为强类型的 Hub）。相比 Hub 类，它多了一个类型参数 T，表示包含客户端方法的接口类型。在访问 Clients 属性时，会自动指向 T 类型，并且可以直接访问 T 的方法成员。

16.3.3 示例：实时更新进度条

本示例将通过服务器端调用客户端方法修改<progress>元素的进度显示。

（1）本示例中，被调用的客户端方法为 SetProcess()。下面的代码定义 ITestClient 接口，其中包含 SetProcess()方法。此接口不需要实现类。

```
public interface ITestClient
{
    Task SetProcess(int p);
}
```

（2）自定义 Hub 类型，T 类型为 ITestClient 接口。

```
public class TestHub : Hub<ITestClient>
{
    public async Task Run()
    {
        int x = 0;
        while(x < 100)
        {
            x++;
            // 调用客户端方法，设置进度
            await Clients.Caller.SetProcess(x);
            // 延时
            await Task.Delay(100);
        }
    }
}
```

Run()方法模拟一个需要在服务器上花费一定时间运行的任务。变量 x 的值从 0 增长到 100，以模拟处理进度，每次进度更新后都会延时 100ms。强类型的 Hub 不再需要调用 SendAsync()方法，而是直接访问 ITestClient 接口的 SetProcess()方法。

（3）将 HTML 和 JavaScript 写在 Razor 文件中（本示例使用 Razor Pages 方式呈现，文件首行要写上@page 指令）。

```
@page
@using MyApp

<!DOCTYPE html>
<html>
<head>
<meta charset="utf-8" />
<title>Demo</title>
</head>
<body>
<p>进度：<span id="txt"></span></p>
<progress id="pr" max="100" value="0"></progress>
<div style="margin-top:15px;">
<button id="btn">开始</button>
</div>
<script src="~/signalr.js"></script>
<script>
        // 获取相关元素的引用
        var btn = document.getElementById("btn");
        var progress = document.getElementById("pr");
        var disp = document.getElementById("txt");
        // 创建连接对象
        var conn = new signalR.HubConnectionBuilder()
            .withUrl("/myhub")
            .build();
        // 注册客户端方法
        conn.on("@(nameof(ITestClient.SetProcess))", (p) => {
            // 设置进度
            progress.value = p;
            disp.textContent = `${p} %`;
        });

        // 建立连接
        try {
            conn.start();
        }
        catch (err) {
            console.log(err);
        }

        btn.addEventListener("click", async (ev) =>{
            if(conn.state !== signalR.HubConnectionState.Connected)
            {
                return;
            }
```

```
            // 禁用按钮，避免重复触发
            ev.target.disabled = true;
            // 调用服务器方法
            await conn.invoke("@(nameof(TestHub.Run))");
            // 恢复按钮为可用状态
            ev.target.disabled = false;
            // 取消默认事件的默认行为
            ev.preventDefault();
        });
    </script>
    </body>
    </html>
```

在调用 conn.on()和 conn.invoke()方法时都可以用@(nameof(...))直接获取各个方法的名称。

（4）运行示例程序，单击页面上的“开始”按钮，服务器开始执行耗时任务，进度条会实时更新，如图 16-8 所示。

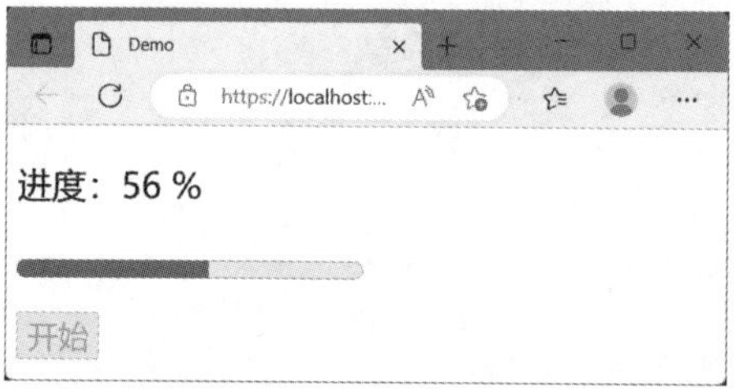

图 16-8　进度条实时更新

16.3.4　示例：记录连接状态

本示例将实现当有客户端成功与服务器连接或断开连接后，服务器会在日志中进行记录。

本示例会用到以下两个知识点。

（1）Hub 类型的依赖注入。从 Hub 派生的类，它的构造函数支持依赖注入，可从服务容器中获取所需的类型实例。

（2）重写 Hub 类的 OnConnectedAsync()方法，当有客户端成功连接后，方法会被调用；重写 OnDisconnectedAsync()方法，当连接断开后被调用。OnDisconnectedAsync()方法有一个 Exception 类型的参数，表示在断开连接时（或之前）所引发的异常。如果无异常发生，此参数为 null。

具体实现步骤如下。

（1）在 ASP.NET Core 应用中（服务器端），需要实现自定义的 Hub——MyHub，代码如下。

```
public class MyHub : Hub
{
    private readonly ILogger _logger;
```

```
    public MyHub(ILoggerFactory logfac)
    {
        _logger = logfac.CreateLogger("SignalR Demo");
    }

    public override async Task OnConnectedAsync()
    {
        _logger.LogInformation("客户端已连接，连接 ID: {0}", Context.
        ConnectionId);
        await base.OnConnectedAsync();
    }

    public override async Task OnDisconnectedAsync(Exception? exception)
    {
        if(exception != null)
        {
            _logger.LogInformation($"连接已断开，但发生以下错误：{exception.
            Message}");
        }
        else
        {
            _logger.LogInformation("连接已断开");
        }
        await base.OnDisconnectedAsync(exception);
    }
}
```

MyHub 类中包含 ILogger 类型的字段_logger，在 OnConnectedAsync()、OnDisconnectedAsync()方法中通过它输出日志消息。

在 MyHub 类的构造函数中，通过依赖注入获得一个 ILoggerFactory 对象，再调用它的 CreateLogger()方法创建一个 ILogger 对象的引用，并赋值给_logger 字段。

（2）在初始化代码中，需要启用 SignalR 相关功能，并映射 MyHub 的终结点。

```
var builder = WebApplication.CreateBuilder(args);
builder.Services.AddSignalR();
var app = builder.Build();

app.MapHub<MyHub>("/hub");
app.Run();
```

（3）客户端是一个 WPF（Windows Presentation Foundation）应用程序。主窗口的 XAML 如下。

```
<Window …
        Title="Demo" SizeToContent="WidthAndHeight">
<Border>
<DockPanel Margin="12" LastChildFill="True">
```

```
<Grid DockPanel.Dock="Bottom">
<Grid.ColumnDefinitions>
<ColumnDefinition Width="*"/>
<ColumnDefinition Width="18"/>
<ColumnDefinition Width="*"/>
</Grid.ColumnDefinitions>
<Button Grid.Column="0" Content="连接" Click="OnConnect"/>
<Button Grid.Column="2" Content="断开连接" Click="OnDisconnect"/>
</Grid>
<Grid Margin="5,5,5,20">
<Grid.ColumnDefinitions>
<ColumnDefinition Width="auto"/>
<ColumnDefinition Width="*"/>
</Grid.ColumnDefinitions>
<TextBlock Text="服务器地址：" Foreground="Blue"/>
<TextBox x:Name="txtServer" MinWidth="200" Grid.Column="1"/>
</Grid>
</DockPanel>
</Border>
</Window>
```

此处要用到面向.NET 的 SignalR 客户端库，即项目要引用 Microsoft.AspNetCore.SignalR.Client 库。

```
<ItemGroup>
<PackageReference Include="Microsoft.AspNetCore.SignalR.Client"…/>
</ItemGroup>
```

（4）在窗口的代码文件中声明 HubConnection 类型的字段。

```
private HubConnection? conn = null;
```

（5）处理“连接”按钮的 Click 事件，与服务器建立连接。

```
private async void OnConnect(object sender, RoutedEventArgs e)
{
    if(txtServer.Text.Length == 0)
    {
        MessageBox.Show("请输入服务器地址");
        return;
    }
    // 创建连接
    conn = new HubConnectionBuilder()
          .WithUrl(txtServer.Text)
          .WithAutomaticReconnect()
          .Build();
    // 连接服务器
    try
    {
        await conn.StartAsync();
        MessageBox.Show("连接成功");
```

```
    }
    catch (Exception ex)
    {
        MessageBox.Show(ex.Message);
    }
}
```

（6）处理“断开连接”按钮的 Click 事件，断开与服务器的连接。

```
private async void OnDisconnect(object sender, RoutedEventArgs e)
{
    if(conn != null && conn.State == HubConnectionState.Connected)
    {
        await conn.StopAsync();
        MessageBox.Show("已断开连接");
    }
}
```

先运行服务器应用程序(ASP.NET Core 应用),此时在控制台窗口中会看服务器的根 URL，如 https://localhost:7159，SignalR 中心的连接地址为 https://localhost:7159/hub。将此地址输入客户端窗口的文本框，如图 16-9 所示。

图 16-9　在客户端输入服务器地址

单击“连接”按钮，在连接建立后，服务器控制台会输出日志信息；接着单击“断开连接”按钮，服务器也会进行记录，如图 16-10 所示。

图 16-10　服务器输出的日志

第 17 章 Blazor

本章要点

- ❑ 服务器托管与 WebAssembly 托管
- ❑ 路由组件
- ❑ 布局组件
- ❑ 组件参数以及级联参数的使用
- ❑ 事件与绑定
- ❑ .NET 与 JavaScript 互操作

17.1 Blazor 概述

Blazor 是 ASP.NET Core 中的一个框架，允许开发人员使用现有的.NET 技术开发 Web 交互，即可以用 C#语言替代 JavaScript 语言编写客户端逻辑。尽管 Blazor 不能完全替代 JavaScript，但它是.NET 开发人员是福音，可以运用自己最熟悉的技术开发 Web UI，免去了学习大量前端框架知识的过程。

Blazor 支持.NET 与 JavaScript 的互操作，即.NET 代码可以调用 JavaScript 代码；反之，JavaScript 代码也可以调用.NET 代码。这样做的好处是可以共享现有代码和第三方类库。

Blazor 有以下两种托管方式。

（1）服务器运行。这种方式与一般 ASP.NET Core 应用一样，.NET 代码在服务器执行，与客户端的交互则由 SignalR 实现。与其他 ASP.NET Core 框架（Razor Pages、MVC）不同的是，Blazor 在客户端 Web UI 更新时不会重新加载整个 HTML 文档，只是通过 SignalR 将要更新的数据发送到客户端，再由客户端代码（主要的脚本文件是 blazor.server.js）更改前端 HTML。若无特殊需求，开发人员应首选服务器托管方式，因为它传输的数据较小，应用加载速度快。.NET 代码在服务器上执行，可减少 Web 浏览器（客户端）的负担。

（2）WebAssembly，简称 wasm。WebAssembly 并不是一种编程语言，而是一种编码方式。它能把 C++、Rust、C#等强类型编程语言编写的代码编译为紧凑的二进制格式（类似于汇编的指令集），性能较高且安全。这使得强类型编程语言编写的代码能够在 Web 浏览器中运行，与 JavaScript 协同工作，大大扩展 Web 应用的功能。使用 WebAssembly 方式托管的 Blazor 应用是在客户端（通常是 Web 浏览器）运行的，并不依赖后端服务器。但其缺点是在初次运行时需要从服务器下载.NET 运行行时和应用程序集，会导致 Web UI 的加载时间变长。当然，它的优点是能减轻服务器的负担，再次运行时执行效率高（毕竟代码都下载到浏览器中，直接启动应用程序）。如果应用程序需要与用户进行较多的交互，可以选择 WebAssembly 方式托管。

不管以何种方式托管应用程序，Blazor 的用户界面逻辑也是使用 Razor 标记完成的（和 Razor Pages、MVC 视图一样），称为 Razor 组件，文件扩展名为.razor。

17.2　服务器托管

Blazor Server 与一般的 ASP.NET Core 应用项目没有区别。项目文件（*.csproj）的 Sdk 属性仍然使用 Microsoft.NET.Sdk.Web。

```
<Project Sdk="Microsoft.NET.Sdk.Web">
…
</Project>
```

在应用程序初始化时，需要在服务器容器上调用 AddServerSideBlazor()扩展方法注册与 Blazor 相关的服务，如添加 SignalR 功能、JavaScript 互操作运行时支持等。

```
builder.Services.AddServerSideBlazor();
```

在 HTTP 管道上还要调用 MapBlazorHub()扩展方法。

```
app.UseStaticFiles();
app.UseRouting();
app.MapBlazorHub();
```

MapBlazorHub()方法用于映射 Blazor 专用的 SignalR 中心（Hub）——ComponentHub。此 Hub 是内部实现类型，用于 Blazor 服务器与客户端之间的数据传输。它的默认访问路径为/_blazor。MapBlazorHub()方法还映射了两个路径：

（1）/_blazor/disconnect/，用于断开 SignalR 连接时调用；

（2）/_blazor/initializers/，返回一组 JavaScript 脚本文件。这些脚本会在 Blazor 初始化过程中执行。这些文件路径是在 CircuitOptions 选项类的 JavaScriptInitializers 属性中配置的。但由于 JavaScriptInitializers 属性被定义为 internal 成员，外部代码无法访问，因此开发人员无法手动添加初始化脚本文件。

Blazor Server 应用程序在启动时要加载 blazor.server.js 脚本，由于此文件是以静态文件的方式提供的，所以在 HTTP 管道上还要调用 UseStaticFiles()方法。blazor.server.js 文件不是项目文件的一部分，它是内嵌到 Microsoft.AspNetCore.Components.Server 程序集中的资源文件。可以在 ASP.NET Core 源代码的项目文件（Microsoft.AspNetCore.Components.Server.csproj）中找到其声明。

```
<PropertyGroup>
<BlazorServerJSFilename>blazor.server.js</BlazorServerJSFilename>
…
</PropertyGroup>
…
<Target …>
<ItemGroup>
<EmbeddedResource Include="$(BlazorServerJSFile)" LogicalName=
"_framework/$(BlazorServerJSFilename)" />
…
</ItemGroup>
</Target>
```

根据上述声明，脚本的加载路径为_framework/blazor.server.js。在 WebHost 初始化过程中，内嵌的资源文件会被合并到 IWebHostEnvironment.WebRootFileProvider 对象中（由 CompositeFileProvider 类封装，该类可以包含 IFileProvider 对象的集合。应用程序可在此集合中查找文件）。当在 HTTP 管道上调用 UseStaticFiles()方法时，内嵌的资源文件也成为静态文件的一部分。

前文曾提到，Blazor 的 UI 部分是通过 Razor 组件实现的。它和 MVC 中的视图组件类似，最终会输出 HTML 片段。但这些 Razor 组件无法自动呈现，它们需要一个容器——HTML 页面。这个 HTML 页面除了承载 Razor 组件外还有一个任务——加载并执行 blazor.server.js 文件，Blazor 应用程序才能真正启动。

Blazor Server 运行在服务器上，它需要执行服务器代码，因此不能使用普通的 HTML 文件，应使用 Razor 文件（*.cshtml）。Razor Pages 和 MVC 视图均可实现，二者只选择一种即可。

17.2.1 示例：使用 Razor Pages 承载 Blazor 应用

要创建 Blazor Server 应用，可以使用以下命令。

```
dotnet new blazorserver …
dotnet new blazorserver-empty …
```

使用 blazorserver 模板创建的 Blazor 项目带有演示代码（天气预报数据、按钮计数器），以及 Bootstrap 前端样式库。如果使用 blazorserver-empty 模板，则不带演示代码和 Bootstrap 样式库。

本示例为了让读者能够更好地理解 Blazor 项目的结构，将使用空白的 Web 项目模板创建应用程序。命令如下。

```
dotnet new web -n <项目名称> -o <项目存放路径>
```

具体实现步骤如下。

（1）在项目中添加 Razor 组件，文件命名为 App.razor。Blazor 启动后将呈现该组件。

```
@using Microsoft.AspNetCore.Components.Web

<div>
<p>请单击下面的按钮</p>
```

```
<button type="button" @onclick="OnClick">请单击</button>
<p>@_message</p>
</div>

@code{
    // 私有字段
    private string? _message;
    // 用于统计按钮被单击的次数
    private int _count;

    void OnClick()
    {
        _count ++;
        // 修改字段内容
        _message = $"你单击了{_count}次按钮";
    }
}
```

第 1 行使用@using 指令引入 Microsoft.AspNetCore.Components.Web 命名空间。在 Razor 组件中，C#代码要写在@code 区域内。App 组件内定义了私有字段_message，该字段的值会显示在<p>元素内。

OnClick()方法用来处理<button>元素的 click（单击）事件。这里要注意语法，如果 onclick 前面没有@符号，表明它是常规的 HTML 元素事件，其响应代码只能用客户端脚本（如 JavaScript）来编写；而加上@符号则表示可以用.NET 代码（如 C#）处理 click 事件。@onclick="OnClick" 将 click 事件与 OnClick()方法绑定，当用户单击<button>元素时会调用 OnClick()方法。

（2）在项目中创建 Pages 目录。

（3）在 Pages 目录下新建 Host.cshtml 文件，即 Razor 页面。

```
@page
@addTagHelper *, Microsoft.AspNetCore.Mvc.TagHelpers
…

<!DOCTYPE html>
<html>
<head>
<title>Demo App</title>
<meta charset="utf-8"/>
</head>
<body>
        @*呈现 App 组件*@
<component type="typeof(App)" render-mode="ServerPrerendered" />
        @*引入 Blazor 脚本*@
<script src="~/_framework/blazor.server.js"></script>
</body>
</html>
```

在 Razor 文件中呈现组件，需要使用 ComponentTagHelper 标记帮助器。它对应的 HTML 元素是<component>。type 属性指定 Razor 组件的类名，本示例中为 App。render-mode 指定组件的呈现方式。ServerPrerendered 表示 App 组件采用“预呈现”策略，即服务器先生成组件的静态 HTML 返回给客户端（不带交互功能），待 Blazor 启动后重新呈现组件内容并添加交互功能。预呈现策略是一个“平衡”方案，可优化组件的加载速度。

在<body>元素的结束标记前需要加载 blazor.server.js 文件，用于启动 Blazor 应用程序。

（4）在初始化应用程序时，向服务容器添加 Razor Pages 和 Blazor 功能。

```
builder.Services.AddRazorPages();
builder.Services.AddServerSideBlazor();
```

（5）在 HTTP 管道上映射 Blazor 应用所需要的 SignalR Hub。

```
app.UseStaticFiles();
app.UseRouting();
app.MapBlazorHub();
```

使用静态文件是因为需要访问 blazor.server.js 文件。

（6）调用 MapFallbackToPage()方法，并指定 Host 页面文件的 URL。

```
app.MapFallbackToPage("/Host");
```

MapFallbackToPage 映射一个“后备”终结点，当请求的 URL 没有匹配的路由时返回 Host 页面。App 组件包含在 Host 页面内，当客户端第 1 次访问时，需要先执行 Host 页面，才能呈现 App 组件。本示例中只有一个 App 组件，并没有定义路由规则。如果项目中有多个 Razor 组件，为了能在不同组件之间切换，每个组件需要指定唯一的路由。

当然，此处改为调用 MapRazorPages()方法也是可以的。

```
app.MapRazorPages();
```

如果调用的是 MapRazorPages()方法，那么访问 Host 页面的 URL 是/host。如果希望通过根地址访问，就要修改 Host 页的 Razor 标记，在@page 指令中指定页面的访问路径。

```
@page "/"
```

运行示例程序，单击页面上的按钮，就会看到统计单击次数的消息，如图 17-1 所示。

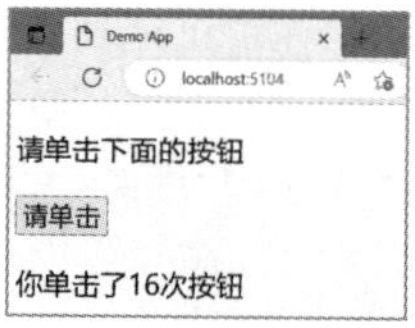

图 17-1 统计按钮被单击次数

17.2.2 示例：在 MVC 视图中承载 Blazor 应用

本示例将演示在 MVC 视图文件中承载 Blazor 应用的方法。具体实现步骤如下。

（1）在项目中添加 Home 控制器（类名为 HomeController）。

```
public class HomeController : Controller
{
    public IActionResult Host()
    {
        return View();
    }
}
```

Host 操作方法将返回同名视图，即视图文件为 Host.cshtml。

（2）在项目中添加 Razor 组件，命名为 App。该组件将用于 Blazor 应用的主界面。

```
<h3 style="color:darkblue">示例组件</h3>
<p>v1.0</p>
```

（3）在项目中新建 Views 目录。

（4）在 Views 目录下新建 Shared 子目录。

（5）在 Shared 目录下添加_Layout.cshtml 布局文件。

```
<!DOCTYPE html>

<html>
<head>
<meta name="viewport" content="width=device-width" />
<title>@ViewBag.Title</title>
</head>
<body>
<div>
      @RenderBody()
</div>
<script src="~/_framework/blazor.server.js"></script>
</body>
</html>
```

在<body>元素结束标记前需要加载 blazor.server.js 文件。

（6）在 Views 目录下新建 Home 目录（与控制器同名）。

（7）在 Home 目录下添加 Host.cshtml 视图文件（与 Host 方法同名）。

```
<component type="typeof(App)" render-mode="Server" />
```

使用<component>元素加载 App 组件。

（8）在 ASP.NET Core 应用程序的初始化代码中添加带视图的 MVC 和 Blazor Server 功能。

```
builder.Services.AddControllersWithViews();
builder.Services.AddServerSideBlazor();
```

（9）在 HTTP 管道上使用静态文件（允许客户端获取 blazor.server.js 文件）。

```
app.UseStaticFiles();
```

（10）映射 Blazor 的 SignalR 中心。

```
app.MapBlazorHub();
```

（11）映射 MVC 终结点。

```
app.MapControllerRoute("app",
"{controller=Home}/{action=Host}");
app.MapFallbackToController("Host",
"Home");
```

MapFallbackToController()方法的作用是在请求的 URL 无效时默认指向 Host 视图，保证 Blazor 应用在 URL 错误的情况下仍可启动。

运行示例程序，结果如图 17-2 所示。

图 17-2　在 MVC 视图中承载 Blazor 应用

17.2.3 初始化脚本

前面曾介绍过，在 HTTP 管道上调用 MapBlazorHub()方法注册 ComponentHub 的访问地址的过程中，会同时注册/_blazor/initializers/路径。该路径将返回一个字符串数组，表示初始化脚本的路径。这些会在 Blazor 应用启动前/启动后执行，完成自定义的初始化操作。

ASP.NET Core 应用程序在初始化配置时，会在 Web 根目录下（如 wwwroot）查找名为{App-Name}.modules.json 的文件（{App-Name}是当前 ASP.NET Core 应用程序的名称），然后读取{App-Name}.modules.json 文件的内容，并添加到 CircuitOptions.JavaScriptInitializers 属性中。具体可参考下面的源代码。

```
public void Configure(CircuitOptions options)
{
   // _environment.ApplicationName 是当前应用程序的名称
    var file = _environment.WebRootFileProvider.GetFileInfo
    ($"{_environment.ApplicationName}.modules.json");
    if (file.Exists)
    {
      // 通过反序列化得到 JSON 文件中的字符串数组实例
       var initializers = JsonSerializer.Deserialize<string[]>
       (file.CreateReadStream());
       for (var i = 0; i < initializers.Length; i++)
       {
          var initializer = initializers[i];
         // 将路径添加到 JavaScriptInitializers 列表中
          options.JavaScriptInitializers.Add(initializer);
       }
    }
}
```

应用程序在生成时并没有在 wwwroot 目录下生成{App-Name}.modules.json 文件，而是生成<项目目录>\obj\{Debug 或 Release}\net{版本号}\jsmodules\ jsmodules.build.manifest.json 文件。在生成应用程序时，在输出目录（一般为\bin）生成静态资源的清单文件{App-Name}.staticwebassets.runtime.json，将 jsmodules.build.manifest.json 文件映射为{App-Name}.modules.json。

```
{
"ContentRoots": [
"<项目目录下的>\\wwwroot\\",
"<项目目录下的>\\obj\\{Dubug|Release}\\net{版本}\\jsmodules\\"
  ],
"Root": {
"Children": {
…
"{App-Name}.modules.json": {
"Children": null,
"Asset": {
"ContentRootIndex": 1,
```

```
"SubPath": "jsmodules.build.manifest.json"
        },
"Patterns": null
      }
    },
...
  }
}
```

只有当应用程序发布时才会在 wwwroot 目录下生成{App-Name}.modules.json 文件。

当客户端请求/_blazor/initializers/路径时，由 CircuitJavaScriptInitializationMiddleware 负责将初始化脚本文件的列表发回给客户端。

```
internal sealed class CircuitJavaScriptInitializationMiddleware
{
    private readonly IList<string> _initializers;

    public CircuitJavaScriptInitializationMiddleware(IOptions
    <CircuitOptions> options, RequestDelegate _)
    {
        _initializers = options.Value.JavaScriptInitializers;
    }

    public async Task InvokeAsync(HttpContext context)
    {
        await context.Response.WriteAsJsonAsync(_initializers);
    }
}
```

初始化脚本中只要包含以下两个方法即可，客户端会自动调用。

（1）beforeStart()：在 Blazor 应用启动前调用。

（2）afterStarted()：在 Blazor 应用启动后调用。

在声明这两个方法时要加上 export 关键字，使其能够从 JavaScript 模块中导出，否则无法被调用。

17.2.4　示例：使用初始化脚本

添加初始化脚本的规则是在 Web 静态资源根目录（默认为 wwwroot）下添加名为{App-Name}.lib.module.js（App-Name 是应用程序的名称）的 JavaScript 脚本文件，然后编写 beforeStart()和 afterStarted()函数。

该脚本文件必须符合上述要求，不能随意命名。除非手动添加{App-Name}.modules.json 文件指定自定义的脚本文件（请参考 17.2.5 节示例）。

本示例实现步骤如下。

（1）执行以下命令，创建一个空白的 Blazor Server 应用（应用名称是 MyApp）。

```
dotnet new blazorserver-empty -n MyApp -o <项目存放路径>
```

（2）在 wwwroot 目录下添加 JavaScript 文件，命名为 MyApp.lib.module.js。

（3）在 MyApp.lib.module.js 文件中定义并导出 beforeStart()和 afterStarted()函数。

```
export function beforeStart() {
    // 创建 HTML 元素
    var el = document.createElement("p");
    // 设置样式
    el.style = "color: blue";
    // 设置文本内容
    el.textContent = "Blazor 应用即将启动…";
    // 将该元素添加到<body>元素中
    document.body.appendChild(el);
}

export function afterStarted() {
    // 创建 HTML 元素
    var elp = document.createElement("p");
    // 设置样式和文本内容
    elp.style = "color: green";
    elp.textContent = "Blazor 应用已启动";
    // 将此元素追加到<body>元素中
    document.body.appendChild(elp);
}
```

在 Blazor 应用启动前，在当前 HTML 文档的<body>元素内追加一个<p>元素，设置其内容为“Blazor 应用即将启动…”；在 Blazor 应用启动后，同样添加一个<p>元素，内容设置为“Blazor 应用已启动”。

（4）运行示例程序，结果如图 17-3 所示。

17.2.5 示例：手动添加 modules.json 文件

本示例将在 wwwroot 目录下手动创建{App-Name}.modules.json 文件，然后写上初始化脚本文件的相对路径。初始化脚本的文件名可以自定义。

在 wwwroot 目录下新建一个目录，命名为 scripts。然后在 scripts 目录下添加 JavaScript 脚本文件，命名为 inits.js。脚本文件内的代码如下。

```
export function beforeStart() {
    console.log("Blazor 应用即将启动");
}

export function afterStarted() {
    console.log("Blazor 应用已启动");
}
```

在 wwwroot 目录下添加 MyApp.modules.json 文件（假设当前示例名为 MyApp），然后以 JSON 数组的格式（路径写在一对中括号内）写入 inits.js 文件的相对路径。

```
[
"scripts/inits.js"
]
```

运行示例程序，打开 Web 浏览器的“开发人员工具”，切换到“控制台”页面，就能看到自定义初始化脚本输出的消息了，如图 17-4 所示。

图 17-3　初始化脚本已修改 HTML 文档

图 17-4　初始化脚本已执行

17.3　WebAssembly 托管

Blazor 托管于 WebAssembly 上，可在 Web 浏览器中运行。其优点是可直接调用.NET 代码，不需要与服务器建立连接。但在首次运行时，需要从服务器下载应用程序以及.NET 运行库，会消耗一定的时间和网络流量。

和服务器托管的 Blazor 应用一样，WebAssembly 托管的 Blazor 应用也有两个项目模板。开发人员可以使用以下 dotnet 命令创建 Blazor 项目。

```
// 带有演示代码的项目
dotnet new blazorwasm -n … -o …
// 不带演示代码的项目
dotnet new blazorwasm-empty -n … -o …
```

Blazor WebAssembly 的项目文件使用的 SDK 名称为 Microsoft.NET.Sdk.BlazorWebAssembly。文件结构如下。

```
<Project Sdk="Microsoft.NET.Sdk.BlazorWebAssembly">

<PropertyGroup>
<TargetFramework>net7.0</TargetFramework>
<Nullable>enable</Nullable>
<ImplicitUsings>enable</ImplicitUsings>
</PropertyGroup>

...

</Project>
```

由于 Blazor WebAssembly 首次运行时需要从服务器下载应用程序和.NET 运行时（Runtime），

因此需要一个服务器提供程序文件。一般可以使用常规的 ASP.NET Core 应用程序。在开发阶段，也可以引用名为 Microsoft.AspNetCore.Components.WebAssembly.DevServer 的 Nuget 库，该库公开一个简单的 ASP.NET Core 应用程序，用于向客户端（浏览器）提供 Blazor 应用的下载服务。

17.3.1 示例：手动创建 Blazor WebAssembly 项目

本示例将手动构建（执行 dotnet new 命令时不使用 blazorwasm 和 blazorwasm-empty 模板）Blazor WebAssembly 项目。通过本示例的演练，读者可以更直观地了解 Blazor WebAssembly 项目的结构。

具体实现步骤如下。

（1）执行 dotnet new 命令，新建一个普通的类库项目，命名为 MyBlazorApp，存放在 Client 目录下。

```
dotnet new classlib -n MyBlazorApp -o Client
```

（2）打开项目文件，将<project>元素的 Sdk 属性修改为 Microsoft.NET.Sdk.BlazorWebAssembly。

```
<Project Sdk="Microsoft.NET.Sdk.BlazorWebAssembly">
…
</Project>
```

（3）执行以下命令，添加对 Microsoft.AspNetCore.Components.WebAssembly 包的引用。

```
dotnet add package Microsoft.AspNetCore.Components.WebAssembly
```

（4）删除项目模板生成的 Class1.cs 文件。

（5）在项目根目录下添加 Razor 组件，命名为 App.razor。

```
<div style="border-block: 2px dashed orange">
<h2>Blazor WASM 示例</h2>
</div>
```

（6）在项目根目录下新建 wwwroot 目录。

（7）在 wwwroot 目录下添加 index.html 文件。

```
<!DOCTYPE html>
<html>
<head>
<meta charset="utf-8" />
<title>Demo</title>
</head>
<body>
<div id="app">正在加载 Blazor 组件…</div>
<!--引用脚本文件-->
<script src="/_framework/blazor.webassembly.js"></script>
</body>
</html>
```

id 为 app 的<div>元素用于呈现 Blazor 组件。注意此处引用的脚本为 blazor.webassembly.js，而不是 blazor.server.js。

（8）添加 Program.cs 文件，定义 Program 类，用于编写程序入口点——Main()方法。

```
using Microsoft.AspNetCore.Components.WebAssembly.Hosting;

namespace MyBlazorApp
{
    public class Program
    {
        static async Task Main(string[] args)
        {
            var hostBuilder = WebAssemblyHostBuilder.CreateDefault(args);
            // 添加 Razor 组件
            hostBuilder.RootComponents.Add<App>("#app");
            // 创建 Host 实例
            var host = hostBuilder.Build();
            // 启动 Host
            await host.RunAsync();
        }
    }
}
```

首先调用 WebAssemblyHostBuilder.CreateDefault()静态方法创建默认的 WebAssemblyHostBuilder 实例。接着通过 RootComponents 属性添加所需的 Razor 组件。本示例只需要添加 App 组件即可。在调用 Add()方法时指定 App 组件将呈现在 id 为 app 的 HTML 元素内——index.html 文件中的<div>元素。

调用 Build()方法构建 WebAssemblyHost 对象。最后调用 RunAsync()方法启动 WebAssemblyHost 对象。

Blazor WebAssembly 客户端已完成。接下来需要创建一个 ASP.NET Core 项目，提供应用程序和.NET 运行时的下载服务。

（9）执行以下命令，创建一个空白的 ASP.NET Core 应用程序。

```
dotnet new web -n ServerApp -o Server
```

（10）进入 Server 目录，执行以下命令，引用 MyBlazorApp 项目。

```
dotnet add reference ../Client/MyBlazorApp.csproj
```

（11）执行以下命令，引用 Microsoft.AspNetCore.Components.WebAssembly.Server 包。

```
dotnet add package Microsoft.AspNetCore.Components.WebAssembly.Server
```

（12）打开 Program.cs 文件，将代码修改为

```
var builder = WebApplication.CreateBuilder(args);
var app = builder.Build();

// 提供 Blazor-WebAssembly 相关文件
```

```
app.UseBlazorFrameworkFiles();
// 使用静态文件
app.UseStaticFiles();
// 通过 index.html 文件加载 Blazor 组件
app.MapFallbackToFile("index.html");

app.Run();
```

注意，非 Blazor Server 版应用程序，不需要调用 MapBlazorHub()方法，只调用 UseBlazorFrameworkFiles()方法即可。此方法允许客户端请求如 blazor.webassembly.js 等文件。为了能访问 index.html 文件，还要调用 UseStaticFiles()方法。index.html 文件位于 MyBlazorApp 项目的 wwwroot 目录下，由于 ServerApp 项目引用了 MyBlazorApp 项目，因此可以访问 wwwroot 目录下的文件。

运行 ServerApp 项目，再用浏览器访问服务器地址，结果如图 17-5 所示。由于 Blazor WebAssembly 需要从服务器下载运行时，所以组件的加载会有些延时。

图 17-5　呈现 App 组件

17.3.2　示例：用 node.js 开发 Blazor WebAssembly 服务器

由于 Blazor WebAssembly 应用是在浏览器中执行的，因此服务器可以使用非 ASP.NET Core 应用程序。本示例将使用 node.js 编写一个简单的 HTTP 服务器，并提供 Blazor WebAssembly 程序文件的下载。具体实现步骤如下。

（1）执行以下命令，新建一个类库应用程序。

```
dotnet new classlib -n MyApp -o
```

（2）打开项目文件，将 Sdk 属性修改为 Microsoft.NET.Sdk.BlazorWebAssembly。

```
<Project Sdk="Microsoft.NET.Sdk.BlazorWebAssembly">

…

</Project>
```

（3）执行以下命令，引用必备的 Nuget 包。

```
dotnet add package Microsoft.AspNetCore.Components.WebAssembly
```

（4）删除模板生成的 Class1.cs 文件。

（5）在项目根目录下添加 Test.razor 组件。

```
@using Microsoft.AspNetCore.Components.Web

<h3>示例组件</h3>
<div style="margin: 20px; height: 45px; width: @(_width)px; background-color: red"></div>
```

```
<div>
<button type="button" @onclick="on_click1">减小宽度</button>
<button type="button" @onclick="on_click2">增大宽度</button>
</div>

@code{
    private int _width = 60;

    void on_click1()
    {
        _width -= 15;
        if(_width < 60)
        {
            _width = 60;
        }
    }

    void on_click2()
    {
        _width += 15;
        if(_width > 300)
        {
            _width = 300;
        }
    }
}
```

<div>元素的宽度（CSS 样式 width）与_width 字段绑定，单击“增大宽度”按钮时，将<div>元素的宽度增大 15；单击“减小宽度”按钮时，将<div>元素的宽度减小 15。

（6）在项目中添加 Program.cs 文件，并编写入口点代码。

```
var builder = WebAssemblyHostBuilder.CreateDefault(args);
// 设置 Test 组件为应用程序的启动组件
builder.RootComponents.Add<Test>("#app");
var host = builder.Build();

await host.RunAsync();
```

（7）执行以下命令，发布应用程序。

```
dotnet publish
```

（8）新建一个目录，命名为 DemoServer（可以位于任意路径下，但路径中最好不要含有非英文字符）。

（9）找到已发布的 Blazor 应用目录（默认在 bin\{Debug|Release}\net{版本号}\publish）中的 wwwroot 目录，将整个_framework 目录复制到 DemoServer 目录下。

（10）在 DemoServer 目录下添加 index.html 文件，此文件将作为主页，用于加载 Blazor 应用程序。

```
<!DOCTYPE html>
```

```
<html>
<head>
<meta charset="utf-8"/>
<title>Demo</title>
</head>
<body>
<div id="app">正在加载…</div>
<script src="_framework/blazor.webassembly.js"></script>
</body>
</html>
```

（11）在 DemoServer 目录下添加 server.js 文件。

```
const http = require("node:http");
const fs = require("node:fs");
const url = require("node:url");
const path = require("node:path");

// 服务器名称
const host = "localhost";
// 服务器端口
const port = 8100;
// 创建服务器
ser = http.createServer((request, response)=>{
    // 获取请求路径
    var thePath = url.parse(request.url).pathname;
    // 去掉路径开头的/
    var file = thePath.substring(1);
    // 如果未指定文件名，默认为 index.html
    if(file.length == 0)
    {
        file = "index.html";
    }
    // 读取文件
    fs.readFile(file, (err, data)=>{
        if(err) {
            response.writeHead(404, {"Content-Type": "text/html"});
        } else {
            let contenttype = {"Content-Type": "text/html"};
            // 处理几个特殊的文件扩展名与 MIME 类型之间的映射关系
            let ext = path.extname(file);
            switch(ext)
            {
                case ".js":
                case ".mjs":
                    contenttype["Content-Type"] = "text/javascript";
                    break;
                case ".json":
                    contenttype["Content-Type"] = "application/json";
```

```
                break;
            case ".htm":
            case ".html":
                contenttype["Content-Type"] = "text/html";
                break;
            case ".css":
                contenttype["Content-Type"] = "text/css";
                break;
            case ".jpg":
            case ".jpeg":
                contenttype["Content-Type"] = "image/jpeg";
                break;
            case ".png":
                contenttype["Content-Type"] = "image/png";
                break;
            case ".gif":
                contenttype["Content-Type"] = "image/gif";
                break;
            default:
                contenttype["Content-Type"] = "application/octet-stream";
        }
        // 发送 HTTP 头部
        response.writeHead(200, contenttype);
        // 发送文件内容
        response.write(data);
    }
    // 响应消息发送完毕
    response.end();
  });
}).listen(port, host);                              // 侦听客户端连接

console.log(`服务器地址: http://${host}:${port}`);
```

首先通过 require 指令导入需要的 node.js 模块。http.createServer()方法将创建 HTTP 服务器，并通过 listen()方法指定要侦听的地址和端口号。

（12）执行以下命令，用 node.js 启动 HTTP 服务器。

```
node server.js
```

（13）在 Web 浏览器中打开 http://localhost:8100，就能看到 Test 组件的呈现效果，如图 17-6 所示。分别单击“减小宽度”和“增大宽度”按钮调整红色长方形的宽度。

图 17-6　Test 组件

17.3.3　示例：初始化脚本

Blazor WebAssembly 应用程序在生成后会产生 blazor.boot.json 文件，自定义的初始化脚本将由 libraryInitializers 属性列出。

```
{
…
"extensions": null,
"lazyAssembly": null,
"libraryInitializers": {
"<项目名称>.lib.module.js": "sha256-XXXXXX"
},
…
}
```

初始化脚本的命名规则与 Blazor Server 项目一致，其格式依旧是<App-Name>.lib.module.js。下面给出一个示例。

```
const _style = "padding:8px; margin:12px 0px; background-color: darkblue;
color: white;";

export function beforeStart() {
    var div = document.createElement("div");
    div.style = _style;
    div.textContent = "Blazor app 正在启动…";
    document.body.appendChild(div);
}

export function afterStarted() {
    var div = document.createElement("div");
    div.style = _style;
    div.textContent = "Blazor app 已启动.";
    document.body.appendChild(div);
}
```

假设项目名称为 MyApp，那么初始化脚本的文件名就是 MyApp.lib.module.js。同样，初始化脚本也要放到 wwwroot 目录下。

示例效果如图 17-7 所示。

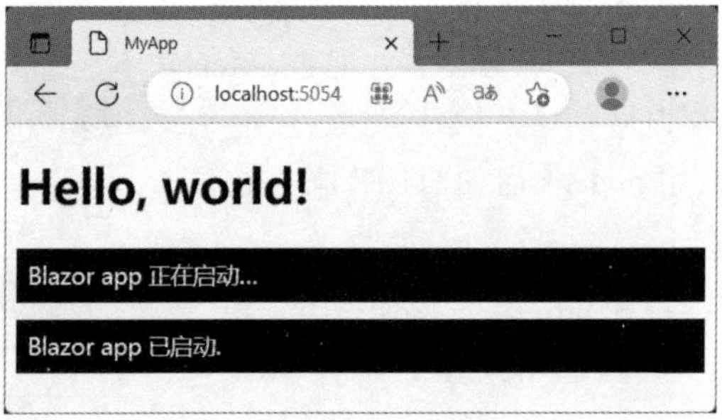

图 17-7　Blazor WebAssembly 应用程序的初始化脚本

17.3.4　DevServer

在创建 Blazor WebAssembly 应用程序时，如果加上--hosted（或-ho）参数，就会生成一个简单的 ASP.NET Core 应用程序，用于提供 Blazor WebAssembly 程序文件的下载。例如：

```
dotnet new blazorwasm -n <项目名称> -o<输出路径> --hosted
```

如果不加--hosted 参数，那么仅生成 Blazor WebAssembly 项目，同时，项目会引用名为 Microsoft.AspNetCore.Components.WebAssembly.DevServer 的 Nuget 包。

```
<Project Sdk="Microsoft.NET.Sdk.BlazorWebAssembly">

...

<ItemGroup>
<PackageReference Include="Microsoft.AspNetCore.Components.WebAssembly"
Version="7.0.3" />
<PackageReference Include="Microsoft.AspNetCore.Components.WebAssembly.
DevServer" Version="7.0.3" PrivateAssets="all" />
</ItemGroup>

</Project>
```

这是一个内置的小型 ASP.NET Core 应用程序，此功能仅在开发阶段使用，方便开发人员调试程序代码，项目发布后不再使用。

17.4　路由组件

Blazor 应用程序（不管是 Server 托管还是 WebAssembly 托管）的项目模板会生成一个名为 App 的 Razor 组件，作为应用程序启动时加载的主界面。例如：

```
<Router AppAssembly="@typeof(App).Assembly">
<Found Context="routeData">
<RouteView RouteData="@routeData" DefaultLayout="@typeof(MainLayout)" />
<FocusOnNavigate RouteData="@routeData" Selector="h1" />
</Found>
<NotFound>
<PageTitle>Not found</PageTitle>
<LayoutView Layout="@typeof(MainLayout)">
<p role="alert">Sorry, there's nothing at this address.</p>
</LayoutView>
</NotFound>
</Router>
```

App 组件的根元素是 Router 组件（路由组件），对应的是 Microsoft.AspNetCore.Components.Routing 命名空间下的 Router 类。Router 组件至少要指定一个程序集（即上述代码中的 AppAssembly 属性，它指定的是 App 组件所在的程序集），并根据 URL 匹配出程序集内的组件并呈现到用户界面上。

Router 组件有以下两个核心属性。

（1）Found：路由匹配成功后要呈现的内容。此处通常会使用 RouteView 组件。该组件会接收来自 Found 属性上下文传递的 RouteData 对象（位于 Microsoft.AspNetCore.Components 命名空间），然后呈现与路由 URL 对应的组件。假设 Abc 组件的 URL 为/test，当导航的地址为 http://host/test 时，RouteView 组件内会呈现 Abc 组件。

（2）NotFound：若请求的 URL 路由匹配失败，就呈现该属性所指定的内容。

Blazor 应用程序只在单个 HTML 页内加载，不同组件之间的切换只是替换相应的 HTML 片段。例如，Index 组件呈现的 HTML 如下。

```
<main>
<h1 tabindex="-1">Hello, world!</h1>
</main>
```

当导航到 Second 组件后，<main>元素内的 HTML 被替换为 Second 组件的内容。

```
<main>
<h3>Musics</h3>
<p>另一个组件</p>
</main>
```

与 Razor Pages 应用程序类似，Razor 组件内可以使用@page 指令定义路由规则。@page 指令写在 Razor 文档的第 1 行，类型为字符串表达式。例如：

```
@page "/listItem"
```

也可以使用 RouteAttribute 特性类（位于 Microsoft.AspNetCore.Components 命名空间）指定路由规则。

```
@attribute [Route("/second")]
```

17.4.1 示例：路由组件的简单应用

本示例将展示路由组件（Router）的功能。示例程序包含以下组件。

（1）App 组件：在 Blazor 应用程序启动时最先呈现的组件。App 组件内使用了 Router 组件。

```
@using Microsoft.AspNetCore.Components.Routing

<Router AppAssembly="typeof(Program).Assembly">
<NotFound>
<p style="color:red">未找到相关的组件</p>
</NotFound>
<Found Context="routeData">
<RouteView RouteData="routeData" />
</Found>
</Router>
```

NotFound 属性指定在找不到指定的组件时显示红色文本“未找到相关的组件”。Found 属性引用一个 RouteView 组件。如果路由匹配成功，目标组件将呈现在 RouteView 组件内部。Context="routeData"用来给上下文对象重命名（默认为 context），它是一个泛型参数，因此具体的数据类型取决于上下文间传递的对象。此处类型为 RouteData。

（2）Default 组件。该组件相当于主页面，当访问根 URL 时，默认呈现此组件。

```
@page "/"

<h3>主页</h3>

<div>
<a href="/comp1">组件 1</a>
<a href="/comp2">组件 2</a>
</div>
```

在 Default 组件中，有两个<a>元素，用于链接到另外两个组件——Comp1 和 Comp2。

（3）Comp1 组件定义的路由模板为/comp1。

```
@page "/comp1"

<h3>组件 1</h3>
```

（4）Comp2 组件定义的路由模板为/comp2。

```
@page "/comp2"

<h3>组件 2</h3>
```

运行示例程序，首先看到的是 Default 组件，如图 17-8 所示。分别单击“组件 1”和“组件 2”链接可以导航到相应的组件，如图 17-9 所示。

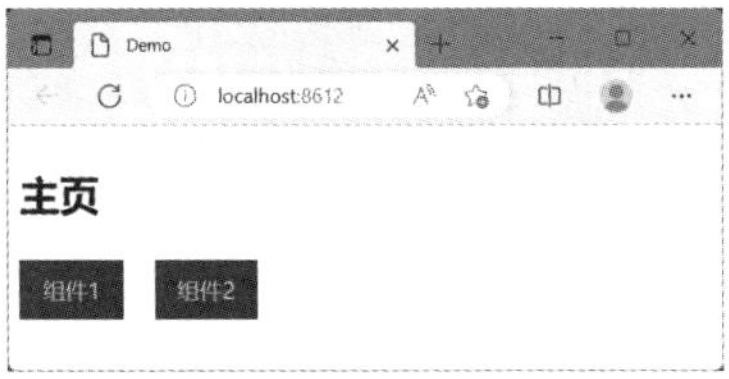

图 17-8　Default 组件

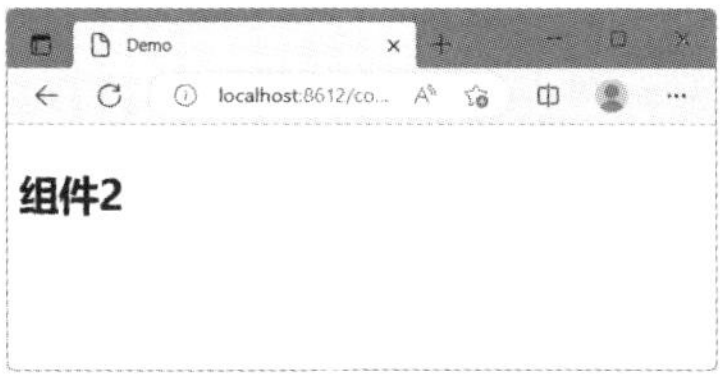

图 17-9　导航到另一个组件

17.4.2　示例：使用路由参数

Razor 组件在定义路由规则时也可以使用路由参数。例如：

```
/index/{test}
```

上述规则中，test 是路由参数。Razor 组件必须声明与参数同名的属性成员，并且要应用 ParameterAttribute 特性。成员名称与路由参数相同即可，不区分大小写。以下声明均有效。

```
[Parameter]
public string test { get; set; }

[Parameter]
public string Test { get; set; }
```

本示例将演示路由参数的使用。示例包含两个 Razor 组件。

（1）第 1 个组件接收名为 message 的路由参数。

```
@page "/{message?}"

<h3>示例组件</h3>
@if(Message is not null)
{
<p>参数：@(nameof(Message)) = @Message</p>
}

@code{
    [Parameter]
    public string? Message{ get; set; }
}
```

message 后面的问号（?）表示此参数为可选，应用程序在访问 URL 时可以省略参数值，即/和/abc 都可以。

（2）第 2 个组件定义了两个路由参数：number1 和 number2。

```
@page "/calc/{number1:int}/{number2:int}"

<h4>计算结果</h4>
<div>
    @Number1 + @Number2 = @Calculate()
</div>

@code{
    [Parameter]
    public int Number1{ get; set; }

    [Parameter]
    public int Number2{ get; set; }

    private int Calculate()
    {
        return Number1 + Number2;
    }
}
```

路由参数默认被解析为字符串类型（string），而该组件内 Number1 和 Number2 属性都需要 int 类型的值，因此在定义路由参数时要加约束条件（:int，路由约束以英文的冒号开始），明确其参数值必须解析为整型数值。

运行示例程序，访问/hello，测试第 1 个组件，如图 17-10 所示。

访问/calc/150/235，测试第 2 个组件，如图 17-11 所示。

图 17-10　测试第 1 个组件

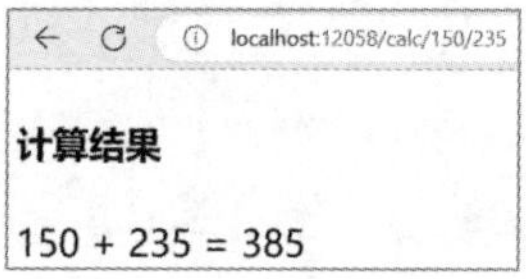

图 17-11　测试第 2 个组件

17.4.3　示例：使用[Route]特性

当路由规则模板需要格式化字符串时，使用 RouteAttribute 要比@page 指令灵活。本示例将演示结合 nameof 运算符，直接用组件属性成员的名称（Title）设置路由参数。

```
@attribute [Route("/{" + nameof(Title) + "?}")]

<h1>@Title</h1>
<hr />
<span>欢迎光临</span>

@code{
    [Parameter]
    public string? Title{ get; set; }

    protected override Task OnParametersSetAsync()
    {
        Title = Title ?? "默认标题";
        return Task.CompletedTask;
    }
}
```

最终产生的路由模板为/{Title?}。上述代码还重写了 OnParametersSetAsync()方法。该方法在组件接收到参数，赋值给属性之后调用。重写此方法可以验证 Title 属性有没有赋值。若为 null，表示 URL 中未提供 Title 路由参数，就为 Title 属性赋值“默认标题”；若已赋值且有效，则值不变。

运行示例程序，直接访问根路径（/），此时 Title 参数无值，显示“默认标题”，如图 17-12 所示。

接着，访问/Hello-App，此时 Title 参数的值有效，就会呈现到页面上，如图 17-13 所示。

图 17-12　Title 参数未赋值

图 17-13　Title 参数已赋值

17.5　布局组件

与 Razor Pages、MVC 一样，Blazor 应用程序也可以将共用的界面元素放到一个组件中，称为布局组件。普通的 Razor 组件默认派生自 ComponentBase 类，而布局组件需要从 LayoutComponentBase 类派生。在 Razor 文档中，可以用@inherits 指令声明组件的基类。例如：

```
@inherits LayoutComponentBase
```

在普通组件中，使用@layout 指令套用布局组件。

```
@layout CustLayout
```

在布局组件内，Body 属性相当于一个占位符。@Body 语句出现的地方将呈现内容组件。如下面的代码所示，内容组件将显示在<div>元素内。

```
<div> @Body </div>
```

在 RouteView 组件上，可以通过 DefaultLayout 属性指定一个默认布局组件。当路由匹配成功后，待呈现的组件如果未使用@layout 指令声明要使用的布局组件，那么将自动套用 DefaultLayout 属性所指定的布局组件。

```
<Router AppAssembly="@typeof(App).Assembly">
<Found Context="routeData">
<RouteView RouteData="@routeData" DefaultLayout="@typeof(MainLayout)" />
</Found>
<NotFound>
…
</NotFound>
</Router>
```

17.5.1 示例：导航栏

本示例将在布局组件中呈现简单的导航栏，然后将此布局应用到其他组件上。单击导航栏中的链接，会呈现相应的内容组件。

将布局组件命名为 CustLayout，代码如下。

```
@inherits LayoutComponentBase

<div id="hd">
<h5>自定义布局</h5>
<nav id="navbar">
<a href="/">主页</a>
<a href="/news">公司新闻</a>
<a href="/list">产品列表</a>
<a href="/about">公司介绍</a>
</nav>
</div>

<div>
    @Body
</div>
```

本示例中有 4 个内容组件，具体如下。

（1）主页（Index），使用根 URL。

```
@page "/"
@layout CustLayout

<h5>主页</h5>
<p>欢迎来到主页</p>
```

@layout 指令将套用 CustLayout 组件，用于布局。

（2）新闻页（News），访问 URL 为/news。

```
@page "/news"
@layout CustLayout

<h5>最近新闻</h5>

<ul>
<li>新闻 A</li>
<li>新闻 B</li>
<li>新闻 C</li>
</ul>
```

（3）产品页（List），访问 URL 为/list。

```
@page "/list"
@layout CustLayout

<h5>产品总览</h5>

<ul>
<li>产品 1</li>
<li>产品 2</li>
<li>产品 3</li>
<li>产品 4</li>
</ul>
```

（4）公司介绍页（About），访问 URL 为/about。

```
@page "/about"
@layout CustLayout

<h5>公司简介</h5>
<hr />
<p>ABCD 公司成立于 2002 年，主要产品线有…</p>
```

运行示例程序，默认显示主页，如图 17-14 所示。

单击“产品列表”链接，将呈现产品列表信息，如图 17-15 所示。

图 17-14　显示主页

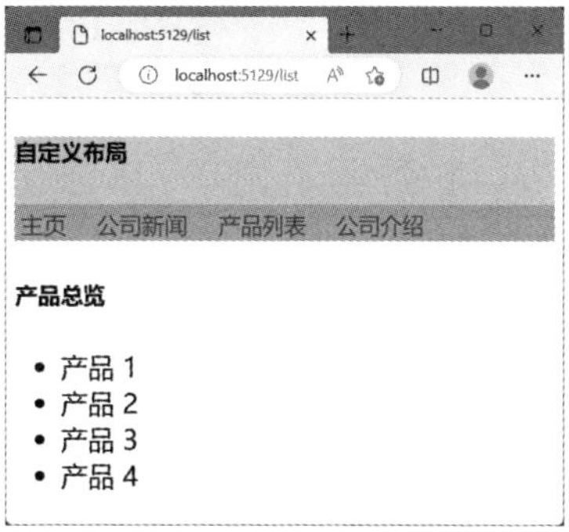

图 17-15　显示产品页

17.5.2 示例：将普通组件用于布局

普通的 Razor 组件（指不是从 LayoutComponentBase 类派生的组件类）也可以当作布局组件使用，方法是在组件中定义名为 Body 的公共属性，类型为 RenderFragment 委托。例如：

```
<div>
<h4>布局组件</h4>
</div>

<main> @Body </main>

@code{
    [Parameter]
    public RenderFragment? Body{ get; set; }
}
```

需要注意的是，Body 属性必须声明为公共成员，并且要应用 ParameterAttribute 后才能正常使用。

17.6 组件参数

组件类的公共属性附加上 ParameterAttribute 后就成为组件参数。在呈现组件时，可以向组件参数赋值。参数的赋值方式与常规的 HTML 元素的属性相同。例如：

```
<myComponent Para1=… Para2=… />
```

在为组件参数赋值时要注意参数值与参数的数据类型匹配。假设 SomeValue 参数为 int 类型，下面的代码却使用了一个浮点数值，会引发错误。

```
<myComponent SomeValue="0.22" />
```

如果组件之间是嵌套关系（如 A 组件内呈现 B 组件，B 组件内呈现 C 组件等），参数传递就比较简单，直接赋值即可。如果组件是 Blazor 应用的根组件（或叫作顶层组件，如项目模板生成的 App 组件），传递参数的方式则相对复杂一些。具体可参考本节的相关示例。

17.6.1 示例：嵌套组件的参数传递

本示例将创建名为 Test 的 Razor 组件，嵌套在项目模板生成的 Index 组件内。在 Index 组件内调用 Test 组件时会传递一个字符串对象。

Test 组件的代码如下。

```
<div>
<span>来自父级组件的参数：</span>
    @if(Data != null)
    {
<span>@Data</span>
    }
    else
```

```
    {
<span>&lt;未传递&gt;</span>
    }
</div>

@code{
    [Parameter]
    public string? Data{ get; set; }
}
```

Data 属性将公开为 Test 组件的参数。若 Data 属性的值有效，则显示其值；若为 null，则显示“<未传递>”（“<” 和 “gt;” 分别是 “<” 和 “>” 的转义字符）。

这里要注意，Test 组件未使用@page 指令，因为它是嵌套在 Index 组件内的，不需要进行路由匹配（无须定义访问 URL）。

在 Index 组件中呈现 Test 组件，并向 Data 参数赋值。

```
@page "/"

<h1>Hello, world!</h1>

<Test Data="一马平川"/>
```

运行示例程序，结果如图 17-16 所示。

图 17-16　在 Test 组件中呈现 Data 参数

17.6.2　示例：顶层组件的参数传递（Blazor Server）

本示例将演示在 Blazor Server 应用程序中，如何向顶层 Razor 组件（即根组件）传递参数。顶层组件没有父级组件，在 Razor Page（或 MVC View）中通过 Component 标记帮助器调用。Blazor Server 项目模块生成的 App 组件就是顶层组件。

顶层组件使用 Component 标记帮助器呈现，即 ComponentTagHelper 类。为了能够传递参数，该类公开了 Parameters 属性，类型为 IDictionary<string,object>，即字典数据结构。在 Razor 文档中可以使用形如 “param-<参数名称>=<参数值>” 的格式设置参数。例如，要为 Title 参数传值，则代码如下。

```
<component type="<组件类型>"…
            param-Title="@("abcdefg")" />
```

由于 Parameters 属性的类型是字典结构，且 Value 为 Object 类型，因此在向 Title 参数传递字符串时，要使用 Razor 表达式，即@(...)。直接使用字符串会导致识别错误。编译器会将参数值误判为组件内某个属性（或字段）成员的名称。例如，若 Title 参数的赋值代码如下。

```
param-Title="abcdefg"
```

编译器会认为这是让 Title 参数引用当前组件中名为 abcdefg 的属性（或字段）的值，而不是将其视为字符串常量。加上 Razor 表达式后，编译器就会知道 abcdefg 是字符串常量了。

本示例创建了名为 Root 的组件，作为顶层组件。

```
<div style="padding:20px; background-color:@(BackColor); color:
@(ForeColor); text-align:center">
```

```
<h3>Blazor 应用</h3>
</div>

@code {
    // 前景颜色
    [Parameter]
    public string ForeColor { get; set; } = "red";
    // 背景颜色
    [Parameter]
    public string BackColor { get; set; } = "white";
}
```

<div>元素的 CSS 样式将引用 ForeColor 和 BackColor 属性的值，使 ForeColor 属性能控制<div>元素的前景颜色；BackColor 属性则控制背景颜色。

在 Razor Page 文件中，使用 Component 标记帮助器呈现 Root 组件，并传递参数值。

```
<component type="typeof(Root)" render-mode="ServerPrerendered"
           param-forecolor="@("yellow")"
           param-backcolor="@("red")"/>
```

最终效果如图 17-17 所示。

图 17-17　通过组件参数设置前景颜色和背景颜色

17.6.3　示例：顶层组件的参数传递（Blazor WebAssembly）

Blazor WebAssembly 应用程序运行于 Web 浏览器内，用于加载 Razor 组件的 HTML 页面（默认为 Index.html）。不执行服务器代码，因此就不能使用像 Component 这样的标记帮助器，只能在程序入口点（Main()方法）通过.NET 代码设置参数。

本示例将创建名为 Root 的顶层组件，代码如下。

```
<div>
<h3>Blazor App</h3>
<hr />
<p>@(nameof(Message1))参数：@Message1</p>
<p>@(nameof(Message2))参数：@Message2</p>
<p>@(nameof(Message3))参数：@Message3</p>
</div>

@code {
    [Parameter]
    public string? Message1{ get; set; }
```

```
    [Parameter]
    public double Message2{ get; set; }

    [Parameter]
    public bool Message3{ get; set; }
}
```

Root 组件公开了 3 个参数：Message1 是字符串类型，Message2 是双精度数值类型，Message3 是布尔类型。

在 Blazor WebAssembly 应用程序入口点代码中，先通过字典对象设置好参数，然后创建 ParameterView 实例，最后将 ParameterView 实例传递给 RootComponents.Add()方法。

```
var builder = WebAssemblyHostBuilder.CreateDefault(args);
// 创建参数字典
var pardic = new Dictionary<string, object?>();
// 为参数赋值
pardic["message1"] = "Demo Data";
pardic["message2"] = 6.000025214d;
pardic["message3"] = false;
// 产生 ParameterView 实例
var parView = ParameterView.FromDictionary(pardic);
// 添加顶层组件，以及传递参数
builder.RootComponents.Add(typeof(Root), "#app", parView);
...
```

ParameterView.FromDictionary()是静态方法，根据已有的字典对象生成 ParameterView 实例。调用 RootComponents.Add()方法添加 Root 为顶层组件，并把 ParameterView 实例也传递进去。在呈现 Root 组件时，应用程序会自动为 Root 组件的参数赋值。

运行示例程序，结果如图 17-18 所示。

图 17-18　显示 3 个参数的值

17.7 级联参数

在 Razor 文档的标记树中使用级联参数（CascadingValue）组件，能够将参数值传递到各个子级组件中。以 CascadingValue 组件为父级，级联参数不仅可以跨越多个层级向下传递参数，也可以同时将值传递给多个子级组件。

CascadingValue 的子级组件需要在属性成员上应用 CascadingParameterAttribute 特性才能接收到级联参数的值。不管是公共属性还是非公共属性，都可以接收级联参数值。级联参数的赋值是自动完成的，不需要编写任何赋值代码。

17.7.1 示例：根据类型接收级联参数

这是级联参数的默认传递方式。在 CascadingValue 组件的子级组件中，只要属性成员满足以下条件，就会自动接收级联参数。

（1）属性的数据类型与 CascadingValue.Value 相同。

（2）属性应用了 CascadingParameterAttribute 特性。

本示例在 Index 组件中使用 CascadingValue 组件，并且 CascadingValue 组件内嵌套 A 组件，A 组件内嵌套 B 组件。A、B 组件都能够接收到级联参数。

A 组件定义 OtherValue 属性，用于接收级联参数。该属性为私有属性。

```
<div style="background-color: yellow; padding:20px; border:2px solid red">
<h3>A 组件</h3>
<p>
      @(nameof(OtherValue)): @OtherValue
</p>
   @ChildContent
</div>

@code{
   [Parameter]
   public RenderFragment? ChildContent{ get; set; }

   // 接收级联参数
   [CascadingParameter]
   private float OtherValue{ get; set; }
}
```

ChildContent 属性只是普通的组件参数，类型是 RenderFragment 委托。它用于呈现 A 组件的子级对象，本示例中是 B 组件。ChildContent 是默认命名，在 Razor 代码中可以省略。

B 组件定义 OuterValue、SomeValue 属性接收级联参数。在本示例中，OuterValue 属性的类型与级联参数匹配，而 SomeValue 属性的类型不匹配。因此，SomeValue 属性不会接收级联参数。

```
<div style="background-color:lightskyblue; padding:12px; border:
2px solid red">
<h3>B 组件</h3>
<p>@nameof(OuterValue): @OuterValue</p>
<p>@nameof(SomeValue): @SomeValue</p>
</div>

@code{

   // 接收级联参数
   [CascadingParameter]
   public float OuterValue{ get; set; }

   // 以下属性与 CascadingValue 组件的值类型不匹配
   [CascadingParameter]
   protected long SomeValue{ get; set; }
}
```

在 Index 组件中，使用 CascadingValue 组件设置一个 float 类型的值作为级联参数，并将 A 组件作为 CascadingValue 的子级，B 组件为 A 组件的子级。

```
@page "/"

<h1>Hello, world!</h1>

<CascadingValue Value="theValue">
<A>
<ChildContent>
<B />
</ChildContent>
</A>
</CascadingValue>

@code{
    private float theValue = 1005.027f;
}
```

其中，A 组件的 ChildContent 属性可以省略。

```
<CascadingValue Value="theValue">
<A>
<B />
</A>
</CascadingValue>
```

运行示例程序，结果如图 17-19 所示。

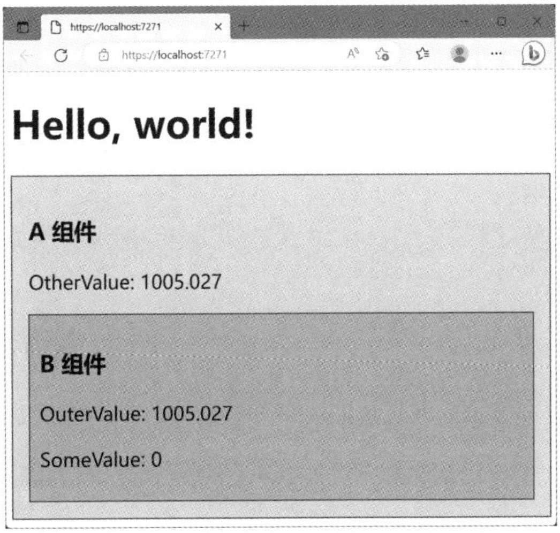

图 17-19　A、B 组件接收到的级联参数

从运行结果可以看到，OtherValue 和 OuterValue 属性都接收到了级联参数的值；SomeValue 属性由于类型与级联参数不匹配，不会接收参数值，因此保留 long 类型的默认值 0。

17.7.2 示例：根据命名接收级联参数

CascadingValue 组件可以嵌套使用——提供多个级联参数值。如果参数的类型相同，或者需要区分不同类型的不同参数，则需要给级联参数命名。子级组件在应用 CascadingParameterAttribute 特性时将通过 Name 属性指定要接收的级联参数的名称。

本示例演示了两个类型相同但值不同的级联参数，子级组件将通过参数名称接收对应的值。

子级组件 ChildTest 的代码如下。

```
<div>
<h3>子级组件</h3>
<p>第一个级联参数：@Value1</p>
<p>第二个级联参数：@Value2</p>
</div>

@code{
    [CascadingParameter(Name = "cascValue1")]
    public int Value1{ get; set; }

    [CascadingParameter(Name = "cascValue2")]
    public int Value2{ get; set; }
}
```

Value1 属性将接收名为 cascValue1 的参数值；Value2 属性将接收名为 cascValue2 的参数值。

在父级组件 ParentRoot 中，使用两个嵌套的 CascadingValue 组件设置级联参数。

```
<h3>父级组件</h3>

<CascadingValue Value="testValue_1" Name="cascValue1">
<CascadingValue Value="testValue_2" Name="cascValue2">
<ChildTest />
</CascadingValue>
</CascadingValue>

@code {
    int testValue_1 = 3000;
    int testValue_2 = 8650;
}
```

第 1 个参数的名称为 cascValue1，其值为 3000；第 2 个参数的名称为 cascValue2，其值为 8650。

运行示例程序，结果如图 17-20 所示。

图 17-20　显示两个级联参数的值

17.8 事件

Blazor 应用中的事件命名与 HTML DOM 原生的事件属性相同，使用时要在事件名称前加上@字符，格式为"@on<事件名称> = <委托>"。例如：

```
<button @onclick="ClickMe">…</button>
```

上述 Razor 标记为<button>元素的单击事件（click）绑定 ClickMe()方法。当按钮被用户单击时，就会调用 ClickMe()方法。当然，也可以使用 Lambda 表达式。

```
<button @onclick="@(e => …)"> …</button>
```

具备以下特点的方法成员或 Lambda 表达式均可赋值给事件处理委托：

（1）方法返回 void；

（2）方法返回 Task；

（3）方法无参数；

（4）方法带有一个事件参数（EventArgs 的派生类）对象。

DOM 事件与事件参数的类型映射被声明在一个名为 EventHandlers 的静态类上。Blazor 框架默认的 EventHandlers 类位于 Microsoft.AspNetCore.Components.Web 命名空间。源代码如下。

```
// Focus events
[EventHandler("onfocus", typeof(FocusEventArgs), true, true)]
[EventHandler("onblur", typeof(FocusEventArgs), true, true)]
[EventHandler("onfocusin", typeof(FocusEventArgs), true, true)]
[EventHandler("onfocusout", typeof(FocusEventArgs), true, true)]

// Mouse events
[EventHandler("onmouseover", typeof(MouseEventArgs), true, true)]
[EventHandler("onmouseout", typeof(MouseEventArgs), true, true)]
[EventHandler("onmouseleave", typeof(MouseEventArgs), true, true)]
[EventHandler("onmouseenter", typeof(MouseEventArgs), true, true)]
[EventHandler("onmousemove", typeof(MouseEventArgs), true, true)]
[EventHandler("onmousedown", typeof(MouseEventArgs), true, true)]
[EventHandler("onmouseup", typeof(MouseEventArgs), true, true)]
[EventHandler("onclick", typeof(MouseEventArgs), true, true)]
[EventHandler("ondblclick", typeof(MouseEventArgs), true, true)]
[EventHandler("onwheel", typeof(WheelEventArgs), true, true)]
[EventHandler("onmousewheel", typeof(WheelEventArgs), true, true)]
[EventHandler("oncontextmenu", typeof(MouseEventArgs), true, true)]

// Drag events
[EventHandler("ondrag", typeof(DragEventArgs), true, true)]
[EventHandler("ondragend", typeof(DragEventArgs), true, true)]
[EventHandler("ondragenter", typeof(DragEventArgs), true, true)]
[EventHandler("ondragleave", typeof(DragEventArgs), true, true)]
[EventHandler("ondragover", typeof(DragEventArgs), true, true)]
```

```
    [EventHandler("ondragstart", typeof(DragEventArgs), true, true)]
    [EventHandler("ondrop", typeof(DragEventArgs), true, true)]

    // Keyboard events
    [EventHandler("onkeydown", typeof(KeyboardEventArgs), true, true)]
    [EventHandler("onkeyup", typeof(KeyboardEventArgs), true, true)]
    [EventHandler("onkeypress", typeof(KeyboardEventArgs), true, true)]

    // Input events
    [EventHandler("onchange", typeof(ChangeEventArgs), true, true)]
    [EventHandler("oninput", typeof(ChangeEventArgs), true, true)]
    [EventHandler("oninvalid", typeof(EventArgs), true, true)]
    [EventHandler("onreset", typeof(EventArgs), true, true)]
    [EventHandler("onselect", typeof(EventArgs), true, true)]
    [EventHandler("onselectstart", typeof(EventArgs), true, true)]
    [EventHandler("onselectionchange", typeof(EventArgs), true, true)]
    [EventHandler("onsubmit", typeof(EventArgs), true, true)]

    // Clipboard events
    [EventHandler("onbeforecopy", typeof(EventArgs), true, true)]
    [EventHandler("onbeforecut", typeof(EventArgs), true, true)]
    [EventHandler("onbeforepaste", typeof(EventArgs), true, true)]
    [EventHandler("oncopy", typeof(ClipboardEventArgs), true, true)]
    [EventHandler("oncut", typeof(ClipboardEventArgs), true, true)]
    [EventHandler("onpaste", typeof(ClipboardEventArgs), true, true)]

    // Touch events
    [EventHandler("ontouchcancel", typeof(TouchEventArgs), true, true)]
    [EventHandler("ontouchend", typeof(TouchEventArgs), true, true)]
    [EventHandler("ontouchmove", typeof(TouchEventArgs), true, true)]
    [EventHandler("ontouchstart", typeof(TouchEventArgs), true, true)]
    [EventHandler("ontouchenter", typeof(TouchEventArgs), true, true)]
    [EventHandler("ontouchleave", typeof(TouchEventArgs), true, true)]

    // Pointer events
    [EventHandler("ongotpointercapture", typeof(PointerEventArgs), true, true)]
    [EventHandler("onlostpointercapture", typeof(PointerEventArgs), true, true)]
    [EventHandler("onpointercancel", typeof(PointerEventArgs), true, true)]
    [EventHandler("onpointerdown", typeof(PointerEventArgs), true, true)]
    [EventHandler("onpointerenter", typeof(PointerEventArgs), true, true)]
    [EventHandler("onpointerleave", typeof(PointerEventArgs), true, true)]
    [EventHandler("onpointermove", typeof(PointerEventArgs), true, true)]
    [EventHandler("onpointerout", typeof(PointerEventArgs), true, true)]
    [EventHandler("onpointerover", typeof(PointerEventArgs), true, true)]
    [EventHandler("onpointerup", typeof(PointerEventArgs), true, true)]

    // Media events
    [EventHandler("oncanplay", typeof(EventArgs), true, true)]
    [EventHandler("oncanplaythrough", typeof(EventArgs), true, true)]
    [EventHandler("oncuechange", typeof(EventArgs), true, true)]
    [EventHandler("ondurationchange", typeof(EventArgs), true, true)]
```

```
    [EventHandler("onemptied", typeof(EventArgs), true, true)]
    [EventHandler("onpause", typeof(EventArgs), true, true)]
    [EventHandler("onplay", typeof(EventArgs), true, true)]
    [EventHandler("onplaying", typeof(EventArgs), true, true)]
    [EventHandler("onratechange", typeof(EventArgs), true, true)]
    [EventHandler("onseeked", typeof(EventArgs), true, true)]
    [EventHandler("onseeking", typeof(EventArgs), true, true)]
    [EventHandler("onstalled", typeof(EventArgs), true, true)]
    [EventHandler("onstop", typeof(EventArgs), true, true)]
    [EventHandler("onsuspend", typeof(EventArgs), true, true)]
    [EventHandler("ontimeupdate", typeof(EventArgs), true, true)]
    [EventHandler("onvolumechange", typeof(EventArgs), true, true)]
    [EventHandler("onwaiting", typeof(EventArgs), true, true)]

    // Progress events
    [EventHandler("onloadstart", typeof(ProgressEventArgs), true, true)]
    [EventHandler("ontimeout", typeof(ProgressEventArgs), true, true)]
    [EventHandler("onabort", typeof(ProgressEventArgs), true, true)]
    [EventHandler("onload", typeof(ProgressEventArgs), true, true)]
    [EventHandler("onloadend", typeof(ProgressEventArgs), true, true)]
    [EventHandler("onprogress", typeof(ProgressEventArgs), true, true)]
    [EventHandler("onerror", typeof(ErrorEventArgs), true, true)]

    // General events
    [EventHandler("onactivate", typeof(EventArgs), true, true)]
    [EventHandler("onbeforeactivate", typeof(EventArgs), true, true)]
    [EventHandler("onbeforedeactivate", typeof(EventArgs), true, true)]
    [EventHandler("ondeactivate", typeof(EventArgs), true, true)]
    [EventHandler("onended", typeof(EventArgs), true, true)]
    [EventHandler("onfullscreenchange", typeof(EventArgs), true, true)]
    [EventHandler("onfullscreenerror", typeof(EventArgs), true, true)]
    [EventHandler("onloadeddata", typeof(EventArgs), true, true)]
    [EventHandler("onloadedmetadata", typeof(EventArgs), true, true)]
    [EventHandler("onpointerlockchange", typeof(EventArgs), true, true)]
    [EventHandler("onpointerlockerror", typeof(EventArgs), true, true)]
    [EventHandler("onreadystatechange", typeof(EventArgs), true, true)]
    [EventHandler("onscroll", typeof(EventArgs), true, true)]

    [EventHandler("ontoggle", typeof(EventArgs), true, true)]
    public static class EventHandlers
    {
    }
```

EventHandlers 类不需要定义任何成员，仅通过 EventHandlerAttribute 特性类创建事件与事件参数的映射关系。例如，onmousemove 表示 mousemove 事件，它的事件处理程序需要一个 MouseEventArgs 类型的事件参数。

Razor 组件的基类（ComponentBase）实现了 IHandleEvent 接口，在处理事件后会自动更新组件的呈现状态，开发人员不需要手动调用 StateHasChanged()方法重新呈现组件。

17.8.1　示例：计数器

本示例将实现简单的计数器。MyCounter 组件的代码如下。

```
@using Microsoft.AspNetCore.Components.Web

<h3>计数器</h3>

<p>当前计数：@_count</p>
<div>
<button @onclick="OnDecrease">递减</button>
<button @onclick="OnIncrease">递增</button>
</div>

@code {
    int _count = 0;

    void OnDecrease()
    {
        _count--;
        if (_count < 0)
        {
            _count = 0;
        }
    }

    void OnIncrease()
    {
        _count++;
        if(_count > 10000)
        {
            _count = 0;
        }
    }
}
```

处理<button>元素的 onclick 事件，以修改_count 字段的值。“递减”按钮的事件处理方法是 OnDecrease(),每次触发都会使_count 字段减 1;“递增”按钮的事件处理方法是 OnIncrease()，每次触发都会使_count 字段加 1。

运行示例程序，单击页面上的按钮进行测试，如图 17-21 所示。

17.8.2　示例：记录鼠标指针的位置

本示例将演示@onmousemove 事件的处理。该事件的参数类型为 MouseEventArgs。此参数提供了以下几种坐标。

（1）OffsetX 和 OffsetY：相对于当前元素的坐标，即以当前元素的左上角为原点进行计算。

（2）PageX 和 PageY：指针位置是相对于整个 HTML 文档。这适合于页面存在滚动条的情况。坐标的参照对象是当前页面，不受滚动影响。

（3）ClientX 和 ClientY：客户区域的坐标，即参照对象是浏览器窗口的可见区域。

（4）ScreenX 和 ScreenY：指针在当前屏幕中的坐标。

本示例仅在一个<div>元素上处理@onmousemove 事件，因此应获取相对于当前<div>元素的坐标，即 OffsetX 和 OffsetY。

```
@page "/"

<h1>Hello, world!</h1>
<div class="x" @onmousemove="OnMove"></div>
<p>指针的当前坐标：@offX，@offY</p>

@code{
    double offX, offY;

    void OnMove(MouseEventArgs me)
    {
        offX = me.OffsetX;
        offY = me.OffsetY;
    }
}
```

offX 和 offY 字段用于记录鼠标指针相对于当前元素的坐标，并在<p>元素中引用它们的值，以显示坐标信息，如图 17-22 所示。

图 17-21　简单的计数器

图 17-22　记录鼠标指针的坐标

17.8.3　EventCallback 结构体

EventCallback 结构体实现了 IEventCallback 接口，用于在组件中声明事件。假设 A 组件公开了 FEvent 事件，其他组件可以订阅 FEvent 事件（绑定某个方法）。当 A 组件引发 FEvent 时，绑定的方法会被调用。

组件事件实际上是以参数属性的形式公开的，即应用 ParameterAttribute。但属性的类型要定义为 EventCallback 或 EventCallback<TValue>。

```
[Parameter]
public EventCallback<string> FEvent{ get; set; }
```

EventCallback<TValue>是 EventCallback 的泛型版本，通过 TValue 类型参数可以设置事件

参数的类型。如果使用 EventCallback，那么事件参数的类型就是 Object。例如，上述代码中 FEvent 事件的事件参数为字符串类型。FEvent 事件可以绑定类似以下方法成员。

```
void SomeMethod();
Task SomeMethod();
void SomeMethod(string args);
Task SomeMethod(string args);
```

在 A 组件外，可以将 FEvent 事件绑定到特定的方法。

```
<A FEvent= "OnFE" … />

@code{
    void OnFE(string args)
    {
        …
    }
}
```

17.8.4 示例：进度条组件

本示例将实现一个进度条组件 ProgressBar，公开 OnProgressChanged 事件。当进度发生改变后会引发此事件。

ProgressBar 组件的代码如下。

```
@implements IDisposable

<div>
<h3>ProgressBar 组件</h3>
<progress max="100" value="@(Progress)" />
</div>

@code {
    private System.Timers.Timer _timer = new(100);
    private int Progress{ get; set; }

    // 公开的事件，可供其他组件使用
    [Parameter]
    public EventCallback<int> OnProgressChanged{ get; set; }

    // 组件初始化时启动定时器
    protected override void OnInitialized()
    {
        _timer.Elapsed += OnTimer;
        _timer.Start();
    }

    // 组件释放时停止定时器
    public void Dispose()
    {
        _timer.Stop();
        _timer.Elapsed -= OnTimer;
```

```
    }

    private async void OnTimer(object? sender, System.Timers.
    ElapsedEventArgs e)
    {
        // 进度递增
        Progress += 1;
        // 如果进度超过 100，就将其重置为 0
        if(Progress > 100)
        {
            Progress = 0;
        }
        await InvokeAsync(async () =>
        {
            // 引发事件
            await OnProgressChanged.InvokeAsync(Progress);
        });
    }
}
```

ProgressBar 组件使用 Timer 类，每 100ms 修改一次 Progress 属性的值。当属性的值到达 100 时，重新设置为 0。要引发 OnProgressChanged 事件，应调用它的 InvokeAsync()方法，并传递事件参数（此处传递的是 Progress 属性的值）。注意上述代码是调用组件实例的 InvokeAsync()方法（继承自 ComponentBase 类），通过 Lambda 表达式调用 OnProgressChanged 事件的 InvokeAsync()方法。这是由于 Timer 对象引发 Elapsed 事件时并不是在主线程中调度代码的，代码上下文并不在主线程上。所以，要通过组件实例的 InvokeAsync()方法“代理”访问组件成员。

在另一个组件中（本示例是 Index 组件）使用 ProgressBar 组件，并将 OnProgressChanged 事件与 OnProgress()方法绑定。当 OnProgressChanged 事件发生时，将进度值实时保存到变量 p 中。

```
@page "/"

<h1>Hello, world!</h1>

<ProgressBar OnProgressChanged="OnProgress" />

<p>当前进度：@p</p>

@code{
    int p;
    void OnProgress(int _p)
    {
        p = _p;
    }
}
```

运行示例程序，结果如图 17-23 所示。

图 17-23 实时显示进度

17.9 CSS 隔离

CSS 隔离，即将样式的作用范围限制在 Razor 组件上，CSS 样式只对当前组件有效，不会影响其他组件。CSS 隔离不能把样式写在 Razor 代码中，而是创建一个与 Razor 组件文件名相匹配的.css 文件，并把所有要隔离的样式全写在此文件中。例如，Index 组件的文件名为 Index.razor，对应的 CSS 隔离文件名就是 Index.razor.css。

读者可以通过以下示例了解 CSS 隔离的使用方法。

（1）在项目中添加 Test1.razor 文件，创建 Test1 组件。

```
<h3>Test1 组件</h3>
```

（2）添加 Test1.razor.css 文件，作为 Test1 组件专用的样式表，将<h3>元素的文本颜色设置为蓝色。

```
h3 {
    color: blue;
}
```

（3）添加 Test2.razor 文件，创建 Test2 组件。

```
<h3>Test2 组件</h3>
```

（4）添加 Test2.razor.css 文件，作为 Test2 组件的专用样式表，并把<h3>元素的文本颜色设置为绿色。

```
h3 {
    color: green;
}
```

（5）在 Index 组件中分别呈现 Test1 和 Test2 组件。

```
@page "/"

<h1>Hello, world!</h1>

<Test1 />
<Test2 />
```

（6）在承载 Blazor 根组件的页面上（Blazor Server 项目中的_Host.cshtml 文件或 Blazor WebAssembly 项目中的 index.html 文件）引用<App-Name>.styles.css 文件。其中，App-Name 是应用程序的名称，如 MyApp。

```
<!DOCTYPE html>
<html lang="en">
<head>
…
<link href="MyApp.styles.css" rel="stylesheet" />
</head>
```

```
<body>
…
</body>
</html>
```

MyApp.styles.css 文件是在生成项目之后生成的。首先，为每个组件单独生成<Component>.razor.rz.scp.css。其中，Component 表示组件名称，本示例中将生成以下两个样式文件。

```
// Test1.razor.rz.scp.css
h3[b-nwxr9mcglr] {
    color: blue;
}

// Test2.razor.rz.scp.css
h3[b-xr31lyzl0g] {
    color: green;
}
```

最后，合并到 MyApp.styles.css 文件中。

```
/* _content/MyApp/Pages/Test1.razor.rz.scp.css */
h3[b-nwxr9mcglr] {
    color: blue;
}
/* _content/MyApp/Pages/Test2.razor.rz.scp.css */
h3[b-xr31lyzl0g] {
    color: green;
}
```

在生成时，Blazor 会为每个组件的隔离样式分配一个 ID，本示例中 Test1 组件的 ID 为 b-nwxr9mcglr，Test2 组件的 ID 为 b-xr31lyzl0g。在最终呈现的 HTML 文档中，将 ID 作为元素的特性。

```
<h3 b-nwxr9mcglr >Test1 组件</h3>
<h3 b-xr31lyzl0g >Test2 组件</h3>
```

也可以自定义这些 ID，方法是打开项目文件（.csproj），添加以下内容。

```
<ItemGroup>
<None Update="Pages/Test1.razor.css" CssScope="cmp-1-id" />
<None Update="Pages/Test2.razor.css" CssScope="cmp-2-id" />
</ItemGroup>
```

CssScope 属性用来指定 CSS 的范围 ID。重新生成项目后，在 obj\{Debug|Release}\net<版本号>\scopedcss\bundle 目录下找到 MyApp.styles.css 文件，此时样式表会变为

```
/* _content/MyApp/Pages/Test1.razor.rz.scp.css */
h3[cmp-1-id] {
    color: blue;
}
/* _content/MyApp/Pages/Test2.razor.rz.scp.css */
```

```
h3[cmp-2-id] {
    color: green;
}
```

最终呈现的 HTML 文档也会改变。

```
<h3 cmp-1-id >Test1 组件</h3>
<h3 cmp-2-id >Test2 组件</h3>
```

（7）运行示例程序，结果如图 17-24 所示。

图 17-24　CSS 隔离

17.10　数据绑定

对于非交互元素（如<p><span>等），因其不需要接收用户输入，通常可以在 Razor 文档中直接引用代码对象（如属性、字段、变量、表达式结果等）。例如：

```
<span>@message</span>

@code{
    private string message = "Test";
}
```

而交互元素（常见于表单元素，如<input><textarea>）除了获取数据，还需要更新数据——将用户输入的内容传输给代码对象（如属性），即实现双向绑定。

在 Blazor 应用程序中，可以处理 onchange 事件，提取用户输入的内容，然后更新代码。

```
<input type="text" value="@data" @onchange="OnChanged" />
<p>你输入的内容：@data</p>

@code{
    private string? data;

    void OnChanged(ChangeEventArgs args)
    {
        data = args.Value?.ToString();
    }
}
```

当<input>元素失去焦点时，会引发 onchange 事件，此时可以通过处理该事件更新代码。

还有一种方法是使用@bind 指令。该指令能自动将输入元素的 value 属性与代码对象进行双向绑定，且默认会识别 onchange 事件。上述示例可以这样修改：

```
<input type="text" @bind="data" />
<p>你输入的内容：@data</p>

@code{
    private string? data;
}
```

显然，在双向绑定场景中，使用@bind 指令会更简洁。应用程序在编译后会生成以下代码。

```
// 写入<input>元素的开始标记
__builder.OpenElement(0, "input");
// 设置 type=text
__builder.AddAttribute(1, "type", "text");
// 设置 value=@data
__builder.AddAttribute(2, "value", global::Microsoft.AspNetCore.
Components.BindConverter.FormatValue(data));
// 生成@onchange 事件的处理代码
__builder.AddAttribute(3, "onchange", global::Microsoft.AspNetCore.
Components.EventCallback.Factory.CreateBinder(this, __value => data =
__value, data));
__builder.SetUpdatesAttributeName("value");
// 结束<input>元素的呈现
__builder.CloseElement();
```

其中比较重要的一行代码是

```
__builder.AddAttribute(3, "onchange", global::Microsoft.AspNetCore.
Components.EventCallback.Factory.CreateBinder(this, __value => data =
__value, data));
```

这里编译器自动生成了 onchange 事件的处理代码。__value 是<input>元素中最新输入的内容，它将最新的值传递给 data 字段。

17.10.1　示例：绑定日期输入元素

本示例将完成 DateTime 类型的字段成员与日期输入元素（<input>元素）之间的绑定。当用户在界面上选择日期后，被绑定字段的值也随之更新。

```
<input type="date" @bind="dateObj" />
<p>你输入的日期是：@dateObj?.ToShortDateString()</p>

@code{
    DateTime? dateObj = new DateTime(2023, 1, 1);

}
```

其中，@bind="dateObj"等价于@bind="@(dateObj)"。这是因为@bind 命令会自动完成非字符串对象的类型转换，不需要添加 Razor 表达式。

运行示例程序，修改<input>元素中的日期，<p>元素中所显示的日期会自动更新，如图 17-25 所示。

17.10.2　示例：使用 oninput 事件

@bind 指令绑定后，默认会关联交互元素的 onchange 事件，以达到更新绑定源（属性、字段成员，或代码表达式）的目的。在用户输入内容过程中，onchange 事件不会发生，只有当输入元素失去焦点时才会发生。

若要实时获取输入的内容，应当处理 oninput 事件，而非 onchange 事件。@bind 指令允许通过添加 event 属性修改默认事件。

下面的代码将使用<input>元素的 oninput 事件实时显示输入字符数量和输入内容。

```
<div>
<label for="txt">请输入：</label>
<input type="text" id="txt" @bind="Content" @bind:event="oninput" />
</div>
<p>你输入的内容：@Content</p>
<p>已输入@(Content?.Length)个字符</p>

@code{
    private string? Content { get; set; }
}
```

@bind 指令与 event 属性之间用英文的冒号连接，并且全部使用小写字母，不能写成@bind:Event。

运行示例程序，在输入元素中随机输入字符。可以看到，就算焦点未离开输入元素，两个<p>元素的内容都会立即更新，如图 17-26 所示。

图 17-25　日期对象的双向绑定

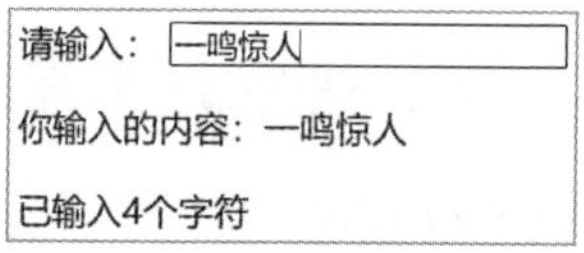

图 17-26　实时更新绑定源

17.10.3　组件之间的绑定

Razor 组件之间也可以使用@bind 指令进行绑定，即组件参数之间的双向绑定。绑定时使用@bind-<参数名>的格式，如

```
<X @bind-Size="@_size" />

@code{
    private int _size;
}
```

上述代码将 X 组件的 Size 参数与_size 字段进行绑定。

当被绑定的组件参数的值更新后，@bind 指令默认查找名为"<参数名>Changed"的事件。例如，上述示例中，X 组件的 Size 参数若被修改，将通过 SizeChanged 事件将参数值传回给_size 字段。

如果 X 组件中 Size 参数对应的更改事件不是 SizeChanged，那么就要用@bind-Size:event 指令指定事件的名称。

```
<X @bind-Size="@_size" @bind-Size:event="SizeUpdated" />
```

17.10.4 示例：Slider 组件

本示例将实现一个滑动条组件（用户可以在界面上拖动滑块来更改组件的值）——Slider。此组件公开 CurrentValue 参数，表示当前的进度值。CurrentValue 参数对应的更改事件为 CurrentValueChanged，其作用是让绑定到 CurrentValue 参数的其他组件能获取到更改通知。

Slider 组件的实现代码如下。

```
<input type="range" list="ticks" min="0" max="30" step="1" @bind=
"@(CurrentValue)" @bind:after="ValueUpdated" />
<datalist id="ticks">
<option value="0" label="0"></option>
<option value="10" label="10"></option>
<option value="20" label="20"></option>
<option value="30" label="30"></option>
</datalist>

@code {
    [Parameter]
    public int CurrentValue { get; set; }

    protected async override void OnInitialized()
    {
        // 在初始化时触发一次事件
        // 这是为了让与此组件绑定的其他组件能获得更改通知
        CurrentValue = 15;
        await CurrentValueChanged.InvokeAsync(CurrentValue);
    }

    [Parameter]
    public EventCallback<int> CurrentValueChanged { get; set; }

    async Task ValueUpdated()
    {
        await CurrentValueChanged.InvokeAsync(CurrentValue);
    }
}
```

<input type="range">元素会在 Web 页面上呈现一个滑动条。max 特性设置滑块的最大值（本例为 30），min 特性设置滑块的最小值（本例为 0）。step 是步长值，1 表示滑块每移动一个单位，<input>元素的值就增加 1。<datalist>元素用于在滑动条下方显示刻度值，本示例只显示 0、10、20、30。

Slider 组件内部也使用了绑定，将<input type="range">元素的值与 CurrentValue 参数绑定，滑块被拖动后会自动更新 CurrentValue 的值。但是，绑定到 CurrentValue 参数的外部组件无法知道值已经被修改，因此，用@bind:after 指令绑定 ValueUpdated()方法，这个方法在完成双向绑定后会调用。在 ValueUpdated()方法内引发一次 CurrentValueChanged 事件，其他绑定的组件就会收到通知。

在 Index 组件中呈现 Slider 组件，并将 CurrentValue 参数与_value 字段绑定。

```
@page "/"

<h1>Hello, world!</h1>

<Slider @bind-CurrentValue="@_value" />

<p>当前值：@_value</p>

@code{
    private int _value;
}
```

运行示例程序，当滑块被拖动时，会在 Slider 组件下方实时显示最新的值，如图 17-27 所示。

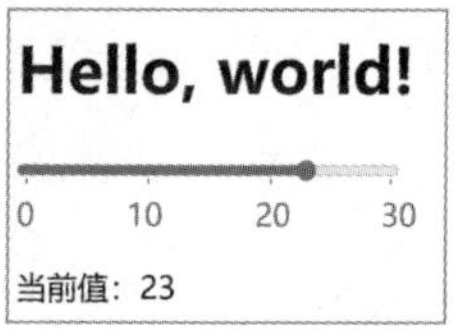

图 17-27　与 Slider 组件绑定

17.11　用.NET 代码编写组件

Razor 组件也可以直接使用.NET 代码（如 C#）来编写。普通组件从 ComponentBase 类派生即可，布局组件应派生自 LayoutComponentBase 类。除此之外，组件类还会用到一些特性类。表 17-1 列出了这些特性类与 Razor 指令之间的关系。

表 17-1　组件特性类与Razor指令对照

功　能	特　性　类	Razor指令
指定路由规则	RouteAttribute	@page
套用布局组件	LayoutAttribute	@layout
从服务容器中获取注入对象	InjectAttribute	@inject

假设有以下 Razor 组件。

```
@page "/index"

<h1>Hello, world!</h1>
<hr />
<span>ID: @MyID</span>

@code {
    private Guid MyID { get; set; }

    protected override void OnInitialized()
    {
        // 为 MyID 属性设置默认值
        MyID = Guid.NewGuid();
    }
}
```

若改为完全使用 C#语言编写，则对应的代码如下。

```
    [RouteAttribute("/index")]
    public class Index : ComponentBase
    {
    protected override void BuildRenderTree(RenderTreeBuilder __builder)
    {
            __builder.AddMarkupContent(0, "<h1>Hello, world!</h1>\r\n<hr>\r\n");
            __builder.OpenElement(1, "span");
            __builder.AddMarkupContent(2, "ID: ");
            __builder.AddContent(3, MyID);
            __builder.CloseElement();
        }

        private Guid MyID { get; set; }

        protected override void OnInitialized()
        {
            // 为 MyID 属性设置默认值
            MyID = Guid.NewGuid();
        }
    }
```

重写 ComponentBase 类的 BuildRenderTree()方法，通过 RenderTreeBuilder 对象的相关方法输出要呈现的内容。

17.11.1 渲染树

Render Tree（可翻译为“渲染树”）指由 HTML 文档中可见元素所构成的树状结构。浏览器在加载 HTML 文档时，会对其进行解析，构造 DOM 树。随后分析相关 CSS 规则，进行样式相关的计算，以确定元素的呈现大小和位置。DOM 树与 CSS 规则组建渲染树。最后浏览器将渲染树中可见的元素绘制出来，展示给用户。

渲染树由一系列指令组成，每个指令可以称为“帧”（RenderTreeFrame），表示渲染树中要处理的最小单位。帧的种类由 RenderTreeFrameType 枚举（位于 Microsoft.AspNetCore.Components.RenderTree 命名空间）定义，如表 17-2 所示。

表 17-2 RenderTreeFrameType枚举成员

枚举成员	说明
Element	表示要呈现一个HTML元素
Text	要呈现的是文本内容
Component	表示要呈现组件
Attribute	元素或组件的特性
Markup	HTML标记片段（纯HTML标记，未包含Razor标记）
Region	将一组帧放到一起，拥有独立的序列编号
ComponentReferenceCapture 或 ElementReferenceCapture	捕捉某个元素或组件的实例引用

通过.NET 代码编写组件，需要开发者手动构建渲染树。其中，核心的类是 RenderTreeBuilder。该类公开一组用于构建渲染树的方法成员。

（1）OpenElement：发出一帧，表示要向渲染树追加 HTML 元素。在设置好元素的特性或文本内容后，一定要调用 CloseElement()方法关闭元素（写入结束标记）。

（2）OpenComponent：发出一帧，表示追加 Razor 组件。同样也要调用 CloseComponent()方法关闭组件。

（3）OpenRegion：表示即将追加一组帧，完成后必须调用 CloseRegion()方法。

（4）AddAttribute：为新追加的 HTML 元素或 Razor 组件添加特性。

（5）AddMarkupContent：追加标记内容。一般是纯 HTML 内容。

（6）AddContent：追加标记内容。此方法可以追加纯 HTML 标记的内容，也可以追加组件的内容。

（7）Clear：清除帧。

重写 ComponentBase 类的 BuildRenderTree()方法，就可以获得 RenderTreeBuilder 对象的实例引用。

```
protected override void BuildRenderTree(RenderTreeBuilder builder)
{
    // 发出渲染帧
}
```

RenderTreeBuilder 对象在发出渲染帧时都需要指定一个整数值，这个数值表示生成渲染树后的代码位置（类似于代码行号）。这个整数可以是连续的，也可以是不连续的。

17.11.2 示例：用.NET 代码实现 App 和 Index 组件

本示例将演示通过编写.NET 代码直接实现 App 和 Index 组件。

（1）实现布局组件，命名为 MyLayout。它的基类是 LayoutComponentBase。

```
public class MyLayout : LayoutComponentBase
{
    protected override void BuildRenderTree(RenderTreeBuilder builder)
    {
        // 追加<main>元素
        builder.OpenElement(0, "main");
        // 呈现 Body 属性
        builder.AddContent(1, Body);
        // 关闭元素
        builder.CloseElement();
    }
}
```

布局组件的排版比较简单。<main>元素内直接设置 Body 属性，Body 属性的内容直接呈现在<main>元素内。

（2）实现 Index 组件。此组件包含<h1>元素，其内容为“Hello, World!”。随后是<button>元素，并处理它的 onclick 事件，在按钮被单击时累加计数（修改_count 字段）。最后，<p>元素内呈现_count 字段的值。

```
[Route("/")]
public class Index : ComponentBase
{
    protected override void BuildRenderTree(RenderTreeBuilder builder)
    {
        // 追加<h1>元素
        builder.OpenElement(0, "h1");
        // 设置内容
        builder.AddContent(1, "Hello, World!");
        builder.CloseElement();
        // 追加<button>元素
        builder.OpenElement(2, "button");
        // 处理 onclick 事件
        builder.AddAttribute(3, "onclick", ()=>{
            // 字段值加 1
            _count++;
        });
        // 设置文本内容
        builder.AddContent(4, "请单击这里");
        builder.CloseElement();
        // 追加<p>元素
        builder.OpenElement(5, "p");
        // 添加文本内容
        builder.AddContent(6, "当前计数：");
        builder.AddContent(7, _count);
        builder.CloseElement();
    }

    // 字段
    private int _count;
}
```

RouteAttribute 特性指定 Index 组件的访问路由为/。

为 HTML 元素设置内容可以调用 AddContent()和 AddMarkupContent()方法。两者的区别是 AddContent()方法所添加的内容仅作为普通文本处理（如 abcd），而 AddMarkupContent()方法所添加的内容会包含 HTML 元素（如<div>abcd</div>）。

处理<button>元素的 onclick 事件是通过 AddAttribute()方法实现的，即设置 onclick 特性，它的值是委托类型，可以传递方法成员引用、匿名委托或 Lambda 表达式。

（3）实现 App 组件。该组件内的根节点是 Router 组件。Router 组件上要设置 AppAssembly 属性，以及重要的 Found、NotFound 属性。

```
public class App : ComponentBase
{
    protected override void BuildRenderTree(RenderTreeBuilder builder)
    {
        // Router 组件
        builder.OpenComponent<Router>(0);
        // 设置待查找组件的程序集
        builder.AddAttribute(1, nameof(Router.AppAssembly), typeof
        (Program).Assembly);
        // 当找到目标组件后（Found 属性）
        builder.AddAttribute(2, nameof(Router.Found), (RenderFragment
        <Microsoft.AspNetCore.Components.RouteData>)(routeData =>
        {
            return _builder =>{
                // RouteView 组件
                _builder.OpenComponent<RouteView>(3);
                // 设置 RouteData 属性
                _builder.AddAttribute(4, nameof(RouteView.RouteData),
                routeData);
                // 设置默认布局组件
                _builder.AddAttribute(5, nameof(RouteView.DefaultLayout),
                typeof(MyLayout));
                _builder.CloseComponent();
            };
        }));
        // 若组件未找到（NotFound 属性）
        builder.AddAttribute(6, nameof(Router.NotFound), (RenderFragment)
        (_builder =>
        {
            // 追加 LayoutView 组件
            _builder.OpenComponent<LayoutView>(7);
            // 设置布局组件
            _builder.AddAttribute(8, nameof(LayoutView.Layout), typeof
            (MyLayout));
            // 设置内容
            _builder.AddAttribute(9, nameof(LayoutView.ChildContent),
            (RenderFragment)(_builder2 =>
            {
                _builder2.AddMarkupContent(10, "<p>未找到组件</p>");
            }));
            _builder.CloseComponent();
        }));
        builder.CloseComponent();
    }
}
```

Router 的 NotFound 属性的类型是 RenderFragment 委托。此委托类型的声明如下。

```
public delegate void RenderFragment(RenderTreeBuilder builder);
```

RenderFragment 委托要求与其关联的方法返回类型为 void，带一个 RenderTreeBuilder 类型的参数。也就是说，RenderFragment 委托绑定的方法与 BuildRenderTree()方法是一样的。这使得 NotFound 属性能够创建渲染树的子树。子树会合并到当前节点中。

Found 属性也是委托类型，但它的声明为

```
public delegate RenderFragment RenderFragment<TValue>(TValue value);
```

这个委托返回 RenderFragment 委托的引用，用于构建渲染子树。但它有一个类型参数 TValue，表示一个上下文数据。在 Router 组件中，TValue 参数的实际类型是 RouteData。这个上下文数据会传递给 RouteView 组件的 RouteData 属性。

（4）新建一个 Host.cshtml 文件（Razor 页面），用于承载 App 组件。

```
@page
@addTagHelper *, Microsoft.AspNetCore.Mvc.TagHelpers

<!DOCTYPE html>
<html lang="zh">
<head>
<meta charset="utf-8"/>
<title>Demo</title>
</head>
<body>
<component type="@typeof(App)" render-mode="@RenderMode.
ServerPrerendered" />
<script src="~/_framework/blazor.server.js"></script>
</body>
</html>
```

（5）在应用程序的初始化代码中，启用 Razor Pages 和 Blazor Server 功能，并映射相关的终结点。本示例的 Blazor 应用是基于服务器的，Blazor WebAssembly 的实现原理相同（组件是通用的）。

```
var builder = WebApplication.CreateBuilder(args);
builder.Services.AddRazorPages().WithRazorPagesAtContentRoot();
builder.Services.AddServerSideBlazor();
var app = builder.Build();

app.UseStaticFiles();
app.UseRouting();
app.MapBlazorHub();
app.MapFallbackToPage("/Host");

app.Run();
```

运行示例程序，结果如图 17-28 所示。

17.11.3　示例：使用依赖注入

本示例演示的是在使用.NET 代码编写 Razor 组件时使用依赖注入。

在 Razor 组件中，不使用构造函数接收注入对象，而是将要接收的对象放到属性成员上。

接收依赖注入的属性成员必须应用 InjectAttribute 特性。

执行以下命令，创建一个空白的 Blazor WebAssembly 应用。

```
dotnet new blazorwasm-empty -n <项目名称> -o <项目存放路径>
```

创建自定义组件，命名为 Test。

```
public class Test : ComponentBase
{
    [Inject]
    private IWebAssemblyHostEnvironment? Env { get; set; }

    protected override void BuildRenderTree(RenderTreeBuilder builder)
    {
        // 追加第 1 个<div>元素
        builder.OpenElement(0, "div");
        // 设置内容
        builder.AddContent(1, "运行环境：");
        builder.AddContent(2, Env?.Environment);
        // 关闭元素
        builder.CloseElement();
        // 追加第 2 个<div>元素
        builder.OpenElement(3, "div");
        // 设置内容
        builder.AddContent(4, "基础地址：");
        builder.AddContent(5, Env?.BaseAddress);
        // 关闭元素
        builder.CloseElement();
    }
}
```

Env 属性的值来自依赖注入，服务类型为 IWebAssemblyHostEnvironment。在构建渲染树时，把当前运行环境名称以及应用程序基础地址输出到 Web 页面。

打开 Index 组件（由项目模板生成的 Index.razor 文件），呈现 Test 组件。

```
@page "/"

<h1>Hello, world!</h1>
<Test />
```

运行示例程序，结果如图 17-29 所示。

图 17-28 用.NET 代码编写组件

图 17-29 输出运行环境和基础地址

17.12 .NET 与 JavaScript 互操作

.NET 代码与 JavaScript 代码支持相互调用。尽管 Blazor 应用程序能够使用.NET 代码替代 JavaScript 代码，但目前来说也做不到完全替代，因此，在特定需求下，两种代码之间相互调用是有意义的。

.NET 代码调用 JavaScript 代码需要用到 IJSRuntime 核心接口。实现此接口的类型实例将通过依赖注入从服务容器中获得。对于 Blazor Server 应用程序，在调用 AddServerSideBlazor()方法时，自动向服务容器注册 RemoteJSRuntime 类（该类是内部类型，未公开）。对于 Blazor WebAssembly 应用程序，默认注册的是 DefaultWebAssemblyJSRuntime 类（该类也是内部类型）。IJSRuntime 接口公开 InvokeAsync()方法，用来调用 JavaScript 代码。另外，JSRuntimeExtensions 类也提供用于调用无返回值 JavaScript 函数（即 undefined）的 InvokeVoidAsync()方法，以及可接收 params 参数的 InvokeAsync()方法。

例如，用.NET 代码调用 JavaScript 中的 console.log()方法。

```
await js.InvokeVoidAsync("window.console.log", txt);
```

txt 是传递给 log()方法的参数。console.log()方法没有返回值，因此调用时可以使用 InvokeVoidAsync()方法。window 对象是全局对象，在访问时可以省略。上述代码也简化为

```
await js.InvokeVoidAsync("console.log", txt);
```

在 JavaScript 代码中，将通过 DotNet 对象的 invokeMethod()或 invokeMethodAsync()方法访问.NET 代码。其中，invokeMethod()方法是同步调用，invokeMethodAsync()方法是异步调用。注意 Blazor Server 应用不能使用同步调用（因为需要通过网络远程调用，有一定的延时，不适合同步调用），而异步调用可兼容 Blazor Server 和 Blazor WebAssembly 应用程序。

在.NET 代码中，必须在方法成员（静态方法或实例方法都可以）上应用 JSInvokableAttribute 特性。该特性类可通过 identifier 参数为方法成员提供自定义标识。如果 identifier 参数未赋值，则使用方法成员的名称作为标识。方法标识在其所在的程序集中必须是唯一的。

例如，在.NET 代码中定义名为 GetNumber()的静态方法。

```
[JSInvokable]
public static Task<int> GetNumber(){
…
}
```

在 JavaScript 代码中通过 DotNet.invokeMethodAsync()方法调用。

```
DotNet.invokeMethodAsync("Test", "GetNumber")
    .then(data => {
    …
    });
```

上述代码中，Test 是.NET 代码所在的程序集名称，GetNumber 是要调用的.NET 方法名称。

17.12.1 示例：调用 JavaScript 中的 alert()方法

本示例将演示在.NET 代码中调用 JavaScript 中的 alert()方法。alert()方法属于 window 全局对象，因此在调用时下面两种引用方式都可以使用。

```
window.alert();
alert();
```

alert()方法的功能是弹出提示框，参数接收要显示在对话框中的文本。

在 Index 组件中添加<button>元素，被单击后弹出提示对话框。

```
@page "/"
@inject IJSRuntime jsCaller

<h1>Hello, world!</h1>

<button @onclick="OnClick">单击这里</button>

@code{
    async Task OnClick()
    {
        // 生成随机数
        double v = Random.Shared.NextDouble();
        // 显示提示框
        await jsCaller.InvokeVoidAsync("alert", $"数值：{v}");
    }
}
```

在组件内需要用@inject 指令从服务容器中获得 IJSRuntime 服务实例，并分配一个名为 jsCaller 的局部变量。由于 window.alert()方法无返回值，调用时可以用 InvokeVoidAsync()方法。

运行示例程序，单击页面上的“单击这里”按钮，Web 浏览器会弹出提示对话框，显示一个随机生成的双精度数值，如图 17-30 所示。

图 17-30 弹出提示信息

17.12.2 示例：调用 QRCode.js 生成二维码

本示例将使用.NET 代码调用 JavaScript 代码，生成二维码。该操作需要用到 QRCode.js 文件，它位于 wwwroot\scripts 目录下。

在承载 Blazor 应用的页面上（Blazor Server 是 Pages_Host.cshtml 文件，Blazor WebAssembly 是 wwwroot\index.html 文件），将以下 JavaScript 代码添加到<body>元素的结束标记之前（即</body>之前）。

```
<script src="~/scripts/qrcode.js"></script>
<script>
    window.makeCode = (element, data) =>
```

```
    {
        // 生成前先清除现有数据
        element.innerHTML = null;
        var qr = new QRCode(element, {
            height: 150,                    // 高度
            width: 150                      // 宽度
        });
        qr.makeCode(data);
    }
</script>
```

上述代码为全局的 window 对象添加 makeCode()方法（这样封装是为了方便调用）。其中，element 参数表示要呈现二维码的 HTML 元素，data 表示二维码所包含的文本数据。

在 Index 组件中，将 Razor 代码修改为

```
@page "/"
@inject IJSRuntime js

<label for="data">请输入文本：</label>
<input id="data" @bind="_data" />

<button @onclick="onClick">生成二维码</button>
<div @ref="_showCodeDiv" style="margin: 20px"></div>

@code{
    // 二维码的文本数据
    string? _data;
    // 特殊对象，用于引用 HTML 元素
    ElementReference _showCodeDiv;

    async void onClick()
    {
        // 调用 makeCode()方法
        await js.InvokeVoidAsync("window.makeCode", _showCodeDiv, _data);
    }
}
```

先通过依赖注入获得 IJSRuntime 对象的引用，变量名为 js。

在该组件中，_data 字段表示二维码中要包含的数据，类型为字符串，与<input>元素绑定，可获取键盘输入的字符；_showCodeDiv 字段用于保存对<div>元素的引用，类型为 ElementReference 结构体，用于在.NET 代码中封装对 HTML 元素的引用。单击按钮后，触发 onclick 事件，生成二维码，并呈现在<div>元素内。

运行示例程序，先在文本框内输入内容，接着单击“生成二维码”按钮，结果如图 17-31 所示。读者可以通过微信等手机 App 的“扫一扫”功能检验所生成的二维码是否正确。

图 17-31 生成二维码图片

17.12.3 示例：阶乘计算器

本示例实现在页面上输入一个整数值，然后计算其阶乘（$1\times2\times3\times\cdots\times n$）。计算阶乘的过程由.NET 代码完成，再由 JavaScript 代码调用。具体步骤如下。

（1）用 C#代码实现 FactHelper 类。

```
public class FactHelper
{
    // 计算阶乘
    [JSInvokable("factor")]
    public static string Run(uint baseNum)
    {
        BigInteger result = new BigInteger(1);
        uint x = 1;
        while(x <= baseNum)
        {
            result *= x;
            x++;
        }
        return result.ToString();
    }
}
```

Run()方法是静态成员，为了让它能被 JavaScript 代码调用，必须应用 JSInvokableAttribute 特性类。上述代码在应用 JSInvokableAttribute 特性时为 Run()方法指定了别名 factor。JavaScript 代码将使用此命名查找 Run()方法。

阶乘的计算结果一般比较大，不能用常见的 32 位或 64 位整数类型，而应当使用 BigInteger 类型。

（2）在 Blazor 应用的承载页面中，将以下 JavaScript 代码添加到<body>元素的末尾。

```
<script>
    window.doFact = async (number, element) => {
        var r = await DotNet.invokeMethodAsync("MyApp", "factor", number);
        element.textContent = `计算结果：${r.toString()}`;
    };
</script>
```

将调用.NET 代码的 JavaScript 代码封装为一个函数，并添加到 window 全局对象中。

（3）对 Index 组件做以下修改。

```
@page "/"

<h2>阶乘计算器</h2>
<label for="the_num">请输入：</label>
<input id="the_num" type="number"/>
<button onclick='window.doFact(parseInt(document.getElementById
("the_num").value), document.getElementById("result"))'>计算</button>
<p id="result" style="word-break:break-all"></p>
```

注意，此处<button>元素在处理 onclick 事件时没有加@字符，这是因为这里使用的是 JavaScript 代码，并非.NET 代码。onclick 事件直接关联到 window.doFact()方法上，单击按钮时会直接调用。

（4）运行示例程序，输入一个整数，然后单击“计算”按钮，计算结果将显示在按钮下方，如图 17-32 所示。

图 17-32　计算输入整数的阶乘

17.12.4　示例：JavaScript 调用.NET 对象的实例方法

本示例将演示如何在 JavaScript 中调用.NET 对象实例的方法成员。由于调用静态.NET 方法时不需要对象引用，因此在 JavaScript 代码中直接调用 DotNet.invokeMethodAsync()方法并传递程序集的名称即可调用。而调用实例方法则需要先获取到.NET 对象的实例引用，大致流程如下。

（1）.NET 代码公开一个静态方法 S，允许 JavaScript 代码调用。S 方法中使用 DotNetObjectReference<TValue>类封装.NET 对象的引用，并返回给 JavaScript 代码。

（2）JavaScript 代码调用 S 方法，返回的.NET 对象引用由 DotNetObject 对象重新封装。

（3）通过 DotNetObject 对象的 invokeMethod()或 invokeMethodAsync()方法调用.NET 实例方法 A。

（4）.NET 的实例方法 A 被调用并返回结果（如果有返回值）。

（5）JavaScript 代码获取调用结果。

在.NET 代码中，对象引用由 DotNetObjectReference 类封装；而在 JavaScript 代码中，对象引用则由 DotNetObject 类封装。

本示例用.NET 代码实现了 Hasher 类，其功能是使用 SHA256 算法计算输入字符串的哈希值，返回计算结果。随后用 JavaScript 代码调用 Hasher 类的 ComputeSHA256()方法。

```
public class Hasher
{
    [JSInvokable]
    public ValueTask<string> ComputeSHA256(string text)
```

```
    {
        SHA256 sha256 = SHA256.Create();
        byte[] data = Encoding.UTF8.GetBytes(text);
        byte[] outdata = sha256.ComputeHash(data);
        sha256.Dispose();
        string result = Convert.ToHexString(outdata);
        return ValueTask.FromResult(result);
    }
}
```

ComputeSHA256()方法需要应用 JSInvokableAttribute 特性，使其允许 JavaScript 代码访问。另外，需要一个静态方法，方便 JavaScript 代码获取 Hasher 实例的引用。

```
public static class Helpers
{
    [JSInvokable]
    public static ValueTask<DotNetObjectReference<Hasher>> GetHasherObject()
    {
        var obj = new Hasher();
        var refobj = DotNetObjectReference.Create(obj);
        return ValueTask.FromResult(refobj);
    }
}
```

DotNetObjectReference.Create 是静态成员，直接调用就能创建 DotNetObjectReference 对象。

本示例未使用 Razor 组件，而是直接在 Razor 页面中用 JavaScript 代码访问 Hasher 类。

```
@page "/"

<!DOCTYPE html>
<html lang="zh">
<head>
<meta charset="utf-8" />
<base href="~/" />
</head>
<body>
<div style="margin-bottom:18px">
<label for="txt">请输入文本：</label><br />
<input type="text" id="txt" />
</div>

<button id="btn">计算 SHA256</button>
<p id="display" style="word-break:break-all"></p>

<script src="_framework/blazor.server.js"></script>
<script>
      // 获取 HTML 元素的引用
      var button = document.getElementById("btn");
      var input = document.getElementById("txt");
```

```
        var showresult = document.getElementById("display");
        // 为按钮添加单击事件的处理代码
        button.addEventListener('click', async ()=>{
            if(input.value === '') return;
            // 调用静态的.NET 方法
            DotNet.invokeMethodAsync("MyApp", "GetHasherObject").then
            (async (obj) => {
                var str = input.value;
                // 调用 Hasher 类的实例方法成员
                var res = await obj.invokeMethodAsync("ComputeSHA256", str);
                // 显示结果
                showresult.textContent = `结果: ${res}`;
                // 释放引用
                obj.dispose();
            });
        });
</script>
</body>
</html>
```

虽然本示例未使用 Razor 组件，但用到了 Blazor 功能，因此 blazor.server.js 文件是必须引用的（Blazor WebAssembly 应用也如此）。

运行示例程序，在文本框中输入测试字符串，再单击“计算 SHA256”按钮，就会显示字符串的 SHA256 哈希值，如图 17-33 所示。

图 17-33　计算输入文本的哈希值

第 18 章

验证与授权

本章要点

- ❑ 验证与授权的关系
- ❑ 验证处理程序与中间件
- ❑ 授权的必要条件
- ❑ 授权策略

18.1 验证与授权的关系

验证与授权是应用程序安全方面的两个独立行为，但在许多应用场景下两者是搭配使用的。而且，它们的英文单词在拼写上非常接近，容易混淆。

验证（Authentication）即“你是谁”，此行为检查用户所提供的各种标识，明确其身份。就像去银行办理相关业务时，办理者需要出示个人身份证一样，必须以有效的方式证明“是你本人”。应用程序常用的验证方案有用户名和密码、指纹识别、人脸识别、电子邮箱或手机号码、进出小区所使用的门卡、汽车钥匙等。

授权（Authorization）即“允许你做什么”。用户通过了身份验证，顺利登录某系统后，系统会根据用户相关信息对其所拥有的权限做评估（分析）。例如，QQ 群的管理员可以删除群内用户，而普通成员是不能删除群内其他成员的。或者将授权进行更细化的配置，如某办公系统中，小李可以查看公司的采购记录，而小张则不能查看。

验证与授权虽然是独立进行的，但它们之间也有着密不可分的依赖关系。如果应用程序根本不知道访问它的是什么人，那么应用程序又拿什么去评估访问者的权限呢？假设小明要去小红家里做客，当小红家的门铃响起时，小红先要确定来人是不是小明，如果是，则开门让他进来（验证）。小明进入小红家后，只能在客厅活动，未经对方允许（授权），是不能随便进入小红的卧室的。

可见，授权是在验证的基础上进行的，缺少验证过程收集的用户信息，就无法进行权限评估。匿名访问也是经过验证的，只是此过程中不需要收集太多用户信息，并在授权阶段不做严格的评估，直接通过。

18.2　与验证有关的核心服务

ASP.NET Core 应用程序要支持验证行为，必须在初始化阶段向服务容器注册相关的服务类型。调用 AddAuthenticationCore()或 AddAuthentication()扩展方法即可完成。通常推荐调用 AddAuthentication()方法。该方法内部也会调用 AddAuthenticationCore()方法，并注册一些额外的功能，如数据保护（DataProtection）、Web 文本编码器（WebEncoders）、系统时钟服务（ISystemClock）等。

验证功能中有 3 个服务类型（接口）比较重要。

（1）IAuthenticationSchemeProvider：该接口公开一组方法，可以获取已注册的验证方案（Authentication Schemes）。一个验证方案会对应一个 IAuthenticationHandler 服务，用于实现具体的验证逻辑。

（2）IAuthenticationHandlerProvider：上文提到过，验证方案与 IAuthenticationHandler 是一一对应的。IAuthenticationHandlerProvider 接口要求公开 GetHandlerAsync()方法，可以根据当前 HTTP 请求关联的 HttpContext 对象以及验证方案的名称来检索出 IAuthenticationHandler 实例。

（3）IAuthenticationService：前面两个接口只有在需要深度自定义验证功能时才会用到，而 IAuthenticationService 接口在常规开发中会用得比较多。不过，ASP.NET Core 已为 HttpContext 类定义了一组扩展方法（如 AuthenticateAsync()、ChallengeAsync()等），它们内部实际调用的是 IAuthenticationService 接口的方法，在能够方便访问到 HttpContext 对象的场景（如中间件、MVC 控制器内）中可以直接调用这些扩展方法。但在编写自定义服务类时，可以通过依赖注入获得 IAuthenticationService 引用，再调用其成员方法完成验证。

18.3　验证处理程序

验证处理程序（即 IAuthenticationHandler 接口）用于实现验证逻辑，要求实现类必须公开以下方法成员。

（1）InitializeAsync()：当验证处理程序被激活时进行初始化工作。通过该方法能获得 HttpContext 对象的引用，方便在验证过程中提取客户端所发送的数据。另外，AuthenticationScheme 类型的参数提供了与验证方案有关的信息。

（2）AuthenticateAsync()：实现验证逻辑。

（3）ChallengeAsync()：若验证未成功，可以调用此方法要求客户端重新发送有效的数据（如凭据）。例如，跳转到登录页面，让用户输入密码。

（4）ForbidAsync()：禁止请求。假设客户端已经验证失败超过 3 次，就可以调用此方法。实现该方法可以实现自定义的响应消息，告知客户端该请求被禁用。

18.3.1 示例：验证 HTTP 消息头

本示例的验证方案名称为 abc，为了方便在代码中引用，可以对其进行硬编码。

```
public static class AuthenSchemes
{
    public static readonly string Default = "abc";
}
```

在需要提供验证方案名称的地方直接使用 AuthenSchemes.Default。

（1）定义 CustAuthenticationHandler 类，实现 IAuthenticationHandler 接口。

```
public class CustAuthenticationHandler : IAuthenticationHandler
{
    #region 私有属性
    private HttpContext HttpContext { get; set; }
    private string SchemeName { get; set; }
    #endregion

    public Task InitializeAsync(AuthenticationScheme scheme, HttpContext
    context)
    {
        // 获取相关对象的引用
        HttpContext = context;
        SchemeName = scheme.Name;
        return Task.CompletedTask;
    }

    public Task<AuthenticateResult> AuthenticateAsync()
    {
        AuthenticateResult result = AuthenticateResult.NoResult();
        // 验证方案是 abc
        if (SchemeName == AuthenSchemes.Default)
        {
            // 查找 x-auth 消息头
            var req = HttpContext.Request;
            if (req.Headers.TryGetValue("x-auth", out var value))
            {
                // 有效值：8174e、6353k
                string headerval = value.ToString();
                if(headerval.Equals("8174e", StringComparison.
                OrdinalIgnoreCase)
                || headerval.Equals("6353k", StringComparison.
                OrdinalIgnoreCase))
                {
                    // 验证成功
                    ClaimsPrincipal myPrincipal = new(new ClaimsIdentity(
                    authenticationType: AuthenSchemes.Default ));
                    var ticket = new AuthenticationTicket(myPrincipal,
```

```
                authenticationScheme: AuthenSchemes.Default);
                result = AuthenticateResult.Success(ticket);
            }
            else
            {
                // 验证失败
                result = AuthenticateResult.Fail("x-auth 消息头的值无效");
            }
        }
    }
    return Task.FromResult(result);
}

public async Task ChallengeAsync(AuthenticationProperties? properties)
{
    // 验证未通过，要求客户端重新发送凭据
    HttpContext.Response.StatusCode = StatusCodes.Status401Unauthorized;
    HttpContext.Response.ContentType = "text/plain;charset=utf-8";
    await HttpContext.Response.WriteAsync("验证未通过，请再次尝试");
}

public Task ForbidAsync(AuthenticationProperties? properties)
{
    HttpContext.Response.StatusCode = StatusCodes.Status403Forbidden;
    return Task.CompletedTask;
}
}
```

CustAuthenticationHandler 类的验证规则是检查客户端请求的消息中是否包含名为 x-auth 的消息头。如果是，就判断它的值是否有效。只有 8174e、6353k 两个值是有效的，其他值均视为验证失败。

AuthenticateAsync()方法需要产生一个 AuthenticateResult 实例描述验证结果。AuthenticateResult.Success 表示验证成功，AuthenticateResult.Fail 表示验证失败，AuthenticateResult.NoResult 表示无验证结果（验证过程中未获取到足够的信息，状态不能确定时，可以返回无结果）。

（2）将 CustAuthenticationHandler 类与验证方案 abc 关联起来。在服务容器上调用 AddAuthentication()扩展方法，并通过 AuthenticationOptions 选项类添加验证方案。

```
builder.Services.AddAuthentication(options =>
{
    options.AddScheme<CustAuthenticationHandler>(AuthenSchemes.Default,
    null);
});
```

（3）在 HTTP 管道上映射一个终结点/test，被访问时会调用 abc 方案进行验证。如果验证通过，就返回文本“验证通过，欢迎使用本服务”。如果验证未通过，则调用 ChallengeAsync()方法要求客户端再次发送请求。

```
app.MapGet("/test", async (HttpContext context) =>
{
```

```
    context.Response.ContentType = "text/plain;charset=utf-8";
    // 验证
    var result = await context.AuthenticateAsync(AuthenSchemes.Default);
    if (result.Succeeded)
    {
        await context.Response.WriteAsync("验证通过，欢迎使用本服务");
    }
    else
    {
        // 未通过验证
        await context.ChallengeAsync(AuthenSchemes.Default);
    }
});
```

（4）在用于测试的 Razor 页面上，通过 JavaScript 代码访问/test 终结点。

```
@page "/"

<!DOCTYPE html>
<html lang="zh-cn">
<head>
<title>Demo</title>
<meta charset="utf-8" />
</head>
<body>
<p>
<label for="hdvalue" style="display:block">请输入 x-auth 消息头的值:</label>
<input type="text" id="hdvalue" />
</p>
<button id="btn">验证</button>
<p id="result" style="color:darkblue"></p>
<script>
            // 获取所需 HTML 元素的引用
            var button = document.getElementById("btn");
            var input = document.getElementById("hdvalue");
            var result = document.getElementById("result");
            // 侦听事件
            button.addEventListener("click", e =>{
                let headerval = input.value;
                // 准备 HTTP 消息头
                let headers = new Headers;
                headers.append("x-auth", headerval);
                // 发送请求
                fetch("/test",{
                    headers: headers,
                    method: "GET"
                }).then(response =>{
                    // 显示验证结果
                    response.text().then(str =>{
```

```
                    result.textContent = str;
                });
            });
        });
</script>
</body>
</html>
```

JavaScript 代码使用 fetch()方法向服务器发出请求，请求方式为 HTTP-GET，并携带名为 x-auth 的头部。

运行示例程序，在文本框中输入 s729，并单击下方按钮。由于 x-auth 消息头的值无效，服务器响应“验证未通过，请再次尝试”，如图 18-1 所示。

输入 8174e，再次提交，服务器响应“验证通过，欢迎使用本服务”，如图 18-2 所示。

图 18-1　验证失败

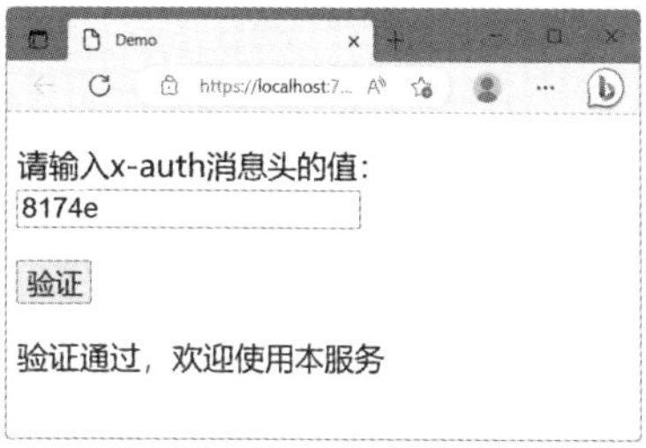

图 18-2　验证成功

18.3.2　示例：多个验证方案共用一个 IAuthenticationHandler 接口

本示例将实现一个 MyAuthenHandler 验证处理程序，并且与 3 个验证方案（s1、s2、s3）关联。这 3 个验证方案的名称使用静态字段声明，便于访问。

```
public class AuthenSchemes
{
    public static readonly string Scheme1 = "s1";
    public static readonly string Scheme2 = "s2";
    public static readonly string Scheme3 = "s3";
}
```

MyAuthenHandler 类只实现 AuthenticateAsync()方法。由于该处理程序仅用于演示，所以验证结果总是 Success。

```
public class MyAuthenHandler : IAuthenticationHandler
{
    #region 私有属性
    private string SchemeName { get; set; }
    private HttpContext HttpContext { get; set; }
    #endregion

    public Task<AuthenticateResult> AuthenticateAsync()
    {
```

```
        // 此处只用于演示，因此总是验证成功
        // 创建用户标识
        ClaimsIdentity id = new(SchemeName);
        // 创建安全实体
        ClaimsPrincipal principal = new ClaimsPrincipal(id);
        // 返回验证结果
        return Task.FromResult(AuthenticateResult.Success(new
        AuthenticationTicket(principal, SchemeName)));
    }

...

    public Task InitializeAsync(AuthenticationScheme scheme,
    HttpContext context)
    {
        SchemeName = scheme.Name;
        HttpContext = context;
        return Task.CompletedTask;
    }
}
```

在服务容器中依次添加 3 个验证方案，并且与它们关联的处理程序都是 MyAuthenHandler 类。

```
builder.Services.AddAuthentication(options =>
{
    options.AddScheme<MyAuthenHandler>(AuthenSchemes.Scheme1, null);
    options.AddScheme<MyAuthenHandler>(AuthenSchemes.Scheme2, null);
    options.AddScheme<MyAuthenHandler>(AuthenSchemes.Scheme3, null);
});
```

在 HTTP 管道上映射一个终结点，路径为/test，根据 URL 中的 scheme 参数的值执行对应的验证方案。例如，访问/test?scheme=s2 表示执行 s2 方案。

```
app.MapGet("/test", async (HttpContext context) =>
{
    var req = context.Request;
    // 读取 URL 查询参数
    if (!req.Query.TryGetValue("scheme", out var value))
    {
        value = AuthenSchemes.Scheme1;
    }
    // 执行验证
    var result = await context.AuthenticateAsync(value);
    if (result.Succeeded)
    {
        string msg = $"方案：{result.Principal.Identity?.
        AuthenticationType ?? "<未知>"}，验证成功";
        context.Response.ContentType = "text/plain;charset=utf-8";
        await context.Response.WriteAsync(msg);
```

```
    }
});
```

在 Razor 页面中，通过<select>元素可选择不同的验证方案，通过 JavaScript 代码构造要访问的 URL，最后用 fetch()全局方法向服务器发出请求。

```
@page "/"
…

<!DOCTYPE html>
<html lang="zh">
<head>
<meta charset="utf-8" />
<title>Demo</title>
</head>
<body>
<div>
<label for="sel">请选择：</label>
<select id="sel">
<option selected value="@AuthenSchemes.Scheme1">方案 1</option>
<option value="@AuthenSchemes.Scheme2">方案 2</option>
<option value="@AuthenSchemes.Scheme3">方案 3</option>
</select>
</div>
<div style="margin-top:17px">
<button id="btn">验证</button>
<p id="result"></p>
</div>

<script>
          // 获取 HTML 元素的引用
          var select = document.getElementById("sel");
          var button = document.getElementById("btn");
          var result = document.getElementById("result");
          // 处理事件
          button.addEventListener("click", ()=>{
              // 获取已选择的验证方案
              var selval = select.value;
              // 拼接请求 URL
              var url = '/test';
              if(selval !== '')
              {
                  url += "?scheme=" + selval;
              }
              // 发送请求
              fetch(url).then(response =>{
                  response.text().then(t => {
                      result.textContent = t;
                  });
```

```
                });
            });
    </script>
    </body>
    </html>
```

运行示例程序，在页面上的下拉列表中选择一个验证方案，然后单击“验证”按钮。服务器会根据不同的验证方案响应不同的信息，如图 18-3 所示。

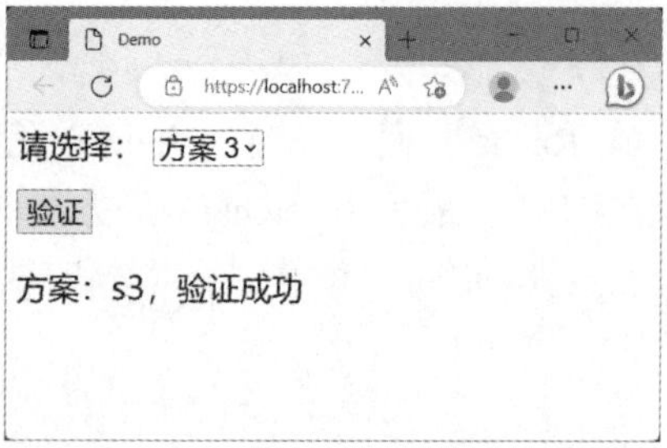

图 18-3　根据所选择的方案返回相应的信息

18.4　IAuthenticationSignInHandler 接口

IAuthenticationSignInHandler 接口用于处理用户的登录逻辑。该接口派生自 IAuthentication-SignOutHandler 接口，而 IAuthenticationSignOutHandler 接口派生自 IAuthenticationHandler 接口。所以，IAuthenticationSignInHandler 接口不仅存在 AuthenticateAsync()、ChallengeAsync() 等方法成员，还包括以下方法成员。

（1）SignInAsync()：处理用户登录，主要用于保存登录状态信息。存储登录状态可以用会话（Session）、Cookies 或数据库。

（2）SignOutAsync()：从 IAuthenticationSignOutHandler 接口继承而来，用于处理用户的登出（注销）逻辑，主要任务是删除在 SignInAsync()方法中所保存的登录状态信息。

综上所述，若实现 IAuthenticationSignInHandler 接口，可以同时实现验证、登录和登出等功能。

下面的示例将实现简单的登录-验证-登出处理。在 Web 页面上输入用户 ID 进行登录，服务器返回一个令牌，随后可以提供用户 ID 和令牌进行验证或取消登录操作。具体步骤如下。

（1）定义 TokenAuthenticationHandler 类，实现 IAuthenticationSignInHandler 接口（同时实现登入/登出和验证功能）。

```
public class TokenAuthenticationHandler : IAuthenticationSignInHandler
{
    #region 私有成员
    #pragma warning disable CS8618
    HttpContext _httpContext;
    string _schemeName;
    TokensManager _tokenMgr;
    #pragma warning restore CS8618

    // 从 URL 查询参数中读取用户 ID 和令牌
    (string? userid, string? token) GetUserAndTokenFromContext()
    {
        var request = _httpContext.Request;
        // 从 URL 查询字符串中提取用户 ID 和令牌
```

```
    _ = request.Query.TryGetValue("userid", out var uid);
    _ = request.Query.TryGetValue("token", out var tk);
    return (uid, tk);
}
#endregion

public Task<AuthenticateResult> AuthenticateAsync()
{
    AuthenticateResult result = AuthenticateResult.NoResult();
    string? userId, token;
    (userId, token) = GetUserAndTokenFromContext();
    // 验证是否提供用户 ID
    if(string.IsNullOrEmpty(userId))
    {
        result = AuthenticateResult.Fail("未提供用户 ID");
    }
    else
    {
        // 检查令牌是否有效
        if(token != null && token == _tokenMgr.GetToken(userId))
        {
            ClaimsIdentity id=new(new[]
            {
                new Claim(ClaimTypes.Name, userId),
                new Claim(ClaimTypes.UserData, token)
            }, _schemeName);
            ClaimsPrincipal principal = new(id);
            AuthenticationTicket ticket = new(principal, _schemeName);
            result = AuthenticateResult.Success(ticket);
        }
        else
        {
            result = AuthenticateResult.Fail("未提供有效的令牌");
        }
    }
    return Task.FromResult(result);
}

public Task ChallengeAsync(AuthenticationProperties? properties)
{
    // 此处不需要实现
    throw new NotImplementedException();
}

public Task ForbidAsync(AuthenticationProperties? properties)
{
    // 此处不需要实现
    throw new NotImplementedException();
}
```

```
    public Task InitializeAsync(AuthenticationScheme scheme, HttpContext
    context)
    {
        _schemeName = scheme.Name;
        _httpContext = context;
        _tokenMgr = context.RequestServices.GetRequiredService
        <TokensManager>();
        return Task.CompletedTask;
    }

    public async Task SignInAsync(ClaimsPrincipal user,
    AuthenticationProperties? properties)
    {
        // 获取用户 ID
        string? userid = user.Identity?.Name ?? null;
        if(string.IsNullOrEmpty(userid))
        {
            return;
        }
        // 生成令牌，表示登录完成
        string newToken = Guid.NewGuid().ToString();
        // 缓存令牌
        _tokenMgr.SetToken(userid, newToken);
        // 将令牌返回给客户端
        await _httpContext.Response.WriteAsync(newToken);
    }

    public Task SignOutAsync(AuthenticationProperties? properties)
    {
        (string? userid, string? token) = GetUserAndTokenFromContext();
        if(userid != null && token != null)
        {
            // 删除令牌
            _tokenMgr.DeleteToken(userid, token);
        }
        return Task.CompletedTask;
    }
}
```

（2）TokenAuthenticationHandler 类中用到了 TokensManager。该类为自定义类型，用于存储/删除用户 ID 和令牌信息。该类将以单个实例的方式注册到服务容器中。

```
// 包含用户和令牌信息的结构体
internal struct TokenEntity
{
    // 用户 ID
    public string? UserId;
    // 用户令牌
```

```
    public string? Token;
    // 过期时间
    public DateTime? Exprire;
}

public class TokensManager
{
    // 存放令牌的集合
    private readonly HashSet<TokenEntity> _tokens;

    // 令牌有效时间
    private readonly TimeSpan _timeout;

    // timeout 表示令牌的有效期
    public TokensManager(TimeSpan timeout)
    {
        _timeout = timeout;
        _tokens = new();
    }

    // 每次访问令牌前都扫描一下
    // 删除已过期的令牌
    private void Scan()
    {
        DateTime _now = DateTime.Now;
        _ = _tokens.RemoveWhere(token => token.Exprire < _now);
    }

    // 根据用户 ID 获取已保存的令牌
    public string? GetToken(string userid)
    {
        if(userid == null)
        {
            throw new ArgumentNullException(nameof(userid));
        }
        Scan();
        var res = (from x in _tokens
                   where x.UserId == userid
                   select x.Token).FirstOrDefault();
        return res;
    }

    // 保存令牌
    public bool SetToken(string userid, string token)
    {
        // 两个参数都不能为 null
        if(userid == null)
        {
            throw new ArgumentNullException(nameof(userid));
```

```
        }
        if(token == null)
        {
            throw new ArgumentNullException(nameof(token));
        }
        TokenEntity entity;
        entity.UserId = userid;
        entity.Token = token;
        entity.Exprire = DateTime.Now.Add(_timeout);
        return _tokens.Add(entity);
    }

    // 删除令牌
    public void DeleteToken(string userid, string token)
    {
        _ = _tokens.RemoveWhere(t => t.UserId == userid && t.Token == token);
    }
}
```

此处仅用于演示，因此令牌只是存放到 HashSet<T>集合中。在实际开发中，可以将用户信息和令牌信息存入数据库或文件中。

（3）将 TokensManager 类注册到服务容器中。

```
builder.Services.AddSingleton<TokensManager>(services => new
TokensManager(TimeSpan.FromSeconds(15)));
```

通过 TokensManager 类的构造函数设置令牌的有效期为 15s。

（4）将 TokenAuthenticationHandler 添加到验证方案集合中。

```
// 验证方案名称
const string AUTHEN_SCHEME="test";

builder.Services.AddAuthentication(opt =>
{
    opt.AddScheme<TokenAuthenticationHandler>(AUTHEN_SCHEME, null);
});
```

（5）在 HTTP 管道上映射登录终结点。

```
app.MapGet("/login", async (HttpContext context) =>
{
    // 获取用户 ID
    bool isOK = context.Request.Query.TryGetValue("userid", out
    StringValues userid);
    if(isOK && !string.IsNullOrEmpty(userid))
    {
        Claim clname = new(ClaimTypes.Name, userid!);
        ClaimsIdentity idt = new(new[]{clname}, AUTHEN_SCHEME);
        ClaimsPrincipal p = new(idt);
        // 登录
        await context.SignInAsync(AUTHEN_SCHEME, p);
```

```
    }
});
```

（6）映射登出（取消登录）终结点。

```
app.MapGet("/logout", async (HttpContext context) =>
{
    await context.SignOutAsync();
    return Results.Ok();
});
```

（7）映射验证终结点。

```
app.MapGet("/authen", async (HttpContext context) =>
{
    var res = await context.AuthenticateAsync(AUTHEN_SCHEME);
    if(res.Succeeded)
    {
        // 验证成功
        return Results.Ok();
    }
    // 验证未通过，返回失败原因
    string reason = res.Failure?.Message ?? "未知原因";
    return Results.Content(reason, "text/plain;charset=utf-8",
    Encoding.UTF8, 499);
});
```

（8）本示例将使用普通的 HTML 文档测试登录和验证。

```
<!DOCTYPE html>
<html lang="zh">
<head>
<title>Demo</title>
<meta charset="utf-8"/>
<link rel="stylesheet" href="app.css" />
</head>
<body>
<div class="card">
<label for="uid">用户：</label>
<input type="text" id="uid" />
</div>
<div class="card">
<label for="tk">令牌：</label>
<input type="text" id="tk"/>
</div>
<p>
<button id="btnSignin">登入</button>
<button id="btnAuth">验证</button>
<button id="btnSignout">登出</button>
</p>
<span class="message" id="msg"></span>
```

```
…
</body>
</html>
```

（9）以下 JavaScript 代码声明相关变量并获取 HTML 元素的引用。

```
var txtUser = document.getElementById("uid");
var txtToken = document.getElementById("tk");
var signInButton = document.getElementById("btnSignin");
var authenButton = document.getElementById("btnAuth");
var signOutButton = document.getElementById("btnSignout");
var showMessage = document.getElementById("msg");
```

（10）处理“登入”按钮的单击（click）事件。

```
signInButton.addEventListener("click", (e)=>
{
    if(txtUser.value !=='')
    {
        fetch("/login?userid="+txtUser.value).then(response =>{
            response.text().then(data =>{
                txtToken.value = data;
                showMessage.textContent = "已获取令牌";
            });
        });
    }
    e.preventDefault();
});
```

（11）处理“登出”按钮的单击事件。

```
signOutButton.addEventListener("click", e=>{
    if(txtUser.value !== '' && txtToken.value !== '')
    {
        let url = `/logout?userid=${txtUser.value}&token=${txtToken.value}
`;
        fetch(url).then(response =>{
            if(response.status == 200)
            {
                showMessage.textContent = "已取消登录";
            }
        });
    }
    e.preventDefault();
});
```

（12）处理“验证”按钮的单击事件。

```
authenButton.addEventListener("click", e=>{
    if(txtUser.value !== '' && txtToken.value !== '')
    {
        let url = `/authen?userid=${txtUser.value}&token=${txtToken.value}`;
        fetch(url).then(response =>{
```

```
            if(response.status == 200)
            {
                showMessage.textContent = "验证成功";
            }
            else
            {
                response.text().then(data =>{
                    showMessage.textContent = data;
                });
            }
        });
    }
    e.preventDefault();
});
```

本示例使用 URL 查询参数来向服务器传递用户 ID 和令牌，如/some/url?userid=<用户ID>&token=<令牌>。

运行示例程序，在“用户”文本框中输入用户 ID（用户名，如 admin），然后单击“登入”按钮。服务器返回的令牌将显示在“令牌”文本框中，如图 18-4 所示。

接着单击“验证”按钮，将显示“验证成功”，如图 18-5 所示。

如果令牌已过期或无效，再次单击“验证”按钮，服务器将返回“未提供有效的令牌”，如图 18-6 所示。

图 18-4　获取令牌

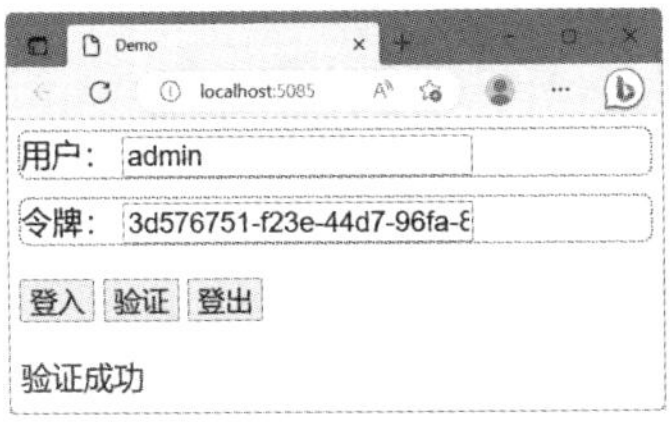

图 18-5　令牌验证成功

图 18-6　令牌验证失败

18.5　验证中间件

在 HTTP 管道需要进行验证的地方插入验证中间件——AuthenticationMiddleware 类（位于 Microsoft.AspNetCore.Authentication 命名空间），将执行默认的验证方案。验证中间件在执行完验证后不会直接处理验证结果，而是完成以下工作。

（1）将验证结果中所包含的用户主体信息（Principal）赋值给 HttpContext.User 属性。

（2）如果验证成功，向 HttpContext.Features 添加 AuthenticationFeatures 对象。对应的访问接口是 IHttpAuthenticationFeature 和 IAuthenticateResultFeature。因此，在

AuthenticationMiddleware 之后的中间件代码中，可以检查 HttpContext.Features 集合，如果包含 IAuthenticateResultFeature 对象则说明验证成功。

以下两种方法均可配置默认的验证方案。

（1）通过 AuthenticationOptions.DefaultAuthenticateScheme 设置默认的验证方案。

（2）应用程序只添加了一个验证方案（视为默认）。

下面的示例将演示使用验证中间件自动执行验证。

CustAuthenHandler 类进行简单的验证处理。从 HTTP 请求的表单（Form）数据中读取用户名和密码，然后判断用户名是否为 Tom，密码是否为 321。若成立，就表示验证成功。

```
public class CustAuthenHandler : IAuthenticationHandler
{
    #region 私有成员
    private HttpContext _context;
    private string _scheme;

    #endregion

    public Task<AuthenticateResult> AuthenticateAsync()
    {
        var result = AuthenticateResult.NoResult();
        var request = _context.Request;
        if(request.HasFormContentType)
        {
            if(request.Form.TryGetValue("user", out var user)
                && request.Form.TryGetValue("pwd", out var passwd))
            {
                // 比较用户名和密码
                if(user == "Tom"&& passwd == "321")
                {
                    // 验证通过
                    ClaimsIdentity id = new(new[]
                    {
                        new Claim(ClaimTypes.Name, user!)
                    }, _scheme);
                    ClaimsPrincipal princ = new(id);
                    AuthenticationTicket ticket = new(princ, _scheme);
                    result = AuthenticateResult.Success(ticket);
                }
                else
                {
                    // 用户名或密码不匹配
                    result = AuthenticateResult.Fail("用户名或密码不正确");
                }
            }
            else
            {
                result = AuthenticateResult.Fail("用户名和密码不完整");
```

```
            }
        }
        return Task.FromResult(result);
    }

    public Task ChallengeAsync(AuthenticationProperties? properties)
    {
        _context.Response.StatusCode = StatusCodes.Status401Unauthorized;
        return Task.CompletedTask;
    }

    public Task ForbidAsync(AuthenticationProperties? properties)
    {
        _context.Response.StatusCode = StatusCodes.Status403Forbidden;
        return Task.CompletedTask;
    }

    public Task InitializeAsync(AuthenticationScheme scheme, HttpContext
    context)
    {
        _scheme = scheme.Name;
        _context = context;
        return Task.CompletedTask;
    }
}
```

在服务容器中启用验证功能并添加验证方案。

```
builder.Services.AddAuthentication(opt =>
{
    opt.AddScheme<CustAuthenHandler>("test", null);
});
```

调用 UseAuthentication()扩展方法，在 HTTP 管道中插入验证中间件。

```
app.UseAuthentication();
```

在验证中间件之后插入一个终结点，检查验证结果。

```
app.Map("/check", (HttpContext context) =>
{
    // 看看是否通过了验证
    IAuthenticateResultFeature? feature = context.Features.Get
    <IAuthenticateResultFeature>();
    string msg = string.Empty;
    if (feature != null && feature.AuthenticateResult != null)
    {
        var authResult = feature.AuthenticateResult;
        // 获取用户名
        string? user = authResult.Principal?.Identity?.Name;
        msg = "验证通过";
        if (user != null)
```

```
        {
            msg += ", 用户名：" + user;
        }
    }
    else
    {
        msg = "未通过验证";
    }
    return Results.Content(msg, "text/plain;charset=utf-8", Encoding.UTF8);
});
```

使用 HTML+JavaScript 代码完成验证测试。

```
<!DOCTYPE html>
<html>
<head>
<meta charset="utf-8" />
<title>Demo</title>
</head>
<body>
<form>
<div style="margin-bottom:10px">
<label for="user">用户名：</label>
<input type="text" name="user" />
</div>
<div style="margin-top:10px">
<label for="pwd">密码：</label>
<input type="password" name="pwd" />
</div>
<div style="margin-top:12px">
<button type="submit">确定</button>
</div>
</form>
<p id="result"></p>

<script>
        // 获取<form>元素的引用
        var form = document.querySelector("form");
        // 处理 submit 事件
        form.addEventListener("submit", e => {
            // 根据<form>元素创建 FormData 实例
            var formdata = new FormData(form);
            fetch('/check', {
                method: "POST",                // 请求方式
                body: formdata                 // 请求正文
            }).then(response => {
                response.text().then(data => {
                    // 显示服务器的响应消息
                    document.getElementById("result").textContent = data;
```

```
            });
        });
        e.preventDefault();                    // 阻止 submit 事件的默认行为
    });
</script>
</body>
</html>
```

运行示例程序，如果输入的用户名是 Tom，密码是 321，那么服务器将返回“验证通过，用户名：Tom”，如图 18-7 所示。

如果输入的用户名是 Bob，密码是 1234，那么服务器将响应“未通过验证”，如图 18-8 所示。

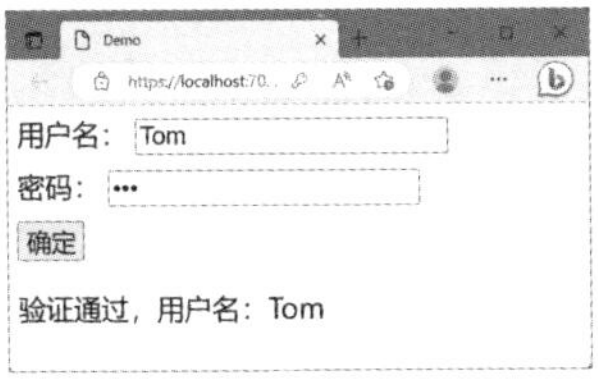

图 18-7　验证通过

图 18-8　验证未通过

18.6　授权处理程序与必要条件

与验证过程类似，授权过程也存在专用的处理程序——IAuthorizationHandler 接口。应用程序可能注册多个 IAuthorizationHandler 服务。可以通过 IAuthorizationHandlerProvider 服务获取已注册的 IAuthorizationHandler 服务集合。

IAuthorizationHandler 接口只声明了一个方法：

```
Task HandleAsync(AuthorizationHandlerContext context);
```

授权上下文对象（AuthorizationHandlerContext）提供了一个 IAuthorizationRequirement 集合，代表授权过程中的必要条件。IAuthorizationRequirement 接口未声明任何成员，仅提供了标记功能，即凡是实现了该接口的类型都被判定为必要条件，用于约束用户权限。如果用户信息满足条件 A，可以调用授权上下文对象的 Succeed()方法向参数传递条件 A 的引用，条件 A 就会从待处理列表（可通过 PendingRequirements 属性获取）中删除。同理，如果条件 B 满足，也会从 PendingRequirements 中删除条件 B，以此类推。当所有 IAuthorizationRequirement 对象都满足授权要求时，HasSucceeded 属性才会返回 true。

需要注意的是，授权上下文对象的实例可以在多个 IAuthorizationHandler 服务实例中共享，所以，必要条件的判定工作可以分布在多个 IAuthorizationHandler 对象中完成。当所有处理程序都执行完毕后，再由 IAuthorizationEvaluator 服务负责汇总并评估用户是否具备访问权限（是否授权给用户）。其中，重要的判断依据是授权上下文对象的 HasSucceeded 属性是否返回 true。

18.6.1 示例：允许指定的部门访问

本示例将演示一种简单的授权方案——只允许特定部门的员工访问某个终结点。具体实现步骤如下。

（1）实现 IAuthorizationRequirement 接口，定义一个必要条件类，用于指定允许用户访问的部门列表。

```
public class PartmentsRequirement : IAuthorizationRequirement
{
    public string[] _parts;

    public PartmentsRequirement(string[] reqPartments)
    {
        _parts = reqPartments;
    }

    // 允许访问的部门列表
    public IEnumerable<string> Partments => _parts;
}
```

（2）自定义授权处理程序，评估访问者是否具备权限。

```
public class CustAuthorizationHandler : IAuthorizationHandler
{
    public Task HandleAsync(AuthorizationHandlerContext context)
    {
        foreach(var req in context.PendingRequirements)
        {
            if(req is PartmentsRequirement partmentsRequirement)
            {
                var principal = context.User;
                foreach(var identity in principal.Identities)
                {
                    // 查找用户信息中的部门信息
                    var claims = identity.FindAll(c => c.Type == "Partment");
                    foreach(Claim cl in claims)
                    {
                        // 用户提供的部门名称是否在允许的部门名单中
                        if(partmentsRequirement.Partments.Contains(cl.Value))
                        {
                            context.Succeed(partmentsRequirement);
                            break;
                        }
                    }
                }
            }
        }
        return Task.CompletedTask;
    }
}
```

用户所属的部门信息是通过 Claim 对象传递的，ClaimType 为自定义的名称 Partment，其值为用户的部门名称。在 HandleAsync()方法中，先读出传递部门信息的 Claim 对象，然后判断 PartmentsRequirement 实例的 Partments 属性是否包含 Claim 对象所提供的部门名称。若包含，表明用户具有访问权限，否则没有访问权限。

（3）在服务容器中注册 CustAuthorizationHandler 类。

```
builder.Services.TryAddEnumerable(ServiceDescriptor.Transient
<IAuthorizationHandler, CustAuthorizationHandler>());
```

（4）注册与授权有关的服务。

```
builder.Services.AddAuthorization();
```

（5）映射根路径的终结点。先通过 URL 查询字符串收集用户提供的信息（此处仅读取用户名和部门名称），再创建 ClaimsPrincipal 实例。最后通过 IAuthorizationService 服务手动执行授权评估。若用户具备访问权限，那么服务器将返回“欢迎光临”。

```
app.MapGet("/", async (HttpContext context) =>
{
    // 收集用户信息
    var req = context.Request;
    ClaimsIdentity idt = new();
    if(req.Query.TryGetValue("name", out var name))
    {
        idt.AddClaim(new Claim(ClaimTypes.Name, name!));
    }
    if(req.Query.TryGetValue("partment", out var partment))
    {
        idt.AddClaim(new Claim("Partment", partment!));
    }
    // 构建用户主体
    ClaimsPrincipal principal = new(idt);

    var authorService = context.RequestServices.GetRequiredService
    <IAuthorizationService>();
    // 授权处理
    var res = await authorService.AuthorizeAsync(principal, null, new
    PartmentsRequirement(new[] { "财务部", "行政部", "法务部" }));
    if(res.Succeeded)
    {
        return "欢迎光临";
    }
    return "抱歉，你没有访问权限";
});
```

运行示例程序，访问以下 URL。

```
http://<主机名:端口>/?name=Jim&partment=生产部
```

由于生产部不在允许访问的部门列表内，因此服务器返回“抱歉，你没有访问权限”。对

URL 做以下修改，再次访问。

```
http://<主机名:端口>/?name=Jim&partment=法务部
```

此时，服务器返回“欢迎光临”。

18.6.2 PassThroughAuthorizationHandler 类

调用 AddAuthorization()扩展方法会向服务容器添加 PassThroughAuthorizationHandler 类。它实现了 IAuthorizationHandler 接口，在 HandleAsync()方法中对所有 IAuthorizationRequirement 对象执行授权处理。源代码如下。

```
public async Task HandleAsync(AuthorizationHandlerContext context)
{
    foreach (var handler in context.Requirements.OfType
    <IAuthorizationHandler>())
    {
        await handler.HandleAsync(context).ConfigureAwait(false);
        if (!_options.InvokeHandlersAfterFailure && context.HasFailed)
        {
            break;
        }
    }
}
```

也就是说，开发人员实际上并不需要自己实现 IAuthorizationHandler 接口，PassThrough-AuthorizationHandler 类会自动完成对必要条件的评估。从上述源代码中可发现，要使必要条件能在 PassThroughAuthorizationHandler 类中进行授权处理，不仅要实现 IAuthorization-Requirement 接口，还要实现 IAuthorizationHandler 接口。

ASP.NET Core 提供了一个名为 AuthorizationHandler 抽象类，可以方便地同时实现 IAuthorizationHandler 和 IAuthorizationRequirement 接口的功能。在派生类中只需要重写以下方法即可。

```
protected abstract Task HandleRequirementAsync(AuthorizationHandlerContext
context, TRequirement requirement);
```

类型参数 TRequirement 表示授权的必要条件（实现 IAuthorizationRequirement 接口的类型）。

下面的示例将演示参与授权评估的两个必要条件。

（1）用户必须提供+86 开头的手机号。

（2）用户年龄在 16 岁以上（含 16 岁）。

下面的代码定义 AgeRequirement 类，用于限制用户的年龄。在类构造函数中通过 age 参数指定年龄，用户年龄必须大于或等于此年龄。

```
public class AgeRequirement : AuthorizationHandler<AgeRequirement>,
IAuthorizationRequirement
{
```

```
    // 最小年龄
    public int Age { get; private set; }

    public AgeRequirement(int age)
    {
        Age = age;
    }

    protected override Task HandleRequirementAsync(
    AuthorizationHandlerContext context, AgeRequirement requirement)
    {
        // 读取用户提供的年龄信息
        foreach(ClaimsIdentity id in context.User.Identities)
        {
            Claim? claim = id.FindFirst("Age");
            if(claim != null)
            {
                int _age = Convert.ToInt32(claim.Value);
                // 用户年龄要大于或等于 Age 属性的值
                if(_age >= requirement.Age)
                {
                    context.Succeed(requirement);
                    break;
                }
            }
        }
        return Task.CompletedTask;
    }
}
```

接着定义 PhoneRequirement 类，用于限制手机号前缀。

```
public class PhoneRequirement : AuthorizationHandler<PhoneRequirement>,
IAuthorizationRequirement
{
    private string _prefix = string.Empty;
    // 手机号前缀
    public string Prefix => _prefix;

    public PhoneRequirement(string prefix)
    {
        _prefix = prefix;
    }

    protected override Task HandleRequirementAsync(
    AuthorizationHandlerContext context, PhoneRequirement requirement)
    {
        // 读取用户提供的手机号
        foreach(var identity in context.User.Identities)
        {
```

```
                Claim? c = identity.FindFirst(ClaimTypes.MobilePhone);
                if(c != null)
                {
                    string phoneNo = c.Value;
                    // 判断是否存在指定的前缀
                    if(phoneNo.StartsWith(requirement.Prefix))
                    {
                        context.Succeed(requirement);
                        break;
                    }
                }
            }
            return Task.CompletedTask;
        }
    }
```

下面的代码先收集客户端提交的用户信息，再使用 IAuthorizationService 手动执行授权评估。

```
app.MapPost("/check", async (HttpContext context) =>
{
    // 读出用户提交的数据
    var request = context.Request;
    _ = request.Form.TryGetValue("name", out var name);
    _ = request.Form.TryGetValue("age", out var age);
    _ = request.Form.TryGetValue("phone", out var phone);
    // 构建用户主体信息
    IList<Claim> claims = new List<Claim>();
    if(!string.IsNullOrEmpty(name))
    {
        claims.Add(new(ClaimTypes.Name, name!));
    }
    if(!string.IsNullOrEmpty(age) && int.TryParse(age, out _))
    {
        claims.Add(new("Age", age!));
    }
    if (!string.IsNullOrEmpty(phone))
    {
        claims.Add(new(ClaimTypes.MobilePhone, phone!));
    }
    ClaimsIdentity id = new(claims);
    ClaimsPrincipal p = new(id);
    // 获取授权服务
    IAuthorizationService authorsv = context.RequestServices.
    GetRequiredService<IAuthorizationService>();
    // 授权必要条件
    AgeRequirement reqm1 = new AgeRequirement(16);
    PhoneRequirement reqm2 = new PhoneRequirement("+86");
    // 执行授权评估
    var result = await authorsv.AuthorizeAsync(
```

```
        p, null, new IAuthorizationRequirement[] { reqm1, reqm2 });
        if (result.Succeeded)
        {
            return "欢迎使用";
        }
        return "你暂时无法访问";
    });
```

上述代码在实例化 AgeRequirement 类时指定用户的最小年龄为 16 岁；在实例化 PhoneRequirement 类时指定手机号的前缀为+86。

假设向/check 路径提交以下数据，服务器将返回“欢迎使用”。

```
name=Tom&age=24&phone=%2B8613550255623
```

其中，%2B 是字符+的转义字符。

18.7　授权策略

在复杂的授权方案中，逐个评估 IAuthorizationRequirement 接口是否满足条件会显得分散且凌乱。此时需要一种机制，可以把一组 IAuthorizationRequirement 接口放到一起，形成一个固定的整体，在执行授权时可以快速调用。这种机制称为授权策略。

授权策略不仅能有效地组织 IAuthorizationRequirement 接口，还可以添加需要的验证方案。当执行授权评估时会自动执行验证，只有成功验证才能进行授权处理。如果身份验证成功但授权失败，授权策略会执行 IAuthenticationHandler.ForbidAsync()方法；如果身份验失败，就会执行 IAuthenticationHandler.ChallengeAsync()方法，要求访问者重新验证。

18.7.1　示例：按用户星级授权

本示例将演示一个依据用户的星级进行授权的策略，当用户星级大于 3 时才允许访问主页。具体实现步骤如下。

（1）定义一个表示用户信息的 UserInfo 类，包括电子邮箱、密码和用户星级。

```
public class UserInfo
{
    public string? Email { get; set; }
    public string? Password { get; set; }
    public int Stars { get; set; }

    public static IEnumerable<UserInfo> GetUsers()
    {
        yield return new UserInfo { Email = "wk001@163.net", Password =
        "12321", Stars = 2 };
        yield return new UserInfo { Email = "chr140x@tse.edu.cn", Password
        = "abcd", Stars = 5 };
        yield return new UserInfo { Email = "gid_32@117.org", Password =
        "5ope", Stars = 3 };
```

```
        yield return new UserInfo { Email = "zero@kict.net", Password =
        "3369", Stars = 4 };
    }
}
```

GetUsers()是一个静态方法，返回一组测试数据。

（2）定义一个 Constants 类，其中包含两个常量成员：验证方案名称和授权策略名称。

```
public class Constants
{
    public const string AUTHEN_SCHEME = "MyAuth";
    public const string AUTHOR_POLICY = "test";
}
```

（3）实现 IAuthenticationSignInHandler 接口，自定义验证方案，包括登入、登出、验证、禁止访问等逻辑。

```
public class MyAuthenticationHandler : IAuthenticationSignInHandler
{
    #region 私有成员
    private HttpContext? _context;
    private string? _scheme;
    #endregion

    public Task<AuthenticateResult> AuthenticateAsync()
    {
        AuthenticateResult result = AuthenticateResult.NoResult();
        // 从 Cookie 中提取__email 的值
        var request = _context!.Request;
        if(request.Cookies.TryGetValue("__email", out string? email)  &&
        email != null)
        {
            // 判断值是否为有效用户名
            var users = UserInfo.GetUsers();
            var q = from u in users
                    where u.Email == email
                    select u;
            if(q.Count() == 1 )
            {
                // 验证通过
                Claim cemail = new(ClaimTypes.Email, email);
                // 收集用户星级
                var cu = q.First();
                Claim cstar = new("Star", cu.Stars.ToString());
                ClaimsIdentity identity = new(new Claim[] { cemail, cstar },
                _scheme);
                ClaimsPrincipal p = new(identity);
                var ticket = new AuthenticationTicket(p, _scheme!);
                result = AuthenticateResult.Success(ticket);
            }
```

```
        else
        {
            // 验证失败
            result = AuthenticateResult.Fail("未找到已登录的 Email");
        }
    }
    return Task.FromResult(result);
}

public Task ChallengeAsync(AuthenticationProperties? properties)
{
    // 跳转到登录页
    _context!.Response.Redirect("/login");
    return Task.CompletedTask;
}

public Task ForbidAsync(AuthenticationProperties? properties)
{
    // 身份验证有效，但无访问权限，即禁止访问
    var response = _context!.Response;
    response.Redirect("/forbid");
    return Task.CompletedTask;
}

public Task InitializeAsync(AuthenticationScheme scheme, HttpContext
context)
{
    _scheme = scheme.Name;
    _context = context;
    return Task.CompletedTask;
}

public Task SignInAsync(ClaimsPrincipal user, AuthenticationProperties?
properties)
{
    string? email = user.Claims.FirstOrDefault(c => c.Type ==
    ClaimTypes.Email)?.Value;

    if(email != null)
    {
        // 设置 Cookie，保存登录状态
        _context!.Response.Cookies.Append("__email", email);
    }
    // 是否有回调 URL
    if(properties != null && properties.RedirectUri != null)
    {
        _context!.Response.Redirect(properties.RedirectUri);
    }
    return Task.CompletedTask;
```

```
    }

    public Task SignOutAsync(AuthenticationProperties? properties)
    {
        var response = _context!.Response;
        // 删除相关的 Cookie，取消登录
        response.Cookies.Delete("__email");
        if(properties != null && properties.RedirectUri != null)
        {
            response.Redirect(properties.RedirectUri);
        }
        return Task.CompletedTask;
    }
}
```

为了让示例更加简单，此处仅仅使用名为__email 的 Cookie 保存已登录的 E-mail。在实际开发中，Cookie 数据应进行加密处理，以提高安全性。

（4）依次实现 AuthorizationHandler 抽象类和 IAuthorizationRequirement 接口，创建一个授权必要条件，可以限制特定星级的用户才能通过授权。

```
public class UserStarRequirement : AuthorizationHandler<UserStarRequirement>,
IAuthorizationRequirement
{
    public int Star { get; private set; }

    public UserStarRequirement(int star)
    {
        Star = star;
    }

    protected override Task HandleRequirementAsync(
    AuthorizationHandlerContext context, UserStarRequirement requirement)
    {
        // 读出在验证阶段收集的用户星级
        foreach(var identity in context.User.Identities)
        {
            var cstar = identity.FindFirst("Star");
            if (cstar != null && int.TryParse(cstar.Value, out int intStar))
            {
                // 用户星级必须大于条件值
                if(intStar > requirement.Star)
                {
                    // 具备权限
                    context.Succeed(requirement);
                }
                else
                {
                    // 不具备权限
                    context.Fail();
                }
```

```
            }
        }
        return Task.CompletedTask;
    }
}
```

（5）在服务容器中启用验证功能，并添加自定义的验证方案。

```
builder.Services.AddAuthentication(opt =>
{
    opt.AddScheme<MyAuthenticationHandler>(Constants.AUTHEN_SCHEME, null);
});
```

（6）在服务容器中启用授权功能，然后添加自定义的授权策略。

```
builder.Services.AddAuthorization(opt =>
{
    opt.AddPolicy(Constants.AUTHOR_POLICY, policybd =>
    {
        // 添加验证方案
        policybd.AddAuthenticationSchemes(new[] { Constants.AUTHEN_SCHEME });
        // 添加授权策略的必要条件
        policybd.AddRequirements(new UserStarRequirement(3));
    });
});
```

AddAuthenticationSchemes()方法向当前策略添加需要的验证方案，在执行授权之前会使用这些验证方案进行身份验证，只有身份验证通过才会执行授权。

AddRequirements()方法向当前策略添加必要的约束条件，本示例限制用户星级必须大于 3 才能拥有访问主页的权限（如 4、5 星级的用户可以访问，而 1、2、3 星级的用户则不能访问）。

在测试的用户数据中，chr140x@tse.edu.cn 是 5 星级用户，因此该用户登录后能顺利进入主页，如图 18-9 所示。

而 wk001@163.net 是 2 星级用户，其登录后被拒绝访问，如图 18-10 所示。

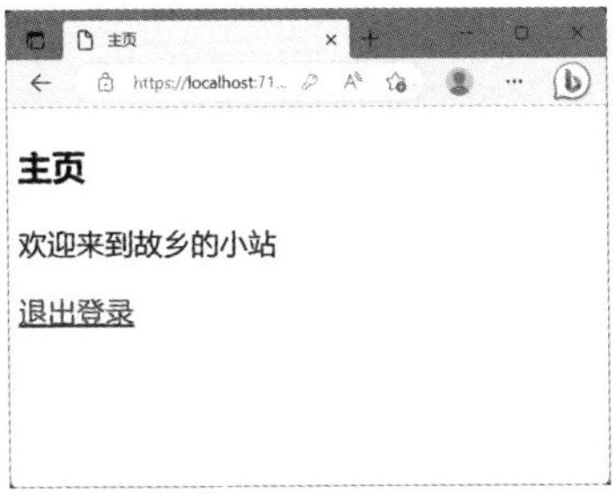

图 18-9　拥有访问权限的用户可以进入主页

图 18-10　2 星级用户无法访问主页

18.7.2　示例：集成内置的 Cookie 验证

ASP.NET Core 内部提供了 Cookie 验证方案，开发人员可以直接使用。登录状态由 Session

和 Cookie 共同存储。此方案可满足常见的身份验证需求。

本示例的授权策略为只允许指定的用户名访问/hello 终结点。与授权策略配合使用的是 Cookie 验证。Cookie 验证默认的方案名称为 Cookies，可通过访问 CookieAuthenticationDefaults.AuthenticationScheme 常量获得。

（1）定义 UserData 类，封装用户信息（用户名和密码）。UserHelper 类用于返回测试数据。

```
public class UserData
{
    public string? UserName { get; set; }
    public string? Password { get; set; }
}

public static class UserHelper
{
    // 生成测试用户数据
    public static IEnumerable<UserData> GetUsers()
    {
        return new UserData[]
        {
            new UserData
            {
                UserName = "Laila",
                Password = "zzxx"
            },
            new UserData
            {
                UserName = "Bob",
                Password = "6565"
            },
            new UserData
            {
                UserName = "Aaron",
                Password = "u3v4"
            },
            new UserData
            {
                UserName = "Maya",
                Password = "02035"
            },
            new UserData
            {
                UserName = "Jack",
                Password = "85a6"
            },
            new UserData
            {
                UserName = "Dallas",
                Password = "pnLf"
            },
```

```
            new UserData
            {
                UserName = "Gale",
                Password = "9728"
            }
        };
    }
}
```

（2）定义 UserNameRequirement 类，它实现 AuthorizationHandler 抽象类和 IAuthorization-Requirement 接口。该类将作为授权的必要条件，限制只允许指定的用户访问相关资源。

```
public class UserNameRequirement : AuthorizationHandler<UserNameRequirement>,
IAuthorizationRequirement
{
    public string[] AllowNames { get; private set; }

    public UserNameRequirement(string[] usernames)
    {
        AllowNames = usernames;
    }

    protected override Task HandleRequirementAsync(
    AuthorizationHandlerContext context, UserNameRequirement requirement)
    {
        // 读取已登录的用户名
        Claim? cuser = context.User.Claims.SingleOrDefault(c => c.Type ==
        ClaimTypes.Name);
        if (cuser != null)
        {
            // 判断当前用户名是否在允许的名单中
            if(requirement.AllowNames.Contains(cuser.Value))
            {
                context.Succeed(requirement);
            }
        }
        return Task.CompletedTask;
    }
}
```

（3）在服务容器中添加验证功能，并进行 Cookie 验证方案的配置。

```
builder.Services.AddAuthentication().AddCookie(options =>
{
    // 登录路径
    options.LoginPath = "/login";
    // 拒绝访问后跳转的路径
    options.AccessDeniedPath = "/denied";
    // Cookie 的名称
    options.Cookie.Name = "my_authen";
    // 退出登录的路径
```

```
        options.LogoutPath = "/logout";
        // 回调 URL 参数的名称
        options.ReturnUrlParameter = "callbackurl";
    });
```

LoginPath 属性指定登录路径，当验证不通过时会自动跳转；AccessDeniedPath 属性指定访问被拒绝后跳转的路径；LogoutPath 属性指定退出登录的路径；Cookie.Name 用于配置登录成功后生成的 Cookie 名称；ReturnUrlParameter 属性设置 URL 查询字符中表示回调 URL 的字段名称，此处设置为 callbackurl，此 URL 用于在登录完成后进行跳转，如/login?callbackurl=/index 表示登录后跳转到/index。

（4）在服务容器中配置授权策略。本示例指定的授权策略命名为 allow_names。

```
builder.Services.AddAuthorization(options =>
{
    options.AddPolicy("allow_names", policybd =>
    {
        // 添加验证方案
        policybd.AddAuthenticationSchemes(CookieAuthenticationDefaults.
        AuthenticationScheme);
        // 添加必要条件
        policybd.AddRequirements(new UserNameRequirement(new[] { "Aaron",
        "Bob", "Jack" }));
    });
});
```

（5）Demo 控制器的代码如下。

```
public class DemoController : Controller
{
    [Route("")]
    [Authorize(Policy = "allow_names")]
    public IActionResult Index()
    {
        return View();
    }

    [Route("login")]
    public IActionResult Login() => View();

    [HttpPost("login")]
    public async Task<IActionResult> Login(UserData user, string?
    callbackUrl)
    {
        // 先检查用户名和密码是否正确
        var userlist = UserHelper.GetUsers();
        var q = from u in userlist
              where u.UserName!.Equals(user.UserName, StringComparison.
              OrdinalIgnoreCase) && u.Password == user.Password
                select u;
```

```
        if (q.Count() == 1)
        {
            // 密码正确，登录
            Claim c = new(ClaimTypes.Name, user.UserName!);
            ClaimsIdentity id = new(new[] { c }, CookieAuthenticationDefaults.
            AuthenticationScheme);
            var principal = new ClaimsPrincipal(id);
            await HttpContext.SignInAsync(CookieAuthenticationDefaults.
            AuthenticationScheme, principal);
            // 跳转回主页
            return Redirect(callbackUrl ?? "/");
        }
        return View(user);
    }

    [Route("denied")]
    public IActionResult Denied() => Content("拒绝访问");

    [Route("logout")]
    public async Task<IActionResult> Logout()
    {
        // 退出登录
        await HttpContext.SignOutAsync(CookieAuthenticationDefaults.
        AuthenticationScheme);
        // 回到登录页
        return RedirectToAction("Login");
    }
}
```

在 Index()方法上应用了 AuthorizeAttribute 特性类，表示执行该方法需要授权。AuthorizeAttribute 特性可以应用到类上，这样类中所有方法成员都需要经过授权后才能访问。若其中需要让某个方法允许匿名访问（不需要授权），则要在此方法成员上应用 AllowAnonymousAttribute 特性，如

```
[AllowAnonymous]
public IActionResult Index()
{
    …
}
```

运行示例程序，首次访问 Index 方法会被服务器拒绝，进而跳转到登录视图，如图 18-11 所示。

使用用户 Jack（密码为 85a6，请参考前文 UserHelper 类的代码）进行登录，由于该用户是允许访问的，所以登录后直接进入 Index 视图，如图 18-12 所示。

若使用用户 Maya（密码为 02035）登录，服务器会拒绝访问，如图 18-13 所示。

图 18-11　登录视图

图 18-12　进入 Index 视图

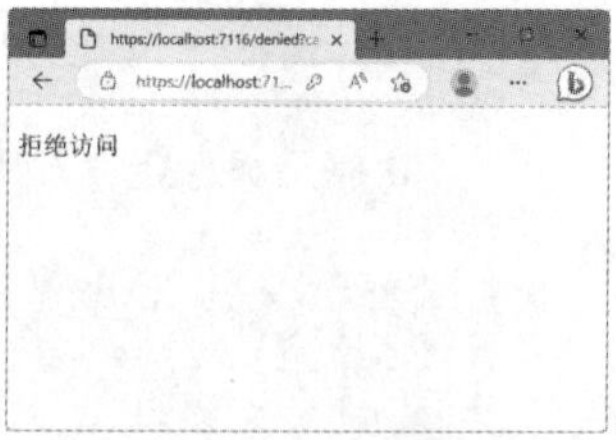

图 18-13　拒绝访问

18.7.3　示例：在终结点上应用授权策略

在调用 MapGet()、MapPut()、MapControllers()等扩展方法添加终结点映射后，可以继续调用 RequireAuthorization()扩展方法指定授权策略。这些终结点需要授权成功后才允许访问。

本示例将演示在 MapGet()方法添加的终结点上使用授权策略，客户端必须提供名为 admin 的用户名才能访问该终结点。用户名将通过自定义消息头 user-name 传递。

（1）定义 HeaderAuthenticationOptions 类，它是一个选项类，派生自 AuthenticationScheme-Options 类。该类包含一个 RequiredHeaders 属性，类型为字符串列表，用于设置一组 HTTP 消息头。在进行身份验证时，要求客户端必须提供 RequiredHeaders 属性所列出的消息头，否则验证无法通过。

```
public class HeaderAuthenticationOptions : AuthenticationSchemeOptions
{
    // 要求客户端请求必须出现的 HTTP 消息头
    public IList<string> RequiredHeaders { get;} = new List<string>();
}
```

（2）定义 HeaderAuthenticationHandler 类，派生自 AuthenticationHandler 类（抽象类），并实现 HandleAuthenticateAsync()抽象方法，处理验证逻辑。

```
public class HeaderAuthenticationHandler : AuthenticationHandler
<HeaderAuthenticationOptions>
{
    // 需要将参数传递给基类的构造函数
    public HeaderAuthenticationHandler(IOptionsMonitor
    <HeaderAuthenticationOptions> options, ILoggerFactory logger,
    UrlEncoder encoder, ISystemClock clock)
        :base(options, logger, encoder, clock)
    {
    }

    protected override Task<AuthenticateResult> HandleAuthenticateAsync()
    {
        var result = AuthenticateResult.NoResult();
        List<Claim> claims = new();
        // 验证请求是否存在必需的消息头
        foreach(var hd in Options.RequiredHeaders)
```

```
        {
            if(!Request.Headers.TryGetValue(hd, out var value))
            {
                result = AuthenticateResult.Fail($"{hd}消息头是必需的");
                break;
            }
            else
            {
                if(StringValues.IsNullOrEmpty(value) == false)
                {
                    // 如果消息头是 user-name，表示用户名
                    if (hd.Equals("user-name", StringComparison.Ordinal
                    IgnoreCase))
                    {
                        claims.Add(new(ClaimTypes.Name, value!));
                    }
                    else
                    {
                        // 其他值将作为 UserData 处理
                        claims.Add(new(ClaimTypes.UserData, value!));
                    }
                }
            }
        }
        if(claims.Count > 0)
        {
            ClaimsIdentity identity = new(claims, Scheme.Name);
            ClaimsPrincipal principal = new(identity);
            AuthenticationTicket ticket = new(principal, Scheme.Name);
            result = AuthenticateResult.Success(ticket);
        }
        return Task.FromResult(result);
    }
}
```

（3）声明两个常量，分别表示验证方案名称和授权策略名称。

```
// 验证方案名称
const string AUTHEN_SCHEME = "test";
// 授权策略名称
const string AUTHORIZ_POLICY = "check_req_name";
```

（4）在服务容器中开启验证功能，添加自定义验证方案。通过 HeaderAuthenticationOptions 选项类配置 RequiredHeaders 属性，要求客户端发送请求时必须携带 user-name 消息头。

```
builder.Services.AddAuthentication().AddScheme<HeaderAuthenticationOptions,
HeaderAuthenticationHandler>(AUTHEN_SCHEME, options =>
{
    options.RequiredHeaders.Add("user-name");
});
```

（5）在服务容器中启用授权功能，添加自定义授权策略。

```
builder.Services.AddAuthorization(options =>
{
    options.AddPolicy(AUTHORIZ_POLICY, bd =>
    {
        // 添加验证方案
        bd.AddAuthenticationSchemes(AUTHEN_SCHEME);
        // 添加必要条件
        bd.RequireUserName("admin");
    });
});
```

RequireUserName("admin")表示由 user-name 消息头提供的用户名必须为 admin。

（6）在 HTTP 管道上调用 MapGet()扩展方法，映射根路径（/）为终结点，返回字符串“Hello World!”。随即调用 RequireAuthorization()方法指定应用自定义的授权策略。

```
app.MapGet("/", () =>"Hello World!").RequireAuthorization(AUTHORIZ_POLICY);
```

运行示例程序，可以用 http-repl 工具进行测试。输入以下命令连接服务器（假设服务器地址为 https://localhost:7218）。

```
httprepl https://localhost:7218
```

向根路径（/）发出 HTTP-GET 请求，并指定消息头 user-name=Lee。

```
get -h user-name=Lee /
```

命令提交后服务器返回的状态码为 403，这是因为用户名不是 admin，授权失败。再次发出请求，此次指定的消息头为 user-name=admin。

```
get -h user-name=admin /
```

由于此次用户名为 admin，授权成功，所以服务器返回的状态码为 200，消息正文为“Hello World!”。